European Yearbook of International Economic Law

Series Editors
Marc Bungenberg, Saarbrücken, Germany
Markus Krajewski, Erlangen, Germany
Christian J. Tams, Glasgow, UK
Jörg Philipp Terhechte, Lüneburg, Germany
Andreas R. Ziegler, Lausanne, Switzerland

Assistant Editor
Judith Crämer, Lüneburg, Germany

Advisory Editors
Armin von Bogdandy, Heidelberg, Germany
Thomas Cottier, Bern, Switzerland
Stefan Griller, Salzburg, Austria
Armin Hatje, Hamburg, Germany
Christoph Herrmann, Passau, Germany
Meinhard Hilf, Hamburg, Germany
John H. Jackson†
William E. Kovacic, Washington, USA
Gabrielle Marceau, Geneva, Switzerland
Ernst-Ulrich Petersmann, Florence, Italy
Hélène Ruiz Fabri, Luxembourg, Luxembourg
Bruno Simma, München, Germany
Rudolf Streinz, München, Germany

More information about this subseries at http://www.springer.com/series/8848

Rafael Leal-Arcas

Solutions for Sustainability

How the International Trade, Energy and Climate Change Regimes Can Help

Rafael Leal-Arcas
Queen Mary University of London
London, UK

ISSN 2364-8392 ISSN 2364-8406 (electronic)
European Yearbook of International Economic Law
ISSN 2510-6880 ISSN 2510-6899 (electronic)
Special Issue
ISBN 978-3-030-23932-9 ISBN 978-3-030-23933-6 (eBook)
https://doi.org/10.1007/978-3-030-23933-6

© Springer Nature Switzerland AG 2019
This work is subject to copyright. All rights are reserved by the Publisher, whether the whole or part of the material is concerned, specifically the rights of translation, reprinting, reuse of illustrations, recitation, broadcasting, reproduction on microfilms or in any other physical way, and transmission or information storage and retrieval, electronic adaptation, computer software, or by similar or dissimilar methodology now known or hereafter developed.
The use of general descriptive names, registered names, trademarks, service marks, etc. in this publication does not imply, even in the absence of a specific statement, that such names are exempt from the relevant protective laws and regulations and therefore free for general use.
The publisher, the authors, and the editors are safe to assume that the advice and information in this book are believed to be true and accurate at the date of publication. Neither the publisher nor the authors or the editors give a warranty, express or implied, with respect to the material contained herein or for any errors or omissions that may have been made. The publisher remains neutral with regard to jurisdictional claims in published maps and institutional affiliations.

This Springer imprint is published by the registered company Springer Nature Switzerland AG.
The registered company address is: Gewerbestrasse 11, 6330 Cham, Switzerland

To Alessandra, Eduardo, and Juan, for teaching me so much about life.

Acknowledgments

This book has been written with the financial support of the Erasmus+ Program of the European Union, which funded my Jean Monnet Chair in EU International Economic Law (project number 575061-EPP-1-2016-1-UK-EPPJMO-CHAIR).

The financial help from two European Union grants is gratefully acknowledged: Jean Monnet Chair in EU International Economic Law (project number 575061-EPP-1-2016-1-UK-EPPJMO-CHAIR) and the WiseGRID project (number 731205), funded by the European Union's Horizon 2020 research and innovation program, for which I am a principal investigator.

The ideas in this book have developed over the years. I am very grateful to the participants of the following events for very insightful discussions that have improved many of the ideas expressed in this book:

- Yale sustainability leadership forum, September 2016 and 2017, Yale University;
- Executive education program on climate change and energy, October 2018, Harvard Kennedy School of Government, Harvard University;
- Oxford Summer School in Ecological Economics, August 2018, Oxford University;
- 35th Round Table on Sustainable Development, OECD Headquarters, Paris, France;
- The Future Sustainability Summit 2019, Abu Dhabi, United Arab Emirates;
- Yale Workshop on Trade and Climate Change, November 2017, Yale University.

While writing this book, I had the fortune to serve as a Visiting Scholar in the following institutions:

- Trade and Environment Division of the World Trade Organization, Geneva, Switzerland;
- Masdar Institute of Science and Technology, Abu Dhabi, United Arab Emirates;
- Sustainable Energy for All Secretariat, Washington, D.C., USA.

I am very grateful to my colleagues in those institutions for hosting me and for helping me shape many of the ideas in this book.

Many concepts addressed in Chaps. 2, 4, and 9 are a further development of R. Leal-Arcas, "Sustainability, Common Concern and Public Goods," The George Washington International Law Review, Vol. 49, Issue 4, pp. 801–877, 2017. Several sections from Chap. 3 are a further development of R. Leal-Arcas and A. Morelli, "The Resilience of the Paris Agreement: Negotiating and Implementing the Climate Regime," Georgetown Environmental Law Review, Vol. 31.1, pp. 1–63, 2018. Several ideas in Chap. 6 are a development from R. Leal-Arcas, "New Frontiers of International Economic Law: The Quest for Sustainable Development," University of Pennsylvania Journal of International Law, Vol. 40, Issue 1, pp. 83–132, 2018. Many ideas in Chaps. 7 and 8 were drawn from R. Leal-Arcas et al., "Smart Grids in the European Union: Assessing Energy Security, Regulation & Social and Ethical Considerations," Columbia Journal of European Law, Vol. 24.2, pp. 311–410, 2018. Parts of Chap. 8 were drawn from the ideas developed in R. Leal-Arcas et al., "Energy Decentralization in the European Union," Queen Mary University of London, School of Law Legal Studies Research Paper No. 307/2019, pp. 1–55.

Rockville, MD, USA   Rafael Leal-Arcas
June 2019

Contents

1 Introduction 1

Part I From the Top Down

2 Cooperation on Issues of Common Concern and Public Goods 19

3 Linking International Trade to Climate Change and Energy 47

4 Using Trade Agreements to Achieve Sustainability: A Counter-Intuitive Conundrum 109

5 Trading Sustainable Energy 183

Part II From the Bottom Up

6 Decentralization and Empowering the Citizen 201

7 Smart Grids and Empowering the Citizen 249

8 Practical Applications of Decentralized Energy in the EU 283

9 Innovation, Research, and Technology 443

Chapter 1
Introduction

Sustainability is a necessity for the twenty-first century that has an inter-generational dimension.[1] Given the urgency[2] of the issue, scientists have proposed concepts such as "planetary boundaries" to define a "safe operating space for humanity"[3] to continue to thrive for years to come.[4] The concept of planetary boundaries is based on scientific research that indicates that, since the Industrial Revolution at the end of the eighteenth century and beginning of the nineteenth century, human activity has gradually become the main driver of global environmental degradation.[5]

A related concept—sustainable development—was coined by the Brundtland Commission[6] in a report titled *Our Common Future*.[7] The concept has three main pillars. First, sustainable development recognizes that part of the environmental challenge is poverty.[8] For example, certain communities need to cut down trees

---

[1] In the case of climate change and energy, the indicators of sustainability are greenhouse gas emissions, primary energy consumption, and the share of renewable energy in gross final energy consumption. For further details on sustainability in the context of economic performance and social progress, see J. Stiglitz, A. Sen and J.-P. Fitoussi, "Report by the Commission on the Measurement of Economic Performance and Social Progress," (2008) available at https://ec.europa.eu/eurostat/documents/118025/118123/Fitoussi+Commission+report.

[2] Jake Sullivan, US national security advisor to US Vice President Biden, once famously said: "Between fatalism and complacency lies urgency." See https://twitter.com/JesseJenkins/status/1062448890543267841.

[3] Rockström et al. (2009), p. 33.

[4] *Planetary Boundaries Research*, Stockholm Resilience Centre, http://www.stockholmresilience.org/research/planetary-boundaries.html [https://perma.cc/HME9-TBNR].

[5] Rockström et al. (2009), p. 33.

[6] Formally known as the World Commission on Environment and Development, the Brundtland Commission was created to persuade countries to aim at sustainable development. World Comm'n on Env't & Dev., *Our Common Future*, Annex 2, U.N. Doc. A/42/427 (1987), available at http://www.un-documents.net/our-common-future.pdf [https://perma.cc/KPK5-CH3Y].

[7] *Id.* at 1.

[8] *Id.* at Overview, ¶ 8.

down for their fuel needs. Second, an integrated approach to sustainable development is important.[9] One of the objectives of the Brundtland Commission was to raise awareness that the various areas (now goals) of sustainable development cannot be addressed in clinical isolation.[10] Rather, sustainable development efforts must integrate economic, environmental, and social considerations. And third, intergenerational ethics apply to sustainable development.[11] Traditionally, however, a short-term approach to issues has been rewarded, as opposed to a long-term approach.

In 2005, one scholar predicted humanity's top ten problems for the next 50 years[12] as follows: (1) energy, (2) water, (3) food, (4) the environment, (5) poverty, (6) terrorism and war, (7) disease, (8) education, (9) democracy, and (10) population.[13] This prediction was based on the fact that in 2004 the world population was 6.5 billion, and in 2050 it is expected to be 10 billion.[14] However, new predictions place the world's population at 11 billion by 2050.[15]

This book explores links and synergies among three of the above challenges: energy,[16] the environment, and poverty (specifically the role of international trade in addressing poverty). It examines international regimes that deal with the three issues, explores possible solutions through multilateralism, and analyzes the potential impact of citizens and prosumers.[17] In this sense, decarbonizing the global economy and achieving energy that is affordable, secure, and clean—in other words, "sustainable energy"[18]—are pressing issues.

---

[9]*Id.* pt. 1, chp. 1, ¶ 42.

[10]*Id.* pt. 1, chp. 2, ¶ 10.

[11]*Id.* pt. 1, chp. 2, ¶ 76. For millennials, values are important. For instance, if they recycle at home, they also want to own or work for firms that recycle.

[12]*See generally* Prashant V. Kamat, *Energy Challenge and Nanotechnology*, Presentation, 1, http://www3.nd.edu/~pkamat/pdf/energy.pdf (referencing humanity's top ten problems for the next 50 years, as identified by Richard Smalley of Rice University) [https://perma.cc/3SSK-NV5E].

[13]*Id.*

[14]*Id.*

[15]*See* Boris Johnson, U.K. Sec'y of State for Foreign and Commonwealth Aff., Speech at Chatham House: Global Britain: UK Foreign Policy in the Era of Brexit, 8 (Dec. 2, 2016), https://www.chathamhouse.org/event/global-britain-uk-foreign-policy-era-brexit [https://perma.cc/HH5K-XSVB].

[16]According to Meghan O'Sullivan, there are ways to see energy: (1) as a driver of domestic development; (2) a shaper of grand strategy; and (3) a determinant of international affairs. Lecture at Harvard University, 3 October 2018.

[17]It is interesting to see the conceptual evolution of this phenomenon of energy actors over time: Initially, one referred to an energy user, then consumer, then customer, and now prosumer. For an analysis of prosumers, see Leal-Arcas et al. (2017), pp. 139–172.

[18]A terminological clarification needs to be made: technically speaking, no energy is sustainable because it disappears as it is consumed; however, for the purposes of this book, we are using the terms 'sustainable energy' and 'energy for sustainable development' interchangeably, bearing in mind that the ultimate goal is energy for sustainable development.

Sustainable energy is crucial in a world where 1.2 billion people still have no access to electricity, which is a serious inequality issue.[19] A solution for sustainable energy is better governance of energy trade.[20] Energy security, or access to energy at an affordable price, is one of the main problems humanity faces.[21] Without access to energy, people and countries cannot develop their potential.[22] Today's environmental challenges are driving a shift from fossil fuels towards clean and renewable energy,[23] i.e., energy from sustainable sources, as opposed to conventional sources such as oil, natural gas,[24] or coal.[25] As the price of oil goes up, there will be a greater incentive for countries to invest in renewables to eventually obtain a cheaper, cleaner and more secure supply of energy. This transition away from fossil fuels will, however, come at a cost.[26] Others argue that the goal of sustainable energy should be "to curb global warming, not to achieve 100% renewable energy."[27]

Two energy developments are happening that explain geopolitical shifts in energy security: (1) a revolution in unconventional energy[28] and (2) a shift to a more environmentally sustainable global energy mix.[29] As a result, we are experiencing a paradigm shift from perceived energy scarcity to energy abundance.

---

[19] See *Energy Access Database*, Int'l Energy Agency, https://www.iea.org/energyaccess/database/.

[20] That said, today, the energy that crosses borders is mostly fossil fuel. See generally Leal-Arcas et al. (2016).

[21] Leal-Arcas (2016).

[22] In the view of the European Commission, there are three main goals in the clean energy transition: putting energy efficiency first, achieving global leadership in renewable energies, and providing a fair deal for consumers. See https://ec.europa.eu/energy/en/news/commission-proposes-new-rules-consumer-centred-clean-energy-transition.

[23] See for instance the growing pressure Cambridge University faced in October 2017 to abandon its investments in fossil fuels because they are incompatible with the Paris Climate Agreement. See A. Mooney, "Pressure grows on £6.3bn Cambridge University fund to drop fossil fuels," *Financial Times*, 8 October 2017.

[24] Nevertheless, one should acknowledge that the abundance of natural gas is transforming the natural-gas market. Prior to 2000, almost all natural gas was either consumed where it was produced or was traded to other countries via pipelines. Growth in supplies and in liquefied natural gas (LNG) is driving growth in gas trade in the world. An increasing number of countries has been importing LNG since 2000, resulting in more fluid, more integrated markets. See lecture by Megan O'Sullivan at Harvard University, 3 October 2018.

[25] Massai (2011).

[26] *100% Renewable Energy: At What Cost?*, The Economist, July 15, 2017, at 58–59. This global concern about environmental challenges is such that there is a United Nations process that could lead to negotiations for an international environmental treaty. See International Centre for Trade and Sustainable Development, "Efforts get underway to consider potential Global Environment Treaty negotiations," *Bridges*, Vol. 22, No. 29, 13 September 2018.

[27] *Renewable-Energy Targets: A Green Red Herring*, The Economist, July 15, 2017, at 10.

[28] For further details, see Leal-Arcas (2018a), pp. 129–142.

[29] See for instance M. O'Sullivan *et al.*, (eds.) "The Geopolitics of Renewable Energy," Harvard Kennedy School Faculty Research Working Paper Series, Working Paper 17-027, June 2017, available at https://www.hks.harvard.edu/publications/geopolitics-renewable-energy. There are several mechanisms where renewables could shape geopolitics: critical materials supply chains,

Therefore, we can expect that the energy transition will have major geopolitical consequences.[30] Equally, it is very likely that, in future, there will be less energy trade in commodities due to a more environmentally sustainable global energy mix. Instead, it is very probable that we will see more renewable energy trade via electricity thanks to technology.

One way to enhance energy security could be through greater energy efficiency, which may prove more effective than the deployment of renewable energy when it comes to reducing greenhouse gas (GHG) emissions.[31] Higher energy efficiency means that less fossil fuel must be burned to get the world economy going. Trade provides another way to enhance energy security: north-eastern Germany is not very industrialized and therefore does not consume much energy, which is needed in south Germany and other more industrialized parts of the country. Here is where trading energy can help enhance energy security. This book explores solutions through better governance of energy trade.[32]

Objectives of the Book

This book challenges the view that trade's only impact on the environment is negative.[33] On the contrary, this research takes the unconventional view that the trading system goes beyond benefiting the economy and society in that it can also contribute to environmental protection, with a specific focus on decarbonization. In this sense, this book proposes a paradigm shift in how we approach trade and develops a new theory based on the triple benefit of trade. This book incorporates the current trend of bottom-up, rather than top-down, solutions to today's global challenges. In my analysis of trade's potential for environmental protection, this book:

- Investigates how trade agreements may be more effective legal instruments than environmental agreements for environmental-protection purposes—a possibility that is both counter-intuitive and surprising;
- Identifies opportunities to promote sustainable energy and environmental protection in future trade agreements; and
- Explores the role of prosumers as new actors on the energy market towards the achievement of energy transition via energy *decentralization, democratization, digitalization, de-regulation,* and *decarbonization.* This approach shifts the

---

technology and finance, new resource curse, electric grids, reduced oil and gas demand, avoided climate change, and sustainable access to energy. Ibid.

[30]Ibid.

[31]*Renewable-Energy Targets: A Green Red Herring*, The Economist, July 15, 2017, at 10. As an example of energy inefficiency, think for instance of air-conditioning in public buses in the US, where, at times, the air-conditioning cannot be turned off, or even down, even when it is unnecessarily cold inside the bus.

[32]Leal-Arcas et al. (2016), p. 40.

[33]See for instance Lilliston, B. "The climate cost of free trade: How the TPP and trade deals undermine the Paris climate agreement," *Institute for Agriculture and Trade Policy*, 2016.

current paradigm from a top-down to a bottom-up perspective in the governance of sustainability.

1.1 State of the Art

Existing literature has taken a comparative approach,[34] but focuses only on some aspects of the problem.[35] Some have carried out their comparative analyses on the trade aspects of energy[36]; others on the external dimension of EU energy law and policy[37] or the internal dimension of EU energy policy[38]; others on the inter-relationships between trade,[39] investment,[40] transit[41] and/or environmental agendas[42] vis-à-vis energy[43]; and others, while having carried out thorough cross-

---

[34]Marin Quemada et al. (2011); Cottier and Delimatsis (2011); Leal-Arcas and Filis (2013), pp. 1–58, Oxford University Press; Sovacool (2011a), pp. 3832–3844.

[35]There are any number of books on the different aspects of globalization. There are many books on the challenges facing the global economy and those facing the global environment. There are only a few books that attempt to address the complex interrelationships between the global economy and the global environment. There are no books that do so in the context of making and re-making international economic and environmental law, and of linking international economic and environmental institutions in the broader context of sustainability.See, for instance, the following books: Hufbauer and Suominen (2010), Victor (2011) and Sachs (2015). These books devote only a few pages to international economic and environmental rules and to the possible clashes between them. None of them addresses the need to erase the lines dividing global economic and environmental governance. None of them offers specific proposals for a future global framework for providing such governance from the bottom up. Moreover, none of these books views these issues through the prism of international and EU law and policy.

[36]Selivanova (2011); UN Conference on Trade and Development (2000); Farah and Cima (2015); Ostry (2010); Sakmar (2008), p. 96.

[37]Dupont and Oberthuer (2015), Goldthau and Sitter (2015) and Talus (2013).

[38]Baumann (2010). The rationale for energy policy is mainly pollution externalities from fossil fuel combustion and in the production of energy, energy security, imperfect competition, government finance, and redistribution of wealth.

[39]Shih (2009), p. 433.

[40]Ghosh (2011); World Energy Council, "World Energy Perspectives – Non-Tariff Measures: Next Steps For Catalysing The Low-Carbon Economy," 2016, available at https://www.worldenergy.org/wp-content/uploads/2016/08/Full-report__Non-tariff-measures_next-steps-for-catalysing-the-low-carbon-economy..pdf.

[41]Yafimava (2011), Pogoretskyy (2017) and Azaria (2015).

[42]Esty (1994); Van de Graaf (2013a), p. 14; Guruswamy (1991), p. 209; Brown Weiss (2016), pp. 367–369; Wettestad (2009), pp. 393–408; Yamarik and Ghosh (2006), p. 15; Leal-Arcas (2014), pp. 11–54; Barrett S, 'Climate Change and International Trade: Lessons on their Linkage from International Environmental Agreements' (Geneva, 2010), available at https://www.wto.org/english/res_e/reser_e/climate_jun10_e/background_paper6_e.pdf.

[43]Pauwelyn (2010).

policy comparative examinations,[44] do not explore the systemic implications of their subject matter for energy security *per se*.[45] Moreover, there is literature on the implications of global[46] and regional[47] systems on energy security,[48] specific to certain structures—e.g., the EU[49] and the North American Free Trade Agreement— and to limited memberships.[50] Other research looks at the relationship of regional or sectoral systems for global energy governance,[51] but does not focus comprehensively on sustainable energy.

1.2 Originality and Ground-Breaking Nature of the Book

This book is unique in that there is no similar comparative study that takes a comprehensive approach over how the trading system can help mitigate climate change and enhance sustainable energy. The book is also unique in that it brings together a top-down and bottom-up approach to the governance of sustainability. Furthermore, it is one of the first monographs that analyze two of the most relevant global regulatory trends in recent international trade agreements, namely climate change and sustainable energy.

The link between international trade law and renewable energy law has been largely neglected by the existing literature. Previous studies have examined specific micro-aspects of the field, e.g., the effect of trade liberalization[52] on energy transit. However, there has been no overarching, cross-disciplinary study that examines the implications of international governance regimes for renewable energy and trade and EU/trans-national energy security. Due to the lack of similar studies, despite the growing research area of sustainable energy, academic literature is limited to the fields of economics and international relations.

---

[44]Barton et al. (2004); McElroy (2009); Goldemberg (2012); Van de Graaf (2013b); Florini and Sovacool (2011), p. 57; Smith and Htoo (2008), p. 217.

[45]Leal-Arcas et al. (2014), Johnston and Block (2012) and Sovacool (2011b).

[46]Birol (2012), p. 184; Sovacool and Florini (2012); Cherp et al. (2011), p. 75; Dubash and Florini (2011), p. 6.

[47]See, *e.g.*, Marketos (2008), Pirani (2009), Andrews-Speed (2012), Tunsjø (2013), Bedeski and Swanström (2012), Fingar (2016), Patnaik (2016), Malashenko (2013), Cooley (2012), Kavalski (2010) and Jonson (2006).

[48]Maniruzzaman (2002), p. 1061; Kalicki and Goldwin (2013); Goldthau and Sitter (2015). See also the contributions in Barton et al. (2004).

[49]Glachant et al. (2012a, b).

[50]de Jong, S. and Wouters, J. "Institutional Actors in International Energy Law," Leuven Center for Global Governance Studies, Working Paper No. 115, July 2013.

[51]Barton et al. (2004); Gunningham (2012), p. 119; Kuzemko (2012); Leal-Arcas (2009).

[52]Hancher et al. (2015), Bjørnebye (2010), Grubb et al. (2008), Roggenkamp et al. (2016), Isser (2015), Talus (2013) and Finger and Künneke (2011).

However, there is an omission of legal analysis in the academic market regarding the topic of this book. In fact, this is the first study to analyze the governance of renewable energy and trade comprehensively, from the perspective of law, political economy and international relations. Looking at renewable energy and trade solely from the perspective of law would not suffice to understand its complexity. Going beyond the law will help us rethink the law. It will then allow us to create an ambitious framework for sustainability.

The book breaks new ground in international relations, international and EU law by assisting in the gradual constitutionalization of international and EU renewable energy trade law in its wider economic context.[53] Its in-depth concentration on EU sustainable energy, and how it can be enhanced by changing the relationships through trade actors, is where this book is innovative and ground-breaking. Taking such a detailed look at the potential of the international trading system to go beyond economic incentives and incorporate the social and environmental aspects of development as well is why this book goes beyond the state of the art, with benefits for a broad spectrum of researchers.

This book answers a very large research question in its entirety and seeks to establish a new interdisciplinary field of study at the juncture of two different but related sectors (i.e., trade and climate change/renewable energy).

1.2.1 Trade As a Vehicle for Climate Action and Sustainable Energy

The first new concept that this book offers is how the trading system can be utilized as a vehicle for climate action and sustainable energy (i.e., energy that is affordable, secure, and clean). The book argues that the trading system can be part of the solution, not the problem, to environmental protection. How can international trade and climate change mitigation work together harmoniously without impeding each other in the context of an emerging decentralized energy system?

The book demonstrates that we are not capitalizing on the international trading system to achieve two crucial twenty-first century goals: sustainable energy and climate change mitigation. The book undertakes a broad analysis of links between the trade, energy, and climate regimes, and examines both bottom-up and top-down approaches in using trade law to tackle climate change and sustainable energy. In other words, it looks at the flagging multilateral trading system and identifies how the international trading system can be used effectively towards environmental protection.

---

[53]Cottier and Hertig (2003), pp. 261–328; Cottier (2011), pp. 495–532.

Very little research has been conducted on the impact of preferential trade agreements (PTAs)[54] in addressing climate change mitigation/environmental protection and energy security.[55] Moreover, the book argues that we can use trade law as a vehicle not only for climate action and sustainable energy, but for many of the other Sustainable Development Goals (SDGs).[56] In other words, trade can serve as an enabler for achievement of the SDGs, which are bottom-up activities, given that it embraces environmental, economic, and social dimensions.[57] This is in line with the commitments of many countries to sustainable energy. However, currently, the governance of trade and renewable energy is fragmented, with many institutions and legal instruments.

There is insufficient research on how the trade and renewable energy regimes can cooperate. No one has conducted the kind of research required to understand trade's role in sustainability issues. New agreements are starting to include sustainable development chapters,[58] and various think tanks and institutes are starting to discuss this issue.[59] But there is no thorough empirical and theoretical study of what impact such sustainable development chapters are having so far, how they can improve, how the two regimes operate and can align better, and how to capitalize on trade to really push forward the renewable energy agenda.[60]

---

[54] Regional trade agreements (RTAs) and preferential trade agreements (PTAs) are used interchangeably throughout this book.

[55] Leal-Arcas and Wilmarth (2015), pp. 92–123; Falkner and Jaspers (2012); Reuveny (2010); Eerola (2006), pp. 333–350, November; Vranes (2009); Carraro and Egenhofer (2007).

[56] Resolution adopted by the General Assembly on 25 September 2015, *Transforming our world: the 2030 Agenda for Sustainable Development*, A/RES/70/1, available at http://www.un.org/ga/search/view_doc.asp?symbol=A/RES/70/1&Lang=E. For further details on the SDGs, see Cutter, A. *et al.*, "Sustainable Development Goals and Integration: Achieving a better balance between the economic, social and environmental dimensions," Stakeholder Forum, available at http://www.stakeholderforum.org/fileadmin/files/Balancing%20the%20dimensions%20in%20the%20SDGs%20FINAL.pdf; Sachs (2016), available at http://worldhappiness.report/ed/2016/.

[57] There is also literature on how environmental agreements can help achieve the sustainable development goals. See Balakrishna Pisupati, UNEP/DELC, '*Role of Multilateral Environmental Agreements (MEAs) in achieving Sustainable Development Goals (SDGs)*', (UNEP Division for Environmental Laws and Conventions 2016) 8.

[58] Charnovitz (2008), pp. 249–251; Falkner and Jaspers (2012), p. 245. Also, environmental issues are starting to play a more prominent role in the WTO dispute settlement system. See for example, United States — Standards for Reformulated and Conventional Gasoline (29 April 1996) WT/DS2/AB/R; United States — Import Prohibition of Certain Shrimp and Shrimp Products (12 October 1998) WT/DS58/AB/R; European Communities — Measures Affecting Asbestos and Products Containing Asbestos (12 March 2001) WT/DS135/AB/R.

[59] Branford (2014); Mattoo, Aaditya and Subramanian, Arvind (04 May 2013) "Four changes to trade rules to facilitate climate change action," VOX CEPR's Policy Portal; Mavroidis and de Melo (2015), pp. 225–236.

[60] This situation raises the question whether climate change demands a normative shift in how we think about the trade regime. For analyses, see Daly (1995), p. 313; Lilliston (2016); Kanemoto et al. (2014), pp. 52–59; Gordon, Kate and Lewis, Matthew, "It's Time to Close the 'Carbon Loophole'," *Wall Street Journal* (13 November 2017), available at https://blogs.wsj.com/experts/2017/11/13/its-time-to-close-the-carbon-loophole/.

1.2.2 Governance from the Bottom Up

A second concept that this book offers is citizens' empowerment, which is an emerging concept in global governance. Empowering citizens has implications for societal change as it provides a human element to governance.[61] More direct participation by citizens is increasingly necessary to reach good governance.

New opportunities are coming up for ensuring energy security. The energy sector is undergoing a large-scale low-carbon transition. What is under-emphasized in this transition is that it involves a major paradigm shift from a supply-driven to a demand-side energy policy. Driven by a mix of geopolitical, economic, climate, and technological considerations, the energy sector is moving towards a new architecture, the principal pillars of which are progressive electrification, a cleaner energy mix, renewable indigenous energy production, increased energy efficiency, and the development of new markets to produce, transmit, and, crucially, manage energy.[62]

The shift in paradigm can be explained as follows: a top-down guidance to sustainable development will come from inter-governmental decisions (i.e., high level of abstraction), whereas a bottom-up approach means that action/implementation will happen from consumers'/citizens' participation (i.e., low level of abstraction). National governments are essential, but are no longer the only key actors, given the rise of (informal) bottom-up approaches to governance. This raises the question whether cities can make effective change if national governments do not deliver. At what point should businesses have to step up if politicians fall short? Cities around the world are demonstrating innovative strategies for advancing solutions to climate change.[63] Via this bottom-up approach to governance, citizens can ask states for reform via referenda.

1.3 Interdisciplinary and Inter-Sectoral Aspects of the Book

The book is interdisciplinary and inter-sectoral, bringing together an analysis of international trade and sustainable energy from the perspective of law, international political economy, and international relations. It takes the novel approach of bringing together law, international political economy, and international relations to explain sustainability as an academic discipline. The book applies methods of legal analysis, namely a comprehensive analysis of treaties, case law and academic

---

[61] Leal-Arcas (2018b), pp. 1–37.

[62] For an overview of the current legal and policy situation in EU energy, see Leal-Arcas and Wouters (2017).

[63] SUSTANIA, "Explore 100 City Solutions for a Greener and Fairer Future"; C40 Cities. *Powering Climate Action: Cities as Global Changemakers.* 2015. https://issuu.com/c40cities/docs/powering_climate_action_full_report.

writings from scholars as well as literature from other social science disciplines, such as international relations and international political economy, to help explore the challenges addressed.[64]

The international political economy of sustainable energy is integrated into this book to identify barriers and success factors that will explain why some countries have been more successful than others when it comes to renewables.[65] The importance of a geopolitical analysis of renewables as part of the national and international energy supply is a relatively new phenomenon.[66]

Legal scholars have used international relations theories and quantitative and qualitative methods from political science, whereas political scientists have tried to understand the causes and consequences of the legalization of international relations. Political Science/International Relations has provided most of the theoretical content and methodological guidance of international legal/political science scholarship, whereas international legal scholarship has contributed with institutional design and processes, as well as dispute settlement mechanisms. This interdisciplinary approach to challenges is in keeping with the direction that legal education has taken in recent decades. To examine the relationship between trade law and sustainable energy from the sole perspective of a legal analysis would result in incomplete insights of little worth to outsiders to the discipline of law. Scholars of international relations and international law have increasingly been involved in approaches that are more problem-driven and less theory-driven, and more receptive to the fact that the realities of international law and international politics may be related to power-politics considerations associated with realism or functional concerns related to institutionalism.[67]

1.4 Structure of the Book

This book is divided into two Parts. After this introduction, Part 1 (Chaps. 2–5) offers a top-down approach to the main themes of economic governance and sustainability (namely trade, energy, and climate change). Such an approach was typical of the twentieth century and remains necessary to provide guidance.

Chapter 2 analyzes the conceptual links between sustainability, common concern and public goods. It also examines incentives for regional and global cooperation on decarbonizing the economy.

The purpose of Chap. 3 is to explain new horizons and perspectives in international economic law in the context of sustainable development. Chapter 3 explores

---

[64]Dunoff and Pollack (2013) and Frodeman et al. (2010).
[65]Moe and Midford (2014).
[66]USAID, (2014) Encouraging Renewable Energy Development: A Handbook for International Energy Regulators.
[67]Katzenstein and Okawara (2001), pp. 153–185; Katzenstein and Sil (2008), pp. 109–130.

1.4 Structure of the Book

the potential of the trading system in helping mitigate climate change and enhancing sustainable energy. The argument is that trade agreements have tremendous potential to help mitigate climate change, which is currently under-explored. Chapter 3 first explains how trade agreements may be a legal instrument to mitigate climate change and enhance sustainable energy. It then provides an analysis of the challenges of mitigating climate change and enhancing sustainable energy. The last section examines the synergistic links between the trading and climate regimes.

Chapter 3 offers a research agenda in international economic law with proposals on how to reach sustainable development. By doing so, it makes the unconventional claim that the trading system can make a great contribution to decarbonisation. Chapter 3 offers a paradigm shift in thinking about international trade. Traditionally, trade has been understood as a stumbling block to sustainable energy. I argue that trade is a building block and that the international community should capitalize on the proliferation of regional trade agreements (RTAs)/bilateral trade agreements to enhance energy security via renewable energy and achieve clean energy. Both can be achieved with the inclusion of strong chapters on trade in goods and services related to sustainable development and renewable energy in RTAs.

Chapter 4 proposes the novel idea of using mega-regional trade agreements (RTAs) to mitigate climate change and enhance sustainable energy. It proposes the argument that only a few major greenhouse gas emitters and just three mega-RTAs can make a great contribution towards climate change mitigation and the enhancement of sustainable energy. Chapter 4 offers forum options that best deal with them with the aim to help mitigate climate change and enhance sustainable energy.

Chapter 5 contributes to the debate on energy security by highlighting various aspects of the energy sector, specifically electricity and gas, which are relevant to the discussion on regional integration and their convergence with sustainable development. Development is not possible without energy and sustainable development is not possible without sustainable energy. Chapter 5 discusses energy sustainability through the trading system and proposes regional trade agreements as means to further enhance peace and the energy security agenda of the region. It makes the case that, to achieve security of energy supply, two main factors are important: diversification to minimize risk and regional cooperation. It suggests that all the regions of the world should reduce or eliminate the technical barriers to energy trade.

Part 2 (Chaps. 6–9) looks at economic governance from the bottom up, a new feature of the twenty-first century for the implementation of projects and for action to happen. Part 2 explores in theoretical and practical terms how this new system of governance and sustainability can be achieved.

Chapter 6 examines the mega-trends of the twenty-first century in the context of sustainability. By doing so, it brings forward the novel idea of how greater participation of citizens can be very promising in helping achieve the Sustainable Development Goals. More specifically, it brings forward the novel approach that the role of citizens in international trade, climate change mitigation, and sustainable energy is crucial to reach sustainability.

Chapter 6 also critically analyzes the new challenges and opportunities that prosumers, as new energy actors, bring to achieving energy security goals.

Following trends in the EU towards new levels of cooperation in energy governance, decentralization, and the emergence of a 'gig' economy, the energy sector is currently undergoing a large-scale transition. One of its core aspects is the progressive top-down diffusion of potential, competences, and leverage across the energy value chain from States and corporate actors towards prosumers.

The novelty of Chap. 6 is that it aims to explain the paradigm shift in the governance of sustainable development: the twentieth century was characterized by a top-down approach to the governance of climate action, energy, and international trade; the twenty-first century, however, offers a bottom-up approach, marking one of the mega-trends of the twenty-first century. Chapter 6 then investigates how the international trading system can be governed from the bottom up so that there is an open, more inclusive trading system in political, legal, and economic terms.

Chapter 7 focuses on how smart grids can contribute to a broader economic transformation. It considers the economic transition occurring globally towards collaborative economics and how the EU aims to incorporate new market exchange models into smart grids energy systems. It considers the potential social and environmental benefits in addition to the challenges that lie ahead in realizing policy goals about the future. Chapter 7 explores the social and ethical dimension of smart grids in the context of the collaborative economy. Chapter 7 also explores how the EU is working towards fostering more flexible, open, transparent, and dynamic policies within the energy sector. To achieve a low-carbon sustainable society that is fair and equitable for all, the new model also has to reduce the use of resources and to use them efficiently. Finally, Chap. 7 then analyzes the role of prosumers in energy security.

The purpose of Chap. 8 is to offer the practical applications of decentralized energy in the EU. It provides an analysis of smart grids in the EU as a way forward to reach sustainable energy. It represents a significant milestone in the upscaling of the various aspects of smart grid technology across the EU. Chapter 8 examines progress on energy decentralization in various EU jurisdictions. It focuses on specific outcomes of decentralization, including deployment of smart grids and smart meters, promoting demand response, the promotion of electric vehicles and greater interconnection with neighboring countries. It also examines any existing barriers. Chapter 8 also provides an analysis of smart grids in the context of the circular economy and digital technology, including cybersecurity and data-management issues.

While the trend of energy decentralization creates ample potential for facilitating and improving the EU's security of supply as well as fulfilling its climate change targets, several caveats exist. These caveats are not confined within energy security prerogatives; they also extend to the critical management of digital security, which the digitalization of energy services brings to the fore. Private and public finance should be effectively attracted and directed to infrastructure schemes that will enable a transition from the traditional centralized power network to the decentralized nexus of smart grids. Technology will play a crucial role in facilitating the role of prosumers in the new market in-the-making.

Finally, Chap. 9 explores sustainability in the context of innovation, research, and technology. Chapter 9 concludes with the expression that there is a knowledge gap on the links between four major global concerns: trade, energy, climate change, and sustainability. With the threat of climate change looming and energy increasingly important to all aspects of human and economic development, learning more about these links is extremely timely.

References

Andrews-Speed P (2012) The governance of energy in China: transition to a low-carbon economy. Palgrave MacMillan, Basingstoke
Azaria D (2015) Treaties on transit of energy via pipelines and countermeasures. Oxford University Press, Oxford
Barton B et al (eds) (2004) Energy security: managing risk in a dynamic legal and regulatory environment. Oxford University Press, Oxford
Baumann F (2010) Europe's way to energy security. The outer dimensions of energy security: from power politics to energy governance. Eur Foreign Aff Rev 15:77–95
Bedeski R, Swanström N (2012) Eurasia's ascent in energy and geopolitics: rivalry or partnership for China, Russia and Central Asia? Routledge, Abingdon
Birol F (2012) Energy for all: the next challenge. Global Policy 3:184
Bjørnebye H (2010) Investing in EU energy security: exploring the regulatory approach to tomorrow's electricity production. Kluwer Law International, Alphen aan den Rijn
Branford D (2014) International trade disciplines and policy measures to address climate change mitigation and adaptation in agriculture. In: Meléndez-Ortiz R, Bellmann C, Hepburn J (eds) Tackling agriculture in the Post-Bali Context. International Centre for Trade and Sustainable Development, Geneva
Brown Weiss E (2016) Integrating environment and trade. J Int Econ Law 19(2):367–369
Carraro C, Egenhofer C (eds) (2007) Climate and trade policy: bottom-up approaches towards global agreement. Edward Elgar, Cheltenham
Charnovitz S (2008) The WTO's environmental progress. In: Davey WJ, Jackson J (eds) The future of international economic law. Oxford University Press, Oxford, pp 249–251
Cherp A, Jewell J, Goldthau A (2011) Governing global energy: systems, transitions, complexity. Global Policy 2:75
Cooley A (2012) Great games, local rules: the new great power contest in Central Asia. Oxford University Press, Oxford
Cottier T (2011) Towards a five storey house. In: Joerges C, Petersmann E-U (eds) Constitutionalism, multilevel trade governance and international economic law. Hart, Oxford, pp 495–532
Cottier T, Delimatsis P (eds) (2011) The prospects of international trade regulation: from fragmentation to coherence. Cambridge University Press, Cambridge
Cottier T, Hertig M (2003) The prospects of 21st century constitutionalism. In: von Bogdandy A, Wolfrum R (eds) Max Planck yearbook of United Nations Law, vol 7. Martinus Nijhoff, Leiden, pp 261–328
Daly HE (1995) Against free trade: neoclassical and steady-state perspectives. J Evol Econ 5:313
Dubash NK, Florini A (2011) Mapping global energy governance'. Global Policy 2:6
Dunoff J, Pollack M (eds) (2013) Interdisciplinary perspectives on international law and international relations: the state of the art. Cambridge University Press, Cambridge
Dupont C, Oberthuer S (2015) Decarbonization in the European Union: internal policies and external strategies. Palgrave Macmillan, Basingstoke

Eerola E (2006) International trade agreements, environmental policy, and relocation of production. Resour Energy Econ 28(4):333–350

Esty D (1994) Greening the GATT: trade, environment, and the future. Peterson Institute Int'l Ec, Washington, D.C.

Falkner R, Jaspers N (2012) Environmental protection, international trade and the WTO. In: Heydon K, Woolcock S (eds) The Ashgate research companion to international trade policy. Ashgate, Farnham

Farah P, Cima E (2015) WTO and renewable energy: lessons from the case law. J World Trade 49(6):1103–1116

Fingar T (2016) The New Great Game: China and South and Central Asia in the era of reform. Stanford University Press, Stanford

Finger M, Künneke R (eds) (2011) International handbook of network industries: the liberalization of infrastructure. Edward Elgar, Cheltenham

Florini A, Sovacool BK (2011) Bridging the gaps in global energy governance. Global Governance 17:57

Frodeman R, Thompson Klein J, Mitcham C (eds) (2010) The Oxford handbook of interdisciplinarity. Oxford University Press, Oxford

Ghosh A (2011) Seeking coherence in complexity? The governance of energy by trade & investment institutions. Global Policy 2:106–119

Glachant J-M, Ahner N, Meeus L (eds) (2012a) EU energy innovation policy towards 2050. Claeys & Casteels, Deventer

Glachant J-M et al (2012b) A new architecture for EU gas security of supply. Claeys & Casteels, Deventer

Goldemberg J (2012) Energy: what everyone needs to know. Oxford University Press, Oxford

Goldthau A, Sitter NA (2015) Liberal actor in a realist world: the European Union regulatory state and the global political economy of energy. Oxford University Press, Oxford

Grubb M, Jamasb T, Pollitt M (2008) Delivering a low-carbon electricity system. Cambridge University Press, Cambridge

Gunningham N (2012) Confronting the challenge of energy governance. Transnational Environ Law 1:119

Guruswamy L (1991) Energy and environment security: the need for action. J Environ Law 3:209

Hancher L, de Hauteclocque A, Sadowska M (eds) (2015) Capacity mechanisms in the EU energy market: law, policy and economics. Oxford University Press, Oxford

Hufbauer G, Suominen K (2010) Globalization at risk: challenges to finance and trade. Yale University Press, New Haven

Isser S (2015) Electricity restructuring in the United States: markets and policy from the 1978 Energy Act to the present. Oxford University Press, Oxford

Johnston A, Block G (2012) EU energy law. Oxford University Press, Oxford

Jonson L (2006) Tajikistan in the New Central Asia: geopolitics great power rivalry and radical Islam. I.B. Taurus, New York

Kalicki J, Goldwin D (eds) (2013) Energy and security: strategies for a world in transition, 2nd edn. Johns Hopkins University Press, Baltimore

Kanemoto K et al (2014) International trade undermines national emission reduction targets: new evidence from air pollution. Global Environ Change 24:52–59

Katzenstein PJ, Okawara N (2001) Japan, Asian-Pacific security, and the case for analytical eclecticism. Int Secur 26(3):153–185

Katzenstein PJ, Sil R (2008) Eclectic theorizing in the study and practice of international relations. In: Reus-Smit C, Snidal D (eds) The Oxford Handbook of international relations. Oxford University Press, New York, pp 109–130

Kavalski E (2010) The New Central Asia: the regional impact of international actors. World Scientific, Singapore

Kuzemko C (2012) Dynamics of energy governance in Europe and Russia. Palgrave Macmillan, Basingstoke

References

Leal-Arcas R (2009) The EU and Russia as energy trading partners: friends or foes? Eur Foreign Aff Rev 14:337–366

Leal-Arcas R (2014) Trade proposals for climate action. Trade Law Dev 6(1):11–54

Leal-Arcas R (2016) The European Energy Union: the quest for secure, affordable and sustainable energy. Claeys & Casteels, Deventer

Leal-Arcas R (2018a) Unconventional sources of fossil fuel in the European Union and China: perspectives on trade, climate change and energy security. In: Hefele P, Palocz-Andresen M, Rech M, Kohler J-H (eds) Climate and energy protection in the EU and China. Springer, Cham, pp 129–142

Leal-Arcas R (2018b) Empowering citizens for common concerns: sustainable energy, trade and climate change. GSTF J Law Soc Sci 6(1):1–37

Leal-Arcas R, Filis A (2013) The fragmented governance of the global energy economy: a legal-institutional analysis. J World Energy Law Bus 6(4):1–58

Leal-Arcas R, Wilmarth CM (2015) Strengthening sustainable development through preferential trade agreements. In: Wouters J et al (eds) Ensuring good global governance trough trade: EU policies and approaches. Edward Elgar, Cheltenham, pp 92–123

Leal-Arcas R, Wouters J (eds) (2017) Research handbook on EU energy law and policy. Edward Elgar Publishing Ltd, Cheltenham

Leal-Arcas R et al (2014) International energy governance: selected legal issues. Edward Elgar, Cheltenham

Leal-Arcas R et al (2016) Energy security, trade and the EU: regional and international perspectives. Edward Elgar Publishing Ltd, Cheltenham

Leal-Arcas R et al (2017) Prosumers: new actors in EU energy security. Netherlands Yearb Int Law 48:139–172

Lilliston B (2016) The climate cost of free trade: how TPP and trade deals undermine the Paris climate agreement. Institute for Agriculture and Trade Policy, Minnesota

Malashenko A (2013) The fight for influence: Russia in Central Asia. Carnegie Endowment for International Peace, Washington, DC

Maniruzzaman A (2002) Towards regional energy co-operation in the Asia Pacific: some lessons from the energy charter treaty. J World Investment 3:1061

Marin Quemada JM et al (2011) Energy security for the EU in the 21st century: markets, geopolitics and corridors. Routledge, Abingdon

Marketos T (2008) China's energy geopolitics: The Shanghai cooperation organization and Central Asia. Routlegde, Abingdon

Massai L (2011) European climate and clean energy law and policy. Routledge, Abingdon

Mavroidis P, de Melo J (2015) Climate change policies and the WTO: greening the GATT, revisited. In: Barrett S, Carraro C, de Melo J (eds) Towards a workable and effective climate regime. Centre for Economic Policy Research Press, London, pp 225–236

McElroy MB (2009) Energy: perspectives, problems, and prospects. Oxford University Press, Oxford

Moe E, Midford P (2014) The political economy of renewable energy and energy security: common challenges and national Responses in Japan, China and Northern Europe. Palgrave, Basingstoke

Ostry S (2010) Energy security and sustainable development: the WTO and the energy charter treaty. In: Kirton J et al (eds) Making global economic governance effective: hard and soft law institutions in a crowded world. Ashgate, Farnham

Patnaik A (2016) Central Asia: geopolitics, security and stability. Routledge, Abingdon

Pauwelyn J (ed) (2010) Global challenges at the intersection of trade, energy and the environment. Centre for Trade and Economic Integration, Geneva

Pirani S (2009) Russian and CIS gas markets and their impact on Europe. Oxford University Press, Oxford

Pogoretskyy V (2017) Freedom of transit and access to gas pipeline networks under WTO law. Cambridge University Press, Cambridge

Reuveny R (2010) On free trade, climate change, and the WTO. J Globalization Stud 1(1):90–103

Rockström J et al (2009) Planetary boundaries: exploring the safe operating space for humanity. Ecol Soc 14:32

Roggenkamp M et al (2016) Energy law in Europe. Oxford University Press, Oxford

Sachs J (2015) The age of sustainable development. Columbia University Press, New York

Sachs J (2016) Happiness and sustainable development: concepts and evidence. In: Helliwell J, Layard R, Sachs J (eds) World happiness report 2016 update, vol I. Sustainable Development Solutions Network, New York. available at http://worldhappiness.report/ed/2016/

Sakmar S (2008) Bringing energy trade into the WTO: the historical context, current status, and potential implications for the middle east region. Indiana Int Comp Law Rev 18:96

Selivanova Y (2011) Regulation of energy in international trade law: WTO, NAFTA, and Energy Charter. Kluwer, Alphen aan den Rijn

Shih W (2009) Energy security, GATT/WTO, and regional agreements. Nat Resour J 49:433

Smith M, Htoo N (2008) Energy security: security for whom? Yale Hum Rights Dev Law J 11:217

Sovacool BK (2011a) An international comparison of four polycentric approaches to climate and energy governance. Energy Policy 39(6):3832–3844

Sovacool BK (ed) (2011b) The routledge handbook of energy security. Routledge, Abingdon

Sovacool B, Florini A (2012) Examining the complications of global energy governance. J Energy Nat Resour Law 30(3):235–263

Talus K (2013) EU energy law and policy: a critical account. Oxford University Press, Oxford

Tunsjø Ø (2013) Security and profit in China's energy policy: hedging against risk. Columbia University Press, New York

UN Conference on Trade and Development (2000) Trade agreements, petroleum and energy policies. United Nations, New York

Van de Graaf T (2013a) Fragmentation in global energy governance: explaining the creation of IRENA. Global Environ Politics 13:14

Van de Graaf T (2013b) The politics and institutions of global energy governance. Palgrave/Mamillan, Basingstoke

Victor D (2011) Global warming gridlock: creating more effective strategies for protecting the planet. Cambridge University Press, Cambridge

Vranes E (2009) Trade and the environment: fundamental issues in international law, WTO law, and legal theory. Oxford University Press, Oxford

Wettestad J (2009) Interaction between EU carbon trading and the international climate regime: synergies and learning. Int Environ Agreements Politics Law Econ 9(4):393–408

Yafimava K (2011) The transit dimension of EU energy security: Russian gas transit across Ukraine, Belarus, and Moldova. Oxford University Press, Oxford

Yamarik S, Ghosh S (2006) Do regional trading arrangements harm the environment? An analysis of 162 countries in 1990. Appl Econometrics Int Dev 6(2):15

Part I
From the Top Down

Chapter 2
Cooperation on Issues of Common Concern and Public Goods

2.1 Introduction[1]

The changing global landscape of the twenty-first century saw the emergence of new challenges which threaten the economic prosperity of states, the well-being of nations, and the human rights of individuals. This chapter takes the view that some of those challenges, which have affected the European Union and its citizens profoundly, can be resolved through an effective and unified system of energy governance.[2] Accordingly, this chapter demonstrates that successful decarbonization through regional and global collective action will both boost the economy and help resolve significant human rights issues and concerns that continue to plague the European Union, such as the current refugee crisis.

---

[1]Many concepts addressed in this chapter are a further development of Leal-Arcas (2017), pp. 801–877.

[2]For instance, according to one study, by removing regulatory barriers to participating in the production of renewable energy, over 180 million Europeans (so-called "energy citizens") could produce their own renewable electricity by 2050. *See* Bettina Kampman *et al.*, *The Potential of Energy Citizens in the European Union* 5, 18, 20 (2016). This approach suggests that a bottom-up approach to renewable energy generation is desirable. *See Communication from the Commission to the European Parliament, the Council, the European Economic and Social Committee, the Committee of the Regions and the European Investment Bank, A Framework Strategy for a Resilient Energy Union with a Forward-Looking Climate Change Policy*, at 2 (stating that the European Commission's vision is "an Energy Union with citizens at its core, where citizens take ownership of the energy transition"). The Author subscribes to this idea.

In February 2015, the European Commission launched the Framework Strategy for a European Energy Union,[3] a project that envisages a resilient "Energy Union"[4] with a forward-looking climate change policy. To achieve greater energy security, sustainability, and competitiveness, the European Commission aims to strengthen and promote solidarity and trust, the full integration of the European market, energy efficiency that will contribute to moderation of demand, the effective decarbonization of the economy, and the promotion of research, innovation, and competitiveness.[5]

Decarbonization[6] is one of the pillars of the European Energy Union because it is a way to achieve both *energy security*[7] and *climate change mitigation*.[8] The latest data indicate that in 2014 the European Union imported 53% of its energy, which makes it the largest energy importer in the world.[9] In addition, six E.U. member states still depend entirely on a single supplier for their gas imports, which makes them vulnerable to supply shocks.[10] The disputes between Ukraine and Russia in 2006, 2009, and 2014 had consequences for the E.U. economy and its citizens' quality of life.[11] Sudden disruptions of energy supply could cripple the European Union and have devastating consequences.

The decarbonization of the economy through the use of renewable energy sources[12] can lead to greater energy security, as the European Union can decrease

---

[3] *Communication from the Commission to the European Parliament, the Council, the European Economic and Social Committee, the Committee of the Regions and the European Investment Bank, A Framework Strategy for a Resilient Energy Union with a Forward-Looking Climate Change Policy*, at 1–4, COM (2015) 80 final (Feb. 25, 2015), http://eur-lex.europa.eu/resource.html?uri=cellar:1bd46c90-bdd4-11e4-bbe1-01aa75ed71a1.0001.03/DOC_1&format=PDF [hereinafter Energy Union Communication] [https://perma.cc/GXR9-CPB2].

[4] *Id.* The European Energy Union is an ambitious project aiming at secure, affordable, and climate-friendly energy in the European Union. *See* European Comm'n, *Energy Union and Climate*, http://ec.europa.eu/priorities/energy-union-and-climate_en [https://perma.cc/AT7P-2ECH]; *see also* Leal-Arcas (2016) (analyzing the goals of the project) [hereinafter *European Energy Union*].

[5] *See* Energy Union Communication, at 4.

[6] Decarbonization refers to the increased use of low-carbon energy sources, such as renewables and nuclear, as well as the act of capping greenhouse gas (GHG) emissions. For the purposes of this chapter, decarbonization refers to the transition to a low-carbon economy through the use of renewable energy sources, unless stated otherwise. *See European Energy Union*, at 93–132.

[7] The International Energy Agency defines energy security as "the uninterrupted availability of energy sources at an affordable price." *See Energy Security*, Int'l Energy Agency, https://www.iea.org/topics/energysecurity/ [https://perma.cc/JLR6-E4HQ]; *see also* Leal-Arcas et al. (2016) (analyzing energy security in the context of international trade).

[8] *European Energy Union*, at 107; *see generally* Leal-Arcas (2013).

[9] *See* Energy Union Communication.

[10] *Id.* Imports are a sign of economic weakness.

[11] *See* Leal-Arcas et al. (2016), p. 1.

[12] These could be tidal, wind, solar (which is one of the big hopes for the future, but it does not work at night), hydro, wave, biomass, to name a few. There are more than 10 countries with near 100% of electricity supplied by renewable energy mostly from hydro. These countries are Albania, Angola, Bhutan, Burundi, Costa Rica, the Democratic Republic of Congo, Lesotho, Mozambique, Nepal,

2.1 Introduction

its reliance on external energy suppliers. This approach will make the bloc less vulnerable to unexpected disruptions of energy supplies. Decarbonization through renewables could also significantly reduce greenhouse gas emissions and contribute to climate change mitigation. The Paris Agreement on Climate Change (Paris Agreement),[13] negotiated in December 2015, sets a goal of keeping global average temperatures below 2 °C above preindustrial levels, as well as pursuing efforts to limit the temperature increase to 1.5 °C above preindustrial levels,[14] "recognizing that this would significantly reduce the risks and impacts of climate change,"[15] which are all local, due to the weather.

After its negotiation, it was said that the Paris Agreement was a success,[16] but real success will come once it is implemented and greenhouse gas emissions are reduced. The ultimate fate of the Agreement rests on developed countries and large emerging economies. The Agreement is a hybrid between a top-down centralized approach (which serves as oversight, guidance, and coordination) and a bottom-up approach

---

Paraguay, Tajikistan, and Zambia. One down-side of large hydro plants is that they are very invasive because they displace large numbers of people. Some sources of renewable energy work well at small-scale and local level. Equally, bio-wastes—created by human beings—can be a major source of energy. Cities produce a lot of bio-wastes, such as food wastes and sewage wastes. It is a form of sustainable biomass, with no extra land-use implications. A way to rectify food waste is via consumer behaviour through education and awareness. Incidentally, food waste has a major impact on the environment because it produces GHG emissions. If it were a country, it would be the third largest emitter of GHGs. See Climate Action Tracker, "Reducing food waste and changing diet could drastically reduce agricultural emissions," 23 January 2018, available at https://climateactiontracker.org/press/reducing-food-waste-and-changing-diet-could-drastically-reduce-agricultural-emissions/.

[13]The Paris Agreement on Climate Change is one of four major legal instruments used to mitigate climate change. The other three are the UN Framework Convention on Climate Change (UNFCCC), the Kyoto Protocol and the Copenhagen Accord. The UNFCCC distinguishes itself because its objective (Article 2) is *qualitative*, not quantitative (namely it does not provide any guidance about temperature reduction in numerical terms). Another feature that makes the UNFCCC a prominent legal document of climate change mitigation is the principle of common but differentiated responsibilities (Article 3.1). They Kyoto Protocol imposes legally binding obligations to reduce greenhouse gas emissions to specific countries (so-called Annex I countries). Unlike the Kyoto Protocol, the Copenhagen Accord is not legally binding, which means that it is a political agreement to mitigate climate change. Moreover, unlike the UNFCCC, the Copenhagen Accord provides a *quantitative* objective, namely 'to hold the increase in global temperature below 2 °C' (paragraph 2). The Paris Agreement on Climate Change is more flexible than the UNFCCC in that it does not create categories of countries, but instead offers nationally determined contributions to mitigate climate change.

[14]Despite common belief to the contrary, more people die because of cold weather than hot weather. *See* Norberg (2016), p. 120. For instance, almost twice as many U.S. citizens died between 1979 and 2006 from excess cold than from excess heat. *See* Goklany (2009), p. 106.

[15]*See* Paris Agreement, art. 2(1), Apr. 22, 2016, https://treaties.un.org/doc/Publication/UNTS/No%20Volume/54113/Part/I-54113-0800000280458f37.pdf [https://perma.cc/UC9R-FS99].

[16]*See* Michael Levi, *Two Cheers for the Paris Agreement on Climate Change*, Council on Foreign Relations: Energy, Security, & Climate (Dec. 12, 2015), http://blogs.cfr.org/sivaram/2015/12/12/two-cheers-for-the-paris-agreement-on-climate-change/ [https://perma.cc/R6JM-9T6R].

(via the nationally determined contributions to the global response to climate change).

Fulfilment of the European Commission's ambitious plan for a resilient Energy Union requires a degree of unity and dedication, as well as enhanced cooperation among member states, both regionally and globally. However, the European Union currently faces serious challenges to its security, sustainability, stability, and ultimately its *legitimacy*. In the wake of raging war on the outskirts of Europe's borders,[17] an unprecedented refugee crisis,[18] an economic debt crisis,[19] and the recent challenges associated with the United Kingdom's decision to leave the European Union,[20] the European Union faces serious integration challenges that threaten not only its legitimacy, but also its very future. This raises two vital questions. First, why would E.U. member states cooperate regionally and globally towards the decarbonization of the economy when they already face serious integration challenges? More importantly, why would E.U. member states concede to speaking with one voice on energy matters when that voice is already fragmented?

This chapter demonstrates that despite the notable integration challenges currently looming over the European Union, E.U. member states have numerous economic, legal, and political incentives to cooperate both regionally and globally. Issues such as climate change and energy supply are matters of common concern that require collaboration at the global level. Climate change mitigation is a global public good, which requires collective action by states and concerted efforts at the regional and global level.[21] This chapter contends that energy security that is achieved through the use of renewable energy sources is a global public good, the type that requires and enables collective action at the global level.

After this introduction, the chapter explores the notion of public goods and matters of common concern in the broader context of international economic law and governance. The chapter then examines possible incentives for regional and global cooperation to decarbonize the economy.

---

[17]Jim Yardley, *Has Europe Reached the Breaking Point?*, N.Y. Times (Dec. 15, 2015), https://www.nytimes.com/2015/12/20/magazine/has-europe-reached-the-breaking-point.html [https://perma.cc/78TJ-FVEF].

[18]*Id.*

[19]*Id.*

[20]*See generally* Rafael Leal-Arcas, *Three Thoughts on Brexit*, Queen Mary School of Law Legal Studies Research Paper No. 249/2016 (2016) (describing future trade relations between the United Kingdom and the European Union as well as the impact on the United Kingdom alone).

[21]See D. King *et al.*, "Climate Change: A risk assessment," available at http://www.csap.cam.ac.uk/media/uploads/files/1/climate-change%2D%2Da-risk-assessment-v9-spreads.pdf.

2.2 What Are Public Goods?

2.2.1 The Concept

Public goods, also known as "collective consumption goods," are defined by economists as the kind of goods that one individual may consume but cannot exclude access to by others.[22] For this reason, economists characterize public goods as "nonrivalrous" and "nonexcludable."[23] Classic examples of public goods include, inter alia, public water supplies,[24] street lighting, lighthouse protection for ships, and national defense services. Unlike private goods, which are usually excludable and rivalrous, public goods are not generally supplied by the private sector, as they cannot be supplied for a profit.[25] The key to why public goods present a challenge for the private sector lies in the potential for unfettered access to the benefits derived from such goods once they are made available, a phenomenon that is known as the "free rider problem."[26]

Thus, the provision of public goods is usually left to governments, which evaluate the social benefits and costs of supplying public goods, usually implementing them through taxation. Apart from the free rider problem, public goods give rise to what some have referred to as "the prisoner's dilemma."[27] The prisoner's dilemma represents a situation in which the lack of information impedes collaboration between two parties.[28] In the context of supplying public goods, the prisoner's

---

[22] See Kaul (2012a), p. 731; see also Global Environmental Commons: Analytical and Political Challenges in Building Governance Mechanisms (Eric Brousseau et al. eds., 2012); Reflexive Governance for Global Public Goods (Eric Brousseau et al. eds., 2012); Kaul (2012b), p. 38.

[23] Kaul et al. (1999), p. 2.

[24] Predications are that water demand will increase by 55% globally between 200 and 2050. See X. Leflaive, "Water Outlook to 2050: The OECD calls for early and strategic action," *Global Water Forum*, 21 May 212, available at http://www.globalwaterforum.org/2012/05/21/water-outlook-to-2050-the-oecd-calls-for-early-and-strategic-action/.

[25] See Elizabeth Hoffman, What Goods and Services are Best Provided by the Public Sector and Which are Best Provided by the Private Sector?, Iowa State University, https://www.econ.iastate.edu/node/710 [https://perma.cc/XJA4-PB4E]; see generally Samuelson (1954) (discussing challenges in determining consumption preferences for collective consumption goods).

[26] The free rider problem leads to underprovision of a good, and thus to market failure. This is so because access to a public good cannot be restricted once it is made available, thus it is difficult to charge people for benefiting from it. See Geoff Riley, *Public Goods and Market Failure*, Tutor2u: Economics, https://www.tutor2u.net/economics/reference/public-goods [https://perma.cc/K76N-4J4R].

[27] See Kaul et al. (1999), pp. 7–9.

[28] Id. The authors explain the prisoner's dilemma using the example of two prisoners who are faced with a choice of denying or confessing to a crime. If one confesses and the other denies, the one who confesses will be granted his freedom, while the other will serve 5 years in prison. If they both confess, they will both serve a reduced term of 3 years. If they both deny, they will both serve 1 year on a lesser charge that can be proven without a confession. As the prisoners are held in separate cells, they cannot communicate and agree on a common story. The authors further describe the prisoner's dilemma as follows:

dilemma could arise where the process is not supported by effective cooperation mechanisms between those who supply the goods and benefit from them, and those who simply benefit as free riders.[29] In line with this, experts and academics have contended that without a mechanism for *collective action*, public goods are at risk of being underproduced.[30]

Finally, even though the list of criteria that define a public good is exhaustive, the list of current public goods is not. Goods that were previously classified as private could later become public, and vice versa. The phenomenon of globalization,[31] technological advancements in recent years, as well as the discovery of new sources of energy, could eventually lead to the reclassification of certain goods and commodities as public, and even the creation of new public goods.

2.2.2 Global Public Goods

In recent years, the notion of a public good has expanded significantly. In an increasingly globalized world, issues such as poverty, war, climate change, blatant abuses of human rights, and market failures have caused ripple effects across the

---

Prisoner A quickly realizes that no matter what prisoner B chooses (deny or confess), he is always better off confessing to the crime. If prisoner B denies the crime, prisoner A can get off with no punishment by confessing. If prisoner B confesses, prisoner A faces 3 years in jail if he also confesses the crime, and 5 years if he denies it. Thus, prisoner A will confess. Prisoner B, facing identical choices, will also confess. The result: both prisoners will confess to the crime and will each serve 3 years in jail. The prisoners' "dilemma" arises from the fact that both would be better off cooperating—by denying the crime—than defecting—by confessing. If they could maintain their silence, they could each serve 1 year rather than three. *Id.* (emphasis added).

The concept of the prisoner's dilemma was originally framed by Merrill Flood and Melvin Dresher. *See Prisoner's Dilemma*, Stan. Encyclopedia Phil., https://plato.stanford.edu/entries/prisoner-dilemma/ [https://perma.cc/QQ38-MMRV].

[29] A good example would be where the government provides street lighting. Street lighting is a public good, thus its supply gives rise to the free rider and prisoner's dilemma problems. The government cannot exclude its citizens from benefiting from the street lighting it provides, as once it is made available, everyone can benefit. In addition, if the government does not communicate to its citizens that without their contribution, the government will not be able to supply street lighting (due to lack of funds and resources), its citizens will make the selfish choice of free riding until lighting is cut off or a cooperation mechanism is established (i.e., agreement to pay taxes). Once the government effectively communicates to its citizens that contributing (by way of taxes) will enable it to keep supplying the street lighting that everyone benefits from, this will give rise to a mutual agreement to collaborate and contribute for the common good. *See* Séverine Deneulin and Nicholas Townsend, *Public goods, Global Public Goods and the Common Good*, ESRC Research Group on Wellbeing in Developing Countries, WeD Working Paper No. 18 at 4 (Sept. 2006), http://www.welldev.org.uk/research/workingpaperpdf/wed18.pdf [https://perma.cc/KB8K-K997].

[30] *See* Stewart and Coleman (2005), p. 121.

[31] The phenomenon of globalization, which was initiated from the top down, has now reached the micro level and seems unstoppable.

globe.[32] As a result, a growing number of experts have written about the rise of the "global public good,"[33] a tangible or intangible commodity that benefits the wider public, not just at the national level, but also at the international. Globalization is good for the fight against climate change because one can make better use of resources, e.g., technology can help in the fight against climate change.

For the purposes of this chapter, a global public good is a tangible or intangible product, the production and supply of which gives rise to the infamous free rider and prisoner's dilemma issues. It is non-excludable and non-rivalrous, and is more or less available worldwide. Consequently, to avoid the underproduction of global public goods, effective mechanisms of collaboration must be established at the global level, including, inter alia, incentives and effective tools that encourage state-to-state cooperation.

2.3 Matters of Common Concern

Matters of common concern represent the worries and issues that drive people to cooperate.[34] The principle of cooperation underlies all national and international efforts to find solutions to common problems, reflected in the proliferation of international treaties and institutions. The very concept of the European Union arose out of a need for consolidated efforts to tackle matters of common concern.[35] Issues such as war, climate change, and economic crises are matters of common concern at the global level. Acknowledging this interdependency, states enter into international agreements, transforming mere desire and willingness to cooperate into

---

[32] See, for instance, the current situation in parts of the Middle East and sub-Saharan Africa. U.N. Sec'y-Gen., Address to the General Assembly (Sept. 24, 2013), https://www.un.org/sg/en/content/sg/statement/2013-09-24/secretary-generals-address-general-assembly-delivered-%E2%80%93-bilingual [https://perma.cc/W9HY-7ELX].

[33] *Defining Global Public Goods*, at 9; *see, e.g.*, Inge Kaul & Raul Mendoza, *What are Global Public Goods*? Nautilus Inst. (May 17, 2008), http://nautilus.org/gps/applied-gps/global-public-goods/what-are-global-public-goods/ [https://perma.cc/WX4X-SBPZ]; *Global Public Goods*, Global Policy Forum, https://www.globalpolicy.org/social-and-economic-policy/global-public-goods-1-101.html [https://perma.cc/N84H-UJQ2].

[34] *See* Shelton (2009), pp. 34–36 ("What makes a concern a common one . . . [is] the importance of the values at stake. This idea is also implicit in the Martens Clause and in the International Court Justice's recognition that *erga omnes* obligations arise by their very nature in view of the importance of the rights involved.") (internal quotations omitted). Issues of common concern are connected to the recognition of erga omnes obligations and the formation of collective compliance institutions and procedures that reinforce the erga omnes obligations imposed in the common interest. *See* Shelton and Kiss (2007), pp. 13–15.

[35] *See generally The EU in Brief*, Europa, https://europa.eu/european-union/about-eu/eu-in-brief_en (describing how the European Union was created to foster economic cooperation and to avoid future conflicts in the post-World War II era) [https://perma.cc/9Z95-XG6F].

legally binding obligations.[36] Thus, when it comes to some matters of common concern, states are not simply *encouraged* to cooperate; they are *obliged* to do so, in line with their responsibilities under international law.[37]

This chapter puts forth the unconventional view that, in addition to climate change, sustainable energy is ultimately a common concern. An example is Sub-Saharan migration to the EU resulting from the consequences of climate change and energy poverty in sub-Saharan Africa.[38] How can trade agreements address such issues?[39] Greater access to sustainable energy could be done regionally via the trading system.

---

[36]For example, the Paris Conference of the Parties, held in December 2015, demonstrated how states could transform the desire to cooperate on common concerns, such as climate change, into a legally binding obligation. *See* Paris Agreement; U.N. Secretary-General, Paris Agreement: Entry into Force, C.N.735.2016.TREATIES-XXVII.7.d (Depositary Notification) (Dec. 12, 2015), https://treaties.un.org/doc/Publication/CN/2016/CN.735.2016-Eng.pdf [https://perma.cc/ASE9-FVH6]. The Paris Agreement on Climate Change (Paris Agreement) came into force in 2016 after the threshold for entry into force was achieved. *See Paris Agreement – Status of Ratification*, U.N. Framework Convention on Climate Change, http://unfccc.int/paris_agreement/items/9444.php. It did so in the form of an internationally legally binding instrument, which was signed and ratified, in accordance with Article 21 of the Paris Agreement. Legally, there is no higher level of commitment at the international level. Treaties are the strongest tool available to states to enhance and solidify their international commitments to each other. *See* Guzman (2005), p. 597.

[37]For example, some international treaties have called for cooperation on environmental issues, such as the 1992 U.N. Framework Convention on Climate Change, U.N. Framework Convention on Climate Change [UNFCCC], art. 4(1), May 9, 1992, S. Treaty Doc. No. 102-38, 1771 U.N.T.S. 107; the 1992 Convention on Biological Diversity, Convention on Biological Diversity, art. 5, June 5, 1992, 1760 U.N.T.S. 79; and the 1994 Convention to Combat Desertification, U.N. Convention to Combat Desertification in Those Countries Experiencing Serious Drought and/or Desertification, Particularly in Africa, art. 4, June 17, 1994, S. Treaty Doc. No. 104-29, 1954 U.N.T.S. 3. In addition, the International Tribunal on the Law of the Sea issued an order on provisional measures on December 3, 2001, in the *MOX Plant Case (Ireland v. U.K.)*, where it indicated that the duty to cooperate may be legally enforceable. The MOX Plant Case (Ir. v. U.K.), Case No. 10, Order of Dec. 3, 2001, ¶ 89, https://www.itlos.org/fileadmin/itlos/documents/cases/case_no_10/Order.03.12.01.E.pdf. Ireland had invoked the U.N. Convention on the Law of the Sea (UNCLOS) Article 123, which requires states to cooperate in exercising their rights and performing their duties with regard to enclosed or semi-enclosed seas. *Id.* ¶ 26. The court held that UNCLOS and general international law make the duty to cooperate a fundamental principle for the prevention of marine pollution (in the Author's view, a matter of common concern), and that certain rights arise from it, which the tribunal can enforce by ordering provisional measures. *Id.* ¶ 82.

[38]Despite the current sub-Saharan emigration, the African continent has tremendous potential as reflected by its young population, economic growth, number of mobiles, population growth, and natural resources (which act as energy supply). The African continent is currently in need for electrification, industrialization, and ending poverty.

[39]The European Union is already committed to referencing its participation in the Paris Agreement on Climate Change in its trade agreements. See for instance the very innovative Japan-EU Economic Partnership Agreement (JEEPA), the first trade agreement that makes explicit reference to the Paris Agreement. See European Commission, "EU and Japan finalise Economic Partnership Agreement," Press release, 8 December 2017, available at http://trade.ec.europa.eu/doclib/press/index.cfm?id=1767. The JEEPA also sets out serious commitments to environmental protection more broadly, labor rights, and sustainable development. Moreover, the EU has clearly stated that

There is a clear growth in energy demand due to population growth, high immigration rates in certain regions of the world, economic growth,[40] and large infrastructure projects. One could make use of economies of scale in energy generation, whether renewable energy or gas. One could also increase economic efficiency, harmonize regional policies, and improve governance. Through energy cooperation, one could have further market integration.

2.4 Public Goods and Matters of Common Concern

This Part deals with public goods such as climate change mitigation and common concerns such as climate change. Who will pay for climate change, which is an inter-regional, inter-generational, inter-national issue? How are we going to pay for it in a world of uncertainty, whether it is economic growth, population growth, technology innovation, or energy supply? The tragedy of the commons[41] is an economic theory used to explain a situation where shared resources exist, and self-interest undermines collective public goods. Such a situation raises questions of who pays the costs and who reaps the benefits. The defining features of what engenders the tragedy of the commons are excludability and rivalry (see Table 2.1). The following Sections explore these indicators in greater detail.

The 'tragedy of the commons'[42] is a well-known example of a collective-action problem. How do we balance the personal interest of costs and benefits for individuals of an outcome against the common good of a collective group? Is the current situation one where we overweigh the value of downside risks because it is direct

---

the US's continued participation in the Paris Climate Agreement is essential if the two sides are to resume formal trade talks on the Trans-Atlantic Trade and Investment Partnership. See "EU Officials: US Participation in Paris Climate Deal Key for Resuming Trade Talks," *Bridges*, Vol. 22, No. 5, 15 Feb. 2018, available at https://www.ictsd.org/bridges-news/bridges/news/eu-officials-us-participation-in-paris-climate-deal-key-for-resuming-trade?utm_source=International%20Centre%20for%20Trade%20and%20Sustainable%20Development%20%28ICTSD%29&utm_campaign=3f37cbf8d6-EMAIL_CAMPAIGN_2018_02_15&utm_medium=email&utm_term=0_7d42ca02cd-3f37cbf8d6-158267977. See also E.A. Crunden, "EU will only make trade deals with nations that ratify Paris climate agreement," *ThinkProgress*, 6 February 2018, available at https://thinkprogress.org/eu-paris-us-decd4aad9145/. Moreover, in November 2017, the Inter-American Court of Human Rights recognized the right to a healthy environment as fundamental to human existence. See Advisory Opinion OC-23/17 of 15 November 2017, available at http://www.aida-americas.org/sites/default/files/oc23_corte_idh.pdf. On how free-trade agreements can enhance environmental cooperation, see Yoo and Kim (2016), pp. 721–738.

[40]But see Costanza and Daly (1992), pp. 37–46.

[41]*See generally* Hardin (1968), pp. 1244–1245 (explaining the "tragedy of the commons" concept and providing an example of its application in respect to population growth and issues of environmental pollution).

[42]Hardin (1968), pp. 1243–1248.

Table 2.1 Indicators in the tragedy of the commons[a]

	Rivalrous	Non-rivalrous
Excludable	*Private goods*	*Club goods*
	Food, fuels, minerals	Cable TV, cinemas
	Car	Some social services
	House	
Non-excludable	*Common pool resources*	*Public goods*
	Forests	Air
	Fisheries	Law enforcement
	Wildlife	Public radio, broadcast TV
	Fossil fuels	Streetlights

[a]Lisa Dale, Lecture at Yale University titled Multiple Scales of Sustainability Governance (Sept. 2016) (on file with author)

and personal? Equally, do we underweigh the value of benefits because they are diffuse and general?[43]

Matters of common concern such as climate change and economic crises have far-reaching and devastating effects, such as loss of sea ice, accelerated sea level rise,[44] more intense heat waves and poverty.[45] There is a need to clarify the relationship between the legal principle of common concern, the economic concept of public goods, and legal scholarship on the governance of global public goods.[46] Most countries are dependent on energy imports, so it is necessary to cooperate to find ways to enhance sustainable energy via the trading system. Issue-linkages provide a way to increase cooperation on global common concerns by increasing participation in regional/global institutions.[47] Yet, this raises a timely question: Why would, say, EU Member States cooperate towards the creation of an Energy Union when the EU is facing integration challenges?[48]

---

[43]See Brian Efird, "The political economy of a regional electricity market," KAPSARC PowerPoint presentation, 10 May 2018, p. 3.

[44]In fact, new evidence and concerns about the fact that countries lying at sea level will be great victims of sea level rise meant that the nations of the world agreed on a new target at the Paris climate summit in 2015: keeping global warming "well below" 2 °C above pre-industrial temperatures. For details of where we stand as of 2018, see UN Environment, "Emissions Gap Report 2018," November 2018, available at http://wedocs.unep.org/bitstream/handle/20.500.11822/26895/EGR2018_FullReport_EN.pdf?isAllowed=y&sequence=1.

[45]See for instance Soltau (2016).

[46]Leal-Arcas (2017), pp. 801–877; Kaul (2016); Petersmann E.-U. (2012) (ed.) *Multilevel Governance of Interdependent Public Goods: Theories, Rules and Institutions for the Central Policy Challenge in the 21st Century* (EUI Working Papers); Cafaggi and Caron (2012), pp. 643–649.

[47]Meyer (2012), pp. 319–348; Low, P. 'Hard Law and 'Soft Law': Options for fostering International Cooperation' (International Centre for Trade and Sustainable Development and the World Economic Forum, October 2015).

[48]For an interesting explanation of environmental cooperation from a game-theory point of view, see Hirsch (2009), pp. 503–510.

2.4 Public Goods and Matters of Common Concern

At the regional level, the EU seems to be one of the most committed regions in the world to climate change mitigation. With great potential for solar energy in the south and wind energy[49] in the north of Europe—two of several sources of renewable energy—the EU is in a position to make use of renewables to mitigate climate change and to gradually become more energy independent and efficient,[50] following the theory of comparative advantage,[51] thereby making energy affordable and reliable. When the wind does not blow in one region or the sun does not shine in another region, regions can exchange each other's renewable energy through interconnectors.[52]

In addition, the EU's Partnership Instrument[53] promises to be pivotal when dealing with cooperation on issues of common concern. Its aim is to support "public diplomacy, people to people contacts, academic cooperation and outreach activities to promote the Union's values and interest."[54] The Partnership Instrument aims at greater interaction between the EU and countries that "play an increasingly prominent role in global affairs, international economy and trade, multilateral fora and global governance and in addressing challenges of global concern."[55]

Climate change and energy supply issues are matters of common concern that give rise to *erga omnes* obligations, due to the value and importance of the rights involved. The destructive impact of climate change must be mitigated through joint efforts and collective action at the global level. Energy supply issues have become more prevalent in recent years, as states become increasingly conscious of the dangers associated with heavy reliance on traditional energy resources.[56] In a world of growing energy demands,[57] the rising scarcity of traditional energy resources[58] and soaring levels of pollution[59] highlight the urgent need for collective

---

[49]To see the various areas of high wind in the word for wind power generation, see Global Wind Atlas, available at https://globalwindatlas.info/.

[50]From an economic perspective, see the views of Ayres (1994), pp. 435–454.

[51]Ricardo (1817).

[52]Since full transition to 100% renewable energy is not realistic in the near future, when the wind is not blowing and the sun is not shining, natural gas can fill those gaps.

[53]European Commission, International Cooperation and Development, "Partnership Instrument," available at https://ec.europa.eu/europeaid/funding/funding-instruments-programming/funding-instruments/partnership-instrument_en.

[54]Ibid.

[55]Ibid.

[56]*See* Leal-Arcas and Minas (2016), p. 647.

[57]*See EIA Projects 48% Increase in World Energy Consumption by 2040*, U.S. Energy Info. Admin. (May 12, 2016), https://www.eia.gov/todayinenergy/detail.php?id=26212 [https://perma.cc/3GXJ-H7UM]. Reducing energy demand is an easy way to mitigate climate change.

[58]*See Energy Scarcity*, ResilientCity.org, http://www.resilientcity.org/index.cfm?id=11897 [https://perma.cc/L9HN-W5GZ].

[59]*See, e.g., Delhi Air Quality Plunges to 'Severe' Category as Pollution Levels Soar*, Hindustan Times (Nov. 12, 2016), http://www.hindustantimes.com/delhi/delhi-air-quality-plunges-to-severe-category-as-pollution-levels-soar/story-5ZWa6MndeqDPf61WgSjYNI.html [https://perma.cc/RR9U-KDN6]. In fact, more than 5.5 million people die prematurely every year as a result of air

global action to mitigate the negative effects of climate change and ensure global energy security.[60]

2.4.1 Climate Change Mitigation Is a Global Public Good That Calls for Collaborative Effort

Climate change mitigation has long been regarded as a public good.[61] "The atmosphere is an international public good, in that all countries benefit from each country's reduction in [GHG] emissions."[62] Climate change mitigation is both non-rivalrous and non-excludable and, because it is available on a worldwide basis, is a global public good.[63] Accordingly, the reduction of GHG emissions presents the same issues and challenges that are commonly associated with the provision of public goods at the national level, such as the lack of economic incentives, and the infamous free rider and prisoner's dilemma issues.[64] So, from an economic perspective, climate change mitigation requires collaborative effort and

---

pollution. See J. Amos, "Polluted air causes 5.5 million deaths a year new research says," *BBC*, 13 February 2019, available at https://www.bbc.co.uk/news/science-environment-35568249. To put this fact in context, it is as if countries with the population of Finland would disappear from the world map every year.

[60] As a result, China has been very active in climate action in recent years and intends to do so in years to come. See, for instance, China's ambition to spend over $360 bill on renewables by 2020, M. Forsythe, "China Aims to Spend $360 billion on renewable energy by 2020," *The New York Times*, 5 January 2017, available at https://www.nytimes.com/2017/01/05/world/asia/china-renewable-energy-investment.html?mcubz=0; on wind energy, China's investment has been remarkable: S. Evans, "Mapped: How China dominates the global wind energy market," 19 April 2016, available at https://www.carbonbrief.org/mapped-how-china-dominates-the-global-wind-energy-market; see also S. Lacey, "China adds more than 5GW of solar PV capacity in the first quarter of 2015," 21 April 2015, available at https://www.greentechmedia.com/articles/read/china-adds-more-than-5gw-of-solar-pv-capacity-in-the-first-quarter-of-2015#gs.pgeEFKg; on solar energy, in 2017 China opened the world's largest floating solar plant (https://www.weforum.org/agenda/2017/06/china-worlds-largest-floating-solar-power/) and built a 250-acre solar farm shaped like a giant panda (www.sciencealert.com/china-just-built-a-250-acre-solar-farm-shaped-like-a-giant-panda).

[61] *See, e.g.*, Hasson et al. (2010), p. 331.

[62] *See* Climate Change 1995: Economic and Social Dimensions of Climate Change 21 (James P. Bruce et al. eds., 1996), https://www.ipcc.ch/ipccreports/sar/wg_III/ipcc_sar_wg_III_full_report.pdf [https://perma.cc/N8CF-RA4S].

[63] *See* Kaul (2012b), pp. 46–47; *see also* Leal-Arcas and Minas (2016), p. 622.

[64] *See* Climate Change 1995: Economic and Social Dimensions of Climate Change, (James P. Bruce *et al.* eds., 1996), at 167. The prisoner's dilemma issue presents itself in the context of climate change mitigation when, in the absence of effective cooperation between states, the negative effects of climate change cannot be mitigated. States must exchange information on emission cuts and other areas of specialized knowledge and expertise to effectively mitigate the effects of climate change.

collective action.⁶⁵ There is a temporal dimension to climate change in that GHGs accumulate in the atmosphere.

All of this raises the question of what is the most effective way to reduce GHG emissions. If half the world's population became vegetarian, it would save 66bn tones of CO_2.⁶⁶ If the international community were to replant two thirds of degraded tropical forests, it would save 61bn tones of CO_2.⁶⁷ If we increased by one third the global bicycle journeys to work, we would save just 2.3bn tones of CO_2.⁶⁸ By contrast, it has been argued that, in fact, "making air-conditioners radically better" may be the answer in that, if we replace refrigerants that damage the atmosphere, it would reduce GHGs by 90bn tones of CO_2 by 2050.⁶⁹ However, until energy can be produced without carbon emissions, these extra air-conditioners will warm the planet. Currently, air-conditioners are part of a vicious cycle: the warmer the Earth, the more people will need them. At the same time, the more energy-inefficient air-conditioners there are, the warmer the planet will be.⁷⁰ Hence, the need to make them drastically better.

2.4.2 Energy Security Through the Lens of a Public Goods Analysis

Energy security has become a significant issue of concern for the European Union, given the region's precarious energy situation.⁷¹ The traditional concept of energy security focuses on the *continual* availability of energy sources at an affordable price, which so far has been associated with a steady and constant availability and supply of traditional energy resources, such as oil and gas.⁷² While it is generally agreed that climate change mitigation, as discussed above, is a global public good,⁷³ the classification of energy security as a public good has divided experts and academics.⁷⁴ The traditional interpretation of a public good cannot be applied to the concept of energy security, as energy security does not fall under the definition of

---

⁶⁵See generally Stavins (2013, 2012), Aldy and Stavins (2010) and Hay et al. (2005).

⁶⁶The Economist, "In praise of air-conditioning: Rebirth of the cool," 25 August 2018, pp. 8–9, at 8.

⁶⁷Idem.

⁶⁸Idem.

⁶⁹Idem.

⁷⁰Ibid, at p. 9.

⁷¹*See* Leal-Arcas (2016), Chapter 1, at 11; Leal-Arcas et al. (2016), p. 1.

⁷²*See* Leal-Arcas et al. (2016), pp. 6–7.

⁷³*See* Leal-Arcas (2012a), p. 892; *see also* Leal-Arcas and Minas (2016), p. 647.

⁷⁴*See, e.g.*, Christopher A. Simon, *Is Energy a Public Good?*, Renewable Energy World (July 2, 2007), http://www.renewableenergyworld.com/articles/2007/07/is-energy-a-public-good-49201.html [https://perma.cc/KG9F-8LJ2].

Fig. 2.1 The attributes of sustainable energy in the energy trilemma

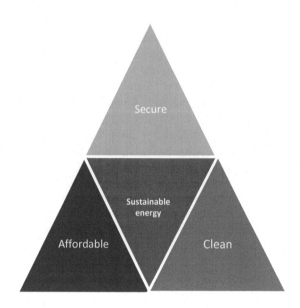

a non-rivalrous and non-excludable good, as defined by economists.[75] The consumption of traditional sources of energy, such as oil and gas, naturally leads to depletion and excludability; hence energy security in this context cannot be classified as a public good. However, by shifting the focus of global efforts towards the creation of a framework that delivers uninterrupted, secure, affordable, clean, and *sustainable* energy through the use of modern technology, states can achieve global renewable energy security, which is a global public good. So, renewable energy may become the engine to obtain the three attributes of sustainable energy in the energy trilemma[76]: namely clean, secure, and affordable energy (see Fig. 2.1).

2.4.2.1 Global Renewable Energy Security

A further paradigm shift is the concept that renewable energy is part of energy security enhancement. The concept of global renewable energy security is rooted in a belief that states—with the help of modern technology—can achieve uninterrupted,

---

[75] A rivalrous good is a good that, once consumed by one consumer, cannot be consumed by other consumers. *See* Weimer and Vining (2005).

[76] "The World Energy Council's definition of energy sustainability is based on three core dimensions–energy security, energy equity, and environmental sustainability. These three goals constitute a 'trilemma,' entailing complex interwoven links between public and private actors, governments and regulators, economic and social factors, national resources, environmental concerns, and individual behaviours." *World Energy Trilemma*, World Energy Council, https://www.worldenergy.org/work-programme/strategic-insight/assessment-of-energy-climate-change-policy/ [https://perma.cc/E4H6-C6P7].

secure, clean, sustainable, and affordable[77] energy through the use of renewable energy resources.[78] The concept of global renewable energy security is better understood through examples *A* and *B*, illustrated below.

2.4.2.1.1 Example *A*

State *X* is rich in sunlight but lacks the technological capacity to process solar energy. State *Y*, on the other hand, possesses the technological capacity to process solar energy, but does not have renewable energy capacity because it is not rich in renewable natural resources such as sunlight. State *X* and state *Y* enter into an agreement whereby state *Y* supplies state *X* with access to the technology it needs to process solar energy and, in turn, state *X* gives state *Y* access to processed renewable energy. As a result, both states *X* and *Y* gain access to uninterrupted, secure, clean, sustainable, and affordable energy.

In the example above, the benefits reaped by states *X* and *Y* also become available to the wider global community, as surplus renewable energy can then be sold to other states.[79] Other states can now gain access to renewable energy generated by states *X* and *Y*, even if those other states do not have the technological capacity to process raw renewable material. Example *B*, below, illustrates this latter condition.

---

[77] Andrew Griffin, *Solar and Wind Power Cheaper than Fossil Fuels for the First Time*, Independent (Jan. 4, 2017), http://www.independent.co.uk/environment/solar-and-wind-power-cheaper-than-fossil-fuels-for-the-first-time-a7509251.html [https://perma.cc/6EWX-JWSH].

[78] *See generally* Leal-Arcas and Minas (2016), pp. 621–666 (reviewing governance structures that govern energy policy and manage energy security).

[79] Scientists are developing new and effective mechanisms that allow for the storage of different types of renewable energy in times of deficit or surplus in production, and the subsequent transportation of any excesses. *See, e.g.*, Gotz et al. (2016), p. 1371 (describing a power-to-gas process chain that could be implemented to store renewable energy). Common forms of renewable energy storage include pumped-storage hydroelectric dams, rechargeable batteries, thermal storage (including molten salts that can store and release large amounts of heat energy), compressed air energy storage, flywheels, cryogenic systems, and superconducting magnetic coils. *See, e.g.*, *Energy Storage Technologies*, Energy Storage Ass'n, http://energystorage.org/energy-storage/energy-storage-technologies [https://perma.cc/W833-XZ9J]. One way to store and transport renewable energy is through the "power-to-gas" method. Gotz et al. (2016), p. 1371. The term "power-to-gas" refers to the new technologies that are used for the storage and transport of regenerative energy in the form of methane or hydrogen. *Id*. For example, renewable electric energy can be transformed into storable methane via electrolysis and subsequent methanation. *Id*. See also F. Reinhardt and S. Hiatt, "Colbrun and the Future of Chile's Power," Harvard Business School, pp. 1–20, 2013.

2.4.2.1.2 Example *B*

State Z, which is not rich in sunlight,[80] does not have the technological capacity to process raw renewable energy resources such as solar energy. Thus, state Z relies on supplies of conventional fossil fuels to meet its energy demands. However, state Z can now enter into an agreement with either state X or state Y to secure its supply of renewable energy through a separate agreement with either or both states.[81]

Example *B* demonstrates just a fraction of the vast potential of renewable energy that can be shared to help meet global energy demands. The model above could be applied to any type of renewable energy resource, such as wind, sunlight, or rain. The agreement between states X and Y in the example above opens the door for trade in renewable energy at the regional and global level, with endless possibilities for states to engage in bilateral, trilateral, plurilateral, and multilateral arrangements for trade in renewable energy. Such arrangements could lead to increased flows of renewable energy throughout the globe, through the use of various mechanisms such as renewable energy trading platforms or intergovernmental agreements on energy trade.

The gradual proliferation of renewable energy around the world metaphorically resembles a spider web, the center of which connects modern technology and renewable energy sources. This gradual spread of renewable energy across the globe—made possible by modern technology and innovation—will ultimately lead to global renewable energy security.

2.4.2.2 Global Renewable Energy Security As a Global Public Good

Rapid changes in technology can alter the categorization of goods, turning previously private goods into public goods, and vice versa. For this reason, this chapter suggests that global renewable energy security is a global public good, as it is

---

[80]Even though all states have access to sunlight, sunlight is not as consistent or easily accessible in all parts of the world. In addition, the amount of energy that can be generated by solar power is unpredictable, as its supply depends on, inter alia, weather conditions. Thus, energy generated by solar power can be produced in excess or deficit and quantities can be quite volatile. One way to resolve issues related to the variability of renewable energy production could be through state-to-state trade in renewable energy. *See Energy Challenges: Reliability*, EDF Energy, https://www.edfenergy.com/future-energy/challenges/reliability [https://perma.cc/5AQF-V536].

[81]Renewable energy trading is a good way for states that do not have renewable energy capacity to secure access to clean, sustainable energy. Directive 2009/28/EC, adopted under the auspices of the European Union's 2020 action plan, encourages states to exchange energy from renewable sources through a combination of domestic production and foreign imports. *See* Directive 2009/28/EC of the European Parliament and of the Council of Apr. 23, 2009 on the Promotion of the Use of Energy from Renewable Sources and Amending and Subsequently Repealing Directives 2001/77/EC and 2003/30/EC, 2009 O.J. (L 140) 16, arts. 1, 6.

nonexcludable[82] and *nonrivalrous*,[83] and it is available, to a greater or lesser extent, on a worldwide scale. Common issues associated with the provision of public goods—such as the free rider issue and the prisoner's dilemma—could arise when renewable energy becomes widely and globally available. For example, where a state secures uninterrupted access to sustainable energy, that energy becomes a common good, from which the wider public can freely enjoy benefits. Because of the non-excludability of global renewable energy security, there is a risk that people will take advantage of the benefits it generates without paying for them.

Finally, achieving global renewable energy security requires collective action and cooperation between the various actors involved in the supply and demand chain. Without effective collaborative mechanisms in place to ensure the free flow of information, technical knowledge, and skills, global renewable energy security cannot be achieved.

2.4.3 *Regional and Global Cooperation on Decarbonizing the Economy Will Contribute to Climate Change Mitigation*

In line with its obligations under the Paris Agreement,[84] the European Union has made a pledge to reduce GHG emissions by at least 40% by 2030,[85] 60% by 2040,[86] and 80% below 1990 levels by 2050.[87] The 2030 climate and energy framework also sets two additional targets for the year 2030: achieving at least a 27% share of renewable energy, and at least a 27% improvement in energy efficiency.[88] A shift

---

[82]When global renewable energy security is achieved, no person in the world can be excluded from consuming the available energy, as it becomes freely and widely available.

[83]One state's agreement to supply another state with renewable energy does not diminish the overall capacity of renewable energy available, and thus, use by one state does not reduce availability for other states. In addition, where an individual consumes renewable energy, her consumption does not reduce the availability for other individuals in the same or other states.

[84]*See* Paris Agreement.

[85]*See* European Comm'n, *2030 Climate and Energy Framework*, http://ec.europa.eu/clima/poli cies/strategies/2030/index_en.htm [https://perma.cc/6SLM-VQEL].

[86]*See* European Comm'n, *2050 Low-Carbon Economy*, http://ec.europa.eu/clima/policies/strate gies/2050/index_en.htm [https://perma.cc/3XA3-K298].

[87]*Id.* ("The European Commission is looking at cost-efficient ways to make the European economy more climate-friendly and less energy consuming . . . . The roadmap suggests that, by 2050, the [European Union] should cut its emissions to [eighty per cent] below 1990 levels through domestic reductions alone (i.e. rather than relying on international credits). This [goal] is in line with [E.U.] leaders' commitment to reducing emissions by [eighty to ninety-five per cent] by 2050 in the context of similar reductions to be taken by developed countries as a group. To reach this goal, the [European Union] must make continued progress towards a low-carbon society. Clean technologies play an important role.").

[88]*See* European Comm'n, *2030 Climate and Energy Framework*.

away from volatile fossil fuels will ensure that the European Union reaches its GHG emission targets, and that it introduces a higher share of renewable energy resources into its economy, in line with its 2030 climate and energy framework and its obligations under the Paris Agreement. The effective decarbonization of the economy, however, cannot occur if E.U. member states act in isolation. If the European Union is to reach its target goals by 2030, its member states must cooperate on decarbonizing the economy, both regionally and globally.[89]

Additionally, concerted action is needed in order to tackle poverty and low standards of living, as developing states that still grapple with such issues are less likely to achieve low-carbon economies within the timeframe set under the Paris Agreement. Cooperation between developed and developing states, for example, could lead to the exchange of technology, skills, expert knowledge, and resources.[90] This exchange, in turn, can stimulate economic growth in developing states, and accelerate the process of decarbonization. Otherwise, developing states may be less willing to cut their emissions, as slowing down the process of industrialization could harm their economies. As deep and successful decarbonization requires profound changes to countries' energy and production systems, the only way to achieve this by 2030, or as soon as possible, is through deep collaborative efforts. By establishing solid collaborative mechanisms that encourage the exchange of renewable energy resources and technology,[91] E.U. member states can become the driving actors in promoting the development of critical low-carbon technologies and making them

---

[89] Article 6 of the Paris Agreement outlines methods by which market mechanisms that were established under the Kyoto Protocol can be developed into mechanisms that allow for the sharing of responsibility for climate action across borders. Paris Agreement, art. 6. Article 6 recognizes the potential of cooperation to promote sustainable development and environmental integrity. *Id.*

[90] Aldy and Stavins (2012), pp. 1043–1044; Stavins, Robert "Paris Agreement: A Good Foundation for Meaningful Progress," December 12, 2015. Blog Post at An Economic View of the Environment.

[91] For example, the United Kingdom and France signed a declaration on nuclear energy and cooperation on climate change action in 2014. *See* Press Release, U.K. Dep't of Energy & Climate, UK and France Sign Declaration on Nuclear Energy and Agree Cooperation on Ambitious Climate Change Action (Jan. 31, 2014), https://www.gov.uk/government/news/uk-and-france-sign-declaration-on-nuclear-energy-and-agree-cooperation-on-ambitious-climate-change-action [https://perma.cc/F3YJ-VVH7]. The declaration paved the way for, inter alia, the successful mitigation of climate change and the development of low-carbon secure electricity, which provides new green jobs and investment. *Id.* In addition, in 2012, the United Kingdom and Iceland signed an agreement to encourage enhanced cooperation between the two states, as well as greater use of interconnectors for the transportation of energy under the sea. *See* Announcement, U.K. Dep't of Energy & Climate, UK and Iceland Sign Energy Agreement (May 30, 2012), https://www.gov.uk/government/news/uk-and-iceland-sign-energy-agreement [https://perma.cc/E37C-JNQM]. Further agreements on cooperation on renewables have been signed between, inter alia, Denmark and China, and the south-west communities of England and the Channel Islands. *See* Press Release, Danish Ministry of Energy, Utilities & Climate, China and Denmark Sign New Cooperation Agreement on Energy Efficiency (May 1, 2014), http://www.efkm.dk/en/news/china-and-denmark-sign-new-cooperation-agreement-on-energy-efficiency [https://perma.cc/6UF5-PU3J]; *Channel Islands' Link with South-West England on Marine Power*, BBC News (Dec. 3, 2013), http://www.bbc.co.uk/news/world-europe-guernsey-25200486 [https://perma.cc/W43X-7GXC].

commercially available and accessible to both developed and developing states. The establishment of collaborative mechanisms can catalyze the process of decarbonization, allowing the European Union to quickly and effectively honor its international responsibilities and obligations on climate change mitigation.

2.4.4 Regional and Global Efforts Towards Decarbonization Could Contribute to the Resolution of Pressing Economic and Human Rights Issues

This Section focuses on the importance of sustainable development in the context of economic growth.[92] A good example of sustainable development is improved access to energy. It is a well-known fact that development leads to an increase in the level of per capita energy consumption.[93] Most of those around the world living without electricity (around 96%) are in sub-Saharan Africa and developing Asia.[94] Eighty per cent of them live in rural areas.[95] Yet, in the case of Africa, the continent receives the least amount of climate finance in the world—around 4%.[96] Controlled energy costs and increased availability will ensure a more efficient use of electricity as well as changes in lifestyle, but would limit economic growth in the developing world. For all these reasons, the energy future should be sustainable, based on renewable energy.[97]

---

[92]Quite inspiring are the words of Aldous Huxley in 1928: '*Progress! You politicians are always talking about it. As though it were going to last. Indefinitely. More motors, more babies, more food, more advertising, more money, more everything, for ever. You ought to take a few lessons in my subject. Physical biology. Progress, indeed!*' See A. Huxley, *Point Counter Point*, The Literary Guild of America, 1928. See also Leal-Arcas (2018a).

[93]*Energy Use (Kg of Oil Equivalent Per Capita)*, The World Bank (2014), http://data.worldbank.org/indicator/EG.USE.PCAP.KG.OE (demonstrating how developed nations consume more energy per capita because their citizens have a higher purchasing power, which translates into buying and using goods that consume energy, namely cars, boats, computers, and cell phones, to name but a few) [https://perma.cc/S3X5-YWP2].

[94]*Id.*

[95]*Rural Population*, The World Bank (2014), http://data.worldbank.org/indicator/SP.RUR.TOTL.ZS [https://perma.cc/BXX6-ESPP].

[96]United Nations Climate Change Conference (@COP22), Twitter (Aug. 31, 2016, 4:10 AM), https://twitter.com/cop22/status/770941942922350593 [https://perma.cc/A7NP-4MBP].

[97]It is interesting to note that most of the renewable energy for electricity production comes from wind energy, not solar, because the wind blows also at night, unlike solar energy, where there is no sun at night, therefore no solar energy during night time. Another reason why most of the renewable energy for electricity production comes from wind energy is that there is greater use of land with solar energy than with wind energy, which is taken from agricultural space.

2.4.4.1 The European Union's Human Rights Crisis

Poverty, war, and repression have driven thousands of people to seek refuge in the European Union.[98] A large number of refugees that attempt to cross the European Union's borders risk their lives and those of their loved ones in order to escape poverty and pitiable living conditions, brought about mostly by conflicts, climate change, and environmental degradation.[99] Energy poverty in particular is a serious issue in sub-Saharan Africa.[100] It has led to an increase in migration to the European Union and is regarded by many as a security problem associated with international crime, terrorism, and trafficking,[101] and has in turn contributed to xenophobia and racism in the European Union.[102]

Related to the notion of refugee status is the concept of (economic or climate) migrant status,[103] often related to energy poverty. Demographic and economic changes push citizens out of poor and middle-income countries and into developed

---

[98] Various reports and articles published in the last 10 years demonstrate that there has been a surge in the influx of refugees from the African continent, particularly from sub-Saharan Africa. *See Key Facts: Africa to Europe Migration*, BBC News (July 2, 2007), http://news.bbc.co.uk/1/hi/world/europe/6228236.stm [https://perma.cc/5PM7-5CYW]. The majority of refugees are forced to seek refuge in European countries due to war, conflict, political upheaval, poverty, and climate change. Isabel Schäfer, *Migration to Europe – Is North Africa Europe's Boarder [sic] Guard?* at 1, German Dev. Inst. (June 8, 2015), https://www.die-gdi.de/uploads/media/German_Development_Institute_Schaefer_08.06.2015.pdf [https://perma.cc/896L-TXB3]. An increasing number of refugees come from sub-Saharan Africa, a region that suffers from energy poverty and where the negative effects of climate change have driven many to relocate in search of a better future. *Id.*

[99] Aryn Baker, *How Climate Change is Behind the Surge of Migrants to Europe*, Time (Sept. 7, 2015), http://time.com/4024210/climate-change-migrants/ [https://perma.cc/D78L-E5U3].

[100] The region has a tremendous energy deficit that is considered by many to be one of the major elements constraining Africa's economic and social development. *See* Matt Timms, *Energy Poverty Stifles Sub-Saharan Africa's Economic Development*, World Fin. (May 3, 2015), http://www.worldfinance.com/markets/energy-poverty-stifles-sub-saharan-africas-economic-development [https://perma.cc/ZH4M-PBMB]; *see* Int'l Energy Charter, "Africa and the Energy Charter: The Bountiful Continent and the Energy Conundrum" (2015), http://www.energycharter.org/fileadmin/DocumentsMedia/Infographics/2015_Energy_Charter_And_Africa.pdf ("According to recent [International Energy Agency] data, less than 300 million Sub-Saharan Africans out of roughly 915 million people living in the region have access to electricity. This means that between [sixty and seventy per cent] of Africans are disconnected. In overall terms, there are about 1.2 billion people in the world with no access to electricity, half of whom live in the African continent.") [https://perma.cc/4LQK-DCNF].

[101] Flahaux and De Haas (2016), pp. 1–2, https://comparativemigrationstudies.springeropen.com/articles/10.1186/s40878-015-0015-6 [https://perma.cc/L2UW-8TSF].

[102] Interestingly, periods of economic progress in the United States and Europe have traditionally been conducive to tolerance and openness because autochthonous populations did not feel that migrants threatened locals' ability to progress economically. *See* Benjamin M. Friedman, *The Moral Consequences of Economic Growth* 7–9 (2005). *A contrario*, whenever economic growth has been low, racism and discrimination have been on the rise, due to local populations feeling pushed down economically as a result of migrants. *Id.*

[103] For a discussion on the controversial concept of "climate migrants," *see generally* Leal-Arcas (2012b), p. 410.

2.4 Public Goods and Matters of Common Concern

countries.[104] The demographic trend of aging population that we observe in developed countries is also happening in growing countries like China[105] and India.[106] In addition, in India and China, for instance, cultural beliefs, sex-selective abortions, and gendercide have caused an excess of boys and men.[107] Many young men, unable to find wives, have great incentives to migrate. Similarly, migrants pushed out of their countries due to energy poverty turn to Western countries for greater opportunity.[108]

In addition, migration flows into the European Union have increased significantly over the past years due to the volatile security situation in North Africa and parts of the Middle East. For example, studies conducted by the Global Migration Data Analysis Centre indicate that the number of asylum seekers has consistently grown since 2011 and was at a record high as of 2015.[109] Moreover, the United Nations estimates that by 2060, fertility in all regions of the world, except for Africa, will have reached the replacement rate of 2.1 children per woman or below, which was already the case in many Western countries as of 2010.[110] Africa will have a birth rate of around 2.7 children per woman by 2060.[111] Many Africans may be tempted to migrate to wealthier Europe so long as they continue to be the victims of climate change and energy poverty, exacerbated by population growth.

The growing number of refugees seeking asylum in the European Union has brought to light the shortages in resources and facilities that would permit the European Union to embrace asylum seekers and meet its obligations under regional and international human rights instruments.[112] The recent readmission agreement of

---

[104] A. Majid and S. Newey, "Defusing the 'demographic timebomb': the world's population challenges in 13 charts," *The Telegraph*, 18 September 2018, available at https://www.telegraph.co.uk/news/0/defusing-demographic-timebomb-worlds-population-challenges-13/.

[105] China Power, "Does China have an aging problem?" *The Economic Times*, 13 August 2018, available at https://www.telegraph.co.uk/news/0/defusing-demographic-timebomb-worlds-population-challenges-13/.

[106] "Demographic time bomb: Young India ageing much faster than expected," available at https://economictimes.indiatimes.com/news/politics-and-nation/demographic-time-bomb-young-india-ageing-much-faster-than-expected/articleshow/65382889.cms.

[107] *Answering for India's 'Missing Girls': Sex-Selective Abortion in India*, Record (Feb. 11, 2014), https://www.newsrecord.co/answering-for-indias-missing-girls-sex-selective-abortion-in-india/ [https://perma.cc/RY8Q-DECL].

[108] Halff et al. (2014); *see also* Isabel Schäfer, *Migration to Europe – Is North Africa Europe's Boarder [sic] Guard?* at 1, German Dev. Inst. (June 8, 2015), https://www.die-gdi.de/uploads/media/German_Development_Institute_Schaefer_08.06.2015.pdf.

[109] Int'l Org. for Migration: Glob. Migration Data Analysis Ctr., 2015 Global Migration Trends Factsheet 9 (2017), http://gmdac.iom.int/global-migration-trends-factsheet [https://perma.cc/FK6F-MW5L].

[110] *See* Max Roser, *Future World Population Growth*, Our World in Data, https://ourworldindata.org/future-world-population-growth/#note-7 [https://perma.cc/AZY5-9ZEP].

[111] *Id.*

[112] The European Union is bound by the Charter of Fundamental Human Rights in the course of implementing E.U. legislation. European Comm'n, *EU Charter of Fundamental Rights*, http://ec.

March 2016 between the European Union and Turkey[113] further highlights these shortages in capacity and serves to undermine the credibility of E.U. institutions, because the agreement calls for the return of asylum seekers to Turkey, a state with a dubious human rights record.[114] Many have questioned the legality of the readmission agreement as its implementation may lead to violations of E.U. and international regulations on the treatment and return of refugees.[115]

Regional and global cooperation on the decarbonization of the economy could help resolve some of the pressing matters that underpin the current human rights crisis described above. The exchange of technology and renewable energy could stimulate economic growth and alleviate energy poverty in Africa, particularly in states where poverty is more prevalent, such as parts of sub-Saharan Africa. Studies

---

europa.eu/justice/fundamental-rights/charter/index_en.htm ("On [December 1, 2009], with the entry into force of the Treaty of Lisbon, the Charter became legally binding on the EU institutions and on national governments, just like the EU Treaties themselves.... The provisions of the charter are addressed to: the institutions and bodies of the [European Union] with due regard for the principle of subsidiarity; and the national authorities only when they are implementing EU law.") [https://perma.cc/VP6L-DEG9]. In addition, E.U. member states are also bound by the Convention for the Protection of Human Rights and Fundamental Freedoms, better known as the European Convention on Human Rights, and the International Covenant of Civil and Political Rights (ICCPR). *See* Convention for the Protection of Human Rights and Fundamental Freedoms, Nov. 4, 1950, 213 U.N.T.S. 221; International Covenant on Civil and Political Rights, Dec. 16, 1966, 999 U.N.T.S. 171; U.N. High Comm'r for Human Rights, The EU and International Human Rights Law 7, 9, http://www.europe.ohchr.org/Documents/Publications/EU_and_International_Law.pdf [https://perma.cc/47H8-F7VB].

[113]*See* Press Release, European Council, *EU-Turkey Statement, 18 March 2016* (Mar. 18, 2016), http://www.consilium.europa.eu/en/press/press-releases/2016/03/18-eu-turkey-statement/ [https://perma.cc/N42M-RXM9].

[114]Human rights have been under attack in Turkey. According to Amnesty International's 2016/2017 Annual Report:

An attempted coup prompted a massive government crackdown on civil servants and civil society. Those accused of links to the Fethullah Gülen movement were the main target. Over [forty thousand] people were remanded in pretrial detention during six months of emergency rule. There was evidence of torture of detainees in the wake of the coup attempt. Nearly [ninety thousand] civil servants were dismissed; hundreds of media outlets and [nongovernmental organizations] were closed down and journalists, activists, and [members of parliament] were detained. Violations of human rights by security forces continued with impunity, especially in the predominantly Kurdish southeast of the country, where urban populations were held under [twenty-four]-hour curfew. Up to half a million people were displaced in the country. The European Union and Turkey agreed to a migration deal to prevent irregular migration to the [European Union]; this led to the return of hundreds of refugees and asylum seekers and less criticism by EU bodies of Turkey's human rights record.

Annual Report: Turkey 2016/2017, Amnesty Int'l, https://www.amnesty.org/en/countries/europe-and-central-asia/turkey/report-turkey/ [https://perma.cc/H9AV-9ECK].

[115]E.U. and international legislation require that there must be no risk of serious harm and no threat that those returned will be sent to another country that is deemed unsafe. *See* Council Directive 2011/95, arts. 32–33, 2011 O.J. (L 337) 9 (EU).

conducted by the International Renewable Energy Agency demonstrate that Africa's economies are currently growing at an average rate of 4% per year.[116] Further, six of the world's ten fastest growing economies over the last decade were found in sub-Saharan Africa.[117] Sustaining the same level of growth, however, will only be possible if supported by a much larger and better-performing energy sector.[118]

As one of the world's major economic powers, the European Union has the capacity and means to invest in research, develop new renewable energy technologies,[119] and encourage innovation. Given its commitment to and investment in clean energy, E.U. states such as Germany may have the capacity to lead the retreat from fossil fuels and initiate the transformation of the global energy sector.[120] Cooperation with, inter alia, African states on the decarbonization of the economy would produce a number of benefits to both Africa and the European Union.

First, such cooperation would facilitate economic growth in the African continent and eradicate energy poverty in sub-Saharan Africa, significantly improving the living conditions of millions of people around the world, including in the European Union. Second, interstate cooperation would reduce the number of economic migrants who travel to the European Union from sub-Saharan Africa: fewer people will feel compelled to undertake the dangerous journey from Africa to Europe. Third, collaborative decarbonization would ensure that the European Union has the capacity to deal with refugees and asylum seekers who enter the European Union to escape persecution and violence due to war and political upheaval. This would remove the current strain on national authorities and reduce the number of refugees that need to be sent away to third countries, such as Turkey.

2.4.4.2 Efforts Towards Decarbonization Will Boost the European Union's Economy

Economic growth is one of the core tenets of the European Union and a powerful incentive for regional and global collaboration. Collaboration on the decarbonization of the economy will benefit individual member states and the overall economy of the European Union by proliferating the spread of renewable energy around the globe

---

[116]Int'l Renewable Energy Agency, "Africa's Renewable Future: The Path to Sustainable Growth" 5 (2013), http://www.irena.org/documentdownloads/publications/africa_renewable_future.pdf [hereinafter Africa's Renewable Future] [https://perma.cc/CP5X-G9F8].

[117]*Africa's Impressive Growth*, The Economist (Jan. 6, 2011), http://www.economist.com/blogs/dailychart/2011/01/daily_chart [https://perma.cc/8BNP-M3US].

[118]*See* "Africa's Renewable Future," available at http://www.irena.org/documentdownloads/publications/africa_renewable_future.pdf.

[119]The main criteria for energy technology are: (1) inter-operability, i.e., appliances that communicate with each other; (2) being modular, i.e., being easy to install; and (3) reliability.

[120]*See* Robert Kunzig, *Germany Could Be a Model for How We'll Get Power in the Future*, Nat'l Geographic (Nov. 2015), http://www.nationalgeographic.com/magazine/2015/11/germany-renewable-energy-revolution/ [https://perma.cc/5URH-7UFS].

and ensuring stable and sustainable global economic growth.[121] Enhanced cooperation also ensures that the European Union will make considerable progress in attaining its objectives under the revised E.U. Sustainable Development Strategy (EU SDS),[122] key among which is the attainment of economic prosperity through the promotion of "a prosperous, innovative, knowledge-rich, competitive and eco-efficient economy which provides high living standards and full and high-quality employment throughout the European Union."[123] Collaboration on the establishment of a fossil fuel-free economy will pave the way for improved trade and diplomatic relations between nations, which can, in turn, reduce tariffs for renewable energy-related goods and services in international trade agreements[124]; expand the Energy Charter Treaty's (ECT) membership; and generate employment.

First, lower tariffs for renewable energy-related goods and services will lead to lower prices for consumers and hence, increased competition. Renewable energy markets will thus soar and make way for new opportunities, increased investment, and economic welfare.[125]

---

[121] As early as the 1960s and 1970s, there was discussion about the limits to growth and the importance of sustainable growth. *See, e.g.*, Ehrlich (1968), rev. ed.; Donella H. Meadows *et al.*, *The Limits to Growth* (1972). More recent works include Sabin (2013), Klein (2014) and Maxton and Randers (2016).

[122] Regional and global cooperation on decarbonization is in line with the European Union's commitment to sustainable development under the E.U. Sustainable Development Strategy (EU SDS). *See* Council of the European Union, Note from Gen. Secretariat to Delegations, *Review of the EU Sustainable Development Strategy (EU SDS) – Renewed Strategy*, SEC (2006) 10917/06 (June 9, 2006), http://www.etuc.org/IMG/pdf/st10117.en06.pdf [https://perma.cc/9VN4-KBSJ]. According to the General Secretariat's Note to the delegations:

Sustainable development means that the needs of the present generation should be met without compromising the ability of future generations to meet their own needs. It is an overarching objective of the European Union set out in the [EU SDS], governing all of the [European] Union's policies and activities. It is about safeguarding the earth's capacity to support life in all its diversity and is based on the principles of democracy, gender equality, solidarity, the rule of law, and respect for fundamental rights, including freedom and equal opportunities for all. It aims at the continuous improvement of the quality of life and well-being on Earth for present and future generations. To that end [the EU SDS] promotes a dynamic economy with full employment and a high level of education, health protection, social and territorial cohesion and environmental protection in a peaceful and secure world, respecting cultural diversity.

Id.

[123] *See id.* at 4.

[124] Such an argument is in line with the theory of economic integration, as analyzed by economist Bela Balassa, which contemplated "degrees" of economic integration, with a free trade agreement as a first step towards integration, harmonization of external tariffs as a step further, and common internal regulations as a step even further. *See generally* Bela Balassa, *The Theory of Economic Integration* (1961).

[125] As an illustrative analog, reduced costs of photovoltaics in recent years have contributed greatly to solar power becoming increasingly competitive. In particular:

2.4 Public Goods and Matters of Common Concern

Second, the expansion of the ECT's[126] membership (for regulation of the energy industry)[127] to countries in the Middle East and North Africa (MENA) region and the Economic Community of West African States will attract investment in the African continent.[128] Collaboration on the decarbonization of the economy—the exchange of renewable energy resources, technology, and expert knowledge—particularly with states in the MENA region and Africa at large, could lead to

---

2015 was a record year for renewable energy, with China, Africa, the [United States], Latin America and India in particular driving forward the global energy transition. A photovoltaics boom is [also] forecast for the United States. . . . While China, Japan and the [United States] dominated the photovoltaics markets in 2015, Europe was also able to reach an important expansion milestone. Total photovoltaics output in [Europe] reached the 100-[gigawatt] mark in 2015.

Solar Market Set to Soar Globally Throughout 2016, Renewable Energy Focus (May 10, 2016), http://www.renewableenergyfocus.com/view/44164/solar-market-set-to-soar-globally-throughout-2016/ [https://perma.cc/2VRT-YSZ5].

[126]The Energy Charter Treaty (ECT) is an international agreement which aims to provide a "multilateral framework for energy cooperation" based on the principles of "open, competitive markets and sustainable development." Energy Charter Secretariat (2004), p. 13. For a commentary on the Energy Charter Treaty, see Leal-Arcas (2018b).

[127]For scholarly work on the topic, see Sioshansi (2013), pp. 227–264; Victor and Heller (2007). See also the book series on energy law published by the Academic Advisory Group to the Energy, Environment, Natural Resources and Infrastructure Law of the International Bar Association with Oxford University Press. Most recently, Godden et al. (2018).

[128]Efforts are already underway to encourage the accession to the ECT of regional organizations such as the Economic Community of West African states, which currently holds observer status. In addition, East African Community (EAC) states such as Burundi, Tanzania, and Uganda have also signed the International Energy Charter 2015, but have not yet acceded to the ECT. States such as, inter alia, Burundi, Kenya, Rwanda, South Sudan, Tanzania, and Uganda face a number of drawbacks within their energy sector, such as limited access to electricity, high costs of electricity generation, and, among others, overdependence on biomass. In relation to the Middle East and North Africa (MENA) region, even though most MENA states already have observer status with the ECT, accession has not yet taken place. Despite the potential for investment in renewable energy, many international developers, investors, and companies are not clear as to how to enter the market. Acceding to the ECT could help resolve some of these regional issues by attracting investment, opening up energy markets, and encouraging international cooperation. However, full accession to the ECT requires that states are able to abide by universal market-based principles, which may require them to undertake further steps before proceeding. The economies of acceding states are assessed against such principles before accession can take place. *See* Victoria Ritah Nalule, Energy Charter Secretariat Knowledge Centre, "Energy in the East African Community: The Role of the Energy Charter Treaty," (2016), http://www.energycharter.org/fileadmin/DocumentsMedia/Occasional/Energy_in_the_East_African_Community.pdf [https://perma.cc/Z9H6-3MG4]; Clean Energy Pipeline, The Future of Renewable Energy in the MENA Region, http://www.cleanenergypipeline.com/Resources/CE/ResearchReports/The%20Future%20for%20Renewable%20Energy%20in%20the%20MENA%20Region.pdf [https://perma.cc/MXW8-9WU8]. For further details, see *Consolidation, Expansion and Outreach*, Int'l Energy Charter, http://www.energycharter.org/what-we-do/conexo/overview/ [https://perma.cc/C4YV-ZWCB]; *see also The International Energy Charter*, Int'l Energy Charter, http://www.energycharter.org/process/international-energy-charter-2015/overview/ [https://perma.cc/798L-6ENC].

stabilization of the energy sectors in these regions[129] and, in turn, facilitate the expansion of ECT membership. This expansion could create reciprocity through technology transfer, while enhancing E.U. energy security by creating an infrastructure that will ultimately boost international, long-distance trade in renewable energy. In addition, it will create a large global renewable energy market where the European Union can compete on a level playing field, and new producers of energy from the MENA region and sub-Saharan Africa can contribute to the energy security of the European Union and the wider global community.

Third, intraregional collaboration on decarbonization will create employment opportunities. Unemployment, and particularly youth unemployment, has been an issue of concern in the European Union.[130] Recent data indicate that 19.750 million adults in the European Union (of whom 15.439 million were in the euro area) were unemployed in February 2017.[131] In addition, in February 2017, 3.905 million young persons (under 25) were unemployed in the European Union, of whom 2.722 million were in the euro area.[132] Regional and global cooperation on the decarbonization of the economy could generate new opportunities for investment and expand the global renewable energy market.[133] Innovation, technological advancement, and research in the field of renewable energy can lead to the creation of new posts and generate employment.

One final point: identifying region-specific incentives to enhance sustainable energy use abroad may, in the long run, contribute towards EU sustainable energy indirectly by making more efficient use of energy resources globally.

References

Aldy JE, Stavins RN (eds) (2010) Post-Kyoto international climate policy: implementing architectures for agreement. Cambridge University Press, New York
Aldy J, Stavins R (2012) Climate negotiators create an opportunity for scholars. Science 337 (6098):1043–1044
Ayres R (1994) On economic disequilibrium and free lunch. Environ Res Econ 4:435–454

---

[129]Many states in parts of Africa and the MENA region can better address energy-related challenges through the exchange of technology, technical knowledge, and skills. Such exchanges will allow for the introduction of relevant compliance mechanisms that will enable these countries to abide by universal market-based principles, and thus lead to speedier accession to the ECT.

[130]*See Unemployment Statistics*, Europa: Eurostat, http://ec.europa.eu/eurostat/statistics-explained/index.php/Unemployment_statistics#Youth_unemployment_trends [https://perma.cc/R3MF-RGAT].

[131]*Id.*

[132]*Id.*

[133]Indeed, various kinds of innovative actions between the private and public sectors are emerging to mitigate climate change. This is the commitment of Mission 2020. *See About*, M2020, http://www.mission2020.global/ [https://perma.cc/F5VF-NXLH].

References

Cafaggi F, Caron DD (2012) Global public goods amidst a plurality of legal orders: a symposium. Eur J Int Law 23:643–649

Costanza R, Daly H (1992) Natural capital and sustainable development. Conservation Biol 6 (1):37–46

Ehrlich PR (1968) The population bomb. Sierra Club, San Francisco

Energy Charter Secretariat (2004) The Energy Charter treaty and related documents: a legal framework for international energy cooperation. Energy Charter Secretariat, Brussels, p 13

Flahaux M-L, De Haas H (2016) African migration: trends, patterns, drivers. Comp Migr Stud 1:1–2

Godden L, Zillman D, Roggenkamp M (eds) (2018) How technological and legal innovation are transforming energy law. Oxford University Press, Oxford

Goklany I (2009) Deaths and death rates from extreme weather events: 1990-2008. J Am Phys Surg 14:102

Gotz M et al (2016) Renewable power-to-gas: a technological and economic review. Renew Energy 85:1371

Guzman AT (2005) The design of international agreements. Eur J Int Law 16:579

Halff A et al (2014) Energy poverty: global challenges and local solutions, vol 1. Oxford University Press, Oxford

Hardin G (1968) The tragedy of the commons. Science 162(3859):1243–1248

Hasson R et al (2010) Climate change in a public goods game: investment decision in mitigation versus adaptation. Ecol Econ 70:331

Hay BL, Stavins RN, Vietor RHK (2005) Environmental protection and the social responsibility of firms: perspectives from law, economics, and business. Resources for the Future, Washington, D.C.

Hirsch M (2009) Game theory and international environmental cooperation. J Energy Nat Resour Law 27(3):503–510

Kaul I (2012a) Global public goods: explaining their underprovision. J Int Econ Law 15:729

Kaul I (2012b) Rethinking public goods and global public goods. In: Brousseau E et al (eds) Reflexive governance for global public goods. The MIT Press, Cambridge, p 37

Kaul I (ed) (2016) Global public goods. Edward Elgar, Cheltenham

Kaul I et al (1999) Defining global public goods. In: Kaul I et al (eds) Global public goods: international cooperation in the 21st century. Oxford University Press, New York, p 2

Klein N (2014) This changes everything: Capitalism vs. the Climate. Simon & Schuster, New York

Leal-Arcas R (2012a) Unilateral trade-related climate change measures. J World Investment Trade 13:875

Leal-Arcas R (2012b) On climate migration and international trade. Vienna J Int Constitutional Law 6:410

Leal-Arcas R (2013) Climate change and international trade. Edward Elgar, Cheltenham

Leal-Arcas R (2016) The European Energy Union: the quest for secure, affordable and sustainable energy. Claeys & Casteels Publishing, Deventer

Leal-Arcas R (2017) Sustainability, common concern and public goods. George Washington Int Law Rev 49(4):801–877

Leal-Arcas R (2018a) New frontiers of international economic law: the quest for sustainable development. Univ Pa J Int Law 40(1):83

Leal-Arcas R (ed) (2018b) Commentary on the Energy Charter Treaty. Edward Elgar Publishing, Cheltenham

Leal-Arcas R, Minas S (2016) Mapping the international and European governance of renewable energy. Oxford Yearb Eur Law 35:621

Leal-Arcas R et al (2016) Energy security, trade and the EU: regional and international perspectives. Edward Elgar Publishing Ltd, Cheltenham, p 1

Maxton G, Randers J (2016) Reinventing prosperity: managing economic growth to reduce unemployment, inequality and climate change. Greystone books, Vancouver

Meyer T (2012) Global public goods, governance risk, and international energy. Duke J Comp Int Law 22:319–348

Norberg J (2016) Progress: ten reasons to look forward to the future. OneWorld, London, p 120

Ricardo D (1817) On the principles of political economy and taxation. John Murray, London

Sabin P (2013) The Bet: Paul Ehrlich, Julian Simon, and our gamble over earth's future. Yale University Press, New Haven

Samuelson PA (1954) The pure theory of public expenditure. Rev Econ Stat 36:387

Shelton D (2009) Common concern of humanity. Iustum Aequum Salutare 1:33

Shelton DL, Kiss A (2007) Guide to international environmental law. Martinus Nijhoff Publishers, Leiden, pp 13–15

Sioshansi F (ed) (2013) Evolution of global electricity markets: new paradigms, new challenges, new approaches. Elsevier, Oxford, pp 227–264

Soltau F (2016) Common concern of humankind. In: Gray K, Tarasofsky R, Carlarne C (eds) The Oxford handbook of international climate change law. Oxford University Press, Oxford

Stavins RN (2012) Economics of the environment: selected readings, 6th edn. W. W. Norton & Company, New York

Stavins RN (2013) Economics of climate change and environmental policy: selected papers of Robert N. Stavins, 2000-2011. Edward Elgar Publishing, Inc., Northampton

Stewart JB, Coleman M (2005) The black political economy paradigm and the dynamics of racial economic inequality. In: Conrad C et al (eds) African Americans in the U.S. economy. Rowman and Littlefield, Lanham, p 121

Victor D, Heller T (eds) (2007) The political economy of power sector reform: the experiences of five major developing economies. Cambridge University Press, Cambridge

Weimer DL, Vining AR (2005) Policy analysis: concepts and practice, vol 72, 4th edn. Pearson Prentice Hall, Upper Saddle River

Yoo IT, Kim I (2016) Free trade agreements for the environment? Regional economic integration and environmental cooperation in East Asia. Int Environ Agreements Politics Law Econ 16(5):721–738

Chapter 3
Linking International Trade to Climate Change and Energy

3.1 Introduction[1]

The argument goes that the causes of environmental problems in a market economy are economics and that the consequences of environmental problems have important economic dimensions that travel from producers to consumers of good and services. Hence, the importance of solutions that are environmentally effective, economically sensible, and politically pragmatic. In the specific case of trade, traditionally, the thinking has been that more trade meant more energy consumption and therefore higher levels of greenhouse gas (GHG) emissions. Economic growth has come at a cost to the environment. But it does not have to be that way. Trade can be part of the solution to reducing GHG emissions by providing preferential treatment to green goods/services in trade agreements, leading consumers to buy green goods such as electric cars.

There is a lot of talk about the fact that, as a result of trade, we have increased social inequality,[2] nationally and internationally, and that the level of carbon dioxide and other greenhouse gas (GHG) emissions has been going up over time, also a

---

[1]Sections from this chapter are a further development of Leal-Arcas and Morelli (2018), pp. 1–63.

[2]See generally World Economic Forum, "The Global Risks Report," 12th edition, Cologny/Geneva: Switzerland, 2017, available at http://www3.weforum.org/docs/GRR17_Report_web.pdf. Regarding arguments on the correlation between trade liberalization, on the one hand, and climate change and inequity, on the other, neither of these problems seems to be driven by economics or the trading system. For instance, trade makes every country richer. But it is not for the WTO to decide who individually (as citizens) gets how much from the benefits of trade. That is for national governments to decide based on national taxation. For a broader analysis of wealth and inequality, see Hughes (2018) and Milanovic (2018). For an inspirational statement on the core principles of development policymaking, see the Stockholm Statement of 2016, available at https://www.wider.unu.edu/sites/default/files/News/Documents/Stockholm%20Statement.pdf.

result of international trade,[3] and what can be done about it.[4] In fact, it has been reported that trade-related GHG emissions account for 26% of global emissions[5] and several investigations conclude that trade agreements undermine climate change mitigation efforts.[6] Moreover, trade has damaged the environment by facilitating access to non-environmental goods, namely goods that did not prevent, reduce or eliminate pollution.[7] Thus, trade is, in many senses, considered to be the competitor of environmental protection. In this chapter, however, I argue that, while all of that may be true, trade can contribute to climate change mitigation.

Yet other investigations conclude that trade agreements undermine climate change mitigation efforts[8] and so they have negative effects. In the words of Cossar-Gilbert, "Trade agreements are often stumbling blocks for action on climate change. Current trade rules limit governments' capacity to support local renewable energy, undermine clean technology transfer and empower fossil fuel companies to attack climate protection in secret courts. Trade policies are preventing a sustainable future."[9] With such a large carbon footprint,[10] it might seem odd to purport that trade can play a role in creating a greener economy. In this chapter, however, I argue that, while all of that may be true, trade can contribute to climate change mitigation.

---

[3]*Is Trade Good or Bad for the Environment?* OECD, http://www.oecd.org/trade/tradeandenvironment.htm [https://perma.cc/4VXK-KTWC]; *see also Environmental Goods and Services Sector*, Europa: Eurostat, http://ec.europa.eu/eurostat/web/environment/environmental-goods-and-services-sector ("The purpose of environmental goods and services is to prevent, reduce and eliminate pollution and any other form of environmental degradation ... and to conserve and maintain the stock of natural resources, hence safeguarding against depletion.") [https://perma.cc/5R27-B7FH].

[4]Some policy suggestions include a tax on the carbon content of imports and refund the tax to companies when they export, as the European Union is doing with cement. See The Economist, "Externalities: The lives of others," pp. 60–61, at 61. Others have studied the effects of a tax policy on greenhouse gas (GHG) emissions. See Nordhaus et al. (2013).

[5]Andrew et al. (2013), p. 23.

[6]Lilliston (2016).

[7]*Is Trade Good or Bad for the Environment?* OECD, http://www.oecd.org/trade/tradeandenvironment.htm [https://perma.cc/4VXK-KTWC]; *see also Environmental Goods and Services Sector*, Europa: Eurostat, http://ec.europa.eu/eurostat/web/environment/environmental-goods-and-services-sector ("The purpose of environmental goods and services is to prevent, reduce and eliminate pollution and any other form of environmental degradation ... and to conserve and maintain the stock of natural resources, hence safeguarding against depletion.") [https://perma.cc/5R27-B7FH].

[8]Lilliston (2016).

[9]"Trade rules trump climate action: U.S. blocks India's ambitious solar plans," *EcoWatch*, 26 February 2016, available at https://www.ecowatch.com/trade-rules-trump-climate-action-u-s-blocks-indias-ambitious-solar-pla-1882181449.html.

[10]The notion of measuring the carbon footprint as part of a sustainable world is even vivid in the 'Clean Label' movement, which aims to provide honest information to the consumer and food professionals on questions such as what there is in our food, who made it, what is the carbon footprint and related issues. See https://gocleanlabel.com/about/.

3.1 Introduction

Why trade? For one thing, trade plays a crucial role in access to and usage of energy resources globally. The international community is currently experiencing a major energy transition,[11] where trade in sustainable energy resources is critical if the international community wishes to move forward cleanly.[12] Sustainable energy, as discussed earlier, is vital for global economic and human development.[13]

Another reason to consider the potential of international trade is its unique cross-sectoral impact. Shocking news such as the fact that the world's eight richest people have the same level of wealth as the poorest 50% makes one wonder about social sustainability and development.[14] Or to put it differently, in the UK, households in the bottom 10% of the population have a disposable income around 10 times less than that of the top 10%.[15] In the past, efforts to achieve the different dimensions of development—economic, social, and environmental—have tended to be obstructed by "silo" mentalities, namely sectors not collaborating with each other to have a holistic view of a multifaceted problem.[16]

Today, however, the international community increasingly recognizes the need to take an integrated approach in addressing global development issues.[17] Trade—an area that every country participates in and, to different degrees, benefits from—cuts

---

[11] *See generally* Robert A. Hefner III, The Grand Energy Transition: The Rise of Energy Gases, Sustainable Life and Growth, and the Next Great Economic Expansion (2015) (detailing the global economic transition to sustainable energy).

[12] As a result, interesting questions would need to be answered, such as could the trade regime be modified to reduce fossil fuel consumption? Should it be? See in this respect the work of Trachtman, Joel, "WTO Law Constraints on Border Tax Adjustment and Tax Credit Mechanisms to Reduce the Competitive Effects of Carbon Taxes," *Resources for the Future* (January 2016); see also Workshop Report: Reforming Fossil Fuel Subsidies through the WTO and International Trade Agreements Monday, 22 May 2017, WTO, Geneva, Switzerland, available at http://climatestrategies.org/wp-content/uploads/2017/04/WTO-Workshop-Report-May-2017.pdf; Cosbey, Aaron *et al.*, "A Guide for the Concerned: Guidance on the elaboration and implementation of border carbon adjustment," *Entwined* (November 2012).

[13] *See* Amie Gaye, *Access to Energy and Human Development*, United Nations Development Programme, Human Development Report 2007/2008, http://hdr.undp.org/sites/default/files/gaye_amie.pdf [https://perma.cc/3JL8-EUT5]; *see also Trade Beyond the Tweet*, Bertelsmann Found., http://www.bfna.org/publication/bvisual-trade-beyond-the-tweet (analyzing how major economies are supported by trade) [https://perma.cc/JES9-G5ZA].

[14] *World's Eight Richest People Have Same Wealth as Poorest 50%*, The Guardian (Jan. 15, 2017), https://www.theguardian.com/global-development/2017/jan/16/worlds-eight-richest-people-have-same-wealth-as-poorest-50 [https://perma.cc/PGP8-XS9L].

[15] *The Scale of Economic Inequality in the UK*, Equality Trust, https://www.equalitytrust.org.uk/scale-economic-inequality-uk [https://perma.cc/G8K5-9W8E].

[16] *Breaking 'Silo' Approach Key in Toppling Barriers to Merging Three Pillars of Sustainable Development, Speaker Tells High-level Political Forum*, U.N. Meetings Coverage ECOSOC/6705 (June 30, 2015), https://www.un.org/press/en/2015/ecosoc6705.doc.htm [https://perma.cc/4YFM-H6DY].

[17] *See, e.g., Sustainable Development Knowledge Platform*, U.N. Dep't of Econ. & Soc. Affairs, https://sustainabledevelopment.un.org/?menu=1300 (identifying the sustainable development goals articulated by the United Nations) [https://perma.cc/U9QU-APJQ].

across almost every aspect of development in its role of reducing poverty, creating jobs,[18] and promoting cross-border cooperation.

This chapter links trade with climate change and energy security in the context of the green economy. Climate change is one of the biggest challenges humanity faces today.[19] The International Monetary Fund has been warning about the devastating economic effects resulting from climate change.[20] As a result of trade, there is increased social inequality[21] as well as more carbon and other GHG emissions in the atmosphere.[22] That said, in the past, trade law has been a very powerful instrument for change, as the following three examples show:

1. First, trade agreements brought around one billion people out of extreme poverty between 1990 and 2010[23];
2. Second, due to trade agreements, more people have access to medicines[24];
3. Third, trade has promoted the protection of human rights. Seventy per cent of countries participate in trade agreements that protect human rights.[25]

In spite of its potential to effect change, trade has been overlooked as a platform to address important global agendas. If the trading system has been instrumental for the above highly complex issues, why not use trade law to mitigate climate change? The trading system can be a powerful tool to fight climate change, provide access to

---

[18]Cynthia D. Crain and Dwight R. Lee, *International Trade Creates More and Better Jobs*, Nat'l Council on Econ. Educ. (2015), http://www.econedlink.org/lessons/docs_lessons/575_international_trade1.pdf [https://perma.cc/4JNP-BXYT]. In the Western world, only thirteen per cent of job losses are the result of trade agreements; the remaining eighty-eight per cent come from technology and innovation. *See* Paul Wiseman, *Why Robots, Not Trade, are Behind so Many Factory Job Losses*, AP (Nov. 2, 2016), https://apnews.com/265cd8fb02fb44a69cf0eaa2063e11d9/mexico-taking-us-factory-jobs-blame-robots-instead [https://perma.cc/F48H-ZQS7].

[19]Archer (2016) and Wagner and Weitzman (2016). Mark Carney, Governor of the Bank of England, has even warned of the catastrophic impact climate change could potentially have on the international financial system. See R. Partington, "Mark Carney warns of climate change threat to financial system," *The Guardian*, 6 April 2018, available at https://www.theguardian.com/business/2018/apr/06/mark-carney-warns-climate-change-threat-financial-system.

[20]International Monetary Fund, "World Economic Outlook Update," January 2019, p. 7.

[21]Gary Burtless, *Worsening American income: Inequality: Is World Trade to Blame?*, Brookings Inst., (Mar. 1, 1996), https://www.brookings.edu/articles/worsening-american-income-inequality-is-world-trade-to-blame/ [https://perma.cc/XB8B-KLGJ].

[22]Stockholm Env't Inst., Peter Erickson *et al.*, International Trade and Global Greenhouse Gas Emissions: Could Shifting the Location of Production Bring GHG Benefits? 2 (Apr. 2013), https://www.sei-international.org/mediamanager/documents/Publications/SEI-ProjectReport-EricksonP-InternationalTradeAndGlobalGreenhouseGasEmissions-2013.pdf [https://perma.cc/RKA7-UBYZ].

[23]*See Towards the End of Poverty*, The Economist (June 1, 2013), http://www.economist.com/news/leaders/21578665-nearly-1-billion-people-have-been-taken-out-extreme-poverty-20-years-world-should-aim [https://perma.cc/WHD7-8TZY].

[24]Wise (2006), p. 342.

[25]Susan Ariel Aaronson, *Human Rights*, The World Bank 443, 443 http://siteresources.worldbank.org/INTRANETTRADE/Resources/C21.pdf [https://perma.cc/P5KD-KZX8].

3.1 Introduction

sustainable energy, and make people and countries richer. This could be achieved through greater cooperation between major emitters of GHGs and more trade liberalization on environmental goods and services. Countries can fight climate change by providing a system that creates incentives to trade in environmental goods. Such a system can stimulate the global economy by creating new jobs, innovative companies, and goods that can be building blocks of a sustainable future.

The international community should conduct more coherent regulation and policy making so that the potential for trade to positively contribute to the climate action effort can be realized.[26] Such actions would also ensure that climate measures do not distort trade and instead promote an open economic system that contributes to an equitable and inclusive sustainable development.

One of the arguments of this chapter is that trade agreements have tremendous potential to help mitigate climate change, which is currently under-explored. This hypothesis raises the following question: How to reconcile progressive trade liberalization with the protection of non-economic interests. Is there an inherent conflict? Trade liberalization can have positive effects.[27]

Trade law can be a tool to help mitigate climate change and enhance sustainable energy. And we all know that, thanks to trade,[28] countries grow economically in an evolutionary manner, as evidenced by empirical and theoretical research.[29] Hence, the triple benefit of trade. This hypothesis may be replicated in other governance issues,[30] such as human rights, where trade in certain goods could be used for capital punishment, torture or other cruel, inhuman or degrading treatment or punishment.[31]

The novelty of this chapter is that it aims to explain the paradigm shift in the governance of sustainable development: The twentieth century was characterized by a top-down approach to the governance of climate change mitigation (e.g., the Kyoto Protocol),[32] energy (e.g., inter-governmental energy agreements), and international trade (e.g., inter-governmental trade agreements). The twenty-first century, however,

---

[26]Prag (2017), Di Leva and Shi (2017), Brandi (2017) and Cosbey (2016).

[27]Esty (1994).

[28]See the views of Gregory Mankiw, stating that trade improves average living standards. See G. Mankiw, "Why Economists Are Worried About International Trade," *The New York Times*, 16 February 2018, available at https://www.nytimes.com/2018/02/16/business/trump-economists-trade-tariffs.html.

[29]See generally the work by the political economist Joseph Schumpeter. But see also Ayres (1996), pp. 117–134 (arguing that "trade was at best a minor contributor to growth in the past and is probably now contributing negatively to both national wealth and equity, hence to welfare").

[30]One can think, for instance, of the argument that, if China and India bring millions of people into the middle class, the world will not be sustainable due to higher levels of consumption (of goods, food, energy) in these two countries. However, Sustainable Development Goal 12 (ensure sustainable consumption and production patterns) is about "promoting resource and energy efficiency, sustainable infrastructure, and providing access to basic services, green and decent jobs and a better quality of life for all." See UN Sustainable Development Goals, Goal 12, available at http://www.un.org/sustainabledevelopment/sustainable-consumption-production/.

[31]See Council Regulation (EC) No 1236/2005 of 27 June 2005, OJ L 200, 30.7.2005, p. 1.

[32]Leal-Arcas (2010a), pp. 17–90.

offers a bottom-up approach. One of the mega-trends[33] of the twenty-first century is the shift to this bottom-up approach[34] in the democratic (in the true sense of the term, namely that power remains with the citizens) implementation of *climate change mitigation plans*[35]—a creation of the Paris Agreement on Climate Change,[36] which has become the locomotive of unilateral[37] climate action.[38] In *climate action*, the implementation of the Paris Agreement on Climate Change is done from the bottom up via citizens, NGOs, mayors, governors, businesses, universities,[39] faith-based organizations or smart cities.[40]

The same is true in *energy governance*, where we are witnessing an energy democratization by decentralizing the governance of energy security and creating new actors such as prosumers and renewable energy cooperatives.

But how about *international trade* governance? How can it be governed from the bottom up so that there is an open trading system in political, legal, and economic terms? How can we have greater involvement of civil society? How can we empower citizens in trade diplomacy? Traditionally, trade policy has been conducted by trade diplomats. Should we not listen to citizens' concerns and those of small and medium enterprises?

---

[33] John Naisbitt popularized the term 'megatrends' with his book *Megatrends: Ten new directions transforming our lives*, New York: Warner Books, 1982.

[34] Daniel Esty of Yale Law School has developed 10 mega-trends of the twenty-first century, one of which is a bottom-up approach to climate action. See Esty (2017), pp. 41–42.

[35] Several factors exacerbate climate change. For instance, increasingly, the world is experiencing frequent cases of floods and they are predicted to increase exponentially. One cause is global warming. Warmer seas evaporate faster and warmer air can retain more water vapour, which provokes the violence of storms and the intensity of heavy rains. See The Economist, "How to cope with floods," p. 11, 2 September 2017. Also, eating meat from animals has negative effects on climate change. See The Economist, "Feed as well as food," pp. 13–14, at 13, 2 September 2017.

[36] D. Victor, "Why Paris worked: A different approach to climate diplomacy," *Yale Environment 360*, 15 December 2015.

[37] By unilateral action, we mean that the Paris Agreement promotes diversity in that countries are free to do unilaterally what they think is best for their own political economy in the fight against climate change. Countries, therefore, agree to collective targets, but can also implement their own goals.

[38] See Environmental Protection Agency, "Climate Change in the United States: Benefits of Global Action," June 2015. Unlike the UN Framework Convention on Climate Change of 1992, which divides countries into Annex I and non-Annex I and makes only Annex I countries be bound to climate change mitigation, the Paris Agreement on Climate Change of 2015 proposes universal goals for climate change mitigation. At the COP 23 in November 2017, the US was rather passive given its intent to withdraw from the Paris Agreement. Without the US's leadership, it might be difficult to reach new climate rules and China may use the potential American absence to lead future negotiations.

[39] Leal and Leal-Arcas (2018).

[40] It is remarkable to see the transformation of climate change agreements in terms of governance structure in such a short period of time: in less than 20 years, the 1997 Kyoto Protocol as an example of a top-down approach to climate change mitigation, and the 2015 Paris Climate Agreement as an example of a bottom-up approach to climate change mitigation. For an analysis of the Paris Climate Agreement, see Bodansky (2016), pp. 288–319; Druzin (2016), pp. 18–23.

3.1 Introduction

So in addition to the top-down process to trade governance, we propose a bottom-up process, with greater citizen participation, which has been a big and unanticipated success since 2016 in the climate change field.

To get there, a fundamental question needs to be answered: How can the trading system increase economic well-being while promoting climate change mitigation and enhancing sustainable energy?[41] This question can be broken into four sub-questions:

A first set of questions focuses on identification of synergies and possible modernization and adjustments within current institutions, regulatory frameworks, and national policies in order to improve sustainable energy.

i. Why Do States Create Overlapping Institutions in a Field Governed by a Specific Institution?

The regulation of renewable energy in international law is fragmented and largely incoherent. This question enables us to map the regulatory competences of all the institutions which play a role in regulating renewable energy and trade so as to identify gaps and overlaps. Specifically, it enables us to map the complexities of EU competences against its energy policy, in accordance with Articles 4(2) and 194 of the Treaty on the Functioning of the EU, to identify governance gaps. For instance, currently, the EU is unable to act the way a sovereign actor such as China[42] or the US would in promoting their energy security. For the EU, therefore, improving energy security will involve taking regulatory and policy measures which address the internal-external cleavage. Once we have identified the high-level normative framework which results from this complex set of regulations, one can propose ways in which gaps could be filled and overlaps eliminated whilst remaining true to the high-level normative framework, concentrating on those measures which would enhance sustainable energy. Given the structure, mandate, membership, regulatory framework and extent of interplay of the various elements of the international trade and renewable energy systems, this question will analyze the specific issues that pose challenges to sustainable energy.

---

[41]'The international trade system – the World Trade Organization (WTO) as well as regional and bilateral trade agreements – has often been criticised from a climate policy perspective, with trade rules perceived by some as a barrier to stronger climate ambition. Yet trade rules can also be looked at as something that could potentially help to achieve transformative change in climate policy.' See Climate Strategies, "The trade system and climate action: Ways forward under the Paris Agreement," Working Paper, October 2016, available at http://climatestrategies.org/wp-content/uploads/2016/01/Trade-and-Paris-Agreement-Summary.pdf, p. 2.

[42]Miller (2017). See also Copper (2016) and Laruelle and Peyrouse (2012).

ii. What Lessons Can the Renewables Governance System Learn from Other Governance Regimes?

The complexities of trade, investment and the environment are different, but the tools used in their management often coincide: taxes, tariffs, regulations, and subsidies. This question will look at other issues addressed by global governance regimes, including investment, climate change, and trade. This question enables us to analyze how these regimes developed, which theories guided their evolution, and find areas of comparison to governance of renewable energy. It also allows us to examine whether similar fragmentation exists within these other regimes, and whether it has impacted their focal point areas. Can governance of renewables find parallels and shape itself accordingly to better promote sustainable energy? How are other global issues governed and can renewable energy extract any lessons in this regard?

A second set of questions brings us to think of the following: From energy transit, to technology transfer, to investment protection, renewable energy and trade present interplays across various fields. For instance, what improvements can be made to the international trading system to ensure sustainable energy and more efficient energy markets?

i. Why would States cooperate regionally/globally on common concerns?
ii. What legal and political-economy instruments can promote sustainable energy?

This chapter explores the potential of the trading system in contributing to decarbonization, focusing on the EU and international level. All of this raises the following questions: Which trade measures would be most effective to achieve a greener society and greener growth? How can the trading system promote sustainable use of natural capital? How do opportunities and challenges to green growth via the trading system differ for developed and developing countries?

This chapter first sets the scene on how trade agreements may be a legal instrument to mitigate climate change and enhance sustainable energy.[43] The chapter provides the state of the art and then offers to go beyond the state of the art in the trade field to help decarbonize the economy. The chapter provides a section on the synergistic links between the trading and climate regimes. The chapter ends with a section on what the future may hold on the links between international trade and renewable energy.

---

[43]There is extensive literature on how trade in environmental goods can support a sustainable future. See for instance Lester, Simon and Watson, K. William (August 19, 2013) "Free Trade in Environmental Goods: The Trade Remedy Problem," Cato Institute Free Trade Bulletin No. 54; Zhang (2013), pp. 673–699; Wu (2014), p. 93; Araya (2016).

3.2 Setting the Scene

This chapter offers a paradigm shift in thinking about international trade. Traditionally, trade has been understood as a stumbling block to sustainable energy.[44] I argue that trade is a building block and that the international community should capitalize on the proliferation of regional trade agreements (RTAs)/bilateral trade agreements (BTAs) to enhance energy security via renewable energy (RE) and achieve clean energy.[45] Both can be achieved with the inclusion of strong chapters on trade in goods and services related to sustainable development and renewable energy in RTAs.

Trade can play a powerful role in achieving two of humanity's most urgent needs—namely, sustainable energy and climate change mitigation—yet trade has been overlooked as a platform to address important global agendas.

Trade has caused harm to the environment because the goods that were traded were not environmental goods, namely goods that prevent, reduce or eliminate pollution.[46] A major aim of the international community is to decarbonize the economy.[47] Renewable energy has many advantages over fossil fuels in terms of health and the environment.[48] With the rise of renewables, international trade in energy and in renewable technologies is likely to increase because renewables are intermittent and the trading system will be necessary to export electricity from where it is plentiful to where it will be necessary. How can the trading system help existing green technologies be implemented at scale? What are the regulatory barriers to doing so? There is a lot of potential for exporting renewables from developed to developing countries. Most of the renewable-energy work is done in developed countries, although China is an exception. In turn, the trading system can be a major vehicle towards moving away from fossil fuels to renewable energy.[49] It can provide fair competition, economies of scale and knowledge transfer.[50]

---

[44]See for instance Peters and Hertwich (2006), pp. 379–387.

[45]On the links between climate change and trade, see Leal-Arcas (2013a).

[46]*Is Trade Good or Bad for the Environment?* OECD, http://www.oecd.org/trade/tradeandenvironment.htm [https://perma.cc/4VXK-KTWC]; *see also Environmental Goods and Services Sector*, Europa: Eurostat, http://ec.europa.eu/eurostat/web/environment/environmental-goods-and-services-sector ("The purpose of environmental goods and services is to prevent, reduce and eliminate pollution and any other form of environmental degradation ... and to conserve and maintain the stock of natural resources, hence safeguarding against depletion.") [https://perma.cc/5R27-B7FH].

[47]Deep decarbonization pathways project (2015) *Pathways to deep decarbonization 2015 report – executive summary*, SDSN. For a comprehensive analysis on the topic, see Bacchus (2018a).

[48]Ottinger (2013).

[49]Tagliapietra (2012).

[50]See a joint statement on Trilateral Meeting of the Trade Ministers of the United States, Japan and the European Union, 31 May 2018, regarding what new WTO rules might look like when it comes to addressing non-market-oriented policies, in the fight against unfair competition. Available at http://trade.ec.europa.eu/doclib/docs/2018/may/tradoc_156906.pdf.

Cleaner shipping[51] and aviation technology exits; in fact, zero-carbon fuels are already entering the market[52] and known technology may make it possible to almost entirely decarbonize shipping by 2035.[53] But it seems that powerful lobbyists are interfering with the process of greening the shipping industry: a report by InfluenceMap noticed that 31% of countries were represented at a 2017 International Maritime Organization meeting by direct business interests.[54] Strong reactions to such a blocking position by lobbyists have come from the highest political levels: the president of the Marshall Islands co-authored an op-ed in the New York Times calling for immediate and determined action.[55]

Today, 80% of the global energy supply comes from fossil fuels.[56] Fossil fuels contribute to climate change and are finite[57]; a situation that leads to energy insecurity.[58] Renewable energy can help here in that it is cleaner than fossil fuels.

---

[51] In the case of shipping, the International Maritime Organization (IMO) has decided to reduce the amount of sulphur permitted in bunker fuel from 3.5% to 0.5% by 2020 because it releases a lot of CO2, thereby trying to fight climate change. An alternative option to bunker fuel is liquefied natural gas, whose burning releases 25% less CO2 than burning bunker fuel. See The Economist, "Marine technology of the future: In need of a clean up," 3 November 2018, pp. 75–77, at 75 and 77. According to Mikhail Sofiev and his colleagues, the IMO's new regulation could prevent thousands of premature deaths a year. See Sofiev et al. (2018), pp. 1–12.

[52] See The Economist, "Shipping: Smoke on the water," 14 April 2018, p. 62.

[53] International Transport Forum, "Decarbonising Maritime Transport: Pathways to zero-carbon shipping by 2035," OECD/ITF, 2018, available at https://www.itf-oecd.org/sites/default/files/docs/decarbonising-maritime-transport.pdf.

[54] InfluenceMap, "Corporate capture of the IMO," available at https://influencemap.org/report/Corporate-capture-of-the-IMO-902bf81c05a0591c551f965020623fda.

[55] H. Heine and C. Figueres, "Polluters on the High Seas," *The New York Times*, 6 April 2018, available at https://www.nytimes.com/2018/04/06/opinion/greenhouse-gases-international-shipping.html.

[56] *World Energy Council Report Confirms Global Abundance of Energy Resources and Exposes Myth of Peak Oil*, World Energy Council (Oct. 15, 2015), https://www.worldenergy.org/news-and-media/press-releases/world-energy-council-report-confirms-global-abundance-of-energy-resources-and-exposes-myth-of-peak-oil/ [https://perma.cc/5ZK6-8LNL].

[57] But see the views of Meghan O'Sullivan, who argues that fears of energy scarcity have given way to the reality of energy abundance. O'Sullivan (2017).

[58] IEA, 'CO2 Emissions from Fuel Combustion: Highlights' (IEA, 2017), available at www.iea.org/publications/freepublications/publication/CO2EmissionsfromFuelCombustionHighlights2017.pdf. Julian Simon questions this statement by arguing that the quantities of natural resources are not limited in the way we think they are. New reserves of natural resources are constantly discovered; others are yet to be discovered; and others are not yet economically viable. *See* Julian Simon, *When Will We Run Out of Oil? Never!*, http://www.juliansimon.com/writings/Ultimate_Resource/TCHAR11.txt [https://perma.cc/F87R-5BNV]. An example that might serve to illustrate Simon's position can be found by comparing predictions concerning copper consumption and dwindling reserves made in the 1970s. "[I]n 1970, identified and undiscovered copper resources were estimated to contain ... reserves of about 280 million metric tons of copper. Since then, almost 480 million metric tons of copper have been produced worldwide, but world copper reserves in 2014 were estimated to be 700 million metric tons of copper, more than double [the estimate] in 1970." U.S. Dep't of the Interior & U.S. Geological Surv., Mineral Commodity Summaries 2015,

3.2 Setting the Scene 57

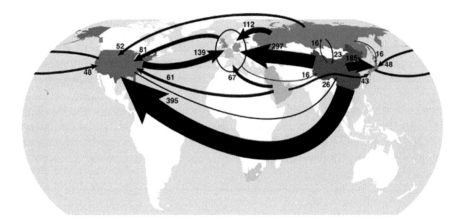

Fig. 3.1 China as the main world exporter of trade-related GHG emissions (source: Steven J. Davis and Ken Caldeira, Consumption-based accounting of CO2 emissions, https://doi.org/10.1073/pnas.0906974107)

It also helps towards geo-political and energy independence and therefore enhances energy security.[59] So renewable energy is the way to peace, as opposed to fossil fuels, which have caused so many wars.

Trade law could be used as a vehicle to achieve this goal. Trade can help everyone lift their living standards by aligning international standards, as opposed to domestic standards, which are low in some countries. Examples of this line of thinking are the policies of Canada, whose trade agreements include strict standards for labor rights and the environment. The European Union (EU) has the same view and concluded the Comprehensive Economic and Trade Agreement with Canada in 2016 along those lines.[60] Other countries such as China,[61] however, want to keep trade deals just to trade issues, avoiding non-economic issues in trade agreements such as labor rights, equity, social welfare, and environmental protection.[62] Including non-economic issues in future trade agreements is an effective manner to reach sustainability. Figure 3.1 depicts China as the main world exporter of trade-related GHG emissions.

---

191 (2015), https://minerals.usgs.gov/minerals/pubs/mcs/2015/mcs2015.pdf [https://perma.cc/FSJ5-HFNA].

[59]Cherp and Jewell (2014), pp. 415–421; Asif and Muneer (2007), pp. 1388–1413.

[60]For a comprehensive analysis of EU trade law, see Leal-Arcas (2019).

[61]Interestingly, China's One Belt One Road Initiative, a massive infrastructure project, will have significant impacts on the energy market, with climate-related consequences. For further details on the initiative, see http://www.bbc.com/news/av/business-39881895/what-is-china-s-one-belt-one-road.

[62]It is interesting to note the separationist approach of mainstream economics, which has mainly focused on efficiency to the detriment of equity and sustainability considerations. Such approach has been heavily criticized by Herman Daly in his prolific work. See Daly (1997).

Renewable energy may become the engine to obtain the three attributes of sustainable energy in the energy trilemma, namely clean, secure and affordable energy. First, renewable energy is clean. Second, efforts are being made to make it increasingly secure via storage systems. The renewable energy trading system can help enhance sustainable energy and, therefore, mitigate climate change. Lastly, regarding affordability, we need to overcome the way in which renewables are viewed in that they are difficult and expensive to produce, attain and transport. Through significant research and many trials, the production of renewable energy will become more and more commonplace. A good analogy is that of computers or cell phones which, initially, used to be (very) expensive, but today are accessible to almost anyone in the world.

I argue that there is great benefit to lowering technical barriers to trade in renewable energy-related goods and services, including in relation to technological goods and services that could encourage the proliferation of renewables and therefore enhance sustainable energy. The Government Procurement Agreement or even the General Agreement on Trade in Services Annex on Telecommunications may be used as WTO model agreements.

3.3 Impediments and Opportunities for Trade and Climate/Renewable Energy

This section proposes as a way forward mapping the macro-level on identification of key impediments and possible adjustments as well as opportunities for cooperation between trade and climate change/renewable energy within current institutions,[63] robust regulatory frameworks for a faster energy transition,[64] and policies to improve access to sustainable energy.[65] There is a large body of institutions and instruments of trade and (sustainable) energy. However, there is insufficient research and researchers are working in silos. I argue that greater (regional) cooperation and greater citizens' participation will lead to climate change mitigation and energy security. But to have regional cooperation, one needs a certain degree of harmonization.

So what are the obstacles and how can we have a win-win situation for trade and climate action? What are the criteria for an effective policy on trade and climate

---

[63] Already in 1990 there was debate on the long-term interaction between the economy and the environment. See Faber et al. (1990), pp. 27–55.

[64] See generally Bodansky et al. (2017), Mayer (2018), Hollo et al. (2013), Dupuy and Vinuales (2018) and Popovski (2018).

[65] See generally Leal-Arcas (2012a), pp. 875–927; Leal-Arcas and Minas (2016), pp. 621–666; Hufbauer and Kim (2009).

action? Is one criterion the enforceability of climate change agreements?[66] Or the harmonization of climate action goals? How can the trading system help with the implementation of the Paris Climate Agreement? In this sense, identifying the gaps and opportunities for cooperation between these two regimes is crucial to have the basis for a new normative framework on how the trading system can help mitigate climate change and enhance energy security.[67] This analytical framework needs to be placed in a political context of non-cooperation by a major international actor, namely the US. The Trump administration's position regarding multilateral trade and the Paris Climate Agreement brings new challenges for the trade-climate nexus.

Currently, governance of energy trade is fragmented, disjointed, with selective membership and guided by State interests. Think for instance of the following institutions and instruments that deal with energy trade: the World Trade Organization, preferential trade agreements, the Energy Charter Conference, the Organization of the Petroleum Exporting Countries or the Gas Exporting Countries Forum. This situation hinders transnational energy flows. Despite apparent overlaps between institutions and regimes involved in renewable energy trade governance, there are significant gaps in the system. The result is a mixed bag of incidental outcomes arising from an array of disjointed energy-related institutions and processes operating at various scales (bilateral, regional, *et cetera*), often each with its own selective membership.

This chapter goes beyond examining the law and governance of international trade in energy and its effects on sustainable energy to identify existing knowledge gaps in energy trade governance in conjunction with non-economic aspects of trade policy. The approach of this chapter is thus interdisciplinary and large-scale, bringing together a holistic analysis of the ever growing and complex interface between trade and renewable energy from the perspective of law, political economy, and international relations,[68] taking into account the fact that the energy field has been traditionally led by economists.

There is evolving literature that tries to explain the complexity of governance in transnational networks and regimes. Although much has been written regarding regimes complexes and transnational networks,[69] and a smaller volume of literature

---

[66]Such a question raises the following issue: do climate change agreements suffer from weak enforcement capacities that ultimately undermine their credibility as instruments of environmental protection? Conversely, are free-trade agreements surprisingly more likely to encourage compliance with environmental commitments than climate change agreements due to a system of encouragement and reward, driven by preferential market access? See Matisoff (2010), pp. 165–186; Morin and Jinnah (2018), pp. 541–565.

[67]See Di Leva and Shi (2016), Leal-Arcas (2018), Leal-Arcas et al. (2016), Leal-Arcas and Wouters (2017) and Leal-Arcas (2016).

[68]Lewis (2014).

[69]See for instance Abbott et al. (2016); Hale and Roger (2014), pp. 59–82; Abbott (2012), pp. 571–590; Gehring and Faude (2013), pp. 119–130; Newman and Zaring (2013), p. 244; Orsini et al. (2013a), pp. 27–39; Raustiala and Victor (2004), pp. 277–309; Slaughter and Zaring (2006); Keohane and Victor (2011), pp. 7–23; Slaughter (2002), p. 1041; Slaughter (2004); Kissinger (2014).

exists regarding the role of private instruments in these transnational regulations[70] and regime complexes[71] and how they relate to social network analysis,[72] this book tries to add to the literature by filling an important gap in the current academic literature related to renewable energy governance at various levels (i.e., international, supranational, and national). Scholars have explained climate change activities in terms of multilevel governance,[73] transnational governance,[74] polycentricity[75] or fragmentation.[76] This chapter will go a step further and look at how the trading system can be used as a vehicle to mitigate climate change, enhance energy security and help grow economically.[77] Is there a transnational legal order for renewables? If yes, what are its boundaries? If not, why did it fail to exist? The book will take the next step and examine the normative framework that relates to renewable energy governance at the inter-State level.

3.3.1 The Challenge of Mitigating Climate Change and Achieving Sustainable Energy

Alongside the crisis of climate change, there remains significant unmet energy demand in the least developed and developing countries as well as significant barriers (including expense, technology and grid capacity) to the growth of renewable energy in many such countries.[78] At the national level, renewable energy law and policy is often characterized by innovation and experimentation. A recent study conducted in 55 developing nations found at least 359 'clean energy-supportive

---

[70] Abbott and Snidal (2010), pp. 315–344; Berliner and Prakash (2014), pp. 217–223; Fransen and Burgoon (2014), pp. 583–619; Dashwood (2014), pp. 551–582; Green (2013), pp. 1–25; Borgatti et al. (1998), pp. 27–36; Borgatti et al. (2009), pp. 892–895; Easley and Kleinberg (2010); Sabel and Zeitlin (2011); Overdevest and Zeitlin (2012); Zeitlin (2011), pp. 187–206; Eberlein et al. (2013).

[71] See for instance Wood et al. (2015); Perez (2012), p. 285.

[72] Richardson (2009), pp. 571–588; Snir and Ravid (2015); Thistlethwaite and Paterson (2015); Beckfield (2010), pp. 1018–1068; David and Westerhuis (2014); Heemskerk et al. (2016), pp. 68–88; Katz and Stafford (2010); Albareda and Waddock (2016); Bodin and Crona (2009), pp. 366–374; Janssen et al. (2006), p. 15; Carpenter and Brock (2008), p. 40; Faure et al. (2015); Kim (2013–14), pp. 980–991; Fenwick et al. (2014), pp. 3–9; Bernstein and Cashore (2004), pp. 33–63; Meidinger (2011), pp. 407–419; Potoski and Prakash (2013), pp. 273–294.

[73] Osofsky (2007), pp. 143–159; Goulder and Stavins (2011), pp. 253–257.

[74] Pattberg and Stripple (2008), pp. 367–388.

[75] Ostrom (2009).

[76] Carlarne (2008), pp. 450–480; Boyd (2010), pp. 457–550.

[77] Bak (2015), Raworth (2017) and Daly (1997) (who criticized mainstream economics for prioritizing efficiency over equity and for wrongly assuming that efficiency and equity are necessarily competing goals).

[78] Yamusa and Ansari (2013), pp. 151–156.

3.3 Impediments and Opportunities for Trade and Climate/Renewable Energy

policies'—almost half of which were introduced in 2012–2013.[79] However, the challenge of effective steering from the international level remains.

Renewable energy governance is in many ways a subset of energy[80] and climate change governance, and has been described 'seriously underdeveloped' at the global level.[81] Former International Atomic Energy Agency director general, Mohamed ElBaradei, has remarked that

> [w]e have a World Health Organization, two global food agencies, the Bretton Woods financial institutions and organizations to deal with everything from trade to civil aviation and maritime affairs. Energy, the motor of development and economic growth, is a glaring exception. Although it cries out for a holistic, global approach, it is actually dealt with in a fragmented, piecemeal way.[82]

Part of the difficulty is that energy, as an issue area, straddles 'highly autonomous' systems of international economic and environmental law.[83] Particular 'institutional and structural diversity' has also been noted in the case of energy governance, as it relates to climate change.[84] A further dimension of complexity comes from the different properties of different sources of renewable energy. Indeed, there is also controversy as to what 'qualifies' for the label of renewable energy. As Bradbrook has observed, 'every energy resource involves a different interface with the law in terms of its exploitation'.[85]

A glance at the multiplicity of processes through which governance of renewable energy is channeled at the international level may leave one with the impression that renewables are indeed a case study in the 'fragmentation' of international law. However, as will become apparent, a countervailing force to fragmentation has emerged in the networking and collaboration taking place across processes and between organizations. The international governance of renewables is increasingly the work of networks. As the International Renewable Energy Agency (IRENA) has averred, 'partnerships remain embedded in every aspect of IRENA's programmatic activities'.[86] The sections below discuss how this constellation of actors and

---

[79]'Climatescope 2014: Mapping the Global Frontiers of Clean Energy Investment - Focus: Asia' (Bloomberg New Energy Finance, 2014) 7.

[80]Interestingly, energy policy has different—and often competing—goals, including security of energy supply, reducing energy poverty and ensuring sustainability. All these goals could be met via renewable energy.

[81]Gunningham (2012), pp. 119, 130–131.

[82]Mohammed ElBaradei, 'A Global Agency is Needed for the Energy Crisis' *Financial Times* (23 July 2008); cited in Gunningham (2012), pp. 130–131.

[83]Perez (2004), pp. 8–9. So autonomous, indeed, that Viñuales has reported that, "[i]nvestment lawyers and environmental lawyers barely speak to each other." Viñuales (2012), p. 1.

[84]Cherp et al. (2011), p. 75.

[85]Bradbrook (1996), pp. 193, 197.

[86]International Renewable Energy Agency, 'Work Program and Budget for 2014–2015: Report of the Director-General' (18 January 2014) A/4/3, para 16.

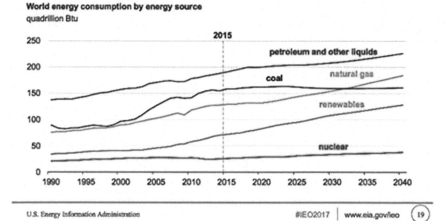

Fig. 3.2 Energy consumption up to 2040. *Source: US Energy Information Administration*

processes shape the development of public international law and governance relating to renewable energy.[87]

Predictions are that the total world energy consumption will increase by 28% from 2015 through 2040.[88] As depicted in Fig. 3.2, with the exception of coal,[89] energy usage from all sources will increase during that period.

As depicted in Fig. 3.3, renewables are projected to supply 31% of world electricity generation in 2040. The great advantage of electricity is that it is difficult to monopolize because it can be produced from several sources of fuel, from natural gas, wind, solar, or biomass. In addition, if the weather conditions are good, anyone can produce electricity: rich and poor countries.

---

[87] Moe and Midford (2014); Ottinger (2013); Tagliapietra (2012); USAID (2014); Van de Graaf (2013); Sovacool and Florini (2012); Florini and Sovacool (2011), p. 57; Florini and Sovacool (2009), pp. 5239–5248; Meyer (2013); Leal-Arcas et al. (2014).

[88] US Energy Information Administration, "International Energy Outlook 2017," p. 19, available at https://www.eia.gov/outlooks/ieo/pdf/0484(2017).pdf. 'The effects of economic growth assumptions on energy consumption are addressed in the High and Low Economic Growth cases. World gross domestic product (GDP) increases by 3.3%/year from 2015 to 2040 in the High Economic Growth case and by 2.7%/year in the Low Economic Growth case, compared with 3.0%/year in the Reference case.' Ibid., p. 8.

[89] Some EU governments are trying to shut down coal plants; the industry has reacted. See "Finland considers speeding up ban on coal; industry reacts strongly," *Xinhuanet*, 7 January 2018, available at http://www.xinhuanet.com/english/2018-01/07/c_136877044.htm; A. Vaugham, "UK government spells out plan to shut down coal plants," 5 January 2018, available at https://www.theguardian.com/business/2018/jan/05/uk-coal-fired-power-plants-close-2025.

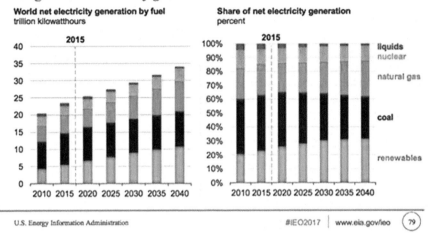

Fig. 3.3 Share of renewables in world electricity generation. *Source: US Energy Information Administration*

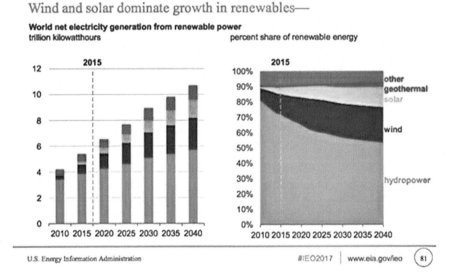

Fig. 3.4 Input of wind and solar in renewable energy. *Source: US Energy Information Administration*

Although wind and solar power will see the most growth among renewables for electricity generation, hydro will remain the largest single source of renewable electricity generation in the world (see Fig. 3.4).

With its extreme reliance on the rest of the world for its energy supply[90] and, consequently, its energy security,[91] it is in the EU's interest to diversify its energy sources and supply channels, and also increase energy efficiency by promoting more sustainable practices and greater energy market integration, which will enhance the EU's renewables potential. In fact, we see efforts in those directions through the promotion of the EU's Internal Energy Market and the Energy Community. Trade policy and regulation can be instrumental in achieving these goals. For example, there is potential to incorporate renewable energy provisions within the EU's numerous regional trade agreements (RTAs)[92]; there are trade incentives to better manage competition and invest in green technologies; and possibilities for exporting cutting-edge EU technologies through the EU's trade and bilateral cooperation agreements.

Access to energy (and renewables more specifically) is a common concern and requires local and global action.[93] Currently, there is no cohesive governance for global energy trade, which has implications for renewable energy trade governance. Governance of energy trade arises by default, rather than design, through the *ad hoc* interplay of different aspects of the international economic system.[94] Many institutions involving different actors and geographical scope address energy security. If combined with climate change mitigation, there are studies that identify more than 60[95] and even hundreds of transnational organizations that deal with its governance.[96] As a result, we have a polycentric and very complex institutional structure.[97] While its polycentric nature is not a problem *per se*—in fact, arguably the complexity of the energy system lends itself to polycentric governance[98]—the situation has resulted in a fragmentation of the global[99] and European energy trade regimes, a lack of cohesiveness of the global and European energy trade systems, divergent national interests, and a diversity of energy sources.[100] For such a polycentric system to succeed, a high level of coordination and trust is necessary between

---

[90]Yergin (2006, 2008, 2011).

[91]Taylor and Van Doren (2008), p. 475; Victor and Yueh (2010), p. 89.

[92]See analytical work by Croquet (2015), pp. 124–157; Jinnah and Morgera (2013), pp. 324–339; Morin, J.F. and Gauthier Nadeau, R. "Environmental Gems in Trade Agreements: Little-known Clauses for Progressive Trade Agreements," *CIGI Papers* No. 148, 2017.

[93]Walljasper, J. "Elinor Ostrom's 8 Principles for Managing a Commons," 2 October 2011, available at http://www.onthecommons.org/magazine/elinor-ostroms-8-principles-managing-commmons#sthash.9MPxgwbH.dpbs.

[94]Florini and Sovacool (2011), p. 57; Florini and Sovacool (2009), pp. 5239–5248.

[95]Bulkeley et al. (2012), pp. 591–612.

[96]Green (2010).

[97]Abbott (2012), pp. 571–590.

[98]Andrews-Speed, P. and Shi, X. (2015) "What Might the G20 under China's Presidency Deliver for Global Energy Governance?" *Energy Studies Institute Policy Brief 8*.

[99]Meyer (2012), pp. 319–348.

[100]Leal-Arcas and Filis (2013a), pp. 1225–1301.

the various actors involved. As it stands, the governance regime for energy trade is not conducive to EU sustainable energy.

How can we explain the proliferation of energy security institutions? Some scholars focused on the international level suggest that the regulatory activity of various international institutions represents a 'regime complex.'[101] This concept is used to refer to 'partially overlapping and nonhierarchical institutions governing a particular issue-area.'[102] Others propose understanding the governance activities by private and public actors in terms of a 'transnational regime complex,' which is composed of civil society organizations, governments and business, which is coined as 'governance triangle.'[103] In the same manner, how can we explain the fragmentation of renewable energy governance? And does it lead to forum-shopping?[104]

At the European level, it is vital that the EU take the right steps and decisions to ensure sustainable energy. A more cohesive governance system for energy trade would facilitate renewable energy usage, avoid unnecessary legal disputes and provide predictability. Achieving this will require a thorough understanding of the elements, workings, and evolution of the current energy trade governance regime and its consequences for European sustainable energy.

Energy trade is a key component of both the global and EU economies, and international trade in renewable energy spans a number of policy areas, including trade, investment, economic development, and environmental protection. The very nature of energy—namely its centrality to almost every field of human endeavor—and the very nature of traditional energy resources—namely finiteness, uneven distribution, and high desirability—lead to the politicization of energy and encourage intense competition for control over energy resources between actors.[105] While energy supply and consumption are important aspects of the global and EU energy economy,[106] they do not exist in an equilibrious relationship. Rather, they are heavily mediated by political considerations and by the very operation of global markets,[107] which dictate the extent to which energy needs are ultimately met.

The dominant opinion is that trade liberalization will increase economic activity[108] and therefore energy consumption. All countries require energy resources, but few possess them, and thus trade in fossil fuels (primarily oil and gas) and renewables is crucial to fulfil global energy needs.[109] Moreover, how energy from traditional sources is governed has an impact on renewable energy governance. Internationally, there is more trade in oil than in anything else. Moreover, half of

---

[101] Keohane and Victor (2011), pp. 7–23; Orsini et al. (2013b), pp. 419–435.

[102] Raustiala and Victor (2004), pp. 277–309.

[103] Abbott and Snidal (2009), pp. 44–88.

[104] Alter and Meunier (2009), p. 13.

[105] Andrews-Speed (2008), Schrijver (1997) and Wenger et al. (2009).

[106] Yergin (2011).

[107] Akerlof and Shiller (2015).

[108] Rodrik (2011).

[109] Luciani (2013).

world trade in services is energy-dependent.[110] Yet, the GATT/WTO has historically not preoccupied itself with energy trade. Very few energy-rich countries saw a need to join the GATT/WTO club, given that the reduction of import restrictions—one of the main goals of the multilateral trading system—is not an issue when it comes to energy. Saudi Arabia, one of the main energy-producing countries in the world,[111] only joined the World Trade Organization (WTO) in 2005 and several energy-producing countries are still not WTO Members.

Wholesale change to institutions requires more time than is available, and so the chapter will focus on the trade aspects of renewable energy governance and on how they could be adjusted to better promote sustainable energy in the EU. EU sustainable energy depends upon institutionalized energy-related internal as well as international cooperation.[112] For instance, arguably, effective systems for energy trade and energy transit enhance energy security for those economies involved in such cooperation. At the domestic level, a single agent—namely the State—is the authority that adopts measures towards the energy security of the territory/economy that it controls.[113] Such an agent is not omnipotent in its attempts towards energy security, given that energy security often relies on factors—e.g., energy commodities' price and availability—over which it has little or no control.

At the EU level, there are numerous actors who have influence over the energy economy, including EU and Member State bodies. This plurality of actors and the variety of interests at play—e.g., interests across the national-regional-international spectrum, the public-private spectrum,[114] and across the policy spectrum—mean that the achievement of EU sustainable energy is a considerably complex challenge.[115] While all sovereign actors/economies have an interest in their respective energy security, global energy security is a concern to none. So global energy security is not currently considered a common concern as is climate change.[116] In that respect, this chapter aims to shed light on incentives for States to cooperate on sustainable energy, highlighting ways in which it is, in fact, a common concern.

At the international level, the EU is one of a patchwork of institutions that may have implications for cross-border energy trade.[117] While the EU lacks the powers of a sovereign actor to diplomatically pursue its energy security in the manner that China or the US may, it does possess a comprehensive energy policy that is multifaceted and that makes good use of the powers that lie within its

---

[110]Gault (2010), p. 9.

[111]Bordoff, Jason, "This Isn't Your Father's OPEC Anymore," *Foreign Policy,* June 26, 2018.

[112]Jegen (2010), p. 73.

[113]Graetz (2011).

[114]Hawken (2010).

[115]Zarrilli (2010) and Anceschi et al. (2011).

[116]Leal-Arcas (2013a).

[117]de Jong, S. and Wouters, J. "Institutional Actors in International Energy Law," Leuven Center for Global Governance Studies, Working Paper No. 115, July 2013.

competences.[118] The WTO also provides governance over trade within its scope, including over energy trade. Many other institutions exist that provide degrees of governance over aspects of trade in energy at the inter-State level. This patchwork of institutions and regimes amounts to a sort of "accidental" energy trade governance, and presents some areas of overlap. For instance, both the WTO and the Energy Charter Treaty have rules that apply to the trade, investment, and environmental-protection aspects of energy. These overlaps in no way amount to cohesive governance of energy trade.

3.3.2 Overlapping Institutions

One of the aims of this book is to map out the governance of renewable energy so as to identify gaps and overlaps and propose ways in which these gaps could be filled and overlaps eliminated.[119] We will do so by looking at the interplay between public international law and European Union (EU) law in the context of renewables,[120] the interconnection between these two legal regimes and how they influence each other. Renewable energy comes from sustainable sources of energy, as opposed to conventional sources of energy such as oil, natural gas, coal, and uranium. Renewable energies, unlike the ones listed, are available in infinite supply, as they have natural forces that continuously replenish them.[121] They include solar, wind (onshore or offshore),[122] wave, hydro, tidal, geothermal, and biomass sources. By renewable energy governance, we understand the ensemble of the various institutions, legal instruments, and processes that deal with renewable energy towards the aim of mitigating climate change, which is a global public good[123] and a collective-action problem that is truly global, with long timescales in the energy systems and CO_2 cycles.[124]

---

[118]Haghighi (2007).

[119]For an initial survey, see Roehrkasten (2015); Kottari and Roumeliotis (2013).

[120]When we refer to *renewables*, unless specifically stated, we refer to renewable sources of energy, as juxtaposed with conventional energy sources such as hydrocarbons/fossil fuels (i.e., gasoline/petrol/oil, gas, and coal), and juxtaposed with other non-conventional sources that are, however, non-renewable—e.g., nuclear power.

[121]Oschmann (2008), p. 19.

[122]Regarding the reliability of both wind and solar energy, there is the concern that the wind does not always blow and the sun does not always shine. This means that energy storage solutions are necessary.

[123]For further details on public goods, see Kaul et al. (1999), Olson (1965), Sandler (2004), Taylor (1987) and Holzinger (2008).

[124]According to Daniel Schrag, "the timescale of decarbonisation is longer than many policy makers and climate change advocates imagine. Deep decarbonisation requires an array of new technologies [...] and a new infrastructure that represents more than 500 years of construction at current rates. It seems unlikely to occur in just a few decades." His own views of a credible timescale for decarbonization are:

I argue in this book that effective renewable energy governance at the international and European level has become a major challenge of public international law and EU law due to the fragmentation of the system and the proliferation of institutions. As a result of the various institutions that govern renewables, interesting questions arise: Where are the governance overlaps between these institutions? What is the interaction between them as a result of this fragmented governance? For instance, is it hierarchical or is it polycentric?

Renewable energy has characteristics of a global public good and requires local and global action. Currently, there is no cohesive governance for global renewable energy. Governance of renewable energy arises by default, rather than design, through the *ad hoc* interplay of different aspects of the international economic system.[125] Several institutions involving different actors and geographical scope address renewable energy. As a result, we have a polycentric and very complex institutional structure.[126] Arguably, the complexity of the renewable energy system lends itself to polycentric governance.[127] The situation has resulted in a fragmentation of the global[128] and European energy regimes, a lack of cohesiveness of the global and European renewable energy systems, divergent national interests, and a diversity of energy sources.[129] For such a polycentric system to succeed, a high level of coordination and trust is necessary between the various actors involved. As it stands, the governance regime for renewable energy is not conducive to sustainable energy.

How can we explain the proliferation of renewable energy institutions? Some scholars focused on the international level suggest that the regulatory activity of various international institutions represents a 'regime complex'.[130] This concept is used to refer to 'partially overlapping and nonhierarchical institutions governing a particular issue-area'.[131] Others propose understanding the governance activities by private and public actors in terms of a 'transnational regime complex', which is composed of civil society organizations, governments and business, which is coined

---

Phase 1 (up to 2030–2050): greater penetration of wind and solar energy, backed by natural gas. Phase 2 (2040 to 2070): continuation of renewables; deployment of storage to manage intermittency of renewables; electrification of electric vehicles. Phase 3 (after 2060): carbon capture and storage for natural gas plants; biofuels. Lecture given by Daniel Schrag on 1 October 2018 at Harvard University.

[125] Similar arguments have been made about developing a framework global (renewable) energy governance more generally. See Bruce (2013); Florini and Sovacool (2009), pp. 5239–5248; Sovacool and Florini (2012); Bradbrook (2011); Bradbook and Wahnschafft (2012).

[126] Abbott (2012), pp. 571–590.

[127] Andrews-Speed, P. and Shi, X. (2015) "What Might the G20 under China's Presidency Deliver for Global Energy Governance?" *Energy Studies Institute Policy Brief 8.*

[128] Meyer (2012), pp. 319–348.

[129] Leal-Arcas and Filis (2013a), pp. 1225–1301.

[130] Keohane and Victor (2011), pp. 7–23; Orsini et al. (2013b), pp. 419–435.

[131] Raustiala and Victor (2004), pp. 277–309.

3.3 Impediments and Opportunities for Trade and Climate/Renewable Energy

as 'governance triangle'.[132] In the same manner, how can we explain the fragmentation of renewable energy governance? And does it lead to forum-shopping?[133]

To understand the various and overlapping layers of renewable energy governance, this book benefits from global legal pluralism and regime theory. The latter argues that regimes or international institutions affect the conduct of States or other international actors in international cooperation. Regime theory is analyzed in the context of idealism[134] and realism[135] to explain the knowledge gap in international renewable energy law and governance, given that it views regimes as intervening variables, such as power, interest and values, on the one hand, and behavior and outcome, on the other.[136] The book also benefits from the concepts of transnational legal order,[137] institutional theory,[138] regime complexes,[139] and systems theory.[140] Given the large choice of renewable energy institutions, legal instruments and processes, we have limited ourselves to some of the most emblematic and representative governance structures of renewable energy.

It is vital that countries take the right steps and decisions to ensure renewable energy. A more cohesive governance system for renewable energy would facilitate its usage, avoid unnecessary legal disputes, and provide predictability. Achieving this will require a thorough understanding of the elements, workings, and evolution of the current renewable energy governance regime.

Renewable energy spans a number of policy areas, including trade, investment, economic development, and environmental protection. The World Trade Organization (WTO) provides governance over trade within its scope, including over renewable energy trade. Many other institutions exist that provide degrees of governance over aspects of renewable energy at the inter-State level. This patchwork of institutions and regimes amounts to a sort of 'accidental' renewable energy governance, and presents some areas of overlap. For instance, both the WTO[141] and the Energy

---

[132] Abbott and Snidal (2009), pp. 44–88.

[133] On forum shopping in global governance, see Alter and Meunier (2009), p. 13.

[134] Idealism placed hope in international law and institutions such as the United Nations.

[135] Realism views international institutions as having no impact on nation-state behaviour, since all international politics could be understood in terms of national interests.

[136] Krasner (1983).

[137] Halliday and Shaffer (2015).

[138] Jupille et al. (2013); Martin and Simmons (1998), pp. 397–419; Krasner (1983); Kingston, C. and Caballero, G. (2008) "Comparing Theories of Institutional Change," available at https://www3.amherst.edu/~cgkingston/Comparing.pdf.

[139] Raustiala and Victor (2004), pp. 277–309; Keohane and Victor (2011), pp. 7–23; Abbott (2014), p. 57.

[140] Von Bertalanffy (1968).

[141] https://www.wto.org/english/docs_e/legal_e/final_e.htm.

Charter Treaty[142] have rules that apply to the trade, investment,[143] and environmental-protection aspects of energy.

Many instruments connect both the trade and climate regimes: the World Trade Organization (WTO) via its case law on environmental protection,[144] the European Union (EU) with its regulation on how trade should protect the environment, the nationally determined contributions,[145] which are now part of the post-2020 Paris climate architecture, and the Environmental Goods Agreement, which is a plurilateral WTO agreement, to name but a few. As a result, it is necessary to understand how WTO rules can contribute to the design of effective climate policies internationally and how climate action can be made compatible with WTO rules.[146] Equally, it is imperative to evaluate trade rules to see how they support climate action without compromising on trade liberalization.[147] In addition, there are overlapping institutions and processes from both the trade and climate regimes. As a result, closer cooperation between these institutions and processes would be necessary, especially when it comes to coherence in the interpretation of environmental agreements included in sustainable development chapters of free-trade agreements (FTAs).

Since the regulation of renewable energy in international law is fragmented and largely incoherent,[148] it is essential to understand the overall trade and renewable energy systems and determine their net effect in terms of sustainable energy. There are competing interests in different jurisdictions, which may explain why new international organizations are created that overlap. This chapter will analyze how the different institutions govern renewables, where are the overlaps, interactions and fragmentation between them. For instance, is the inter-institutional relationship hierarchical or is it polycentric? This section will focus on mapping the regulatory

---

[142]Energy Charter Treaty (opened for signature 17 December 1994, entered into force 16 April 1998) 2080 UNTS 95 ('ECT').

[143]De Brabandere and Gazzini (2014) and Cameron (2010).

[144]For instance China – Measures Concerning Wind Power Equipment (2010) (raised by the US against China in relation to subsidies for wind turbines); United States – Countervailing Duty Measures on Certain Products from China (2012) (launched by China against the US in relation to the price of Chinese solar panels and wind towers); European Union and certain Member States – Certain Measures Affecting the Renewable Energy Generation Sector (2012) (in which China requested WTO consultations with the EU, Greece, and Italy on several feed-in-tariff programs in support of solar energy generation that allegedly contained local content requirements (LCRs)); India – Certain Measures Relating to Solar Cells and Solar Modules (2013) (which was initiated by the US against Indian LCR provisions pertaining to solar cells and solar modules).

[145]The nationally determined contributions are an example of a bottom-up approach to climate change governance: it is up to countries to decide what is best for them in the fight against climate change and how to do it.

[146]Morosini (2010), pp. 713–748.

[147]S. Droege et al., "Mobilising Trade Policy for Climate Action under the Paris Agreement: Options for the European Union," Stiftung Wissenschaft und Politik Research Paper, 2018.

[148]For suggestions on how to promote renewable energy effectively, see Haas et al. (2004), p. 833; Haas et al. (2011), pp. 2186–2193; Held et al. (2006), p. 849; Dubash and Florini (2006), pp. 6–18.

3.3 Impediments and Opportunities for Trade and Climate/Renewable Energy   71

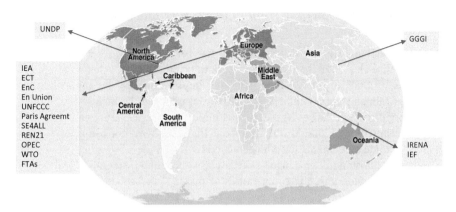

Fig. 3.5 Institutions and Instruments of International Trade and Renewable Energy. Source: The author (Rafael Leal-Arcas); what follows is an explanation of each acronym used in Fig. 3.5. UNDP: United Nations Development Program; IEA: International Energy Agency; ECT: Energy Charter Treaty; EnC: Energy Community; En Union: Energy Union; UNFCCC: UN Framework Convention on Climate Change; SE4All: Sustainable Energy for All; REN21: Renewable Energy Policy Network for the 21st Century; OPEC: Organization of the Petroleum Exporting Countries; WTO: World Trade Organization; IRENA: International Renewable Energy Agency; IEF: International Energy Forum; GGGI: Global Green Growth Institute

competences of all the institutions which play a role in regulating renewable energy and trade so as to identify gaps and overlaps (see Fig. 3.5). Given the proliferation of institutions and instruments dealing with international trade and renewable energy, better governance and coordination is necessary for a better transition of energy.

Since the EU is currently unable to act the way a sovereign actor such as China[149] or the US would in promoting their energy security,[150] improving energy security will involve taking regulatory and policy measures that address the internal-external cleavage. This section aims to propose ways in which gaps could be filled and overlaps eliminated whilst remaining true to the high-level normative framework, concentrating on those measures which would enhance EU sustainable energy.

Addressing the roots and ramifications of, say, energy trade governance fragmentation requires advancing the existing theoretical groundwork towards a comprehensive understanding that will pave the way for further research.[151] Global legal pluralism seems to be an appropriate theory in that it helps to describe and explain comprehensively the various institutions and processes of trade and renewable energy.

---

[149]Leal-Arcas (2012b), pp. 67–81.

[150]Leal-Arcas and Filis (2013a), pp. 1225–1301.

[151]Leal-Arcas and Filis (2013b), pp. 348–405.

3.3.3 Learning Lessons from Other Governance Regimes

Since the beginning of the 1990s, the EU has been making use of human-rights conditionality clauses in its trade agreements to protect human rights or democratic principles.[152] Such a practice has had major implications for the application and respect of human rights,[153] albeit some countries opposed the inclusion of human-rights conditionality clauses in the negotiation of trade agreements on the grounds that trade agreements should only be about trade. So why not use conditionality clauses in trade agreements for climate action?

When establishing the balance between global economic integration and domestic regulatory autonomy, the complexities of trade,[154] investment[155] and the environment are different.[156] However, the tools used in their management often coincide: taxes,[157] tariffs, regulations, and subsidies,[158] all of which are instruments to mitigate climate change and move forward the energy transition.[159] This section proposes as a future research agenda to look at various issues addressed by global

---

[152]Bartels (2005).

[153]McKenzie (2018), pp. 255–271.

[154]Leal-Arcas (2008).

[155]Leal-Arcas (2010b).

[156]See for instance the proposal by Sugiyama and Sinton (2005), pp. 65–88. See also Barrett, S. "Climate Change and International Trade: Lessons on their linkage from international environmental agreements," 2010, available at https://www.wto.org/english/res_e/reser_e/climate_jun10_e/background_paper6_e.pdf. On investment, see Leal-Arcas (2009), pp. 33–135.

[157]In the case of the environment, one can think of a carbon tax, where the consumer, and not the producer, is taxed. Such a tax would place the burden on the West (which is the main consumer of goods), as opposed to the rest. One could also design carbon taxes that exempt renewable energy. For an analysis of how taxes protect the environment, see Baumol and Oates (1971), pp. 42–54; Baumol (1972), pp. 307–322; Ekins (1994), pp. 571–579; Ekins (1999), pp. 39–62; Ekins (2009); Turner et al. (1998), pp. 121–136.

[158]In the fight against climate change, something as basic as transferring subsidies from the fossil-fuel industry to the renewables industry would be a very effective way to mitigate climate change and invest public funding intelligently. For an analysis of the funding spent on fossil fuel subsidies by the G7, see S. Whitley, et al., "G7 fossil fuel subsidy scorecard: tracking the phase-out of fiscal support and public finance for oil, gas and coal" June 2018, (where the authors argue that 'On average per year in 2015 and 2016 the G7 governments gave at least $81 billion in fiscal support and $20 billion in public finance, for both production and consumption of oil, gas and coal at home and overseas'), available at https://www.odi.org/publications/11131-g7-fossil-fuel-subsidy-scorecard.

[159]See Bordoff (2009); Kendall (2012), p. 51ff; J. Hillman, 'Changing Climate for Carbon Taxes. Who's Afraid of the WTO?,' The German Marshall Fund of the United States, Climate & Energy Policy Paper Series (July2013) at 1ff; J. Stiglitz 'A New Agenda for Global Warming' *The Economists' Voice*, July 2006; Goh (2004), pp. 395–423; Babiker and Rutherford (2005), pp. 99–125; Pauwelyn (2012); J.P.M. Sijm and A. van Dril, "The Interaction between the EU Emissions Trading Scheme and Energy Policy Instruments in the Netherlands: Implications of the EU Directive for Dutch Climate Policies," (INTERACT), 2003; Peat (2012), pp. 3–10; Sovacool (2017), pp. 150–163.

3.3 Impediments and Opportunities for Trade and Climate/Renewable Energy

governance regimes, including investment, climate change, and trade.[160] Its purpose is to analyze how these regimes developed, determine which theories guided their evolution, and find areas of comparison to governance of renewable energy. It further proposes an examination of whether similar fragmentation exists within these other regimes and whether it has impacted their focal point areas. Can the governance of renewables find parallels and shape itself accordingly to better promote sustainable energy?[161] How are other global issues governed and can renewable energy extract any lessons in this regard?

That said, commentator Varun Sivaram brings forward three reasons that might impede a long-term clean-energy transition:

1. Some environmental groups in different countries are putting pressure on governments to close down nuclear reactors, which are a source of clean energy;
2. There seems to be less support in the US regarding innovation in solar energy; and
3. There is lobbying for barriers to free trade of solar components, which will make the deployment of solar power more expensive.[162]

As for promoting the use of renewable energy,[163] it is one of the most pressing concerns for climate change and long-term sustainability at a global level. Analyzing how international trade law can contribute to promoting renewable energy addresses an important social, political and legal challenge.[164] We need a deep understanding of the current systemic aspects of energy trade governance and their implications for sustainable energy to achieve effective change.

Our world faces two major challenges when it comes to energy. First, as of 2016, one person in five on the planet still lacked access to electricity,[165] and almost three billion people still use wood, coal, charcoal or animal waste for cooking and

---

[160] See for instance Bacchus (2018b).

[161] See for instance IRENA, "Untapped potential for climate action: Renewable energy in nationally determined contributions," International Renewable Energy Agency, 2017, available at http://www.irena.org/-/media/Files/IRENA/Agency/Publication/2017/Nov/IRENA_Untapped_potential_NDCs_2017.pdf.

[162] V. Sivaram, "The dark side of solar: How the rising solar industry empowers political interests that could impede a clean energy transition," *Brookings Institution*, April 2018, available at https://www.brookings.edu/wp-content/uploads/2018/04/fp_20180416_dark_side_solar.pdf.

[163] Morocco is building a large solar-power plant in the Saharan Desert. See A. Neslen, "Morocco to switch on first phase of world's largest solar plant," *The Guardian*, 4 February 2016, available at https://www.theguardian.com/environment/2016/feb/04/morocco-to-switch-on-first-phase-of-worlds-largest-solar-plant. Dubai is doing the same. See D. Debusmann Jr, "Dubai's $3.8bn solar park continues to break world records," Arabian Business, 20 March 2018, available at http://www.arabianbusiness.com/energy/392315-dubais-38bn-solar-park-continues-to-break-world-records. These investments will help increase the electricity supply and cut energy subsidies.

[164] On the role of the law for the promotion of sustainable development, see Omorogbe (2008), p. 39.

[165] Int'l Energy Agency, WEO 2016 Electricity Access Database, http://www.worldenergyoutlook.org/resources/energydevelopment/energyaccessdatabase/ [https://perma.cc/MZP7-BU5Z].

heating.¹⁶⁶ The other main global energy challenge is that, in places with access to modern energy services, the lion's share of energy usage stems from environmentally damaging fossil fuels.¹⁶⁷ We use fossil fuels because there is demand for it: the situation is demand-driven. Proposals for an alternative way forward have been discussed.¹⁶⁸

Lastly, the environmental regime can learn much from the WTO system (for instance, the trade policy review mechanism) and also from the rest of the UN system (such as the International Labour Organization and the Food and Agriculture Organization). Key factors to integrate as elements of good governance include efficiency, coherence, predictability and transparency.

3.4 Making Greater Use of the Trading System

3.4.1 Filling the Gaps

Very little theoretical work exists on renewable energy law and even less on trade in renewables law. So, this section will address an important knowledge gap from an interdisciplinary perspective. It therefore aims to open up new horizons for research in three ways:

1) currently, we do not have a good understanding of the links between trade and renewable energy, of how trade can promote renewable energy and, therefore, help mitigate climate change¹⁶⁹;
2) through the application of our findings to other areas of sustainable development, e.g., trade and gender issues¹⁷⁰; and
3) how trade is underutilized as a tool to promote issues of global common concern.

---

¹⁶⁶World Health Organization, "Fuel for life: Household energy and health," p. 4, 2006, available at http://www.who.int/indoorair/publications/fuelforlife.pdf.

¹⁶⁷This view is in contrast with that of Scott Pruitt, head of the Environmental Protection Agency under President Donald Trump. Mr Scott believes in true environmentalism, namely 'using natural resources that God has blessed us with.' See The Economist, "Lexington: Salting the Earth," 27 January 2018, p. 40. A similar example is the planning of extraction of coal from Pakistan's Thar Desert, which has the financial help of the China-Pakistan Economic Corridor. See The Economist, "Engro: Thar's coal in the desert," 3 February 2018, pp. 55–56. On the other hand, one would need to extract natural resources, many of which will come from developing countries, to make green technologies and therefore tech-based decarbonization efforts. This means that there will be an increase in the demand for such minerals. Therefore, the challenge is extracting the essential minerals and leaving fossil resources in the ground, and doing so in a sustainable manner. For an opposite view, see Abu Gosh and Leal-Arcas (2013), pp. 480–531.

¹⁶⁸See Skymining, which proposes to replace fossil fuels. Available at https://skymining.com/index.html.

¹⁶⁹For some preliminary work, see Leal-Arcas and Alvarez Armas (2018).

¹⁷⁰Women Watch, "Gender Equality & Trade Policy" Resource Paper, 2011, available at http://www.un.org/womenwatch/feature/trade/gender_equality_and_trade_policy.pdf. But see also Dine (2005).

3.4 Making Greater Use of the Trading System

Since time immemorial, trade has played a role in foreign relations, poverty reduction, and societal advancement, which explains why we should aim at deeper, stronger, and larger international trade. Trade has had many positive impacts on the sustainability agenda, as mandated by the preamble to the WTO Agreement. We aim to fill the theoretical and empirical gap for how trade law can help mitigate climate change and enhance sustainable energy.

As a result of this knowledge gap, we have missed crucial opportunities for cooperation between trade and climate change. Here is an example: In the 1990s, two major agreements were concluded (one on climate change—the United Nations Framework Convention on Climate Change (UNFCCC)[171]—and one on international trade—the WTO Agreement).[172] It is surprising that the WTO Agreement only briefly mentions in its preamble the importance of 'sustainable development' in the context of international trade since the trade community already knew of the danger of climate change from the existence of the UNFCCC.[173] I argue this was a missed opportunity for trade law to play a bigger role in mitigating climate change. Years later, in 2015, a new global climate agreement came into existence—the Paris Agreement on Climate Change—which does not even mention the term 'trade.' These are examples of missed opportunities to cooperate between the trade and climate regimes.

---

[171]The UNFCCC was adopted at the Rio Earth Summit. For an analysis of what went wrong at that summit, see Palmer (1992), p. 1005. Margatet Thatcher, former Prime Minster of the UK, already in 1989, three years before the creation of the UNFCCC, highlighted the danger posed by the increase of CO2 entering the atmosphere at a speech she gave at the UN General Assembly. See https://www.youtube.com/watch?v=VnAzoDtwCBg. Since that speech at the UN, the concentration of CO2 in the Earth's atmosphere is much higher. One happy example of a country whose CO2 emissions have been falling annually since 1990 is the UK. This was due to removing coal drastically from the grid in recent years. See Z. Hausfather, "Analysis: Why the UK's CO2 emissions have fallen 38% since 1990," *Carbon Brief*, 4 February 2019, available at https://www.carbonbrief.org/analysis-why-the-uks-co2-emissions-have-fallen-38-since-1990.

[172]Interestingly, some twentieth-century trade agreements already contained reference to climate change or greenhouse gas (GHG) emissions reduction even before the UNFCCC was signed. Below are two excerpts from pre-1992 (the year when the UNFCCC was signed) agreements referring to GHGs effects and climate change:

"The Parties recognize the value of exchanging views, using existing consultation mechanisms under this Convention, on major ecological hazards, whether on a planetary scale (such as the greenhouse effect, the deterioration of the ozone layer, tropical forests, etc.), or of a more specific scope resulting from the application of industrial technology". See Fourth ACP-EEC Convention signed at Lomé on 15 December 1989, *Official Journal L 229, 17/08/1991 p. 0003 – 0280,* Article 41.

"2. Cooperation shall centre on: [. . .] - global climate change [. . .]. 3. To these ends, the Parties plan to cooperate in the following areas: - [. . .] - development of strategies, particularly with regard to global and climatic issues [. . .]." See Europe Agreement between the European Communities and their Member States, of the one part, and the Republic of Hungary, of the other part, 1991, Official Journal L 347, 31/12/1993 p. 0002 – 0266, Article 79.

[173]On the interaction between the WTO and sustainable development, see the views of Marceau, G. and Morosini, F. "The Status of Sustainable Development in the Law of the World Trade Organization," (8th November 2011), available at SSRN: https://ssrn.com/abstract=2547282.

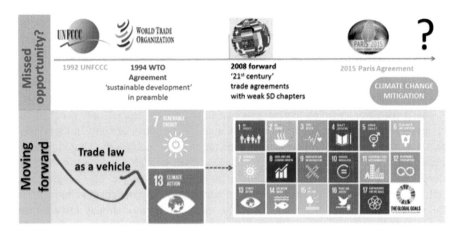

Fig. 3.6 From insufficient cooperation to effective synergies

There is not enough cooperation or consistency between trade and climate change policies. Greater cooperation between the secretariats of the U.N. Framework Convention on Climate Change (UNFCCC) and the World Trade Organization (WTO) is necessary to fill the gap between the theoretical potential for trade law to help mitigate climate change and getting empirical results. This gap is potentially catalytic because it paves the way for using trade to solve other sustainability challenges. As a result of this knowledge gap, we have missed crucial opportunities for cooperation between trade and climate change.

As Fig. 3.6 depicts, in the 1990s, two major agreements were concluded: one on climate change—the UNFCCC—and one on international trade—the WTO Agreement.[174] The WTO Agreement only briefly mentions in its preamble the importance of sustainable development in the context of international trade.[175] Still, considering that sustainable development is a tenet of the WTO Agreement, the multilateral trading system should be more effective at climate change mitigation and promotion of sustainable energy. The WTO Agreement missed the opportunity for trade law to play a bigger role in mitigating climate change by failing to emphasize this purpose.

From 2008, so-called "twenty-first-century trade agreements"[176] with chapters on sustainable development, started to emerge.[177] However, these chapters are often

---

[174]Marrakesh Agreement Establishing the World Trade Organization, Apr. 15, 1994, 1867 U.N.T.S. 154 (1994).

[175]*Id.* pmbl.

[176]This locution refers to trade agreements that touch upon environmental and social issues. It was first used to refer to the Trans-Pacific Partnership. *See* Everett Rosenfeld, *Who Wins and Loses in '21st Century Trade Agreement'*, CNBC (Nov. 13, 2015), http://www.cnbc.com/2015/11/13/who-wins-and-loses-in-21st-century-trade-agreement.html [https://perma.cc/8PCM-NUCB].

[177]Bartels (2015).

3.4 Making Greater Use of the Trading System

vague.[178] In 2015, a new global climate agreement came into existence—the Paris Agreement—which fails to mention the term "trade." These agreements are missed opportunities to cooperate between the trade and climate regimes.

However, the 22nd session of the Conference of the Parties (COP 22) in Marrakesh[179] made some progress towards deciding how the trading system can help achieve the SDGs. The WTO, the U.N. Conference on Trade and Development (UNCTAD), and the International Trade Center (ITC), in collaboration with the secretariats of the UNFCCC and the International Fund for Agricultural Development, came up with a tool box of trade measures that can help mitigate GHG emissions.[180] These are:

1) reducing costs and deploying key climate technologies quickly to places where they will have the biggest impact;
2) stimulating investment in energy, infrastructure, transport, information technology, and other key sectors of the new climate economy; and
3) fostering competitive markets that encourage individuals, enterprises, and entire industries to learn from past experience, innovate, and do better in the future.[181]

Greater cooperation between the trade and climate regimes will lead to climate change mitigation and energy security. It is encouraging to note that some countries are part of the International Solar Alliance,[182] a group of sunshine countries based

---

[178] *See, e.g.*, Trans-Pacific Partnership, chp. 19, Off. U.S. Trade Rep., https://ustr.gov/trade-agreements/free-trade-agreements/trans-pacific-partnership/tpp-full-text [https://perma.cc/5HVL-GSUJ]; *see also* Laura Puccio and Krisztina Binder, *Trade and sustainable development chapters in CETA* 11, European Parliament Research Service (Jan. 2017), http://www.europarl.europa.eu/RegData/etudes/BRIE/2017/595894/EPRS_BRI%282017%29595894_EN.pdf [https://perma.cc/GVT8-JMQ4].

[179] The Conference of the Parties (COP), described in Article 7 of the U.N. Framework Convention on Climate Change (UNFCCC), is the supreme decision-making body of the UNFCCC which meets on a yearly basis unless the parties decide otherwise. *See* UNFCCC, art. 7. The COP's role is to promote and review the implementation of the UNFCCC. *See id.* It periodically reviews existing commitments in light of the convention's objective, new scientific findings, and the effectiveness of national climate change programs, and can adopt new commitments through amendments and protocols. *See id.* In December 1997, at its third session (COP-3), it adopted the Kyoto Protocol, containing stronger emissions-related commitments for developed countries in the post-2000 period. *See* Kyoto Protocol to the United Nations Framework Convention on Climate Change, Dec. 11, 1997, 2303 U.N.T.S. 214 (1998). In 2015, at COP-21, the Paris Agreement was adopted. *See* Paris Agreement.

[180] *COP22: Geneva-Based Agencies Highlight Important Role of Trade in Addressing Climate Change*, U.N. Conf. on Trade and Dev. (Nov. 12, 2016), http://unctad.org/en/pages/newsdetails.aspx?OriginalVersionID=1379 [https://perma.cc/XE92-2QTZ].

[181] *Id.*

[182] Key Information About International Solar Alliance, Int'l Solar Alliance (Nov. 30, 2015), http://isolaralliance.org/projects.html [https://perma.cc/Y7LW-ADWK]. For a list of prospective countries of the International Solar Alliance, see *id.*

between the Tropic of Cancer and the Tropic of Capricorn,[183] where the potential of solar energy is phenomenal. Furthermore, a strong link between energy security and climate change mitigation is the use of renewable energy to diversify energy sources and, therefore, enhance energy security. The use of renewable energy, in turn, is a way to mitigate climate change. Linking all of this to the international trading system as a catalyst will only help mitigate climate change and enhance energy security.

Much is taking place in major developing countries to make this happen. For instance, India plans to reduce its GHG emissions relative to its GDP by 33 to 35% by 2030 from the 2005 level.[184] It intends to do so through policies on the promotion of clean energy, enhancement of energy efficiency,[185] development of less carbon-intensive[186] and more resilient urban centers, as well as the promotion of a sustainable green transportation network.[187] India also pledged to achieve around 40% of its electric power from non-fossil fuel-based energy resources by 2030 with the help of technology transfer and low-cost international finance from the Green Climate Fund.[188] All of this is largely possible if there is greater cooperation between the trade and climate change regimes because this is an area where far too little attention has been given in global policymaking to make both regimes "mutually consistent, supportive, and reinforcing."[189] Therefore, identifying the gaps and opportunities for

---

[183] For a map of the so-called global sunbelt, *see Annual Solar Irradiance, Intermittency and Annual Variations*, Green Rhino Energy, http://www.greenrhinoenergy.com/solar/radiation/empiricalevidence.php [https://perma.cc/95BK-23EW].

[184] Press Release, GOV'T OF INDIA, *India's Intended Nationally Determined Contribution is Balanced and Comprehensive: Environment Minister*, Ministry of Environment, Forest and Climate Change (Oct. 2, 2015), http://pib.nic.in/newsite/PrintRelease.aspx?relid=128403 [https://perma.cc/3T5L-3TKY].

[185] In fact, according to the International Energy Agency (IEA), doubling world GDP by 2040 would require only a small rise in energy demand if countries adopted strict standards like Japan's for the case of vehicle-fuel efficiency. The great winners of this prediction would be consumers and the climate. Greater efficiency makes energy cheaper. Therefore, consumers want more and benefits them. See The Economist, "Energy efficiency: Waste not, want more," 27 October 2018, p. 72 (citing the prediction of the IEA).

[186] An unrelated but interesting fact to note here is that "a study conducted by Cranfield University found that 12,000 rose stems grown in Kenya incurred a carbon footprint of 2,200kg CO2, while the equivalent supply from Holland generated 35,000kg CO2. The Kenyan roses thrived outside in the sunshine, while the Dutch ones were grown in greenhouses heated by fossil fuels." See C. Fry, "Tread lightly: Stop buying famed flowers," *The Guardian*, 25 April 2008, available at https://www.theguardian.com/environment/ethicallivingblog/2008/apr/25/treadlightlystopbuyingfarm.

[187] Press Release, GOV'T OF INDIA, *India's Intended Nationally Determined Contribution is Balanced and Comprehensive: Environment Minister*, Ministry of Environment, Forest and Climate Change (Oct. 2, 2015), http://pib.nic.in/newsite/PrintRelease.aspx?relid=128403.

[188] India, Submission of India's Intended Nationally Determined Contribution to the U.N. Framework Convention on Climate Change, 29 (Oct. 1, 2015), http://www4.unfccc.int/submissions/INDC/Published%20Documents/India/1/INDIA%20INDC%20TO%20UNFCCC.pdf [https://perma.cc/68YU-SG43].

[189] James Bacchus, "Global Rules for Mutually Supportive and Reinforcing Trade and Climate Regimes," 4 (*Int'l Ctr. for Trade and Sustainable Dev.* and *World Econ. Forum* 2016).

cooperation between these two regimes is crucial to create a new normative framework on how the trading system can help mitigate climate change and enhance energy security.

How can the trading system help? How should the trading system deal with climate change mitigation? Very few trade agreements contain sustainable development chapters. Moreover, hardly any scholarly work exists that can inform practice.[190] Trade agreements can be a vehicle to address common concerns. But if all of the possible outcomes are positive, why are countries and their citizens not reacting to them? Are the trade rules preventing the energy transition? What needs to be changed to make the energy transition happen faster? The following Subsections address these concerns.

We stand to achieve considerable gains when trade law becomes a tool for change.[191] As discussed in the introduction to this chapter, the international community can use trade law as a vehicle not only for climate action and sustainable energy,[192] but also to achieve many of the SDGs. And it is well known that, thanks to trade, countries grow economically.[193] Hence, the "triple benefit of trade" (see Fig. 3.7), in that trade can have a positive economic, environmental, and social impact, bringing together all three dimensions of sustainable development. But can countries grow sustainably?[194] This hypothesis that trade law can help mitigate climate change and enhance sustainable energy may be replicated in other governance issues.[195]

On the links between trade and climate action, the French government has made its position clear to the US government: "No Paris Agreement, no trade

---

[190] There is very little scholarship that analyzes how trade agreements can enhance sustainable development, and more specifically climate change mitigation and sustainable energy. *See* Lee and Kirkpatrick (2001), p. 395; Leal-Arcas (2015a), p. 248; Leal-Arcas and Wilmarth (2015).

[191] WTO, "Harnessing trade for sustainable development and a green economy," available at https://www.wto.org/english/res_e/publications_e/brochure_rio_20_e.pdf.

[192] WTO Speeches – DG Pascal Lamy 'The "Greening" of the WTO Has Started' <https://www.wto.org/english/news_e/sppl_e/sppl79_e.htm>.

[193] This notion is in line with Sustainable Development Goal 8, available at https://sustainabledevelopment.un.org/sdg8. See also the views of Gregory Mankiw, stating that trade improves average living standards. See G. Mankiw, "Why Economists Are Worried About International Trade," *The New York Times*, 16 February 2018, available at https://www.nytimes.com/2018/02/16/business/trump-economists-trade-tariffs.html.

[194] Jacobs and Mazzucato (2016) and Summers (2016).

[195] One can think, for instance, of the argument that, if China and India bring millions of people into the middle class, the world will not be sustainable due to higher levels of consumption (of goods, food, energy) in these two countries. However, Sustainable Development Goal 12 (ensure sustainable consumption and production patterns) is about "promoting resource and energy efficiency, sustainable infrastructure, and providing access to basic services, green and decent jobs and a better quality of life for all." See UN Sustainable Development Goals, Goal 12, available at http://www.un.org/sustainabledevelopment/sustainable-consumption-production/.

Fig. 3.7 The triple benefit of trade

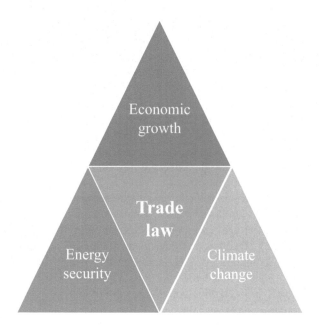

agreement,"[196] following the US's decision to withdraw from the Paris Climate Agreement[197] and making the point that, as a result, there will be no conclusion of the Trans-Atlantic Trade and Investment Partnership (TTIP), a proposed and not yet concluded FTA between the US and the EU. These initiatives of the French government are in line with the views of the EU commissioner for trade, who, in February 2018, twitted: "Paris deal reference needed in all EU trade agreement today."[198]

Commentators Mathilde Dupre and Samuel Lere propose a two-stage process of integrating the Paris Climate Agreement with recent and future trade agreements:

- First, new provisions could be inserted in trade agreements to allow for these agreements (or parts thereof) to be suspended if a Party fails to meet its GHG emissions reduction commitments or to regularly upgrade them. Such a provision would be a very concrete step towards subjecting trade law to environmental law[199];

---

[196]https://www.greenoptimistic.com/france-paris-agreement-no-trade-agreement-20180206/#.WpaYLWrFLX4.

[197]The withdrawal will not be effective until November 2020. Without the participation of the US in the Paris Climate Agreement, the Agreement will, nevertheless, cover around 80% of global GHG emissions. The US is responsible for around 16% of global GHG emissions. See http://www.wri.org/blog/2014/11/6-graphs-explain-world%E2%80%99s-top-10-emitters.

[198]https://twitter.com/Mathilde_Dupre_/status/959112642429423616.

[199]See Petersmann (2000), pp. 1363–1382 (arguing for a shift from negative integration found in the GATT 1947 (such as the elimination of technical barriers to trade) to positive integration found

3.4 Making Greater Use of the Trading System

- Second, to make such a provision operational, trade agreements could create an ad hoc committee composed of scientists and NGO representatives to assess to what extent countries have met their climate commitments (based on information provided to the UN Framework Convention on Climate Change) and propose appropriate trade sanctions, if necessary.[200]

Mathilde Dupre and Samuel Lere further suggest that FTAs remove all provisions that damage the environment.[201]

Recent EU FTAs[202] contain stronger and more detailed provisions on the links between trade and climate change[203] than those EU FTAs negotiated before the 2015 Paris Climate Agreement.[204] Such is the case of the 'trade and sustainable development' chapters in the EU-Singapore,[205] EU-Vietnam,[206] and EU-Japan FTAs.[207]

---

in the TRIPs Agreement, when it comes to integrating human rights into WTO law). This concept of positive integration based on the TRIPs Agreement can be emulated in the WTO context by setting global standards for CO2 emissions-reduction incentives.

[200]M. Dupre and S. Lere, "Trade and climate: How the EU can protect the Paris Agreement," *Euractiv*, 28 February 2018, available at https://www.euractiv.com/section/climate-environment/opinion/trade-and-climate-how-the-eu-can-protect-the-paris-agreement/.

[201]Ibid.

[202]As of 2018, the EU was negotiating an FTA with Mercosur. The result of that negotiation may mean deforestation in Brazil. The EU should try to avoid that as well as the potential abuse of that future FTA regarding forest-risk commodities such as coco.

[203]Non-paper of the Commission services, "Feedback and way forward on improving the implementation and enforcement of Trade and Sustainable Development chapters in EU Free Trade Agreements," 26 February 2018, available at http://trade.ec.europa.eu/doclib/docs/2018/february/tradoc_156618.pdf.

[204]For analyses of trade and sustainable development prior to the conclusion of the Paris Climate Agreement, see Leal-Arcas (2015b), pp. 248–264.

[205]EU-Singapore FTA, Chapter 13, available at http://trade.ec.europa.eu/doclib/docs/2013/september/tradoc_151766.pdf. On 13 February 2019, the trade and investment agreements between the EU and Singapore received the approval of the European Parliament. This means that the trade agreement could enter into force once Singapore concludes its own internal procedure and both parties complete the final formalities. See http://europa.eu/rapid/press-release_IP-19-906_en.htm.

[206]EU-Vietnam FTA, Chapter 15, available at http://trade.ec.europa.eu/doclib/docs/2016/february/tradoc_154229.pdf.

[207]On 6 July 2018, the EU Council of Ministers adopted a package of decisions on the Economic Partnership Agreement between the EU and Japan (EPA), including a decision on the signature of the Agreement and a decision to request the consent of the European Parliament for the conclusion of the Agreement. See Council Decision (EU) 2018/966 of 6 July 2018 at https://eur-lex.europa.eu/legal-content/EN/TXT/?uri=CELEX:32018D0966. The Council also adopted a decision on the signing and provisional application of a Strategic Partnership Agreement (SPA) between the EU and Japan on 26 June 2018. "The SPA is the first-ever framework agreement between the EU and Japan. It seeks to strengthen cooperation and dialogue across a broad range of bilateral, regional and multilateral issues. It highlights the shared values and common principles that constitute the basis for the deep and long-lasting cooperation between the EU and Japan as strategic partners, including democracy, the rule of law, human rights and fundamental freedoms." See Consilium, "EU-Japan: Council adopts decision to sign trade agreement," 6 July 2018, available at http://www.consilium.

The EU-Singapore FTA expressly aims to facilitate trade in climate-friendly goods and services.[208] Moreover, Chapter seven of the Agreement is dedicated to non-tariff barriers to trade and investment in renewable energy generation. Specifically, Article 7.4(a) in Chapter seven of the Agreement states that Parties shall "refrain from adopting measures providing for local content requirements or any other offset affecting the other Party's products, service suppliers, investors or investments." In addition, Chapter thirteen of the Agreement deals with trade and sustainable development. In it, Article 13.11.3 states that "the Parties share the goal of progressively reducing subsidies for fossil fuels," thereby including a provision about the reduction of trade distortions as a consequence of fossil fuel subsidies.[209]

3.4.2 A Potential Agreement on Trade in Energy[210]

This section argues that an Agreement on Trade in Energy will enhance energy flows. So this section proposes as a future research agenda for international economic law to examine whether a General Agreement on Trade in Energy would promote sustainable energy worldwide.[211] Equally interesting would be to see to what extent it would be possible to create a Sustainable Energy Trade Agreement, as suggested by the International Centre for Trade and Sustainable Development.[212] Such an agreement would cover the liberalization of trade in climate-friendly goods and services and would be a plurilateral agreement under Annex 4 of the WTO Agreement that would include a critical mass of major economies and GHG emitters, either in the context of the WTO or outside of the WTO.

---

europa.eu/en/press/press-releases/2018/07/06/eu-japan-council-adopts-decision-to-sign-trade-agreement/.

[208] Not yet signed; not yet in force. As of June 2018, and based on the findings in Opinion 2/15 of Advocate General Sharpston of the Court of Justice of the EU with respect to the EU-Singapore FTA, the EU is currently deciding whether all EU Member States should ratify the Agreement or whether to have a separate agreement with all the provisions that are outside the scope of the EU's exclusive competence. For further analysis on mixed agreements, see Leal-Arcas (2008), chapter 5.

[209] For an analysis of this issue, see Condon (2017), pp. 675, 690; Van de Graaf and van Asselt (2017), p. 313; Wold et al. (2012), pp. 635,694; Burniaux and Chateau (2014), p. 71; Coady, D., Flamini, V., and Sears, L. 'The Unequal Benefits of Fuel Subsidies Revisited: Evidence for Developing Countries' (*IMF Working Paper 15/250*, November 2015), available at http://www.imf.org/external/pubs/ft/wp/2015/wp15250.pdf; David Coady et al. (2015), "How Large Are Global Energy Subsidies?" *IMF Working Paper*, WP/15/105.

[210] Parts of this section were written by Edoard Alvares.

[211] See Cottier et al. (2011); T. Cottier, O. Nartova, L. Rubini, S. Z. Bigdeli, S. Matteotti, and Y. Selivanova, "Background Note for the Second Biennial Global Conference of the Society of International Economic Law (SIEL), 8.07-10.07.2010 - Panel VII: Towards a WTO Framework Agreement on Trade in Energy", Online Proceedings of the Society of International Economic Law, Working Paper No. 2010/40, p. 2.

[212] ICTSD (2011).

3.4 Making Greater Use of the Trading System

All forms of energy should be subject to the same rules. Energy may become part of the WTO agenda in the near future. Given that current WTO rules are far from addressing all the needs of energy trade today, is it necessary to have an Agreement on Trade in Energy in the WTO context? If so, can and should the Energy Charter Treaty be used as a model? Could one invoke GATT Article XX(b) and (g)[213] as exceptions for renewables-related trade in goods and services? Should one have an evolutionary interpretation of GATT Article XX(g) so that the locution 'exhaustible natural resources' encapsulates environmental resources?

Moreover, now that Russia has joined the WTO and that energy is one of its greatest assets in economic terms, would this be the right time to include energy trade as part of the WTO Agreements? The few energy-rich countries that are not yet WTO Members (i.e., Iraq, Iran, Sudan, Azerbaijan, Equatorial Guinea, Algeria, and Libya), but wish to become WTO Members, will most likely follow Russia. These countries should prioritize the conclusion of negotiations to enter the WTO in order to integrate fully into the global trading system and protect their growing interests on world markets. WTO membership will certainly help eliminate any discrimination against them in their trade in sustainable energy.

As energy-rich countries try to diversify their economies away from the fossil-fuel industry, what is the role of the green energy? An impressive example of an energy-rich country that is trying to aggressively diversify its economy by investing $45 billion in solar energy is the Kingdom of Saudi Arabia.[214] Once built, this would be the world's largest solar project, 'about 100 times larger than both the recently announced Solar Choice Bulli Creek PV Plant in Australia and the Helios PV Plant Phase 1 in Greece and more than double what the photovoltaic industry supplied [in 2017].'[215] The project will triple Saudi Arabia's electricity generation capacity,[216] it will create around 100,000 jobs and reduce power costs by $40 billion.[217]

In parallel to the developments presented in previous sections, a further proposal is being put forward in academic circuits: the emergence of a specialized regime within the WTO system. In other words, the elaboration of a specific agreement on

---

[213] GATT Article XX(g) reads as follows:

Subject to the requirement that such measures are not applied in a manner which would constitute a means of arbitrary or unjustifiable discrimination between countries where the same conditions prevail, or a disguised restriction on international trade, nothing in this Agreement shall be construed to prevent the adoption or enforcement by any contracting party of measures:

[...]

(g) relating to the conservation of exhaustible natural resources if such measures are made effective in conjunction with restrictions on domestic production or consumption;

[214] The Economist, "The $100 billion bet," 12 May 2018, p. 11.
[215] "The World's Largest Solar Project," *Leaders K.S.A.*, Issue 17, May 2018, pp. 8–9, at 8.
[216] Idem.
[217] Idem.

trade in energy, covering both energy goods and services.[218] This proposal takes as its point of departure the observation that the submission of energy to the core of the WTO regime is tributary from the international political and geopolitical landscape of the time that the original 1947 General Agreement on Trade and Tariffs (GATT) was enacted: "[...] *at that time liberalising trade in energy was not a political priority. The industry was largely dominated by state run monopolies and thus governed by strict territorial allocation. International trade in energy resources and products was heavily concentrated, cartelised and controlled by a few multinational companies.*"[219] The idea, therefore, would be to strive for the enactment of a "*comprehensive sectoral agreement on energy*"[220] which would help tackle the aforementioned necessary subsidies reform (as well as a commitment-facilitating list of core and related energy services "*and redrafting of Article X:6 of the revised GPA 2007 to include a more explicit recognition of climate-related measures*").[221]

This would not necessarily mean a radical departure from the current architecture of the WTO regimes. A future specific agreement on trade in energy could sit well within the system as it stands nowadays and there are sufficient and adequate means to provide a satisfactory articulation.[222] Allegedly, there are only two necessary conditions to be fulfilled in order to guarantee the coherence of the whole legal construction: (i) "*Energy requires an integrated approach and does not lend itself to sectoral negotiations, depending upon different forms of energy applied to competing energy sectors*"[223]; and (ii) a comprehensive sectoral agreement on energy would need to simultaneously cover both goods and services.[224]

3.5 Finding Synergistic Links Between the International Trading System and the Climate Change Regime

Well-known benefits of trade are cheap imports and higher productivity. From energy transit, to technology transfer,[225] to investment protection, renewable energy and trade present interplays across various fields. This section proposes an

---

[218] Cottier et al. (2011).

[219] Idem, p. 1.

[220] Idem, p. 21.

[221] Idem.

[222] T. Cottier, O. Nartova, L. Rubini, S. Z. Bigdeli, S. Matteotti, and Y. Selivanova, "Background Note for the Second Biennial Global Conference of the Society of International Economic Law (SIEL), 8.07-10.07.2010 - Panel VII: Towards a WTO Framework Agreement on Trade in Energy", Online Proceedings of the Society of International Economic Law, Working Paper No. 2010/40, p. 2.

[223] Idem, p. 2.

[224] Idem., p. 2.

[225] See for instance Brewer (2011), pp. 516–526.

investigation of the basis for how new trade agreements can better address issues of common concern. For instance, this section proposes looking at what improvements can be made to the international trading system to promote climate action,[226] sustainable energy, and more efficient energy markets.[227] There are very few trade agreements with sustainable-development chapters, even fewer with strong and meaningful sustainable-development chapters.[228] Moreover, there is a lack of scholarship that can inform practice. We claim that trade agreements can be a vehicle to address global common concerns. How can the trading system create a regulatory framework that encourages environmental innovation and shifts consumption patterns towards sustainable goods and services?

A few decades ago, few people would have thought of placing the environment in the trade debate. Today, many people recognize that trade and the environment, as an institutional bureaucratic issue, needs to be integrated. Given the importance of greater cooperation between the trading and climate change regimes for a sustainable future, relevant questions arise: What are the trade instruments that best deal with climate action, bearing in mind that there is much public opposition to trade agreements?[229] What kind of trade regime do we need to create for the great transformation of decarbonization? What should be the role of the trading system in reducing fossil-fuel consumption? In accordance with Article 4.2 of the Paris Agreement on Climate Change,[230] how do the nationally determined contributions (NDCs)[231] integrate with the trading system,[232] taking into account that, unlike

---

[226]Leal-Arcas, R. 'International Trade for Climate Action and Inclusive Green Growth' (*Green Growth Knowledge*, 1 February 2018) <www.greengrowthknowledge.org/blog/international-trade-climate-actionand-inclusive-green-growth>.

[227]Leal-Arcas (2013b), pp. 34–42; Leal-Arcas (2014), pp. 11–54; Leal-Arcas (2013c), pp. 1–8.

[228]That said, the number of multilateral environmental agreements (MEAs) referred to in trade agreements is increasing. This is done for various reasons: to determine hierarchy between agreements or for purposes of interpretation, ratification or implementation of MEAs. See J.-F. Morin and C. Bialais, "Strengthening Multilateral Environmental Governance through Bilateral Trade Deals," *Centre for International Governance Innovation*, Policy Brief No. 123, February 2018.

[229]Gonzalez, A. "Trade Agreements Under Attack: Can They Be Salvaged And Is It Worth It?" 24 October 2016, *Huffpost*, available at https://www.huffingtonpost.com/anabel-gonzalez/trade-agreements-under-at_b_12619314.html.

[230]Article 4.2 of the Paris Agreement reads: 'Each Party shall prepare, communicate and maintain successive nationally determined contributions that it intends to achieve. Parties shall pursue domestic mitigation measures, with the aim of achieving the objectives of such contributions.'

[231]Nationally determined contributions (NDCs) "embody efforts by each country to reduce national emissions and adapt to the impacts of climate change." See United Nations Framework Convention on Climate Change "Nationally Determined Contributions (NDCs)," available at http://unfccc.int/focus/items/10240.php. The idea with the NDCs is to make sure that the increase in the global average temperature is "well below 2 °C above pre-industrial levels." See Paris Agreement on Climate Change, 12 December 2015, FCCC/CP/2015/L.9/Rev.1, p. 2. The NDCs are not legally binding.

[232]On the links between trade and the NDCs, see C. Brandi, "Strengthening climate-friendly trade elements in future nationally determined contributions," Opinion, ICTSD, 10 May 2017, available

trade, they are a global public good?[233] What will be the impact for investment flows, say, to expand carbon sinks?[234] How can trade policy attract investments in climate action?[235] How can trade policy help climate action? How can fossil fuel subsidies be reconciled with attempts to integrate the trade and climate change regimes? What is the role of fossil fuel subsidies and border tax adjustments[236] in this equation? Is there a deficiency in WTO law regarding fossil fuel subsidies?[237] Since the trading system should support sustainable development, has globalization (personified by the world trading system) been de-legitimized? Will greater cooperation between both regimes be possible in light of the current political threats to internationalism?[238] Is the fact that the Paris Agreement offers a territorial approach[239] to climate mitigation an issue for international trade? How will emerging technologies and business models affect both the trade and climate change regimes in the future?

---

at https://www.ictsd.org/opinion/strengthening-climate-friendly-trade-elements-in-future-nationally-determined.

[233] For further details on global public goods in the context of sustainability, Leal-Arcas (2017), pp. 801–877.

[234] The UNEP argues that cutting GHG emissions will not be enough to keep global warming within desired expectations and that GHGs must also be removed from the air. See UNEP, "The emissions gap report 2017," chapter 7, available at https://wedocs.unep.org/bitstream/handle/20.500.11822/22070/EGR_2017.pdf.

[235] A case in point here would be a climate-investment facilitation agreement.

[236] On border tax adjustments, see Odell (2018).

[237] Arguably, the possibilities for addressing subsidies under WTO law in general have major limitations. See generally Peat (2012), p. 3.

[238] Examples of deglobalization and political threats to internationalism are Brexit, the US's withdrawal from the Trans-Pacific Partnership, and President Trump's idea of leaving the Paris Agreement on Climate Change, the Intermediate-range Nuclear Forces Treaty, and the world trading system. President Trump is an example of someone de-stabilizing international cooperative structures in environment, climate change, and trade. See for instance Donnan, S. "WTO faces an identity crisis as Trump weighs going it alone," *Financial Times*, 6 December 2017, available at https://www.ft.com/content/38c56f52-d9a5-11e7-a039-c64b1c09b482?desktop=true&conceptId=a208e921-65cb-31b7-8a7c-3e4d0ddcdb53&segmentId=7c8f09b9-9b61-4fbb-9430-9208a9e233c8#myft:notification:daily-email:content:headline:html. It is interesting to observe that today anti-globalizers are in the North (developed world), whereas pro-globalizers are in the South (developing world). The losers of globalization are the poor of the rich world. For an analysis of winners and losers of globalization, see N. Lamp, "How Should We Think about the Winners and Losers from Globalization? Three Narratives and their Implications for the Redesign of International Economic Agreements," Queen's University Legal Research Paper No. 2018-102, 2018, available at https://papers.ssrn.com/sol3/papers.cfm?abstract_id=3290590.

[239] 'Each climate plan reflects the *country's* ambition for reducing emissions, taking into account *its domestic circumstances and capabilities.*' [emphasis added]. See http://unfccc.int/focus/items/10240.php.

3.5 Finding Synergistic Links Between the International Trading System...

There are deficiencies and potential tensions when trying to work out synergistic links between the trade and climate regimes.[240] For instance, the term 'trade' does not appear in the Paris Agreement on Climate Change. In fact, the Paris Agreement incorporates few trade-related elements, such as technology transfer[241] and renewable energy.[242] Moreover, updating[243] the NDCs may lead to tensions based on process and production methods and to intergovernmental trade disputes at the WTO. Furthermore, the US may withdraw from the Paris Agreement. In addition, it is often the case that officials at trade ministries do not have a mindset for the environment, which makes synergies between trade and climate change more challenging.[244]

Arguably, much of the difficulty with greater cooperation between trade and climate change is that the so-called 'trade-in' issues (such as trade in goods or trade in services) are about the reduction or elimination of barriers to trade, whereas the so-called 'trade and' issues (such as trade and climate change) are about regulatory coherence. Any future climate action will need to be WTO-consistent. In this sense, former chairman of the WTO Appellate Body James Bacchus has argued that "to combine the most benefit for the climate with the least risk to trade, a WTO climate waiver is urgently needed"[245] so that "WTO members revise and realign WTO rules in accordance with the objectives of sustainable development."[246] In addition, adding environmental and social issues to the WTO agenda might give the WTO more legitimacy,[247] presenting the WTO as an evolutionary body of law that may go beyond trade, classifying it as 'regime extension.'[248] Other possible solutions are the removal of impediments in the existing law as well as the gradual phase-out and

---

[240] More broadly, for an analysis of integrating sustainable development with other themes, see French (2006), pp. 103–117; Sax (2000), pp. 2375–2402; Sands and Peel (2012); Brown Weiss (2011), pp. 1–27.

[241] Article 6.8 of the Paris Agreement.

[242] Paris Agreement, p. 2.

[243] 'The Paris Agreement requests each country to outline and communicate their post-2020 climate actions, known as their Nationally Determined Contributions (NDCs).' See http://unfccc.int/focus/items/10240.php.

[244] Peru has a ministry for the environment as a result of the 2009 US-Peru Free Trade Agreement. See https://ustr.gov/trade-agreements/free-trade-agreements/peru-tpa.

[245] James Bacchus, "The Case for a WTO Climate Waiver," *Center for International Governance Innovation*, November 2017, available at https://www.cigionline.org/publications/case-wto-climate-waiver?utm_source=climate-l&utm_medium=social&utm_campaign=cop23.

[246] Ibid.

[247] Esty (2002), pp. 7–22.

[248] Adding non-economic issues to the WTO agenda creates a legitimacy-enhancing effect of addressing these problems that clearly matter to the people who are now challenging the validity of the WTO's objectives. The counterargument, however, would be whether economists and trade law professionals have the necessary tools to tackle environmental and social concerns.

eventual prohibition of fossil fuel subsidies.[249] Litigation at the WTO will most likely not move forward many of these difficulties.

Rather, constructive regulation through plurilateral agreements to develop regulation on emerging technologies, carbon capture and storage, batteries, and similar topics may be a more fruitful way forward. Much of the analysis thus far on how the trade and climate regimes can better work together has been done at the multilateral level. There is a need to think about the regime complex too.[250] To do so, political will is necessary, which one can create through leadership. Ultimately, nature, technology, and science will make a difference.[251] Moreover, tensions between the trade regime and the environmental regime are also due to the fact that they do not share a common language. Trade is mainly concerned with the market and is a field where governments are strongly influenced by business and industry groups. In contrast, the environmental regime is concerned mainly with regulation.

Another way forward would be the support of the green infant industry. Doing so would generate green jobs and would translate into climate action without having to import green goods. Furthermore, creative avenues for international dialogue between finance and climate change ministers, instead of trade and climate ministers, is very promising, especially as countries start to feel the economic effects of climate change and extreme weather. Moreover, since the WTO is a member-driven institution, it is relevant in the trade-climate debate to pay close attention to the domestic politics and policies of WTO member states that are major GHG emitters, followed up with individualized action plans on how to address the issue in each of these countries and link the debate to the investment system. The investment regime is relevant in the broader political economy of this debate to address the clean-energy agenda, climate infrastructure, company sustainability, and eco-system protection.

---

[249]There is the so-called Friends of Fossil Fuel Subsidy Reform, which is "a group of countries committed to supporting the reform of inefficient fossil-fuel subsidies." See Global Subsidies Initiative, available at https://www.iisd.org/gsi/about/who-we-work-with/friends-fossil-fuel-subsidy-reform. The International Monetary Fund has studied the cost of energy subsidies, available at https://www.imf.org/en/News/Articles/2015/09/28/04/53/sonew070215a. The Overseas Development Institute has also studied the social cost of fossil-fuel subsidies: Ipek Gencsu et al., "Phase-out 2020: monitoring Europe's fossil fuel subsidies," 2017, available at https://www.odi.org/publications/10939-phase-out-2020-monitoring-europes-fossil-fuel-subsidies.

[250]Leading scholars of international governance, Robert Keohane and David Victor, argue that the diverse range of institutions involved in climate change governance constitutes a regime complex, which has advantages and disadvantages compared to a unitary international regime. Keohane and Victor (2011), pp. 7–23.

[251]Wuebbles, D.J., D.W. Fahey, K.A. Hibbard, B. DeAngelo, S. Doherty, K. Hayhoe, R. Horton, J.P. Kossin, P.C. Taylor, A.M. Waple, and C.P. Weaver, 2017: Executive Summary of the Climate Science Special Report: Fourth National Climate Assessment, Volume I [Wuebbles, D.J., D.W. Fahey, K.A. Hibbard, D.J. Dokken, B.C. Stewart, and T.K. Maycock (eds.)]. U.S. Global Change Research Program, Washington, DC, USA, 26 pp.

In addition, carbon pricing[252] is one of the main issues in the climate change mitigation debate, not trade, to reach the social cost of GHG emissions.[253] Carbon pricing policies, which are necessary but not sufficient, promote GHG cost-effective abatement due to several factors: innovation, carbon pricing is technology-neutral, and thirdly, the polluter-pays principle.

Below is a list of recommendations for how the trade and climate regimes can better cooperate[254]:

3.5.1 Minimize Conflicts Between the Trade and Climate Regimes

Making greater use of existing fora, such as the Committee on Trade and Environment at the WTO secretariat, is very promising. One could also explore the role for a mandatory climate dispute settlement mechanism in the UN Framework Convention on Climate Change and in other international climate agreements. Equally, it may be worth proposing tough environmental trade regulations given that very few dispute settlement cases related to environmental restrictions have been successful and such cases take years to adjudicate.

---

[252] Putting a price on carbon (whether a carbon tax or an emissions trading scheme, i.e., tradable permits) is a way to combat climate change by making people pay for the environmental damage created. A carbon tax provides an added cost to the cost of the product. A tradable permit system sets a cap on the amount of GHG emissions. Firms must buy a permit to emit and there is only a limited number of permits. The cost of the permit is an added cost to the cost of the product. The price is based on the carbon content of the product. Doing so provides an incentive to find low-cost ways to reduce GHG emissions. If a measure costs less than the price, it would make sense to implement the measure, rather than paying the price. If a measure costs more than the price, it would make sense to pay the price. Conversely, people should be rewarded for protecting the environment. There are ethical considerations with putting a price on carbon because it affects the poor the most. Ideally, there should be harmonized carbon taxes, i.e., have the same carbon tax in all countries. See Nordhaus (2007).

[253] Even the US Supreme Court ruled in 2007 that GHGs are a risk to human health. See Massachusetts v. EPA, 549 U.S. 497 (2007). There is a vast body of literature on the links between human rights and environmental protection: Edward Cameron, "Development, climate change and human rights: From the margins to the mainstream?" 123 *World Bank Social Development Working Papers* [2011]; Handl (2001); Padilla (2002), pp. 69–83; Rosemary Rayfuse and Emily Crawford, "Climate Change, Sovereignty and Statehood," *University of Sydney Legal Studies Research Paper* 11/59 (2011); Barnett and Neil Adger (2007), pp. 639–655; Bosello et al. (2006), pp. 579–591; Pontin (2002), p. 21; Dam and Tewary (2005), p. 383; Kravchenko and Bonine (2008), Ch. 12.

[254] These recommendations draw from Bacchus, James, Leal-Arcas, R., et al. (2016) "Global Rules for Mutually Supportive and Reinforcing Trade and Climate Regimes," E15 Expert Group on Measures to Address Climate Change and the Trade System – Policy Options Paper. E15Initiative. Geneva: International Centre for Trade and Sustainable Development (ICTSD) and World Economic Forum.

Moreover, one could make an exception to the most-favored-nation (MFN)[255] and national-treatment (NT)[256] principles of non-discrimination for policy objectives, e.g., climate change mitigation. This could be the drafting of WTO norms to not impose duties on renewables. For instance, one could provide preferential treatment to green goods and services such as electric cars that could be sold preferentially in the world market.

In addition, one could establish a "peace clause" for climate action. Similar to the WTO's Agriculture Agreement, the peace clause would enable countries to protect themselves from action taken against each other's fossil fuel subsidies.[257] In general, although the trade regime does not seem to be an obstacle to climate change action per se, it is also not driving such action. Therefore, the international community should re-gear the trade regime to incentivize innovation and capital flow toward sustainability initiatives.

3.5.2 Revisit the Concept of 'Like' Products

One could establish an international standard for calculating the amount of carbon used in the making of products. In addition, it would be desirable to agree on a "waiver" from WTO obligations for all trade-restrictive climate measures that are based on the amount of carbon used in making a product.

3.5.3 Climate Action Via the Formation of Climate Clubs

An option to be explored is plurilateral action to promote green technology transfer among club members—e.g., G20, G7, the Major Economies Forum, participation of cities (C40), namely a group of 90+ of the largest cities in the world for the implementation of the Paris Agreement on Climate Change. In addition, climate measures pursuant to a climate agreement should fall within the scope of Article XX (b) and (g) of the GATT and of Article XIV of the GATS (i.e., the general exceptions). Moreover, one could envisage granting a waiver for a climate club

---

[255] Most-favored-nation treatment (GATT Article I, GATS Article II, and TRIPs Article 4) is the principle of not discriminating between one's trading partners.

[256] National treatment (GATT Article III) is the principle of giving others the same treatment as one's own nationals.

[257] For an overview of global energy subsidies, see IEA, 'Energy Subsidies' (*IEA*, 2016), available at www.iea.org/statistics/resources/energysubsidies/. See also Benjamin K Savacool, 'Reviewing, Reforming, and Rethinking Global Energy Subsidies: Towards a Political Economy Research Agenda' (*Sussex*, 2017), available at http://sro.sussex.ac.uk/66350/3/Sovacool-Subsidies-070616.pdf.

organized outside the WTO framework to become a plurilateral agreement under the WTO Agreement.

A club approach would overcome free-riding issues of climate change mitigation (e.g., the EU and NATO as examples of clubs where only members have benefits). This area is of great importance because, despite all the progress in scientific and economic understanding of climate change, achieving international agreements on climate change has proven difficult because of the threat of free-riding. In addition, a carbon price, rather than carbon emission reduction requirements, should be a core element of this climate-club approach because states are more likely to reach agreement on a carbon price than on carbon emission reduction levels. Ideally, carbon prices should reach a level up to that of the social cost of GHG emissions.[258]

3.5.4 *An Agreed Framework for a Global Emissions Trading System and Border Measures*

The idea is to ensure that WTO rules apply to a global emissions trading system and to ensure that a carbon tax (whether borne by the producer or the consumer) is an indirect tax on a product that is eligible for a "border tax adjustment" under Article II:2(a) of the GATT.[259] Border tax adjustments can be perceived as a carrot to avoid or minimize a loss of economic competitiveness and to avoid carbon leakage, or as a stick to spur climate action. When deciding who should be taxed, one could think of taxing the consumers because their countries typically outsource the production of the goods they consume to producing countries like China and other Asian countries, typically in Southeast Asia. That would imply counting carbon consumption, instead of production.

When deciding which option is more effective (whether a carbon tax or an emissions trading system)[260] in terms of raising revenue and cutting GHG emissions, a working paper by the International Monetary Fund finds that taxes raise around twice as much revenue as today's cap-and-trade schemes[261] and are about 50% better at cutting GHG emissions.[262] This raises the question how to make taxes attractive to help mitigate climate change. One way to get there is by calling taxes 'contributions' or 'fees.' Another way is by returning the money raised to citizens. A

---

[258]See generally World Bank and Ecofys. 2018. "State and Trends of Carbon Pricing 2018 (May)", by World Bank, Washington, DC.

[259]A carbon tax may not be so easy to implement, as seen in the case of France in 2018 with the *gilets jaunes*, who refused to accept a high fuel tax proposed by President Macron.

[260]The idea of a cap-and-trade system or a carbon tax in countries with high levels of corruption is ludicrous because the authorities would take the revenue.

[261]The cap-and-trade system caps the GHG emissions to gradually reduce them.

[262]Ian Parry, Victor Mylonas, and Nate Vernon, "Mitigation policies for the Paris agreement: An assessment for G20 countries", IMF Working Paper, 2018. See also Haites et al. (2018), pp. 109–182.

case in point is Switzerland, where two thirds of the money raised is returned to its citizens and firms. The remaining amount goes to green investments.[263]

In the US, a group of senior Republicans has come up with a plan that includes duties on dirty imports and tax credits on clean exports.[264] Such a border-adjustment proposal is WTO-consistent in the context of GATT Article XXb, which favors products that are "necessary to protect human, animal or plant life or health."

Arguably, a tax on carbon is regressive in that the less well-off end up paying a higher proportion of their income on energy. This notion explains the *gilets jaunes* movement in France in November 2018 against rising fuel prices.

3.5.5 Shift Subsidies from Fossil Fuels to Renewable Energy

The international community is not penalizing pollution; rather, it is subsidizing it via fossil fuel subsidies.[265] In fact, a G20 Leaders statement clearly recognizes the way forward: *"To phase out and rationalize over the medium term inefficient fossil fuel subsidies while providing targeted support for the poorest.* Inefficient fossil fuel subsidies encourage wasteful consumption, reduce our energy security, impede investment in clean energy sources and undermine efforts to deal with the threat of climate change."[266] So the international community should agree on the gradual phase-out and eventual prohibition of fossil-fuel subsidies because subsidies distort the allocation of resources.[267]

The rationalization of energy prices provides economic and environmental benefits, but could potentially harm the incumbent industries and consumers benefiting from low prices. The rationalization of energy prices also stimulates energy efficiency and low-carbon energy, thereby setting the foundation for effective climate change policy. The impacts of this rationalization on world energy markets could harm energy exporters.

Allegedly, most developed economies are simply financing fossil energy schemes, despite being publicly outspoken against these kinds of sources of energy. This schizophrenia was denounced by a group of 15 environmental NGOs[268] during

---

[263] The Economist, "Free exchange: When the levy breaks," 18 August 2018, p. 62. See also Stern et al. (2018), pp. 669–677.

[264] Climate Leadership Council, "The conservative case for carbon dividends," February 2017, available at https://www.clcouncil.org/wp-content/uploads/2017/02/TheConservativeCaseforCarbonDividends.pdf.

[265] The countries that subsidize energy (whether natural gas, electricity, or oil) the most are Iran, Saudi Arabia, Russia, and Venezuela.

[266] http://www.g20.utoronto.ca/2009/2009communique0925.html.

[267] A more efficient allocation of resources promotes economic growth.

[268] The movement was largely led by Oil Change International, Friends of the Earth U.S., the Sierra Club, and WWF European Policy Office, which are the organizations listed as "publishers" of the report mentioned below.

3.5 Finding Synergistic Links Between the International Trading System...

the wake of the 2017 G20 summit in Hamburg (Germany).[269] These organizations complain that G20 governments provide (in advantageous loans, guarantees, and similar means of preferential financing) almost four times more public finance to fossil fuels than to clean energy.[270] Not only would a shift in the distribution of financing have the double benefit of simultaneously making fossil fuels more expensive and renewable energies cheaper, but it would correspond with the notion of sustainable development. Moreover, it would be an important back-up to the evolution of a sector which, with careful support, can reach maturity status even earlier than projected, as demonstrated by the Danish experience: Denmark is preparing to suppress renewable energy subsidies, for the sector is almost ready to reap profits on its own.[271]

Moreover, I argue that fossil fuels subsidies discourage investments in renewables, but a multilateral scheme for renewables subsidies may fall under permissible subsidies under the WTO's Agreement on Subsidies and Countervailing Measures.[272] Furthermore, since the price of oil is low, countries should have no reason to subsidize the fossil fuel industry.[273]

If we are serious about climate change mitigation, fossil fuel subsidies must come to an end and green industrial policy tools (i.e., renewable energy subsidies, specific performance requirements, technical standards, and local content requirements) must take priority.[274] We should also allow only non-discriminatory purchases of climate-friendly environmental goods and services under the WTO Government

---

[269]Via the publication of a report entitled "Talk is cheap: How G20 Governments are financing climate disaster", available online on October the 4th 2017 (http://priceofoil.org/content/uploads/2017/07/talk_is_cheap_G20_report_July2017.pdf).

[270]Oil change international, "Talk is cheap: How G20 Governments are financing climate disaster", http://priceofoil.org/2017/07/05/g20-financing-climate-disaster/, available online on October the 4th 2017.

[271]Green Tech Media, "Denmark Preparing to End Renewable Energy Subsidies: 'We're Now Very Close'", https://www.greentechmedia.com/articles/read/denmark-preparing-to-end-renewable-energy-subsidies#gs.BRG0_xo, available online on October the 4th 2017.

[272]For a discussion on subsidies and renewable energy, see Peat (2012).

[273]The International Monetary Fund and the Organization for Economic Cooperation and Development have done useful work on the question of defining and measuring fossil fuel subsidies. See for instance Coady, D. et al., "How large are global energy subsidies?" IMF Working Paper WP/15/105, 2015. A useful tool on questions of defining and categorizing various types of subsidies is the WTO Agreement on Agriculture.

[274]The international community should find ways to reduce both the strong influence wielded by industries benefiting from fossil fuel subsidies and the hesitance of political leaders, even those generally committed to environmental issues, to take a stand publicly despite growing awareness of such measures' economic and environmental repercussions. For a study on the benefits of fuel subsidies, see David Coady, Valentina Flamini, and Louis Sears, 'The Unequal Benefits of Fuel Subsidies Revisited: Evidence for Developing Countries' (*IMF*, November 2015), http://www.imf.org/external/pubs/ft/wp/2015/wp15250.pdf. See also Peter Wooders and Cleo Verkuijl, 'Making the International Trade System Work for Climate Change: Five Ways to Address Fossil Fuel Subsidies through the WTO and International Trade Agreements' (*International Institute for Sustainable Development*, 20 June 2017).

Procurement Agreement (GPA) while encouraging more WTO Members to accede to that Agreement.[275] Finally, subsidies should go to the renewable-energy industry to make the transition to a green economy a reality.[276] The need is great, but the science is nascent and economic incentives in the market are still missing. Here is where governments can play a role through subsidies and green taxes.[277]

3.5.6 Trade in Electric Vehicles

The role of electric vehicles (EVs) in decarbonizing the transport sector is another key area to be explored.[278] Research shows that, if all new cars were electric, they would make up 90% of the world's two billion cars by 2040, thereby saving 11 billion barrels of oil every year (or almost half of annual global production) and 4.7 billion tons of CO2 (this figure excludes emissions and oil used to make electric cars).[279] This plausible reality raises questions such as: how can consumers influence the vehicle industry to make them go electric?[280] How can mobility become renewable?[281]

Some European governments seem to be moving firmly in the direction of EVs: in July 2017, the United Kingdom (UK) government announced that it would ban

---

[275] At the moment, the GPA comprises only 47 WTO members out of 164 members. See https://www.wto.org/english/tratop_e/gproc_e/gp_gpa_e.htm.

[276] A case in point is California, whose officials upheld a mandate in December 2018 to require rooftop panels on new homes from 2020. A 30% federal tax credit was used as an example of a generous subsidy. See The Economist, "Distributed energy: Solar eclipsed," 22 December 2018, pp. 91–94, at 91.

[277] That said, Germany has invested $1trn on low-carbon electricity and yet still depends at 50% on fossil fuels for its power. See "Negative emissions: What they don't tell you," *The Economist*, 18 November 2017, p. 12. One simple way to get to negative emissions is by planting more forests. According to The Economist, "even with negative emissions carbon-dioxide release still needs to fall by 45% or thereabouts by 2030. [...] two-thirds of coal use must be phased out in little more than a decade. By the middle of the century virtually all electricity must come from carbon-free sources (up from a quarter today), and all cars will need to run on electric motors [...], as will trains and most ships." The Economist, "Global Warming: War was is better than jaw jaw," 13 October 2018, pp. 76–77, at 77.

[278] See for instance the electric bicycles produced by BYD Auto Co., a Chinese automobile manufacturer.

[279] *See A Flash in the Sky*, The Economist, July 15, 2017, at 16–17.

[280] All of this said, in the case of cars, their sales are falling because better cars and roads mean longer car life, which means fewer new-car sales, and it is a headwind for electric vehicles. *See* Kyle Stock, *The Real Reason Car Sales Are Falling*, Bloomberg, (Aug. 2, 2017), https://www.bloomberg.com/news/articles/2017-08-02/the-real-reason-car-sales-are-falling.

[281] The idea of becoming renewable goes beyond the mobility sector. See M. Safi, "India plans nearly 60% of electricity capacity from non-fossil fuels by 2027," *The Guardian*, 22 December 2016.

3.5 Finding Synergistic Links Between the International Trading System...

the sale of new cars that run solely on petrol or diesel by 2040.[282] The French government spoke in similar terms in its own announcement, planning to end the sales of gas and diesel cars by 2040.[283] Carmakers are heading in the same direction: Volvo announced in 2017 that all Volvo cars will be electric or hybrid as of 2019.[284] BMW, Porsche, and Audi have electric models that will enter the market by 2020.[285] Outside of Europe, although no timeline has been suggested, China's government would like to move towards a ban on gas vehicles, which will have profound implications for global carmakers, given China's market size.[286] This Chinese move is quite promising as China has some of the world's biggest battery producers and is very active in electronics manufacturing.[287]

Morgan Stanley, an investment bank, expects that, of the one billion cars on the road, half will be powered by battery by 2050, since the price of batteries is decreasing.[288] Moreover, when it comes to GHG emissions, aviation and shipping are two key players in the transportation[289] sector—they are responsible for GHG emissions equivalent to those of some countries that are major GHG emitters.[290] For the mitigation of climate change, electric or hybrid engines in aviation and shipping would be very effective. For instance, hybrid airplanes, with a capacity of 100 passengers, could take off and land using jet engines but, during the cruise, they could make use of electrically powered engines.[291] Similarly, lighter electric engines for aviation have been developed. For airplanes to keep flying, we would need a new development of novel aviation biofuel to counterbalance the aircraft emissions.

---

[282] *Business*, The Economist, July 29, 2017, at 8.

[283] *See id.*

[284] Adam Vaughan, *All Volvo Cars to be Electric or Hybrid from 2019*, The Guardian, (July 5, 2017), https://www.theguardian.com/business/2017/jul/05/volvo-cars-electric-hybrid-2019.

[285] *Cleaning Up Cars*, The Economist, Sept. 30, 2017, at 31.

[286] *Electric Cars in China: Zooming Ahead*, The Economist, Sept. 30, 2017, at 68.

[287] *See id.*

[288] However, battery production is not emissions free. *See Charge of the Battery Brigade*, The Economist, Sept. 9, 2017, at 63–64.

[289] The irony with transportation is that tickets for some flights are cheaper than train tickets, which does not help to mitigate climate change.

[290] Leal-Arcas (2013a).

[291] *Let's Twist Again*, The Economist, Sept. 16, 2017, at 82.

3.5.7 Foster Sectoral Approaches Such as Aviation and Shipping

Climate agreements affecting trade made by organizations such as the International Maritime Organization (IMO)[292] and the International Civil Aviation Organization (ICAO) should be upheld in WTO dispute settlement to WTO Members that are parties to those agreements. In addition, since the demand for aviation[293] and trade via shipping will only grow, one should improve the fuel efficiency of new and existing ships and planes via electric engines.[294] Moreover, electric or hybrid engines in aviation and shipping would be very effective for climate change mitigation. For instance, hybrid planes, with a capacity of 100 passengers, could take off and land using jet engines but, during the cruise, they could make use of electrically powered engines. Similarly, lighter electric engines for aviation have been developed. So an electrification program in transport is on its way.

In fact, aviation could be an interesting sector for the fight against climate change if we manage to use battery power in planes. The main issue is the weight of planes because planes have to lift themselves into the sky, which is currently hard to do with battery power, as opposed to jet engines.[295] But there are advantages to electric motors, rather than jet engines: they are lighter and more energy-efficient.[296] Currently, there are trials to see whether a hybrid-electric airliner with a capacity of 100 passengers might enter the market in 2030.[297]

---

[292]One should add that, in April 2018, the IMO pledged to halve annual emissions from ships by 2050 compared with the 2008 levels, thereby making a statement that the IMO is serious about environmental protection. See IMO, "UN body adopts climate change strategy for shipping," available at http://www.imo.org/en/mediacentre/pressbriefings/pages/06ghginitialstrategy.aspx.

[293]"World fleet se to double in next 20 years," 26 July 2018, *Aviation job search*, available at https://blog.aviationjobsearch.com/world-fleet-set-to-double-in-next-20-years/.

[294]In 2010, transportation (namely the movement of people and goods by cars, trucks, trains, ships, airplanes, and other vehicles) represented 14% of global greenhouse gas emissions. See US Environmental Protection Agency, "Sources of Greenhouse Gas Emissions," available at https://www.epa.gov/ghgemissions/global-greenhouse-gas-emissions-data.

[295]That said, there are already companies that make two-seater electric training planes. See for instance the E-fan, available at https://www.siemens.com/innovation/de/home/pictures-of-the-future/mobilitaet-uns-antriebe/die-zukunft-der-mobilitaet-e-fan-x.html.

[296]"Commercial aviation: The electric-flight plan," *The Economist*, 2 December 2017, pp. 69–70, at 70.

[297]Ibid., at 69. According to The Economist, Samsung Electronics said that by incorporating graphene into a lithium-ion battery, it managed to maximize its energy capacity by 45%. Ibid., at 70. It will be interesting to see whether Chile, a very rich country in lithium, will end up a new Saudi Arabia as a result of large amounts of lithium.

3.5.8 Tariffs

Everyone wants a world that is clean, safe, and prosperous, with no poverty. The answer to many of these issues is a trading system that facilitates the movement of goods and services in a way that will help achieve a cleaner, sustainable, and richer world. The necessary change in the trading system is possible via the reduction or elimination of tariff and nontariff barriers to environmental goods and services. For instance, there are countries that charge tariffs as high as 35% on environmental goods.[298] If countries eliminate or reduce technical barriers to trade in environmental goods and services,[299] they would not only help to mitigate climate change, but also provide greater access to sustainable energy and boost the economy through increased trade and jobs.[300] Reducing these barriers will be beneficial to trade, the environment, and sustainable development.

3.6 Outlook

There is still much to do about how the trading system can be supportive of clean energy and climate action. Investing in renewable energy makes economic sense, not just environmental sense, given that solar energy is the fastest growing energy source. There are obvious social and economic benefits in renewable energy in general and solar energy in particular. Renewables can provide energy for all. They can increase the quality of health by reducing the level of pollution.[301] They can be scalable: citizens can decide what type of energy they want (micro-grids or big solar panels). Renewables can enable students to study where there is not light after dark. The key drive of solar energy is its business case: the price has dropped dramatically in recent years. For renewable energy to be produced at large scale, international cooperation will be necessary. One could also have a more decentralized and democratized energy system.[302]

---

[298]*Environmental Goods Agreement*, Off. U.S. Trade Rep., https://ustr.gov/trade-agreements/other-initiatives/environmental-goods-agreement [https://perma.cc/P8NX-8Z8T].

[299]For a list of fifty-four environmental goods on which leaders of the Asia-Pacific Economic Cooperation (APEC) member-States have committed to reduce or eliminate tariffs, see *Annex C – APEC List of Environmental Goods*, APEC (Sept. 8, 2012), http://www.apec.org/Meeting-Papers/Leaders-Declarations/2012/2012_aelm/2012_aelm_annexC.aspx [https://perma.cc/WL98-T5XT].

[300]U.N. Env't Program, Green Jobs: Towards Decent Work in a Sustainable, Low-Carbon World 3–4, 33–34, 284 (2008), http://digitalcommons.ilr.cornell.edu/cgi/viewcontent.cgi?article=1057&context=intl [https://perma.cc/CQ49-62SJ].

[301]For views on the negative impact of GHG emissions on human health due to air pollution, see World Health Organization, 'Air Pollution' (World Health Organization, 2017), available at www.who.int/airpollution/en/.

[302]For a thorough analysis, see Leal-Arcas et al. (2018).

The way to become a rich country and to economic development usually happens through industrialization, open trade, and the development of export industries. Because international trade facilitates technology transfer, the role of international trade (agreements) in renewable energy should be enhanced in the future. For instance, wind and sunshine in northern Europe is not the same as in southern Morocco. The price of northern European renewable energy will therefore be higher. International trade will therefore play a major role to reduce prices. For that to happen, free and fair solar trade in the global value chain of trade in goods and services is a must. The elimination of trade barriers damaging the global value chain is necessary, whether it is local content requirements or fossil fuel subsidies,[303] as is the promotion of green growth for a sustainable future.[304] Equally, greater cooperation will be necessary between developed countries that possess the solar-energy technology and sunshine countries between the Tropic of Cancer and the Tropic of Capricorn, who are fortunate to have tremendous solar-energy potential. All of this will take place with new thinking and innovation, instead of more of the same.

The reduction of GHG emissions is possible if we switch to 100% renewable energy and if we stop using fossil fuels in the energy sector, transportation and agriculture.[305] Green buildings, reforestation of large areas, and organic agriculture[306] will be necessary elements to reach this goal. To all this, one should add the engagement the U.S., the EU, and China (as the major GHG emitters), helping navigate the transition to renewable energy, and the promotion of carbon pricing.

Moreover, politicians suffer from short-termism for obvious electoral reasons. This phenomenon, however, is not the case of entrepreneurs, who have proven time and time again that they have a long-term approach to their vision and actions. This approach will expedite the necessary change to mitigate climate change and enhance international trade.

This brings us to an analysis in the next chapter of the role of trade agreements in getting to a sustainable world.

---

[303] On climate change and the elimination of trade barriers, see Frey (2016). On fossil fuel subsidies in the maritime and aviation realms, see Michael Keen, Ian Parry, and Jon Strand "The (non-)taxation of international aviation and maritime fuels: Anomalies and possibilities," 2014, available at http://voxeu.org/article/non-taxation-international-aviation-and-maritime-fuels-anomalies-and-possibilities.

[304] See for instance the sustainability views of the Chinese Communist Party regarding the belt and road initiative: "Guidance on Promoting Green Belt and Road," 2017, available at https://eng.yidaiyilu.gov.cn/zchj/qwfb/12479.htm. See also Parking, B. "China assumes green power mantle, leaving Germany, U.S. behind," *Bloomberg*, 5 December 2017, available at https://www.bloomberg.com/news/articles/2017-12-05/china-assumes-green-power-mantle-leaving-germany-u-s-behind.

[305] Huang et al. (2011), pp. S9–S13.

[306] According to The Economist, 'Mr Springmann and his colleagues have calculated that in 2050 GHG emissions from agriculture in a vegan world would be 70% lower than in a world where people ate as they do today; in the "healthy global diet" world they would be 29% lower.' See The Economist, "The retreat from meat," pp. 21–23, at p. 23, 13 October 2018.

References

Abbott KW (2012) The transnational regime complex for climate change. Environ Plan C: Gov Policy 30(4):571–590

Abbott K (2014) Strengthening the transnational regime complex for climate change. Transnatl Environ Law 3:57

Abbott K, Snidal D (2009) The governance triangle: regulatory standards institutions and the shadow of the state. In: Mattli W, Woods N (eds) The politics of global regulation. Princeton University Press, pp 44–88

Abbott KW, Snidal D (2010) International regulation without international government: improving io performance through orchestration. Rev Int Organ 5(3):315–344

Abbott KW, Green J, Keohane R (2016) Organizational ecology and institutional change in global governance. Int Organ

Abu Gosh E, Leal-Arcas R (2013) The conservation of exhaustible natural resources in the GATT and WTO: implications for the conservation of oil resources. J World Invest Trade 14 (3):480–531

Akerlof G, Shiller R (2015) Phishing for phools: the economics of manipulation and deception

Albareda L, Waddock S (2016) Networked CSR governance: a whole network approach to meta-governance. Bus Soc

Alter K, Meunier S (2009) The politics of international regime complexity. Perspect Polit 7(1):13

Anceschi L et al (eds) (2011) Energy security in the era of climate change: the Asia-Pacific experience. Palgrave

Andrew R, Davis S, Peters GP (2013) Climate policy and dependence on traded carbon. Environ Res Lett 8:23

Andrews-Speed P (ed) (2008) International competition for resources: the role of the law, the state, and of markets. Dundee University Press

Araya M (2016) The relevance of the environmental goods agreement in advancing the Paris Agreement Goals and SDGs. A Focus on Clean Energy and Costa Rica's Experience. International Centre for Trade and Sustainable Development (ICTSD), Geneva

Archer D (2016) The long thaw: how humans are changing the next 100,000 years of Earth's climate. Princeton University Press, Princeton, NJ

Asif M, Muneer T (2007) Energy supply, its demand and security issues for developed and emerging economies. Renew Sustain Energy Rev 11(7):1388–1413

Ayres R (1996) Limits to the growth paradigm. Ecol Econ 19:117–134

Babiker M, Rutherford T (2005) The economic effects of border measures in subglobal climate agreements. Energy J 26(4):99–125

Bacchus J (2018a) The willing world: shaping and sharing a sustainable global prosperity. Cambridge University Press

Bacchus J (2018b) Triggering the trade transition: the G20's role in reconciling rules for trade and climate change. International Centre for Trade and Sustainable Development

Bak C (2015) Growth, innovation and trade in environmental goods. Center for International Governance Innovation

Barnett J, Neil Adger W (2007) Climate change, human security and violent conflict. Polit Geogr 26:639–655

Bartels L (2005) Human rights conditionality in the EU's international agreements. Oxford University Press

Bartels L (2015) In: Lester S, Mercurio B, Bartels L (eds) Social issues in regional trade agreements: labour, environment and human rights, in bilateral and regional trade agreements, 2nd edn

Baumol W (1972) On taxation and the control of externalities. Am Econ Rev 62(3):307–322

Baumol W, Oates W (1971) The use of standards and prices for protection of the environment. Swedish J Econ 73(1):42–54

Beckfield J (2010) The social structure of the world polity. Am J Sociol 115(4):1018–1068

Berliner D, Prakash A (2014) The United Nations global compact: an institutionalist perspective. J Bus Ethics 122(2):217–223

Bernstein S, Cashore B (2004) Non-state global governance: is forest certification a legitimate alternative to a global forest convention? Hard choices, soft law: voluntary standards in global trade, environment and social governance, pp 33–63

Bodansky D (2016) The Paris Climate Change Agreement: a new hope? Am J Int Law 110(2):288–319

Bodansky D et al (2017) International climate change law. Oxford University Press

Bodin Ö, Crona BI (2009) The role of social networks in natural resource governance: what relational patterns make a difference? Global Environ Change 19(3):366–374

Bordoff JE (2009) International trade law and the economics of climate policy: evaluating the legality and effectiveness of proposals to address competitiveness and leakage concerns. In: Brainard L, Sorkin I (eds) Climate change, trade and competitiveness: is a collision inevitable? Brookings Institution Press

Borgatti SP, Jones C, Everett MG (1998) Network measures of social capital. Connections 21(2):27–36

Borgatti SP, Mehra A, Brass DJ, Labianca G (2009) Network analysis in the social sciences. Science 323(5916):892–895

Bosello F, Boson R, Tol RSJ (2006) Economy-wide estimates of the implications of climate change: human health. Ecol Econ 58:579–591

Boyd W (2010) Climate change, fragmentation, and the challenges of global environmental law: elements of as post-copenhagen assemblage. Univ Pa J Int Law 32(2):457–550

Bradbook A, Wahnschafft R (2012) International law and global sustainable energy production and consumption. In: Bradbrook A et al (eds) The law of energy for sustainable development. Cambridge University Press

Bradbrook AJ (1996) Energy law as an academic discipline. J Energy Nat Resour Law 14(2):193, 197

Bradbrook A (2011) Creating law for next generation energy technologies. Colo J Int Environ Law Policy 22:251

Brandi C (2017) Trade elements in countries' climate contributions under the Paris Agreement. International Centre for Trade and Sustainable Development (ICTSD), Geneva

Brewer T (2011) Climate change technology transfer: a new paradigm and policy agenda. Climate Policy 8(5):516–526

Brown Weiss E (2011) The evolution of international environmental law. Jpn Yearb Int Law 54:1–27

Bruce S (2013) International law and renewable energy: facilitating sustainable energy for all? Melbourne J Int Law 18

Bulkeley H et al (2012) Governing climate change transnationally: assessing the evidence from a database of sixty initiatives. Environ Plan C: Gov Policy 30(4):591–612

Burniaux JM, Chateau J (2014) Greenhouse gases mitigation potential and economic efficiency of phasing-out fossil fuel subsidies. Int Econ 140:71

Cameron P (2010) International energy investment law: the pursuit of stability. Oxford University Press, Oxford

Carlarne C (2008) Good climate governance: only a fragmented system of international law away? Law Policy 30(4):450–480

Carpenter SR, Brock WA (2008) Adaptive capacity and traps. Ecol Soc 13(2):40

Cherp A, Jewell J (2014) The concept of energy security: beyond the four as. Energy Policy 75:415–421

Cherp A, Jewell J, Goldthau A (2011) Governing global energy: systems, transitions, complexity. Global Policy 2(1):75

Condon BJ (2017) Disciplining clean energy subsidies to speed the transition to a low-carbon world. J World Trade 51:675, 690

Copper J (2016) China's foreign aid and investment diplomacy, vol 1. Palgrave McMillan, Basingstoke

Cosbey A (2016) The trade implications of the Paris COP21 Agreement. International Trade Working Paper, No. 2016/17. Commonwealth Secretariat, London

Cottier T, Malumfashi G, Matteotti-Berkutova S, Nartova O, de Sépibus J, Bigdeli SZ (2011) Energy in WTO law and policy. In: Cottier T, Delimatsis P (eds) The prospects of international trade regulation: from fragmentation to coherence. Cambridge University Press, Cambridge

Croquet N (2015) The climate-change norms under the EU-Korea Free Trade Agreement: between soft and hard law. In: Wouters J, Marx A, Geraerts D, Natens B (eds) Global governance through trade EU policies and approaches. Edward Elgar Publishing, pp 124–157

Daly H (1997) Beyond growth: the economics of sustainable development. Beacon Press

Dam S, Tewary V (2005) Is a 'polluted' constitution worse than a polluted environment? J Environ Law 17:383

Dashwood H (2014) Sustainable development and industry self-regulation: developments in the global mining sector. Bus Soc 53(4):551–582

David T, Westerhuis G (2014) The power of corporate networks: a comparative and historical perspective, vol 26. Routledge

De Brabandere E, Gazzini T (eds) (2014) Foreign investment in the energy sector: balancing private and public interests. Brill Nijhoff, Leiden/Boston

Di Leva C, Shi X (2016) The Paris Agreement and the international trade regime: considerations for harmonization. Sustain Dev Law Policy Brief XVII(1)

Di Leva CE, Shi X (2017) The Paris Agreement and the international trade regime: considerations for harmonization. Sustain Dev Law Policy 17(1)

Dine J (2005) Companies, international trade, and human rights. Cambridge University Press, Cambridge

Druzin BH (2016) A plan to strengthen the Paris Climate Agreement. Fordham Law Rev Res Gestae 84:18–23

Dubash N, Florini A (2006) Mapping global energy governance. Global Policy 2:6–18

Dupuy PM, Vinuales J (2018) International environmental law. Cambridge University Press

Easley D, Kleinberg J (2010) Networks, crowds, and markets: reasoning about a highly connected world. Cambridge University Press

Eberlein B, Abbott KW, Black J, Meidinger E, Wood S (2013) Transnational business governance interactions: conceptualization and framework for analysis. Regul Gov

Ekins P (1994) The impact of carbon taxation on the UK economy. Energy Policy 22(7):571–579

Ekins P (1999) European environmental taxes and charges: recent experience, issues and trends. Ecol Econ 31:39–62

Ekins P (2009) Resource productivity, environmental tax reform and sustainable growth in Europe. Anglo-German Foundation for the Study of Industrial Society

Esty D (1994) Greening the GATT: trade, environment, and the future. Washington, D.C., Peterson Institute for International Economics

Esty DC (2002) The World Trade Organization's legitimacy crisis. World Trade Rev 1(1):7–22

Esty D (2017) Red lights to green lights: from 20th century environmental regulation to 21st century sustainability. Environ Law Rev 47(1):41–42. https://law.lclark.edu/live/files/23903-47-1estypdf

Faber M et al (1990) Economy-environment interactions in the long-run: a neo-Austrian approach. Ecol Econ 2:27–55

Faure M, De Smedt P, Stas A (2015) Environmental enforcement networks: concepts, implementation and effectiveness. Edward Elgar Publishing

Fenwick M, Van Uytsel S, Wrbka S (2014) Introduction: networks and networked governance. Networked governance, transnational business and the law. Springer, pp 3–9

Florini A, Sovacool BK (2009) Who governs energy? The challenges facing global energy governance. Energy Policy 37(12):5239–5248

Florini A, Sovacool BK (2011) Bridging the gaps in global energy governance. Global Gov 17:57

Fransen L, Burgoon B (2014) Privatizing or socializing corporate responsibility: business participation in voluntary programs. Bus Soc 53(4):583–619

French D (2006) Supporting the principle of integration in the furtherance of sustainable development: a sideways glance. Environ Law Manage 18(3):103–117

Frey C (2016) Tackling climate change through the elimination of trade barriers for low-carbon goods: multilateral, plurilateral and regional approaches. In: Mauerhofer V (ed) Legal aspects of sustainable development. Springer International Publishing, Switzerland

Gault J (2010) A word of introduction from the energy industry perspective. In: Pauwelyn J (ed) Global challenges at the intersection of trade, energy and the environment. The Graduate Institute, Geneva, p 9

Gehring T, Faude B (2013) The dynamics of regime complexes: microfoundations and systemic effects. Global Gov: Rev Multilateralism Int Organ 19(1):119–130

Goh G (2004) The World Trade Organization, Kyoto and energy tax adjustments at the border. J World Trade 38(3):395–423

Goulder LH, Stavins RN (2011) Challenges from state-federal interactions in U.S. climate change policy. Am Econ Rev Pap Proc 101(3):253–257

Graetz MJ (2011) The end of energy: the unmaking of America's environment, security, and independence. MIT Press

Green J (2010) Private standards in the climate regime: the greenhouse gas protocol. Bus Polit 12(3)

Green JF (2013) Order out of chaos: public and private rules for managing carbon. Global Environ Polit 13(2):1–25

Gunningham N (2012) Confronting the challenge of energy governance. Transnatl Environ Law 1:119, 130–131

Haas R, Eichhamer W, Huber C, Langniss O, Lorenzoni A, Madlener R, Menanteau P, Morthorst PE, Martins A, Oniszk A, Schleich J, Smith A, Vass Z, Verbruggen A (2004) How to promote renewable energy systems successfully and effectively. Energy Policy 32(6):833

Haas R, Panzer C, Resch G, Ragwitz M, Busch S, Held A (2011) Efficiency and effectiveness of promotion systems for electricity generation from renewable energy sources – lessons from EU countries. Energy 36(4):2186–2193

Haghighi S (2007) Energy security: the external legal relations of the European Union with major oil and gas supplying countries. Hart Publishing

Haites E et al (2018) Experience with carbon taxes and greenhouse gas emissions trading systems. Duke Environ Law Policy Forum XXIX:109–182

Hale T, Roger C (2014) Orchestration and transnational climate governance. Rev Int Organ 9(1):59–82

Halliday T, Shaffer G (eds) (2015) Transnational legal orders. Cambridge University Press

Handl G (2001) Human rights and protection of the environment. In: Eide A, Krause C, Rosas A (eds) Economic, social and cultural rights: a textbook, 2nd edn. Martinus Nijhoff Publishers, New York

Hawken P (2010) The ecology of commerce: a declaration of sustainability

Heemskerk E, Fennema M, Carroll WK (2016) The global corporate elite after the financial crisis: evidence from the transnational network of interlocking directorates. Global Netw 16(1):68–88

Held A, Ragwitz M, Haas R (2006) On the success of policy strategies for the promotion of electricity from renewable energy sources in the EU. Energy Environ:849

Hollo E et al (eds) (2013) Climate change and the law. Springer

Holzinger K (2008) Transnational common goods: strategic constellations, collective action problems, and multi-level provision. Palgrave Macmillan

Huang H et al (2011) Climate change and trade in agriculture. Food Policy 36(1):S9–S13

Hufbauer GC, Kim J (2009) The World Trade Organization and climate change: challenges and options. Peterson Institute for International Economics, Working Paper Series, WP 09-9. Available at https://piie.com/sites/default/files/publications/wp/wp09-9.pdf

Hughes C (2018) Fair shot: rethinking inequality and how we earn. St. Martin's Press

ICTSD (2011) Fostering low carbon growth: the case for a sustainable energy trade agreement. International Centre for Trade and Sustainable Development, Geneva

Jacobs M, Mazzucato M (eds) (2016) Rethinking capitalism: economics and policy for sustainable and inclusive growth. Wiley-Blackwell

Janssen MA et al (2006) Toward a network perspective of the study of resilience in social-ecological systems. Ecol Soc 11(1):15

Jegen M (2010) Two paths to energy security: the EU and NAFTA. Int J 66:73

Jinnah S, Morgera E (2013) Environmental provisions in American and EU Free Trade Agreements: a preliminary comparison and research agenda. Rev Eur Comp Int Environ Law 22(13):324–339

Jupille J, Mattli W, Snidal D (2013) Institutional choice and global commerce. Cambridge University Press

Katz DM, Stafford D (2010) Hustle and flow: a social network analysis of the American federal judiciary. Ohio State Law J 71(3)

Kaul I, Grunberg I, Stern MA (1999) Defining global public goods. In: Kaul I, Grunberg I, Stern MA (eds) Global public goods: international cooperation in the 21st century. Oxford University Press

Kendall K (2012) Carbon taxes and the WTO: a carbon charge without trade concerns? Arizona J Int Comp Law 29:49 at 51ff

Keohane RO, Victor DG (2011) The regime complex for climate change. Perspect Polit 9(1):7–23

Kim RE (2013–14) The emergent network structure of the multilateral environmental agreement system. Global Environ Change 23(5):980–991

Kissinger H (2014) World order: reflections on the character of nations and the course of history. Allen Lane

Kottari M, Roumeliotis P (2013) Renewable energy governance: challenges within a "puzzled" institutional map. In: Michalena E, Maxwell Hills J (eds) Renewable energy governance: complexities and challenges. Springer

Krasner S (ed) (1983) International regimes. Cornell University Press

Kravchenko S, Bonine JE (2008) Human rights and the environment: cases, law and policy. Carolina Academic Press, Durham, NC

Laruelle M, Peyrouse S (2012) The Chinese question in Central Asia: domestic order, social change, and the Chinese factor. Hurst Publishers, London

Leal W, Leal-Arcas R (eds) (2018) University initiatives in climate change mitigation and adaptation. Springer

Leal-Arcas R (2008) Theory and practice of EC external trade law and policy. Cameron May

Leal-Arcas R (2009) The multilateralization of international investment law. N C J Int Law Commer Regul 35(1):33–135

Leal-Arcas R (2010a) Kyoto and the COPs: lessons learned and looking ahead. Hague Yearb Int Law 23:17–90

Leal-Arcas R (2010b) International trade and investment law: multilateral, regional and bilateral governance. Edward Elgar

Leal-Arcas R (2012a) Unilateral trade-related climate change measures. J World Invest Trade 13(6):875–927

Leal-Arcas R (2012b) The role of the European Union and China in global climate change negotiations: a critical analysis. J Eur Integrat Hist 18(1):67–81

Leal-Arcas R (2013a) Climate change and international trade. Edward Elgar, Cheltenham

Leal-Arcas R (2013b) Climate change mitigation from the bottom up: using preferential trade agreements to promote climate change mitigation. Carbon Climate Law Rev 7(1):34–42

Leal-Arcas R (2013c) Working together: how to make trade contribute to climate action. International Centre for Trade and Sustainable Development, Information Note No. 18, pp 1–8, Geneva, November

Leal-Arcas R (2014) Trade proposals for climate action. Trade Law Dev 6(1):11–54

Leal-Arcas R (2015a) Mega-regionals and sustainable development: the transatlantic trade and investment partnership and the trans-pacific partnership. Renew Energy Law Policy Rev 4:248

Leal-Arcas R (2015b) Mega-regionals and sustainable development: the Transatlantic Trade and Investment Partnership and the Trans-Pacific Partnership. Renew Energy Law Policy Rev 6(4):248–264

Leal-Arcas R (2016) The European Energy Union: the quest for secure, affordable and sustainable energy. Claeys & Casteels Publishing

Leal-Arcas R (2017) Sustainability, common concern and public goods. George Wash Int Law Rev 49(4):801–877

Leal-Arcas R (ed) (2018) Commentary on the Energy Charter Treaty. Edward Elgar

Leal-Arcas R (2019) EU trade law. Edward Elgar

Leal-Arcas R, Alvarez Armas E (2018) The climate-energy-trade nexus in EU external relations. In: Minas S, Ntousas V (eds) EU climate diplomacy: politics, law and negotiations. Routledge

Leal-Arcas R, Filis A (2013a) Conceptualizing EU energy security through an EU constitutional law perspective. Fordham Int Law J 36(5):1225–1301

Leal-Arcas R, Filis A (2013b) The fragmented governance of the global energy economy: a legal-institutional analysis. J World Energy Law Bus 6(4):348–405, Oxford University Press

Leal-Arcas R, Minas S (2016) Mapping the international and European governance of renewable energy. Oxford Yearb Eur Law 35(1):621–666

Leal-Arcas R, Morelli A (2018) The resilience of the Paris Agreement: negotiating and implementing the climate regime. Georgetown Environ Law Rev 31(1):1–63

Leal-Arcas R, Wilmarth C (2015) Strengthening sustainable development through preferential trade agreements. In: Wouters J et al (eds) Ensuring good global governance through trade: EU policies and approaches, vol 92

Leal-Arcas R, Wouters J (eds) (2017) Research handbook on EU energy law and policy. Edward Elgar

Leal-Arcas R et al (2014) International energy governance: selected legal issues. Edward Elgar

Leal-Arcas R et al (2016) Energy security, trade and the EU: regional and international perspectives. Edward Elgar

Leal-Arcas R, Lasniewska F, Proedrou F (2018) Smart grids in the European Union: assessing energy security, regulation & social and ethical considerations. Columbia J Eur Law 24(2)

Lee N, Kirkpatrick C (2001) Methodologies for sustainability impact assessments of proposals for new trade agreements. J Environ Assess Policy Manage 3:395

Lewis JI (2014) The rise of renewable energy protectionism: emerging trade conflicts and implications for low carbon development. Global Environ Polit

Lilliston B (2016) The climate cost of free trade: how the TPP and trade deals undermine the Paris climate agreement. Institute for Agriculture and Trade Policy

Luciani G (2013) Security of oil supplies: issues & remedies. Claeys & Casteels

Martin L, Simmons B (1998) Theories and empirical studies of international institutions. Int Organ 52(2):397–419

Matisoff D (2010) Are international environmental agreements enforceable? Implications for institutional design. Int Environ Agreements: Polit Law Econ 10(3):165–186

Mayer B (2018) The international law on climate change. Cambridge University Press

McKenzie L (2018) Overcoming legacies of foreign policy (dis)interests in the negotiation of the European Union-Australia free trade agreement. Aust J Int Aff 72(3):255–271

Meidinger E (2011) Forest certification and democracy. Eur J Forest Res 130(3):407–419

Meyer T (2012) Global public goods, governance risk, and international energy. Duke J Comp Int Law 22:319–348

Meyer T (2013) The architecture of international energy governance. Am Soc Int Law Proc 106

Milanovic B (2018) Global inequality: a new approach for the age of globalization. Harvard University Press

Miller T (2017) China's Asian Dream: empire building along the new silk road. Zed Books, London

Moe E, Midford P (2014) The political economy of renewable energy and energy security: common challenges and national responses in Japan, China and Northern Europe. Palgrave

Morin J-F, Jinnah S (2018) The untapped potential of preferential trade agreements for climate governance. Environ Polit 27(3):541–565

Morosini F (2010) Trade and climate change: unveiling the principle of common but differentiated responsibilities from the WTO Agreements. George Wash Int Law Rev 42:713–748

Newman A, Zaring DT (2013) Regulatory networks: power, legitimacy, and compliance. In: Dunoff J, Pollack M (eds) Interdisciplinary perspectives on international law and international relations: the state of the art. Cambridge University Press, p 244

Nordhaus W (2007) To tax or not to tax. Rev Environ Econ Policy

Nordhaus W, Merrill S, Beaton P (eds) (2013) Effects of U.S. tax policy on greenhouse gas emissions. The National Academy Press. Available at https://smarterfuelfuture.org/assets/content/NRC_GHG_-_July_2013.pdf

O'Sullivan M (2017) Windfall: how the new energy abundance upends global politics and strengthens America's power. Simon & Schuster

Odell J (2018) Our alarming crisis demands border adjustments now. International Centre for Trade and Sustainable Development, Geneva

Olson M (1965) The logic of collective action: public goods and the theory of groups. Harvard University Press

Omorogbe YO (2008) Promoting sustainable development through the use of renewable energy: the role of the law. In: Zillman D, Redgwell C, Omorogbe Y, Barrera-Hernández LK (eds) Beyond the carbon economy: energy law in transition, p 39

Orsini A, Morin J-F, Young O (2013a) Regime complexes: a buzz, a boom, or a boost for global governance? Global Gov: Rev Multilateralism Int Organ 19(1):27–39

Orsini A et al (2013b) Regime complexes: a buzz, a boom, or a boost for global governance? Global Gov 14:419–435

Oschmann V (2008) Introduction to European law on renewable energy sources. In: Parker L, Ronk J, Maxwell R (eds) From debate to design: issues in clean energy and climate change law and policy. Yale School of Forestry and Environmental Studies, p 19

Osofsky H (2007) Local approaches to transnational corporate responsibility: mapping the role of subnational climate change litigation. Pac McGeorge Global Bus Dev Law J 20:143–159

Ostrom E (2009) A polycentric approach for coping with climate change. Policy Research Working Paper 5095. World Bank, Washington, DC

Ottinger R (2013) Renewable energy law and development: case study analysis. Edward Elgar

Overdevest C, Zeitlin J (2012) Assembling an experimentalist regime: transnational governance interactions in the forest sector. Regul Gov

Padilla E (2002) Intergenerational equity and sustainability. Ecol Econ 41:69–83

Palmer G (1992) The earth summit: what went wrong at Rio? Wash Univ Law Q 70:1005

Pattberg P, Stripple J (2008) Beyond the public and private divide: remapping transnational climate governance in the 21st century. Int Environ Agreements: Polit Law Econ 8(4):367–388

Pauwelyn J (2012) Carbon leakage measures and border tax adjustments under WTO law. In: Prevost D, Van Calster G (eds) Research handbook on environment, health and the WTO

Peat D (2012) The wrong rules for the right energy: the WTO SCM agreement and subsidies for renewable energy. Environ Law Manage 24:3–10

Perez O (2004) Ecological sensitivity and global legal pluralism: rethinking the trade and environment conflict. Hart Publishing, pp 8–9

Perez O (2012) International environmental law as a field of multi-polar governance: the case of private transnational environmental regulation. Santa Clara J Int Law 10:285

Peters G, Hertwich E (2006) Pollution embodied in trade: the Norwegian case. Global Environ Change 16:379–387

Petersmann E-U (2000) From 'negative' to 'positive' integration in the WTO: time for 'mainstreaming human rights' into WTO law? Common Market Law Rev 37:1363–1382

Pontin B (2002) Environmental rights under the UK's 'intermediate constitution'. Nat Res Environ 17:21
Popovski V (ed) (2018) The implementation of the Paris Agreement on climate change (law, ethics and governance). Routledge
Potoski M, Prakash A (2013) Do voluntary programs reduce pollution? Examining ISO 14001's effectiveness across countries. Policy Stud J 41(2):273–294
Prag A (2017) Trade and SDG 13 – Action on Climate Change. Asian Development Bank Institute Working Paper 735. Asian Development Bank Institute, Tokyo. Available at https://www.adb.org/publications/tradeand-sdg-13-action-climate-change
Raustiala K, Victor DG (2004) The regime complex for plant genetic resources. Int Organ 58(2):277–309
Raworth K (2017) Doughnut economics: seven ways to think like a 21st-century economist. Chelsea Green Publishing
Richardson A (2009) Regulatory networks for accounting and auditing standards: a social network analysis of Canadian and international standard-setting. Account Organ Soc 34(5):571–588
Rodrik D (2011) The globalization paradox: democracy and the future of the world economy. W.W. Norton & Co, NY
Roehrkasten S (2015) Global governance on renewable energy: contrasting the ideas of the Germand and Brazilian governments. Springer
Sabel CF, Zeitlin J (2011) Experimentalism in transnational governance: emergent pathways and diffusion mechanisms. Paper presented at the International Studies Association Conference, Montreal
Sandler T (2004) Global collective action. Cambridge University Press
Sands B, Peel J (2012) Principles of international environmental law, 3rd edn. Cambridge University Press, Cambridge
Sax J (2000) Environmental law at the turn of the century: a reportorial fragment of contemporary history. Calif Law Rev 88(6):2375–2402
Schrijver N (1997) Sovereignty over natural resources: balancing rights and duties. Cambridge University Press, Cambridge
Slaughter A-M (2002) Global government networks, global information agencies, and disaggregated democracy. Mich J Int Law 24:1041
Slaughter A-M (2004) A new world order. Princeton University Press
Slaughter A-M, Zaring DT (2006) Networking goes international: an update. Annu Rev Law Soc Sci 2
Snir R, Ravid G (2015) Global nanotechnology regulatory governance from a network analysis perspective. Regul Gov
Sofiev M et al (2018) Cleaner fules for ships provide public health benefits with climate tradeoffs. Nat Commun 9(406):1–12
Sovacool B (2017) Reviewing, reforming, and rethinking global energy subsidies: towards a political economy research agenda. Ecol Econ 135:150–163
Sovacool B, Florini A (2012) Examining the complications of global energy governance. J Energy Nat Resour Law 30(3):235
Stern N et al (2018) Making carbon pricing work for citizens. Nature Climate Change 8:669–677
Sugiyama T, Sinton J (2005) Orchestra of treaties: a future climate regime scenario with multiple treaties among like-minded countries. Int Environ Agreements: Polit Law Econ 5(1):65–88
Summers L (2016) The age of secular stagnation: what it is and what to do about it. Foreign Aff
Tagliapietra S (2012) The geoeconomics of sovereign wealth funds and renewable energy. Claeys & Casteels
Taylor M (1987) The possibility of cooperation. Cambridge University Press
Taylor J, Van Doren P (2008) The energy security obsession. Georgetown J Law Public Policy 6:475
Thistlethwaite J, Paterson M (2015) Private governance and accounting for sustainability networks. Environ Plan C: Gov Policy

References

Turner R et al (1998) Green taxes, waste management and political economy. J Environ Manage 53:121–136

USAID (2014) Encouraging renewable energy development: a handbook for international energy regulators

Van de Graaf T (2013) The politics and institutions of global energy governance. Palgrave/Mamillan

Van de Graaf T, van Asselt H (2017) Introduction to the special issue: energy subsidies at the intersection of climate, energy, and trade governance. Int Environ Agreements: Polit Law Econ 17(3):313

Victor, Yueh (2010) The new energy order: managing insecurities in the twenty-first century. Foreig Aff:89

Viñuales JE (2012) Foreign investment and the environment in international law. Cambridge University Press, p 1

Von Bertalanffy L (1968) General system theory: foundations, development, applications. George Braziller, New York

Wagner G, Weitzman ML (2016) Climate shock. Princeton University Press, Princeton, NJ

Wenger A et al (eds) (2009) Energy and the transformation of international relations. Oxford University Press, Oxford

Wise J (2006) Access to AIDS medicines stumbles on trade rules. Bull World Health Org 84:342. http://www.who.int/bulletin/volumes/84/5/news.pdf [https://perma.cc/QH9U-6CNW]

Wold C, Wilson G, Foroshani S (2012) Leveraging climate change benefits through the world trade organization: are fossil fuel subsidies actionable? Georgetown J Int Law 43:635, 694

Wood S, Abbott K, Black J, Eberlein B, Meidinger E (2015) The interactive dynamics of transnational business governance: a challenge for transnational legal theory. Transnatl Leg Theory

Wu M (2014) Why developing countries won't negotiate: the case of the WTO environmental goods agreement. Trade Law Dev 6:93

Yamusa SU, Ansari AH (2013) Renewable energy development in two selected African countries: an overview and assessment. Renew Energy Law Policy Rev 4(2):151–156

Yergin D (2006) Ensuring energy security. Foreign Aff 85(2)

Yergin D (2008) The prize: the epic quest for oil, money and power. Free Press, New York

Yergin D (2011) The quest: energy, security, and the remaking of the modern world. Penguin Press, New York

Zarrilli S (2010) Development of the emerging biofuels market. In: Goldthau A, Witte JM (eds) Global energy governance: the new rules of the game. Brookings Instit Press, chapter 4

Zeitlin J (2011) Pragmatic transnationalism: governance across borders in the global economy. Socio-Econ Rev 9(1):187–206

Zhang ZX (2013) Trade in environmental goods, with focus on climate-friendly goods and technologies. In: Van Calster G, Prevost D (eds) Research handbook on environment, health and the WTO, chapter 19. Edward Elgar Publishing, pp 673–699

Chapter 4
Using Trade Agreements to Achieve Sustainability: A Counter-Intuitive Conundrum

4.1 Introduction[1]

There is no question that climate change is one of the biggest challenges humanity faces today. Today, 80% of the global energy supply comes from fossil fuels.[2] Fossil fuels contribute to climate change and are finite, which leads to energy insecurity.[3] We cannot use all the fossil reserves we have without seriously disrupting the climate system. Renewable energy can help here in that it is cleaner than fossil fuels. It also helps towards energy independence and therefore enhances energy security.[4] Renewable energy sources are the only long-term energy supply solution we have at present.[5] Trade law could be used as a vehicle to achieve these goals because trade rules can promote environmental goods and services.[6]

---

[1] Some of the ideas developed in this chapter drawn on Leal-Arcas (2017), pp. 801–877.
[2] *World Energy Council Report Confirms Global Abundance of Energy Resources and Exposes Myth of Peak Oil*, World Energy Council (Oct. 15, 2015), https://www.worldenergy.org/news-and-media/press-releases/world-energy-council-report-confirms-global-abundance-of-energy-resources-and-exposes-myth-of-peak-oil/ [https://perma.cc/5ZK6-8LNL].
[3] Schrag (2007), pp. 171–178.
[4] On the governance of renewable energy, see generally Leal-Arcas and Minas (2016), pp. 621–666.
[5] In June 2011, an Ipsos public opinion survey in 24 countries showed the following global public support for energy sources: solar power 97%; wind power 93%; hydro power 91%; natural gas 80%; coal 48%; nuclear 38%. See https://commons.wikimedia.org/wiki/File:Global_public_support_for_energy_sources_(Ipsos_2011).png. The 24 countries were the following: Argentina, Australia, Belgium, Brazil, Canada, China, France, UK, Germany, Hungary, India, Indonesia, Italy, Japan, Mexico, Poland, Russia, Saudi Arabia, South Africa, South Korea, Spain, Sweden, Turkey and the USA.
[6] Some proponents have gone even further to suggest that "trade must be an engine of growth for all." *See* WTO, IMF and World Bank leaders: *"Trade must be an engine of growth for all"*, WTO (Oct. 7, 2016), https://www.wto.org/english/news_e/news16_e/dgra_07oct16_e.htm [https://perma.cc/Y8D6-LT8D].

I argue that trade agreements can help mitigate climate change since, in the past, they have been a very powerful instrument for change, as the following two examples demonstrate:

1. poverty reduction[7]: due to trade agreements,[8] one billion people have come out of poverty between 1990 and 2010[9]; and
2. the protection of human rights[10]: 75% of countries use trade agreements to protect human rights.[11]

So why not use trade agreements as a novel tool to solve one of the most important challenges of today, namely climate change, given that they are enforceable, unlike traditional environmental agreements?[12]

One of the purposes of this chapter is not just to thoroughly describe the current state of the climate-energy-trade nexus in the external relations of the European Union (EU) and beyond, but to explore, in normative terms, the potential of the EU's external trade[13] towards contributing to global decarbonization. As the EU has done in the case of human rights, EU free trade agreements (FTAs) may be used to help decarbonize the economy by bundling the fate of these agreements and international climate-change obligations, i.e., the breach of the latter would give rise to legal consequences under the dispute settlement mechanisms of the former (in other words, there would be no free-trade-benefits without effective fulfillment of inter-

---

[7]See for instance World Bank Group and World Trade Organization, (2015), "The Role of Trade in Ending Poverty," World Trade Organization: Geneva, available at https://www.wto.org/english/res_e/booksp_e/worldbankandwto15_e.pdf; Santos-Paulino, A. "Trade, Income Distribution and Poverty in Developing Countries: A Survey", UNCTAD Discussion Papers, No. 207, July 2012, available at http://unctad.org/en/PublicationsLibrary/osgdp20121_en.pdf; Das (2011); OECD (2011), available at https://doi.org/10.1787/9789264098978-en; Hayashikawa, M. "Trading Out of Poverty: How Aid for Trade Can Help," available at https://www.oecd.org/site/tadpd/41231150.pdf; The World Bank (2016), available at https://openknowledge.worldbank.org/bitstream/handle/10986/25078/9781464809583.pdf; United Nations, "Rethinking Poverty: Report on the World Social Situation," New York, 2009, available http://www.un.org/esa/socdev/rwss/docs/2010/fullreport.pdf. On the links between international trade and global justice, see Suttle (2014, 2017).

[8]Some voices question that trade can ease poverty. See Hassoun, N. "Free Trade, Poverty, and Inequality," available at http://repository.cmu.edu/cgi/viewcontent.cgi?article=1354&context=philosophy; Cornia (2004).

[9]See *Towards the End of Poverty*, The Economist (June 1, 2013), http://www.economist.com/news/leaders/21578665-nearly-1-billion-people-have-been-taken-out-extreme-poverty-20-years-world-should-aim [https://perma.cc/WHD7-8TZY].

[10]See Cottier et al. (2005); Lang (2011); Gammage (2014), pp. 779–792.

[11]Susan Ariel Aaronson, *Human Rights*, The World Bank 443, 443 http://siteresources.worldbank.org/INTRANETTRADE/Resources/C21.pdf [https://perma.cc/P5KD-KZX8]. See also Cassimatis (2007).

[12]Abbott and Snidal (2000), pp. 421–456; Boyle (2014), pp. 118 ff; Mensah (2008), pp. 50–56.

[13]For a comprehensive analysis of EU external trade law and policy, see Leal-Arcas (2008a, 2010).

4.1 Introduction

national climate-change-related obligations).[14] The feasibility of this idea will be assessed via the textual analysis of the current wording and content of environmental/climate-change-related elements in a sample of relatively recent and/or upcoming FTAs.

The concept of using the trading system to mitigate climate change and enhance sustainable energy will transform our understanding of trade in the context of environmental protection. It will shift the current paradigm from trade as a major cause of environmental harm (think for instance of aviation[15] and shipping[16]) to trade as a tool for environmental protection (for instance, via the inclusion of legally binding and enforceable provisions on sustainable development and clean energy in RTAs).[17] Doing so will enable us to avoid lowering environmental standards in FTAs. There is literature that reports that countries engaged in the largest RTAs/FTAs and international investment agreements are the ones with the highest levels of GHG emissions.[18] Equally, there is literature that analyzes the impacts of climate change on international trade.[19]

This chapter proposes using mega-RTAs to mitigate climate change and enhance sustainable energy. In other words, this chapter makes the claim that trade agreements can be a tool to promote decarbonization. The evidence for this claim is that RTAs have often served as laboratories for covering new disciplines that do not exist in the WTO context.[20] Moreover, RTAs today cover many topics well beyond trade: competition, investment, environmental protection, natural resources, intellectual

---

[14] See European Commission, "Trade for all: Towards a more responsible trade and investment policy," (2015) p. 23 (stating that "as FTAs enter into force, the EU will have to make sure that the provisions on trade and sustainable development are implemented and used effectively."

[15] It is the case that, with the rise of the middle class in heavily populated countries such as China and India, Chinese and Indians are travelling internationally more than ever as tourists. As a result, the levels of GHG emissions from aviation/shipping will only go up. See The Economist, "Tourism in South-East Asia," 14 April 2018, pp. 48–49.

[16] Shipping, like aviation, was not included in the Paris Climate Agreement, surprisingly, since both industries combined are responsible for 8% of global GHG emissions. The aim of the International Maritime Organization is for the shipping industry to cut its GHG emissions by 50% by 2050 based on the 2008 emission levels. See The Economist, "Shipping: Smoke on the water," 14 April 2018, p. 62.

[17] See for instance Leal-Arcas and Alvarez Armas (2018), pp. 153–168. For further analysis, see Anthony VanDuzer (2016).

[18] Gallagher (2017), p. 7, available at https://www.bu.edu/gdp/files/2017/11/Trade.In_.Balance.11-17.final_.pdf.

[19] Dellink et al. (2017), https://doi.org/10.1787/9f446180-en.

[20] For an analysis of the link between regional trade agreements (RTAs) and the WTO, see Leal-Arcas (2011a), pp. 597–629.

property rights, labor rights,[21] and so forth.[22] Since most of the contracting parties to these three mega-regional agreements are also the main GHG emitters, and since RTAs have provisions that bind countries to mitigate climate change, then RTAs may potentially become a very effective solution to climate change mitigation.[23] This hypothesis raises the following question: How can trade agreements be used to enforce environmental standards? This chapter will also look at the landscape of trade and climate change governance to identify the main emitters of GHGs and who the Contracting Parties to the three mega-RTAs *par excellence* are.

The chapter aims to show tangible ways in which the EU can, through its network of PTAs, move towards greater energy independence as renewable energy becomes increasingly economically viable. For instance, by including chapters on renewable energy, EU trade agreements will enhance secure and clean energy. This is a way to promote trade in green goods and services. Another way is via subsidies for renewables[24]: green taxes to increase the price of oil. Renewables could then become popular by increasing the price of oil to such an extent that countries would find renewables-investment more attractive.[25] A further option is for governments to limit the wholesale supply of fossil fuels.[26]

This chapter first explains the case for regionalism/plurilateralism in the context of trade and climate change. The chapter provides an analysis of forum options for dealing with the convergence of the trade and climate regimes with the aim to help mitigate climate change and enhance sustainable energy. It then provides an analysis of the major GHG emitters and their links to mega-regional trade agreements (RTAs). It also offers an analysis on how we can capitalize on RTAs. It then provides an analysis of how trade agreements can be enhancers of sustainable development and of the transition towards a decarbonized economy.[27] It does so

---

[21]Examples of labor-related trade cases are: US-Guatemala FTA (http://www.worldtradelaw.net/document.php?id=labor/cr/cafta-guatemala-labor-2010-cr.pdf) and US-Bahrain FTA (http://www.worldtradelaw.net/document.php?id=labor/cr/us-bahrain-labor-2013-cr.pdf).

[22]*See* Rafael Leal-Arcas, *Brexit and the Future of UK Trade* (Nov. 25, 2016), http://www.qmul.ac.uk/media/news/items/hss/190227.html [https://perma.cc/AS3M-VPQM].

[23]The same argument applies to sustainable energy. *See, e.g.*, Leal-Arcas et al. (2015), pp. 472–514.

[24]Hildreth (2014), pp. 702–730. In 2018, China was implementing a policy that encourages automakers to produce longer range electric vehicles. See Dixon, T. "Chinese Electric Vehicle Subsidy Changes In 2018 — The Details," *EV Obsession*, 6 January 2018, available at https://cleantechnica.com/2018/01/06/chinese-electric-vehicle-subsidy-changes-2018-details/.

[25]However, the opposite seems to be the case as of late 2017 in the US, where the energy secretary, Rick Perry, has a plan to subsidize coal-fired and nuclear plants. See The Economist, "Energy subsidies in America: Abuse of power," 16 December 2017, pp. 11–12.

[26]Höök and Tang (2013), pp. 797–809.

[27]It is interesting to note that, in the negotiations to the Paris Agreement on Climate Change, there were countries that opposed the term 'decarbonization' in the text. See Moosmann, L. *et al.*, "Implementing the Paris Agreement – Issues at Stake in View of the COP 22 Climate Change Conference in Marrakesh," IP/A/ENVI/2016-11, pp.73–74.

4.2 The Case for Regionalism/Plurilateralism in Trade and Climate Change

Multilateralism is embodied in international trade agreements. International trade and the rapidly proliferating network of trade agreements have sparked controversy for decades. Agreements are signed when countries cannot solve a problem domestically. For instance, climate change is an area in which countries give up some sovereignty to help solve domestic problems. While some blame trade agreements for exporting jobs, sowing poverty, furthering illegal migration, and stealing national sovereignty, others praise them as lynchpins of growth, pillars of peace, guarantors of security, and engines of globalization.[29] Still others view them as useful instruments for fostering global trade and investment.[30]

Arguably, multilateralism is in crisis,[31] whether in the field of trade,[32] investment,[33] energy governance,[34] or climate change mitigation.[35] In the case of trade negotiations, the Doha Round[36] of trade negotiations at the WTO has clearly reached an impasse.[37] One of the reasons for this crisis is that citizens were absent from the

---

[28] On this issue, see WTO, UN Environment, "Making trade work for the environment, prosperity and resilience," 2018, available at https://www.wto.org/english/res_e/publications_e/unereport2018_e.pdf.

[29] Leal-Arcas (2010), p. 39.

[30] *Id.*

[31] It is surprising to see the position of the US which, rather than defend liberal institutions it created after World War II, has been neglecting them and even, more recently under President Trump, attacking them.

[32] European Commission, "Concept note on WTO modernisation: Introduction to future EU proposals," 18 September 2018, available at http://trade.ec.europa.eu/doclib/docs/2018/september/tradoc_157331.pdf.

[33] Leal-Arcas (2010), pp. 58–63.

[34] *See* Leal-Arcas et al. (2014), pp. 489–494.

[35] *See* Leal-Arcas (2013), pp. 291–293. The outcome of the COP24 in December 2018 brings a success to multilateral climate change negotiations with its rule book.

[36] If ultimately successful, the Doha Round, with more than 164 countries at the negotiating table as of January 2017, would be the ninth round since World War II. See Leal-Arcas (2008a), pp. 486–487. The previous rounds were, in chronological order: Geneva Round (1948), with twenty-three countries; Annecy Round (1949), with thirteen countries; Torquay Round (1951), with thirty-eight countries; Fourth Round (1956), with twenty-six countries; Dillon Round (1962), with twenty-six countries; Kennedy Round (1967), with sixty-two countries; Tokyo Round (1979), with 102 countries; and Uruguay Round (1994), with 123 countries. *See* ibid, at 486–87 n.7.

[37] *See* Pakpahan (2012), https://www.wto.org/english/forums_e/public_forum12_e/art_pf12_e/art19.htm [https://perma.cc/PZ8X-5XPV]; Lester (2016), https://www.cato.org/publications/free-

process of decision making. Therefore, in addition to the top-down process, this book proposes a bottom-up process with greater citizen participation to improve problematic trade agreements.

An alternative method of governance to multilateralism in a multipolar world[38] is regionalism,[39] which is a method of economic and political integration.[40] While multilateralism has its advantages, regionalism as an alternative to multilateral governance has not been fully explored and appropriately tapped when it comes to climate change mitigation and the enhancement of sustainable energy. Regionalism is the form that perhaps best describes the supranationalism of the integration of European states into a community and union[41]: sovereign states bound themselves both legally and politically into a single entity in which national and supranational institutions share governance and answer to a court that protects not only the institutions of the system, but also the rights of the individual citizens.[42] Specifically relating to regional trade, there are at least four main trends identified in RTAs that serve to remedy impasses in multilateral trade: movement from most favored nation[43]; liberalization to RTAs[44]; a geographical shift to the Asia-Pacific region; and increases in cross-regional RTAs and mega-RTAs.[45]

This chapter focuses on how mega-RTAs can serve as a platform for climate change mitigation and sustainable energy enhancement. While the multilateral trade

---

trade-bulletin/doha-round-over-wtos-negotiating-agenda-2016-beyond [https://perma.cc/6NUM-DYBQ].

[38] For analyses of multipolarity, see Kennedy (2017) and Walton (2007).

[39] A recent example of regionalism is the Lima Group, composed of representatives of countries from the Western Hemisphere in favor of restoring democracy in Venezuela.

[40] Leal-Arcas (2010), p. 72.

[41] Idem.

[42] *Id.* On supranationalism in the European Union, see Leal-Arcas (2007), pp. 88–113.

[43] The most favored nation treatment is the principle of not discriminating between one's trading partners. *See* General Agreement on Tariffs and Trade art. I –II, Oct. 30, 1947, 61 Stat. A-11, 55 U.N.T.S. 194 [hereinafter GATT]; Agreement on Trade-Related Aspects of Intellectual Property Rights art. 4, Apr. 15, 1994, Marrakesh Agreement Establishing the World Trade Organization, Annex 1C, 1869 U.N.T.S. 299, 33 I.L.M. 1197 (1994) [hereinafter TRIPS Agreement].

[44] According to GATT Article XXIV, it is possible to deviate from GATT Article I and therefore give preferential treatment to parties to an RTA, provided doing so does not raise barriers to trade for third countries. GATT Article XXIV requires that duties be eliminated on "substantially all the trade" between the parties of a customs union or free trade area, or at least with respect to substantially all the trade in products originating in such territories. Regarding the locution "substantially all the trade," there is neither an agreed upon definition of the percentage of trade to be covered by a WTO-consistent agreement nor common criteria against which the exclusion of a particular sector from the agreement could be assessed. For more information, see submissions by Australia (TN/RL/W/173/Rev.1 and TN/RL/W/180), European Communities (TN/RL/W/179), China (TN/RL/W/185), and Japan (TN/RL/W/190).

[45] For an analysis of the main trends and characteristics of regional trade agreements, in force and under negotiation, see Roberto V. Fiorentino, Luis Verdeja & Christelle Toqueboeuf, *The Changing Landscape of Regional Trade Agreements: 2006 Update*, WTO Secretariat Discussion Paper No. 12 (2007).

4.2 The Case for Regionalism/Plurilateralism in Trade and Climate Change

system has the potential to help mitigate climate change, amending the WTO rules requires consensus among WTO members.[46] This Subsection tests an alternative means to multilateral trade by which regional trade can facilitate climate change mitigation, namely through mega-RTAs such as the Comprehensive and Progressive Agreement for Trans-Pacific Partnership (CPTPP).

From a climate change point of view, it is easier and more manageable to negotiate among a small number of large players than it is among a large number of small players, which explains the creation of climate change clubs[47] or coalitions of the willing.[48] The concept of a climate change club refers to a relatively small number of countries that produce the large majority of GHG emissions.[49] This concept could entail an agreement on technology transfer or on product efficiency standards. The same argument holds true for trade negotiations. The multilateral trading system's single undertaking[50] is no longer feasible because the WTO has more members than ever—WTO membership is an ongoing process, with more members to come in the near future[51]—and covers increasingly more topics, which, in turn, are more complex than ever, namely trade and climate change or trade-related energy issues.[52] This explains RTA proliferation as the modus operandi for trade liberalization. Trade liberalization means more trade, trade means economic growth, and economic growth means that every country is better off.

RTAs, and regionalism at large, are a more effective way to combat climate change than multilateralism via the Paris Agreement because the nationally determined contributions to the global response to climate change—Article 3 of the Paris Agreement—are not legally binding under that document.[53] Alternatively, at the regional level, the CPTPP—the only of the three mega-RTAs par excellence concluded to date—makes climate action legally binding in the form of a commitment to a low-emissions economy.[54] So, one option for the trading system to help mitigate

---

[46] Marrakesh Agreement Establishing the World Trade Organization art. IX, Apr. 15, 1994, 1867 U.N.T.S. 154 [hereinafter Marrakesh Agreement].

[47] Leal-Arcas (2011b), pp. 39–41.

[48] *Id.*

[49] *See* Leal-Arcas (2013), pp. 298–300.

[50] A single undertaking provision is a provision that requires countries to accept all the agreements reached during a round of multilateral trade negotiations as a single package, as opposed to on a case-by-case basis. It effectively means that nothing is agreed until everything is agreed by all parties. *See How the Negotiations Are Organized*, WTO, https://www.wto.org/english/tratop_e/dda_e/work_organi_e.htm [https://perma.cc/6VVZ-PM4E].

[51] For a list of observer governments, see *Members and Observers*, WTO, https://www.wto.org/english/thewto_e/whatis_e/tif_e/org6_e.htm [https://perma.cc/HG8D-UMKJ].

[52] For a list of trade topics in the WTO, see *WTO Trade Topics*, WTO, https://www.wto.org/english/tratop_e/tratop_e.htm [https://perma.cc/9Y84-S8WT].

[53] Article 4(2) of the Paris Agreement reads: "Each Party shall prepare, communicate and maintain successive nationally determined contributions that it intends to achieve." Such weak wording does not imply that the nationally determined contributions are legally binding on the parties. Paris Agreement, art. 4(2).

[54] Trans-Pacific Partnership, art. 20.15, ¶¶ 1–2.

climate change would be via mega-RTAs such as the CPTPP, and not necessarily via the multilateral (trading/climate change) system. Considering that the United States (which is the only country that, after having signed the agreement, has not ratified the Kyoto Protocol)[55] negotiated and concluded the TPP before President Trump decided to withdraw from it in January 2017, it is significant that the TPP recognizes climate change, albeit not expressly, as a global concern and that transition to a low-emissions economy requires collective action. The U.S. counterproposal of 2014 removed the term "climate change," substituting it with the locution "low-emissions economy" in the final version.[56] Moreover, it removed any reference to the UNFCCC.[57] Nevertheless, the spirit of decarbonization remains present with the wording "low-emissions economy."

Furthermore, the Section calls into question the assumption that only (or mainly) multilateralism will solve collective-action problems such as climate change, although it acknowledges that (international) cooperation is crucial to tackle collective-action problems. Further, economic regionalism has proven to be more effective than multilateralism at liberalizing trade[58] (and arguably can do the same for climate change mitigation and sustainable energy enhancement) and therefore there is no imperative need for a universal treaty that aims to liberalize trade, mitigate climate change, and enhance sustainable energy. Conversely, if there are merits to multilateralism, then the question is: How can we gradually multilateralize plurilateralism to make it more inclusive in membership? What (economic) incentives would be necessary to make this happen? Would economic incentives for clean goods and services be acceptable to help mitigate climate change?

Thus, variable geometry,[59] as opposed to a single undertaking approach, seems to be a plausible way to move the multilateral trade agenda forward because the single undertaking approach is too ambitious. The variable geometry approach has the advantage of removing the current frustration at the WTO negotiating table—and at violent protests organized by civil society[60]—about the WTO's slow negotiating pace. Regionalism/plurilateralism moves faster than multilateralism.

---

[55] *See generally* Leal-Arcas (2013), chapter 5 (discussing the position of the United States regarding the Kyoto Protocol).

[56] *See* U.S. Counterproposal to the TPP Environment Chapter (Feb. 14, 2014), http://www.redge.org.pe/sites/default/files/20140218%20biodiversity%20climate%20change%20TPP.pdf [https://perma.cc/2C5C-CRZP].

[57] *Id.*

[58] *See generally* Leal-Arcas (2010), p. 84 (discussing the rise and fall of multilateralism and the rise of bilateralism/regionalism).

[59] Variable geometry refers to a situation where some but not all WTO members would conclude trade agreements. The benefit of this concept is that those WTO members who wish to undertake deeper integration or trade liberalization may do so irrespective of the unwillingness of other WTO members to go along.

[60] *See 30 Frames a Second: The WTO in Seattle*, Bullfrog Films, http://www.bullfrogfilms.com/catalog/30fr.html [https://perma.cc/8SQP-SU9P].

4.2 The Case for Regionalism/Plurilateralism in Trade and Climate Change

Finally, it seems that trade agreements are stricter on environmental protection (see, for instance, the CPTPP's chapter on environment in relation to a low-emissions economy[61]) than climate change agreements such as the Paris Agreement. This is both surprising and counter-intuitive. That the CPTPP is enforceable on the reduction of GHG emissions,[62] whereas the Paris Agreement is not,[63] enhances those stricter provisions. It is also notable that, even if the Trump administration in the United States decided to withdraw from the Paris Agreement in June 2017,[64] it will take 4 years to do so, in accordance with Article 28(1) and (2).[65] This situation could be an opportunity for China to lead in the geopolitics of climate change globally,[66] especially because the Paris Agreement is far from offering a dispute settlement mechanism similar to that of the WTO or other multilateral treaties—rendering it to the category of soft law. Nevertheless, and despite the view of the Trump administration on climate change, a group of seventeen Republican members of the U.S. Congress signed a resolution in March 2017 to seek economically viable ways to fight climate change.[67]

There are several options to push forward the trade agenda for climate action and sustainable energy.

4.2.1 Multilateralism

Multilateralism has been in decline for some time,[68] and certainly when it comes to trade liberalization. A case in point is the never-concluded Doha Development Agenda, which was initiated in 2001. The international order seems to be fragmenting and, instead, big powers seem to be doing what they want. It is ironic that the US did more than any other nation in the 1940s to create and maintain the values of western liberal democracies multilaterally by pushing for the Marshall Plan for economic assistance to help rebuild Western European economies after the end of World War II, the creation of the World Bank, the International Monetary Fund, the

---

[61] *Compare* CPTPP, Article 20.15 *with* Paris Agreement, Art. 4(2).
[62] CPTPP, Art. 20.15, ¶ 2.
[63] Paris Agreement, art. 6(4).
[64] *UNFCCC Statement on the US Decision to Withdraw from the Paris Agreement*, United Nations Framework Convention on Climate Change (June 1, 2017), http://newsroom.unfccc.int/unfccc-newsroom/unfccc-statement-on-the-us-decision-to-withdraw-from-paris-agreement [https://perma.cc/5EC2-EPFD].
[65] Paris Agreement, art. 28(1)–(2).
[66] China's performance will be crucial when deciding whether climate change mitigation will be successful.
[67] Emily Flitter, *17 House Republicans Just Signed a Resolution Committing to Fight Climate Change*, Reuters (Mar. 15, 2017, 10:58 AM), http://www.businessinsider.com/r-group-of-17-republicans-sign-us-house-resolution-to-fight-climate-change-2017-3 [https://perma.cc/C8T8-XSUV].
[68] See Leal-Arcas (2010).

General Agreement on Tariffs and Trade and NATO. Yet, today's America under President Trump seems to want to reverse all this multilateral success of the 1940s.

In the case of multilateral trade, the Trump administration questions its validity and the WTO's way of settling disputes,[69] where 'frustration has turned to aggression.'[70] In fact, Mr Robert Lighthizer, the US trade representative, has suggested a return to the pre-WTO system, where might was right,[71] rather than making use of the dispute settlement system to resolve trade disputes.[72] One great difficulty for the WTO process of creating new agreements is that decisions are made by consensus. In a WTO of 164 members, and more to come, such structure does not make decision-making easy.

In the climate change front, there is a new global agreement to mitigate climate change, the Paris Agreement on Climate Change of 2015.[73] Two years later, more than 50 world leaders met in Paris for a One Planet Summit to launch an ambitious project to win the battle against climate change.[74] The World Bank announced that it would stop funding fossil fuel explorations by 2019.[75] The consequences of climate change are well known and most climate scientists agree that, if the global temperature continues to increase, there is a serious risk of catastrophic sea level rises and more inundations in various parts of the world. It is therefore imperative that CO_2 emissions be reduced. In fact, the 2017 Emissions Gap report of the UN Environment Program states that the climate pledges submitted by 164 countries represent one third of the cut in GHG emissions needed to meet the Paris Agreement target of keeping global warming below $2°$[76] above the pre-industrial levels.[77] Complications arise due to the Trump administration's announcement that it will not honor the nationally determined contributions submitted by the Obama administration and it

---

[69] "The WTO: The art of the impossible," *The Economist*, 16 December 2017, p. 64.

[70] "The World Trade Organization: Disaster management," *The Economist*, 9 December 2017, p. 19.

[71] Similar arguments have been made by Brink Lindsey and Steven Teles regarding America's political dysfunction. See Lindsey and Teles (2017).

[72] "The World Trade Organization: Situations vacant," The Economist, 9 December 2017, pp. 68–69, at 68. Also, for an analysis of the history of US trade policy, see Irwin (2017) (arguing that trade is neither dull nor as bad as people think).

[73] Michael A. Levi, 'Two Cheers for the Paris Agreement on Climate Change' (Council on Foreign Relations, 12 December 2015), https://www.cfr.org/blog/two-cheers-paris-agreement-climate-change.

[74] https://www.oneplanetsummit.fr/en/.

[75] "Climate summitry: New life for the Paris deal," *The Economist*, 16 December 2017, pp. 51–52, at 51.

[76] The 2-degree Celsius target was first put forward by William Nordhaus, an environmental economist who shared the Nobel economics prize in 2018. It was then adopted by policy-makers, but it is not economics- or science-based, only a political decision.

[77] UNEP, "The emissions gap report 2017," available at https://wedocs.unep.org/bitstream/handle/20.500.11822/22070/EGR_2017.pdf.

will not pay into the UN's Green Climate Fund, whose purpose is to transfer $100bn per year by 2020 to poor countries.[78]

The 2016 Kigali Amendment[79] to the Montreal Protocol[80] will also serve as a catalyst for climate action. Furthermore, the international community agreed on the establishment of a global market-based measure to offset international aviation CO_2 emissions in late 2016 in the framework of the International Civil Aviation Organization. Moreover, the so-called Paris Agreement rulebook, which will establish the necessary rules to provide guidance to fulfil the ambition of the agreement, was adopted in 2018.[81] All these developments show that the Paris Agreement and related legal instruments are the start of a process towards decarbonization of the global economy in the second half of the twenty-first century.

4.2.2 Plurilateralism

Plurilateralism is an option to be explored in order to ease barriers to trade in environmental goods and services.[82] For instance, climate clubs (i.e., getting countries to join various regimes) could serve as a way to promote technology transfer within the club members.[83] In terms of configuration of clubs of major countries (e.g. G-8, G-8+5, G-20, APEC), they could serve as a platform to conclude RTAs in green energy technologies.[84] In the context of trade, the Environmental Goods

---

[78]"Climate summitry: New life for the Paris deal," *The Economist*, 16 December 2017, pp. 51–52, at 51.

[79]In October 2016, in Kigali, Rwanda, 197 countries adopted an amendment to phase down hydrofluorocarbons (HFCs) under the Montreal Protocol on Substances that Deplete the Ozone Layer (Montreal Protocol), "commit[ing] to cut[ting] the production and consumption of HFCs by more than [eighty] percent over the next [thirty] years." *Recent International Developments under the Montreal Protocol*, U.S. Envtl. Prot. Agency, https://www.epa.gov/ozone-layer-protection/recent-international-developments-under-montreal-protocol [https://perma.cc/UL3P-2C4A]. The Kigali Amendment will come into force in 2019.

[80]*See* Montreal Protocol on Substances that Deplete the Ozone Layer, Sept. 16, 1987, S. Treaty Doc. No. 100-10 (1987), 1522 U.N.T.S. 3. The Montreal Protocol's objective was to phase out consumption of replaceable chemical products that harmed the ozone layer but entailed profits for the chemical industry. For an analysis of the lessons from the Montreal Protocol for climate change negotiators, see Richard J. Smith, *The Road to a Climate Change Agreement Runs through Montreal*, Peterson Institute for International Economics Policy Brief No. PB10-21(Aug. 2010), https://piie.com/publications/policy-briefs/road-climate-change-agreement-runs-through-montreal [https://perma.cc/JKV5-VJND].

[81]Government Offices of Sweden, "Paris Agreement 'rulebook' adopted at climate conference in Katowice," available at https://www.government.se/press-releases/2018/12/paris-agreement-rulebook-adopted-at-climate-conference-in-katowice/.

[82]Leal-Arcas (2013).

[83]For literature on climate clubs, see Brewer (2015); Nordhaus (2015), pp. 1339–1370; Hovi et al. (2016), pp. 1–9.

[84]For views on regional integration and climate change, see Fujiwara and Egenhofer (2007).

Agreement (EGA)[85] is a plurilateral[86] trade agreement currently under negotiation between 18 WTO Members to reduce tariffs on environmentally beneficial goods[87] so as to stimulate sustainable practices within global supply chains.[88] The idea is that "there could be a benefit for the multilateral trading system in lowering technical barriers to trade in energy-related goods and services, including in relation to technological goods and services that could encourage the proliferation of renewables."[89] The advantage of the EGA is that it is a type of plurilateral agreement that extends concessions to all WTO members on an MFN basis once adopted.

Combined, the parties represented at the EGA discussions produce 90% of environmental goods.[90] Since tariffs are already very low in many countries, the issue with the EGA is mainly about non-tariff barriers. Much of the issue with the EGA is defining what an environmental good is. Even if a trade agreement can be formed with explicit environmental objectives, arguably, the fact that the EGA has stalled is detrimental in that it is evidence that the international trade regime has little to offer sustainability efforts and is thus disconnected from the Paris Agreement.

Also in the trade context, RTAs are examples of trade clubs, offering the advantage of preferential treatment in trade without penalties, in accordance with GATT Article XXIV. A joint statement by the US, the EU, and Japan at the 11th WTO Ministerial Conference in Buenos Aires in December 2017 pledging "to enhance trilateral cooperation in the WTO" shows that plurilateralism is a strong

---

[85] *Environmental Goods Agreement*, Off. U.S. Trade Rep., https://ustr.gov/trade-agreements/other-initiatives/environmental-goods-agreement. The EGA ended up being an open plurilateral agreement because one needs the consensus of the WTO membership for a closed plurilateral agreement in the context of WTO decision-making.

[86] A plurilateral approach to trade agreements means that the agreements are optional and not binding on those WTO members who do not engage in them. In the WTO context, multilateral negotiations, as opposed to plurilateral negotiations, imply the participation of all WTO members. The nature of the consequent multilateral agreements from these multilateral negotiations implies that commitments are taken by all the WTO members. The idea behind plurilateral negotiations is to make the WTO deliver again on progressive liberalization. See Leal-Arcas (2008b), pp. 9–104.

[87] Lower environmental goods tariffs would allow for cheap procurement of, say, solar panels and other renewable technologies, thus facilitating a transition to a clean energy future as fast as possible. On the other hand, temporarily higher tariffs may allow a country to build up, say, a solar industry that will then be able to compete on the global market, bringing prices down overall and leading to further innovation globally.

[88] The 18 WTO members are: Australia, Canada, China, Costa Rica, the European Union, Hong Kong, Iceland, Israel, Japan, South Korea, New Zealand, Norway, Singapore, Switzerland, Liechtenstein, Chinese Taipei (Taiwan), Turkey, and the United States. *Environmental Goods Agreement*. All the E.U. member states are represented by the European Union in the negotiations, which means that there is a total of 46 WTO member states represented in the Environmental Goods Agreement (EGA).

[89] Leal-Arcas (2015a), pp. 202–219.

[90] "Environmental Goods Agreement," Transport and Environment, 2015, p. 1, available at https://www.transportenvironment.org/sites/te/files/publications/2015%2009%20TE_EGA%20briefing%20note_FINAL.PDF.

4.2 The Case for Regionalism/Plurilateralism in Trade and Climate Change 121

alternative to multilateralism.[91] Another example of plurilateralism being on the rise is the fact that a coalition of countries has signed up for the negotiation of new rules on e-commerce plurilaterally.[92] So long as there is critical mass, such a deal would be possible if there is no discrimination against other WTO members.

In the context of climate change, there is hardly any international cooperation in that countries do what they think is best for them, as opposed to what is good for the world as a whole. International climate agreements offer no incentives for countries to go beyond what is in their self-interest, which explains the lack of international climate cooperation. An example is Canada's withdrawal from the Kyoto Protocol in 2011 without legal consequences, which weakened both the environmental effectiveness and legitimacy of the Kyoto Protocol regime.[93] The current legal instruments are not enough for what the international community needs to solve the climate change issue.

A solution would be to find a mechanism where countries want to join the club and no country wants to leave. That would mean offering benefits to the club members,[94] where the negatives become positives, and where the members can exclude others, who themselves do not wish to join the club.[95] To be in the club, one would need participation and compliance.[96] Such a situation would create stable coalitions. Economic theory and empirical evidence show that stable coalitions with substantial emissions abatement are not likely to form without sanctions against non-participants.

Arguably, in the case of climate change agreements, they are doomed to failure because there is no incentive to remain a Contracting Party to the agreements, as there is no penalty if a country chooses to withdraw from the agreement. Equally, there is no punishment if a Contracting Party does not comply with the agreement. So a future club for climate mitigation could be construed as one that offers benefits for joining, but there would be no punishment if countries wish not to join.

Three characteristics appear evident for the creation of a successful climate club:

1. Most big GHG emitters need to be members of the club;
2. Membership benefits are a must, and they should outweigh obligations; and

---

[91] USTR, "Joint Statement by the United States, European Union and Japan at MC11," available at https://ustr.gov/about-us/policy-offices/press-office/press-releases/2017/december/joint-statement-united-states.

[92] Reuters, "Some WTO members to push for e-commerce rules as broader deal fails," 13 December 2017, available at https://www.cnbc.com/2017/12/13/reuters-america-some-wto-members-to-push-for-e-commerce-rules-as-broader-deal-fails.html.

[93] The Guardian, "Canada pulls out of Kyoto protocol," 13 December 2011, available at https://www.theguardian.com/environment/2011/dec/13/canada-pulls-out-kyoto-protocol.

[94] M. Tomz, J. Goldstein, and D. Rivers, "Membership Has Its Privileges: The Impact of the GATT on International Trade," 2007, available at https://web.stanford.edu/~tomz/pubs/TGR2007.pdf.

[95] For further details on the economic theories of clubs, see Buchanan (1965), pp. 1–14.

[96] A simple example of compliance would be a speeding ticket: if the speeding ticket is very high, the driver will be very careful not to go beyond the speed limit and would therefore comply with the law.

3. The club would need to be related to sanctions for non-compliance.[97]

Who might be the right institution to host such a climate club? The WTO? The Organization for Economic Co-operation and Development (OECD)? The Major Economies Forum on Energy and Climate (MEF) could well be a good platform to link clean energy, climate action and international trade. It was initiated in 2007 by the Bush administration under the name "Major Emitters Forum"[98] and launched by the Obama administration on March 28, 2009.[99] It could be a good candidate for effective and efficient climate change mitigation through cooperation, given its membership.

The MEF is intended to facilitate a candid dialogue among major developed and developing economies, help generate the political leadership necessary to achieve a successful outcome at future UN climate change conferences, and advance the exploration of concrete initiatives and joint ventures that increase the supply of clean energy while cutting GHG emissions.[100]

The MEF partners include: Australia, Brazil, Canada, China, the EU, France, Germany, India, Indonesia, Italy, Japan, Korea, Mexico, Russia, South Africa, the UK, and the U.S.[101] Bringing together these major emitters, which were responsible for around 75% of GHG emissions in the world as of 2009 (these numbers include land-use change),[102] will increase the likelihood of mitigating climate change more quickly, as the MEF is a more efficient and effective negotiating forum than the UN Framework Convention on Climate Change (UNFCCC).[103] Even bringing together the six major emitters, responsible for 60% of global GHG emissions, will be very beneficial for climate action.[104] As a platform for action, the MEF carries legitimacy: it represents 80% of global GHG emissions, 80% of the world's GDP, and 60% of the world's population.[105]

---

[97] One would need to make sure that such sanctions would not violate international law and/or WTO legal rules.

[98] The name was changed because, according to the members, the initial name sounded like an oligopoly of polluters.

[99] The MEF has gone through a number of name changes. It was previously called the Major Emitters Forum and the Major Economies Process on Energy Security and Climate Change.

[100] Stewart et al. (2013), pp. 273–305.

[101] See Major Economies Forum on Energy and Climate, available at http://www.majoreconomiesforum.org/about/descriptionpurpose.html. For an analysis, see Leal-Arcas (2011c), pp. 25–56; Leal-Arcas (2018), pp. 305–317.

[102] Broder, J. "Clinton Says U.S. is Ready to Lead on Climate," *The New York Times*, 27 April 2009, available at http://nyti.ms/huEbYb.

[103] Oye (1985), p. 21.

[104] D. Cappiello, "These 6 Countries Are Responsible For 60% Of CO2 Emissions," *Business Insider*, 5 December 2014, available at http://www.businessinsider.com/these-6-countries-are-responsible-for-60-of-co2-emissions-2014-12.

[105] See generally Leal-Arcas (2013), pp. 337–338.

The MEF fosters technological innovation, brought about by increased trade.[106] To avoid the obstacles faced by the UNFCCC machinery, the MEF should focus on each member's economic weight, GHG emissions reduction responsibilities, and the calculation of responsibility for GHG emissions such as sharing the burden equally between producers and consumers, in order to fairly decide who should reduce GHG emissions and by how much. For instance, most GHGs are emitted because countries do not have clean sources of energy. They have no choice but to use available technologies. If energy-producing countries have to pay 50% of the cost, there would be a greater incentive to shift energy production from fossil fuels to clean energy. Investing in clean energy would then enhance innovation, which is beneficial for the economy.[107]

4.2.3 Bilateralism

Bilateralism, which should be understood as complementary to multilateralism and not mutually exclusive, has been very constructive in clean-energy terms, e.g., US-China relations (the G2). Since a few years now, the US and China[108] have made remarkable advancement on energy- and climate-related cooperation. The United States and China are already cooperating on a number of joint efforts over clean technology, which plays a major role in the relations of the two countries. Below are a few[109]:

1) The United States-China Clean Energy Research Center
2) The United States-China Energy-Efficient Buildings,
3) The United States-China Electric Vehicles,
4) The twenty-first Century Coal Program
5) The China Greentech Initiative
6) The United States Alliances in Chinese Cleantech Industry
7) The United States-China Renewable Energy Partnership
8) The United States-China Energy Cooperation Program,
9) The U.S.-China Regional Cooperation Initiatives

---

[106]Major Economies Forum on Energy and Climate, "Technology Action Plan—Executive Summary," December 2009, available at http://www.majoreconomiesforum.org/images/stories/documents/MEF%20Exec%20Summary%2014Dec2009.pdf.

[107]Leal-Arcas (2013), p. 338.

[108]In fact, China leads the world in clean energy. In recent years, it has spent more on cleaning up its energy system than the US and the EU combined. See The Economist, "The geopolitics of energy," Special report, 17 March 2018, p. 8. A case in point is the fact that China sells more electric vehicles than the rest of the world. See International Energy Agency, "Global EV Outlook 2017," available at https://www.iea.org/publications/freepublications/publication/GlobalEVOutlook2017.pdf, p. 5.

[109]For further details, see Leal-Arcas (2013), pp. 333–336.

In the trade front, bilateralism seems to be beneficial for the big party in a bilateral trade negotiation because it is able to bully the other country. However, higher tariffs means that prices would rise for the consumers of the country that raises the tariffs. For instance, exports that depend on imported components would become less competitive. This new trend of raising tariffs in bilateral deals seems to be at odds with the promise among WTO members not to raise tariffs above agreed levels in the multilateral context and to apply the most-favored nation principle. Therefore, the bilateral approach of raising tariffs is at odds with the multilateral approach of countries not being able to discriminate between their economic friends and enemies, given that lower tariffs granted to one WTO member means granting the same treatment to all.[110]

4.2.4 Unilateralism

There have also been unilateral, sector-specific attempts to mitigate climate change using trade as a tool. For example, the EU has tried to include aviation in the Emissions Trading System.

4.2.5 Vertical Policy-Making

Very promising is the idea of going beyond a horizontal (inter-governmental) approach to policy-making so that the international community starts exploring vertical approaches[111] too (i.e., the involvement of cities through their mayors, NGOs, states through their governors,[112] companies).[113] At the Conferences of the Parties (COPs), there is the issue of physical design if the international community is serious about the engagement of non-state actors during the negotiations.[114] At the

---

[110] The Economist, "Trade blockage," 21 July 2018, pp. 17–20, at 17.

[111] See for instance America's Pledge, where a 'number of U.S. cities, states, businesses, and universities have reaffirmed their commitment to helping America reach its Paris climate goals.' Available at https://www.americaspledgeonclimate.com/.

[112] See for instance the US Climate Alliance, which is a bipartisan coalition of states in the US committed to implementing the objectives of the Paris Agreement on Climate Change within their borders. For further details, see https://www.usclimatealliance.org/.

[113] See Victor (2017).

[114] See "Mike Bloomberg Delivers Remarks on America's Pledge at COP23, Bonn, Germany, on Saturday, November 11th, 2017," available at https://www.mikebloomberg.com/news/mike-bloomberg-delivers-remarks-americas-pledge-cop23-bonn-germany-saturday-november-11th-2017/. To watch the remarks of former mayor of New York Michael Bloomberg and California Governor Jerry Brown regarding the first America's Pledge report detailing how U.S. cities, states, and businesses can continue to make progress on climate change mitigation, regardless of what happens at the US federal level of governance, see https://www.youtube.com/watch?v=gXyFW9_EJ_U.

moment, the structure is such that only governments are participants in the climate change negotiations. In this respect, we have recently seen the relevant work of C40 Cities[115] and R20 regions of climate action outside the COPs.[116]

In fact, in December 2017, city leaders from different parts of the world met in Chicago, invited by its mayor, to discuss how cities can implement the Paris Agreement.[117] In September 2018, California's governor Jerry Brown passed an executive order asking state agencies to start planning for the mitigation of climate change at the sub-national level to make California's economy carbon-neutral by 2045.[118] All these sub-national green efforts are most welcome and show a commitment to the Paris Agreement, even if some countries may not be committed at the federal level.

4.2.6 Regionalism

Finally, regionalism seems to be on the rise,[119] certainly since the decline of multilateralism.[120] The number of RTAs has increased considerably since right before the creation of the WTO in 1995. This phenomenon has created the so-called spaghetti bowl.[121]

To conclude this Section, multilateralism might be the best way forward in terms of legal certainty, predictability, and transparency. The second best option seems to be regionalism/plurilateralism.

The next Section examines in more detail the potential of regionalism by analyzing specific RTAs. It also explores the role of the major emitters.

---

[115]C40 Cities, http://www.c40.org/ [https://perma.cc/J53Q-WNQ9].

[116]Regions of Climate Action, http://regions20.org/ [https://perma.cc/8EPB-X2QN]. A further platform of a coalition of regions/states that fights climate change sub-nationally is Under2 Coalition, at https://www.under2coalition.org/.

[117]"Climate summitry: New life for the Paris deal," *The Economist*, 16 December 2017, pp. 51–52, at 51.

[118]The Economist, "Climate change: Local government v global warming," 15 September 2018, pp. 67–68, at 67.

[119]Bhagwati, J. 'US Trade Policy: The Infatuation with FTAs' [1995] Discussion Paper Series No. 726, Columbia University, available at http://www.columbia.edu/cu/libraries/inside/working/Econ/ldpd_econ_9495_726.pdf; Baier et al. (2008), pp. 461–497.

[120]T. Risse, "The Diffusion of Regionalism, Regional Institutions, Regional Governance," paper presented at the EUSA 2015 Conference, March 5-7, 2015, available at http://www.file:///C:/Users/lcw197/Downloads/Risse%20Diffusion%20EUSA2015.pdf; Leal-Arcas (2011a), pp. 597–629.

[121]Bhagwati (1992), pp. 535–556.

4.3 Major Emitters and Mega-Regional Trade Agreements

By making use of mega-RTAs with binding and enforceable provisions on environmental protection, there will be economic growth and mitigation of climate change. In a world where states have built bridges connecting themselves through trade and technology, the production and supply of public goods has far-reaching, global implications.

The North American Free Trade Agreement (NAFTA)[122] of 1994 was the first preferential trade agreement (PTA) to include a side-agreement to that effect, namely the North American Agreement on Environmental Cooperation.[123] The European Union's RTAs have been incorporating environmental provisions since the mid-1990s.

This section brings forward the novel idea of using mega-regional trade agreements (RTAs) to mitigate climate change and enhance sustainable energy. Adjustments by just a few major GHG emitters[124] and just three mega-RTAs (namely the Trans-Atlantic Trade and Investment Partnership (TTIP), the Comprehensive and Progressive Agreement for Trans-Pacific Partnership (CPTPP),[125] and the Regional Comprehensive Economic Partnership (RCEP))[126] can make a great contribution towards climate change mitigation and the enhancement of sustainable energy.[127]

Since most of the contracting parties to these three mega-regional agreements are also the main GHG emitters, and since RTAs have provisions that bind countries to mitigate climate change and these provisions are enforceable, then RTAs may potentially become a very effective solution to climate change mitigation and the

---

[122]In 2018, the NAFTA Parties concluded an updated version of NAFTA, namely the United States-Mexico-Canada Agreement (USMCA).

[123]http://www.cec.org/about-us/NAAEC. For further details, see Grossman and Krueger (1993).

[124]Based on cumulative emissions (which describe a country's total historic emissions and is really what matters to climate change), according to the World Resources Institute, almost half of global emissions of greenhouse gases comes from just four parties to the UN Framework Convention on Climate Change, namely China, the United States, the European Union, and India). See https://wri.org/blog/2014/11/6-graphs-explain-world%E2%80%99s-top-10-emitters.

[125]Australia became the sixth country to ratify the CPTPP, thus triggering the Treaty's entry into force, which took place on 30 December 2018.

[126]As of early 2018, the former Trans-Pacific Partnership has a new name [Comprehensive and Progressive Agreement for Trans-Pacific Partnership] and the US withdrew from it, the RCEP has come to a halt, and the TTIP negotiations are frozen. On TTIP negotiations, see Philip Blenkinsop, "U.S. trade talks in deep freeze after Trump win, says EU," *Reuters*, 11 November 2016, available at https://www.reuters.com/article/us-usa-election-eu-trade/u-s-trade-talks-in-deep-freeze-after-trump-win-says-eu-idUSKBN1361UN.

[127]See Leal-Arcas et al. (2014, 2016); World Economic Forum, "Mega-regional Trade Agreements Game-Changers or Costly Distractions for the World Trading System?" July 2014, available at http://www3.weforum.org/docs/GAC/2014/WEF_GAC_TradeFDI_MegaRegionalTradeAgreements_Report_2014.pdf.

4.3 Major Emitters and Mega-Regional Trade Agreements

promotion of clean energy.[128] In addition, one can think of even multilateralizing RTAs.[129]

The three concluded or ongoing negotiations for mega-regional trade agreements par excellence based on their percentage of global GDP are:

4.3.1 The Regional Comprehensive Economic Partnership

The Regional Comprehensive Economic Partnership (RCEP) is a free-trade agreement (FTA) negotiation that has been developed among 16 countries in Asia and Oceania: the 10 members of the Association of Southeast Asian Nations (ASEAN) (Brunei, Cambodia, Indonesia, Laos, Malaysia, Myanmar, the Philippines, Singapore, Thailand, and Vietnam) and the six countries with which ASEAN has existing FTAs (Australia, China, India, Japan, South Korea, and New Zealand).[130] In relation to RCEP, these six non-ASEAN countries are known as the ASEAN Free Trade Partners.[131] RCEP countries have a population of more than three billion and a total GDP of around 23 trillion dollars, which is about 30% of global GDP.[132]

4.3.2 The Comprehensive and Progressive Agreement for Trans-Pacific Partnership

The Comprehensive and Progressive Agreement for Trans-Pacific Partnership (CPTPP) is an almost 6000-page long FTA concluded among 11 Asia-Pacific nations, namely Japan, Mexico, Canada, Australia, Malaysia, Chile, Singapore, Peru, Vietnam, New Zealand, and Brunei.[133] Its predecessor (the Trans-Pacific Partnership (TPP)) was concluded on October 5, 2015 after several years of secretive

---

[128] The same argument applies to sustainable energy. *See, e.g.*, Leal-Arcas et al. (2015), p. 472.

[129] Pauwelyn (2009), p. 368; Richard Baldwin, 'Multilateralising Regionalism: Spaghetti Bowls as Building Blocs on the Path to Global Free Trade' NBER Working Paper Series No 12545 (National Bureau of Economic Research 2006) http://www.nber.org/papers/w12545.

[130] *Regional Comprehensive Economic Partnership (RCEP)*, N.Z. Foreign Aff. & Trade, https://www.mfat.govt.nz/en/trade/free-trade-agreements/agreements-under-negotiation/rcep/ [https://perma.cc/KD8W-LRLW].

[131] *Id.*

[132] *Id.*

[133] *DG Azevedo Congratulates TPP Ministers*, WTO (Oct. 5, 2015), https://www.wto.org/english/news_e/news15_e/dgra_05oct15_e.htm [https://perma.cc/C8D3-7SKM]; Roger Yu, *TPP, Explained: What is the Trans-Pacific Partnership that President Trump is Withdrawing From?*, USA Today (Jan. 23, 2017), https://www.usatoday.com/story/money/2017/01/23/what-tpp/96949608/ [https://perma.cc/5T7G-7BSM].

negotiations.[134] The TPP negotiations were conducted with a level of secrecy not witnessed in any previous trade agreement.[135] Even the U.S. Congress was critical about the opaqueness surrounding it.[136] Only 600 "cleared advisors" representing corporations and trade blocs were privy to the negotiating process at the expense of the general public and civil society.[137] The TPP represents 11% of world population,[138] 26% of world trade,[139] and nearly 40% of global GDP.[140] In January 2017, U.S. President Donald Trump signed an executive order for the United States to withdraw from the TPP.[141] The U.S. withdrawal has only a minor effect since the TPP went ahead without the United States and became the CPTPP; moreover, the United States has never been a party to the TPP because the U.S. Congress had not ratified the TPP before the United States' withdrawal from the trade agreement in January 2017[142] and because the TPP was never in force with its entire original membership of 12 countries.[143]

---

[134]Eric Bradner, *How Secretive is the Trans-Pacific Partnership?*, CNN (June 12, 2015), http://www.cnn.com/2015/06/11/politics/trade-deal-secrecy-tpp/ [https://perma.cc/X3ED-SD5P].

[135]*Secret Trans-Pacific Partnership Agreement (TPP) - IP Chapter*, WikiLeaks (Nov. 13, 2013), https://wikileaks.org/tpp/pressrelease.html [https://perma.cc/S4VE-QHBP].

[136]*See* Press Release, *Congressional Democrats Escalate Criticism of Substance, Process of Obama's First Trade Pact – the Trans-Pacific Partnership*, Public Citizen (June 27, 2012), http://www.citizen.org/documents/release-congressional-democrats-escalate-criticism-6-27-12.pdf [https://perma.cc/SC7B-4PNJ].

[137]*See* William Mauldin, *U.S Says Not 'Rushing' Asia-Pacific Trade Deal*, Wall St. J. (Sept. 26, 2013), http://online.wsj.com/news/articles/SB10001424052702303796404579099632713091994 [https://perma.cc/XFN8-CWTP].

[138]*Trans-Pacific Partnership Agreement*, Austl. Gov't: Dep't of Foreign Aff. & Trade (Feb. 7, 2017), http://dfat.gov.au/trade/agreements/tpp/Pages/trans-pacific-partnership-agreement-tpp.aspx [https://perma.cc/VAD5-XMSQ].

[139]*Id.*

[140]This figure represents gross domestic product (GDP) prior to the U.S. withdrawal from the Trans-Pacific Partnership (TPP). *See Overview of the Trans Pacific Partnership*, Off. of the U.S. Trade Representative, https://ustr.gov/tpp/overview-of-the-TPP [https://perma.cc/9PSE-MVC9].

[141]*Trump Executive Order Pulls Out of TPP Trade Deal*, BBC (Jan. 24, 2017), http://www.bbc.co.uk/news/world-us-canada-38721056 [https://perma.cc/Z34R-T44K].

[142]*See* James McBride, *The Trans-Pacific Partnership and U.S. Trade Policy*, Council on Foreign Relations (Jan. 31, 2017), https://www.cfr.org/backgrounder/trans-pacific-partnership-and-us-trade-policy [https://perma.cc/KBY8-VV5F]. In legal terms, this means that the United States has never been a party to the agreement. The U.S. president's authority to withdraw from an international agreement is summarized in Restatement (Third) Foreign Relations Law of the United States § 339 (Am. Law Inst., 1987). The question has been litigated in the context of withdrawal from a treaty, particularly in the case of *Goldwater v. Carter*, 617 F.2d 697 (D.C. Cir. 1979), which concerned termination of the U.S.-Taiwan Mutual Defense Treaty.

[143]Catherine Putz, *TPP: The Ratification Race is On*, Diplomat (Feb. 5, 2016), http://thediplomat.com/2016/02/tpp-the-ratification-race-is-on/.

4.3.3 The Trans-Atlantic Trade and Investment Partnership

The Trans-Atlantic Trade and Investment Partnership (TTIP) is a proposed RTA between the United States and the European Union and its member states.[144] The TTIP was first conceived in November 2011, following a U.S.-E.U. summit and the sixth meeting of the Transatlantic Economic Council.[145] Leaders requested that the U.S.-E.U. High Level Working Group on Jobs and Growth identify "policies and measures to increase U.S.-E.U. trade and investment to support mutually beneficial job creation, economic growth, and international competitiveness."[146] The High Level Working Group concluded that the development of a comprehensive bilateral trade and investment agreement would provide the most benefits for the parties.[147] The TTIP represents nearly 50% of global GDP.[148]

4.3.4 Analysis

Leaving aside the overlapping membership in these three mega-RTAs (Malaysia, Vietnam, Brunei, Japan, Singapore, Australia, and New Zealand are parties to both RCEP and CPTPP), the total aggregate of global GDP that the three mega-RTAs represent is around 80–85%.[149] This means that most of global GDP is represented by these three mega-RTAs. Equally, the 10 major emitters of GHGs are responsible for about 70% of global GHG emissions, out of 196 countries (see Table 4.1).[150]

---

[144] *About TTIP*, European Comm'n, http://ec.europa.eu/trade/policy/in-focus/ttip/about-ttip/ [https://perma.cc/FZ5T-UWN7].

[145] *Fact Sheet: United States to Negotiate Transatlantic Trade and Investment Partnership with the European Union*, Off. of the U.S. Trade Representative (February 13, 2013), https://ustr.gov/about-us/policy-offices/press-office/fact-sheets/2013/february/US-EU-TTIP [https://perma.cc/HWM8-5Y55].

[146] U.S.-European Union High Level Working Grp. on Jobs and Growth, Final Report 1, (Feb. 11, 2013), www.ustr.gov/sites/default/files/02132013%20FINAL%20HLWG%20REPORT.pdf [hereinafter Final Report: High Level Working Group on Jobs and Growth] (citations omitted) [https://perma.cc/RJ4H-XW7J].

[147] *Id.*

[148] *What You Need to Know About TTIP*, Eur. Am. Chamber of Com.: N.Y., https://www.eaccny.com/international-business-resources/what-you-need-to-know-about-ttip/ [https://perma.cc/LX5D-29S2].

[149] This percentage range reflects the author's estimate. This figure represents GDP prior to the United States' withdrawal from the TPP. *See id.*; *see also Overview of the Trans Pacific Partnership*, Off. of the U.S. Trade Representative, https://ustr.gov/tpp/overview-of-the-TPP [https://perma.cc/R75Q-TFK3]; *Regional Comprehensive Economic Partnership (RCEP)*, N.Z. Foreign Aff. & Trade, https://www.mfat.govt.nz/en/trade/free-trade-agreements/agreements-under-negotiation/rcep/ [https://perma.cc/A358-4YCE].

[150] Friedrich et al. (2017), http://www.wri.org/blog/2017/04/interactive-chart-explains-worlds-top-10-emitters-and-how-theyve-changed.

Table 4.1 Major GHG emitters and contracting parties to the three mega-RTAs

Top 10 GHG Emitters[a] (≈70% of global GHG emissions)[b]	RCEP (ASEAN + 6) (≈30% of global GDP)	CPTPP (≈40% of global GDP)[c]	TTIP (≈50% of global GDP)
China	✓		
United States			✓
European Union (28 countries)			✓[d]
India	✓		
Russia			
Indonesia	✓		
Brazil			
Japan	✓	✓	
Canada		✓	
Mexico		✓	
	RCEP parties that are not top 10 GHG emitters: Australia, Brunei, Cambodia, Laos, Malaysia, Myanmar, New Zealand, Philippines, Singapore, South Korea, Thailand, Vietnam	CPTPP parties that are not top 10 GHG emitters: Australia, Brunei, Chile, Malaysia, New Zealand, Peru, Singapore, Vietnam	

[a]The list takes into account emissions deriving from land use change and forestry
[b]Ge et al. (2014), https://wri.org/blog/2014/11/6-graphs-explain-world%E2%80%99s-top-10-emitters [https://perma.cc/9X33-ZAZL]
[c]This figure represents GDP prior to the U.S. withdrawal from the Trans-Pacific Partnership (TPP), the predecessor of the CPTPP. See *Overview of the Trans Pacific Partnership*, available at https://ustr.gov/tpp/overview-of-the-TPP
[d]The E.U. member states will most likely be part of the Trans-Atlantic Trade and Investment Partnership (TTIP) given that some issues of TTIP are shared competence between the E.U. and its member states. For an analysis of trade agreements between the E.U. and a third party where E.U. member states ended up being a party of trade agreements, see Leal-Arcas (2008a)

If one analyzes the table above by considering the European Union as a single economic entity and discounting the E.U. member states that are among the 10 major economies in the world (i.e., Germany, the United Kingdom, France, and Italy), notably, Indonesia and Mexico are the only two emitters in the top 10 that are not among the 10 major economies.[151] Indonesia is the only country in the top

---

[151] According to the International Monetary Fund, these are the ten major economies, excluding any E.U. member state and including the European Union as a single entity: the United States, the European Union, China, Japan, India, Brazil, Canada, South Korea, Russia, and Australia. See *Report for Selected Countries and Subjects*, IMF, http://bit.ly/2dQKeno [https://perma.cc/F4SU-YMDR].

4.3 Major Emitters and Mega-Regional Trade Agreements

10 emitters which is not among the top world economies. This means that its levels of GHG emissions are disproportionately high.

When comparing the membership of mega-RTAs with the major GHG emitters in Table 4.1, it becomes clear that eight out of the ten major GHG emitters are contracting parties to at least one of the three mega-RTAs (in the case of Japan, it is a party to the CPTPP and RCEP). The only two major emitters which are not parties to any of the three mega-RTAs are Brazil and Russia. Two other major GHG emitters (Australia and South Korea), which are not in the top ten major GHG emitters, are contracting parties to at least one of the three mega-RTAs (namely RCEP and CPTPP).

Therefore, by having these three mega-RTAs with legally binding and enforceable provisions on climate change mitigation and low-emissions economy, eight of the ten major GHG emitters could effectively solve most of the climate change problem. Although climate change is a global problem of collective action, mega-RTAs could be an effective way to tackle climate change.

In addition to those three mega-RTAs, there are three concluded or ongoing trade initiatives that are worth mentioning regarding the role of international trade in climate change mitigation and sustainable energy. The first, also a mega-RTA, is the Comprehensive Economic and Trade Agreement between Canada and the European Union and its member states (CETA).[152] Since both Canada and the European Union are parties to some of the three mega-RTAs mentioned above (Canada is a party to the CPTPP, and the European Union to the TTIP), this chapter omits CETA from the table above to avoid repetition of the participation of the top GHG emitters in mega-RTAs.[153] The second trade agreement is the Environmental Goods Agreement (EGA), currently under negotiation.[154] The third agreement is the Information Technology Agreement (ITA), which is relevant for trade in clean energy technologies.[155]

4.3.5 The Comprehensive Economic and Trade Agreement

The Comprehensive Economic and Trade Agreement (CETA) was signed in October 2016.[156] Both Canada and the European Union are in the top 10 GHG

---

[152] *CETA Explained*, European Comm'n, http://ec.europa.eu/trade/policy/in-focus/ceta/ceta-explained/ [https://perma.cc/X6FZ-28VZ].

[153] Within the three chosen mega-RTAs, there is repetition in membership. For instance, there are seven CPTPP signatories that are included in the Regional Comprehensive Economic Partnership (RCEP) negotiations: Australia, Brunei, Japan, Malaysia, New Zealand, Singapore, and Vietnam.

[154] *Environmental Goods Agreement*, WTO, https://www.wto.org/english/tratop_e/envir_e/ega_e.htm [https://perma.cc/8XCV-W86Y].

[155] *See* World Trade Organization, Ministerial Declaration of 13 December 1996, WTO Doc. WT/MIN(96)/16, 36 I.L.M. 375 (1997).

[156] Dominic Webb, House of Commons Library, Briefing Paper No. 7492, CETA: The EU-Canada Free Trade Agreement 5 (2017).

emitters and are among the major economies of the world, and therefore key actors in this area. The parties to CETA agree that economic growth supports their social and environmental goals.[157] CETA has two chapters relevant to the relationship between trade and environmental concerns: Chapter 22 (on trade and sustainable development) and Chapter 24 (on trade and environment).[158] CETA's Chapter 22 recognizes that economic growth, social development, and environmental protection are interconnected. Chapter 24 commits the parties to putting into practice international environmental agreements.[159] More specifically, Chapter 24 protects the rights of the parties to regulate on environmental matters, requires the parties to enforce their domestic environmental laws, and prevents the parties from relaxing their laws to boost trade.[160]

4.3.6 The Environmental Goods Agreement

The EGA is a plurilateral[161] trade agreement currently under negotiation between eighteen WTO Members. Five of the 10 major GHG emitters listed in Table 4.1 above are participating in the EGA.[162] This agreement aims to encourage green growth[163] and sustainable development by liberalizing trade in environmental goods and by reducing or eliminating tariffs in environmental goods[164] such as renewable and clean energy technology.[165] Arguably, a broad liberalization of services could

---

[157] *See* Comprehensive Economic and Trade Agreement (CETA) between Canada and the European Union and its Member States, at 3, Jan. 1, 2017, 2017 O.J. (L 11) 23 [hereinafter CETA].

[158] CETA arts. 22, 24.9.

[159] *Id.* art. 24.

[160] *CETA Chapter by Chapter*, European Comm'n, http://ec.europa.eu/trade/policy/in-focus/ceta/ceta-chapter-by-chapter/ [https://perma.cc/5QZD-CWXG].

[161] A plurilateral approach to trade agreements means that the agreements are optional and not binding on those WTO members who do not engage in them. In the WTO context, multilateral negotiations, as opposed to plurilateral negotiations, imply the participation of all WTO members. The nature of the consequent multilateral agreements from these multilateral negotiations implies that commitments are taken by all the WTO members. The idea behind plurilateral negotiations is to make the WTO deliver again on progressive liberalization. See Leal-Arcas (2008b), p. 28.

[162] *See Environmental Goods Agreement*.

[163] The term 'green growth' can be linked to 'prosperity without growth' or even a 'socially sustainable economic de-growth.' Is economic growth compatible with environmental sustainability? See Jackson (2011).

[164] *The Environmental Goods Agreement (EGA): Liberalising Trade in Environmental Goods and Services*, European Comm'n (Sept. 8, 2015), http://trade.ec.europa.eu/doclib/press/index.cfm?id=1116 [https://perma.cc/C8A6-RP6Z].

[165] On the link between renewables and the trading system, see Leal-Arcas and Filis (2014), p. 3; Leal-Arcas and Filis (2015), p. 482.

4.3 Major Emitters and Mega-Regional Trade Agreements

also be beneficial for sustainable development, as would an expansion of the EGA to services trade.[166] Moreover, a great added value of the EGA is that the "benefits of this new agreement will be extended to the entire WTO membership, meaning all WTO members will enjoy improved conditions in the markets of the participants to the EGA."[167] The extension of these benefits will multilateralize this plurilateral agreement. This agreement is an example of the relevant intersection between international economic law and the SDGs. Such plurilateral agreements could have the potential of most favored nation application and therefore serve as a platform for climate change mitigation worldwide.[168] In sum, once the EGA is in place, it will add traditional products (not just environmental goods), more WTO Members, and nontechnical barriers to trade in environmental services.

A core strategic movement in the promotion of sustainable energy and climate-change mitigation from an international trade perspective would be the removal of barriers to trade "*in products that are crucial for environmental protection and climate change mitigation*".[169] Following this logic, since July 2014, the EU has been engaged, with 17 other partners,[170] in the negotiations that, under WTO auspices, intend to bring to life a multilateral "*Environmental Goods Agreement*" (EGA).[171] As explained by the Directorate General for Trade (DG Trade) of the European Commission: "*Eliminating* [...] *customs duties can boost trade in "green goods", and this in turn can help the European Union and its partners to protect environment and meet their climate and energy targets.*"[172]

The parties' approach so far has consisted in attempting to identify possible lists of "green goods" that would deserve the described barrier-removal privilege. To do so, each of the negotiating parties has submitted lists of goods they would like to see freed of duties. The EU's listing includes "products *used for: generation of renewable energy, control of air pollution, management of solid and hazardous waste, management of waste water and water treatment, environmental remediation and*

---

[166] Services trade includes, for instance, clean water filtration services and the movement of people via mode four of the General Agreement on Trade in Services. *See* General Agreement on Trade in Services, April 15, 1994, Marrakesh Agreement Establishing the World Trade Organization, Annex 1B, 1869 U.N.T.S. 183.

[167] *Environmental Goods Agreement*.

[168] *See* Bryce Baschuk, *Environmental Goods Negotiators Make Incremental Progress,* Bloomberg Int'l Trade Daily, Sept. 26, 2016.

[169] European Commission, "Environmental Goods Agreement: Promoting EU environmental objectives through trade", http://trade.ec.europa.eu/doclib/press/index.cfm?id=1438, available online on October the 3rd 2017.

[170] Australia, Canada, China, Costa Rica, Hong Kong, Iceland, Israel, Japan, Korea, New Zealand, Norway, Singapore, Switzerland, Liechtenstein, Chinese Taipei, Turkey, United States. See World Trade Organization, "Environmental Goods Agreement (EGA)", https://www.wto.org/english/tratop_e/envir_e/ega_e.htm, available online on October the 9th 2017.

[171] European Commission, "Environmental Goods Agreement: Promoting EU environmental objectives through trade", http://trade.ec.europa.eu/doclib/press/index.cfm?id=1438, available online on October the 3rd 2017.

[172] Idem.

clean up, noise and vibration abatement, resource and energy efficiency, environmental monitoring and analysis."[173] Overall, the idea is that *"there could be a benefit for the multilateral trading system in lowering technical barriers to trade in energy-related goods and services, including in relation to technological goods and services that could encourage the proliferation of renewables."*[174]

4.3.7 The Information Technology Agreement

Finally, the ITA, whose relevance for trade in clean energy technologies is crucial, was concluded by 29 parties at the Singapore Ministerial Conference in December 1996.[175] Today, there are 82 parties to the ITA, which represents 97% of international trade in information technology (IT) products.[176] In December 2015, over 50 parties to the agreement concluded an expansion of the ITA, which covers an additional 201 products.[177]

As we have seen, several of these so-called mega-RTAs are incorporating environmental provisions: Chapter 20 of the Comprehensive and Progressive Agreement for Trans-Pacific Partnership,[178] Chapter 24 of the Comprehensive Economic and Trade Agreement, and there is a chapter on trade and sustainable development in the Trans-Atlantic Trade and Investment Partnership negotiations.[179] RTAs of emerging economies are also converging to this 'green' race, especially when the agreements involve countries that belong to the Organization for Economic Cooperation and Development.[180] A case in point is the Peru-Korea FTA, whose Article 19.8.1 recognizes that climate change is a common concern and states that "the Parties agree to promote joint measures to limit or reduce the adverse effects of the climate change."[181] The RTAs-trend seems irreversible and is likely to persist, given the current crisis in the multilateral trading system.

---

[173]Idem.

[174]Leal-Arcas (2015a), pp. 202–219.

[175]Ministerial Declaration of 13 December 1996; *Information Technology Agreement*, WTO, https://www.wto.org/english/tratop_e/inftec_e/inftec_e.htm [https://perma.cc/VEH5-7QZ3].

[176]*Information Technology Agreement.*

[177]*Id.*

[178]Similar to Chapter 20 of the earlier Trans-Pacific Partnership.

[179]See for instance Leal-Arcas (2015b), pp. 248–264.

[180]See Berger et al. (2017).

[181]Article 19.8.2 of the Peru-Korea FTA is more specific on how to tackle climate change by stating:

...each Party, within its own capacities, shall adopt policies and measures on issues such as:

(a) improvement of energy efficiency;

4.4 Capitalizing on RTAS[182]

4.4.1 The Rationale

It is worth exploring the potential of incorporating strong and meaningful chapters addressing climate change mitigation and promoting renewable energy within PTAs, for which major trade actors could make use of their vast network of PTAs.

In fact, trade agreements may be more effective legal instruments than environmental agreements for environmental-protection purposes, which is both counter-intuitive and surprising.[183] This is due to the fact that environmental agreements lack a dispute settlement system that trade agreements offer regarding enforcement. As depicted in Fig. 4.1, this lack of enforceability in environmental agreements raises a very interesting question: could trade agreements be a solution to decarbonization?

There are several ways to remedy this deficit in environmental agreements:

1. A 'name and shame' approach used in environmental agreements could be interpreted as an enforcement mechanism;
2. A cooperative approach, as opposed to sanctions, for the enforcement of agreements;
3. A sanctions-based approach;
4. Invoking Article 60 of the Vienna Convention on the Law of Treaties on the termination or suspension of the operation of a treaty as a consequence of its breach.[184]

An example of how one could promote sustainability[185] through trade is the Comprehensive Economic and Trade Agreement (CETA), a free-trade agreement (FTA) between the EU and Canada, under provisional application since 2017.[186] In October 2017, the French government approved an action plan on the Comprehensive Economic and Trade Agreement (CETA), a free-trade agreement (FTA) between the EU and Canada, following a report by independent experts on

---

(b) research, promotion, development and use of new and renewable energy, technologies of carbon dioxide capture, and updated and innovative environmental technologies that do not affect food security or the conservation of biological diversity; and
(c) measures for evaluating the vulnerability and adaptation to climate change.

[182] For the purposes of this sections, the use of regional trade agreements (RTAs) and free-trade agreements (FTAs) is interchangeable.

[183] See Leal and Leal-Arcas (2018). As a result, the environment may benefit from being addressed properly in FTAs. See also Leal-Arcas and Alvarez Armas (2018), pp. 153–168.

[184] For access to the Vienna Convention on the Law of Treaties, see https://treaties.un.org/doc/publication/unts/volume%201155/volume-1155-i-18232-english.pdf.

[185] There are studies on how to measure sustainability. See Munda (2005), pp. 117–134.

[186] European Commission, "EU-Canada trade agreement enters into force," 20 September 2017, available at http://europa.eu/rapid/press-release_IP-17-3121_en.htm.

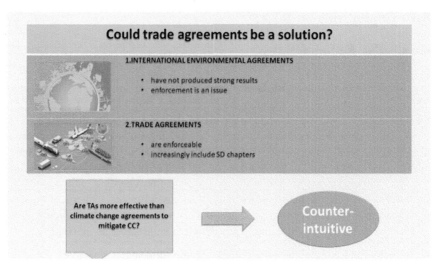

Fig. 4.1 Trade agreements as a potential solution to decarbonization (In Fig. 4.1, 'SD' stands for sustainable development, 'TAs' stands for trade agreements, and 'CC' stands for climate change)

CETA's impact on the environment, climate change, and health.[187] Although the report "did not identify any immediate risks in the provisions of CETA which were likely to stand in the way of the provisional application of the agreement," concerns remain on how the Agreement will work in practice.[188] CETA seems to lack ambition regarding transnational cooperation on climate issues. The action plan has three main objectives:

1. Ensuring that climate regulations are protected from any abusive challenges made by foreign investors;
2. Strengthening international cooperation on climate issues; and
3. Making sure that trade agreements are fully consistent with European policies that contribute to sustainable development.[189]

However, an innovative feature of CETA appears in Article 24.5.1[190] and Article 24.5.3,[191] which state that the contracting Parties may not reduce levels of environmental protection and may not fail to effectively enforce their environmental law to

---

[187]"An action plan for the robust and ambitious implementation of CETA," *Gouvernement.fr*, 26 October 2017, available at http://www.gouvernement.fr/en/an-action-plan-for-the-robust-and-ambitious-implementation-of-ceta.

[188]Ibid.

[189]Ibid.

[190]CETA Article 24.5.1 reads: The Parties recognise that it is inappropriate to encourage trade or investment by weakening or reducing the levels of protection afforded in their environmental law.

[191]CETA Article 24.5.3 reads: A Party shall not, through a sustained or recurring course of action or inaction, fail to effectively enforce its environmental law to encourage trade or investment.

encourage trade or investment. These declarations are promising features for the protection of the environment via international trade, demonstrating that the protection of the environment comes first.[192]

4.4.2 Coherence Between Trade and Climate Change Actions

Two fora seem the most appropriate for creating coherence between trade and climate change policies. First, the establishment of the WTO incorporated the creation of its Committee on Trade and Environment (CTE).[193] The goal of the CTE is to identify and understand the relationship between trade and the environment to promote "sustainable development."[194] The other forum for discussion of trade measures and their links with climate change is the UNFCCC's response measures forum.[195] To avoid the proliferation of climate measures that adversely impact international production and trade, Article 3.5 of the UNFCCC states explicitly that "[m]easures taken to combat climate change, including unilateral ones, should not constitute a means of arbitrary or unjustifiable discrimination or a disguised restriction on international trade."[196] Interestingly, Article 3.5 of the UNFCCC reads conceptually along the same lines as the chapeau of GATT Article XX.

Moreover, some of the WTO agreements under Annex 1 contain provisions that recognize the right of WTO Members to regulate the protection of human, animal, and plant life or health, or the environment.[197] In addition, the Doha Round encompasses specific negotiations concerning various aspects of trade and the

---

[192] See also a non-paper of the European Commission services, "Feedback and way forward on improving the implementation and enforcement of Trade and Sustainable Development chapters in EU Free Trade Agreements" (proposing ways to enforce the commitments under the trade and sustainable development chapters in EU FTAs), 26 February 2018, available at http://trade.ec.europa.eu/doclib/docs/2018/february/tradoc_156618.pdf.

[193] See *The Committee on Trade and Environment ('Regular' CTE)*, WTO, https://www.wto.org/english/tratop_e/envir_e/wrk_committee_e.htm [https://perma.cc/7ZXP-PGGB].

[194] Id.

[195] *Impact of the Implementation of Response Measures*, UNFCCC, http://unfccc.int/cooperation_support/response_measures/items/4908.php [https://perma.cc/R9HT-M9BV].

[196] UNFCCC, art. 3(5).

[197] See Agreement of the Application of Sanitary and Phytosanitary Measures art. 2, Apr. 15, 1994, Marrakesh Agreement Establishing the World Trade Organization, Annex 1, 1867 U.N.T.S. 493; Agreement on Technical Barriers to Trade art. 2, April 15, 1994, Marrakesh Agreement Establishing the World Trade Organization, Annex 1, 1868 U.N.T.S. 120; General Agreement on Trade in Services art. XIV, April 15, 1994, Marrakesh Agreement Establishing the World Trade Organization, Annex 1B, 1869 U.N.T.S. 183.

environment which emphasize the increase in environmental values[198] in the trade sphere.[199] Overall, the WTO seeks to ensure that environmental policies are not barriers to trade liberalization and that trade policies are not an obstruction to environmental protection.[200] However, all the changes that have occurred during the WTO era have not substantially influenced the ongoing interaction between trade and climate change.[201]

The benefit of climate change regulation is reducing the risk of climate change. Reducing carbon dioxide (CO_2) without reducing economic growth or energy security is possible thanks to coherence between trade and climate change regulation and policy. For instance, as pro-globalization scholar Johan Norberg points out, it is possible to develop more efficient production processes, construction that is less energy-consuming, and new sources of energy that are cleaner than using CO_2.[202] Despite the high levels of CO_2 in the United States, thanks to technology, the United States has been able to emit three times less CO_2 than it would have if its technology had been kept at the 1900 level.[203]

It is important to recognize that, in the relationship between WTO rules and multilateral environmental rules, environmental rules should be drafted in a manner that is not in conflict with WTO law. Whenever a conflict between the two disciplines arises, clarification of WTO rules should be made in a manner that puts the environment first.

4.4.3 Trade Agreements As Enhancers of Sustainability[204]

4.4.3.1 Introduction

This section will introduce the role of trade policy in the transition towards a decarbonized economy. Since climate change is one of the biggest challenges

---

[198] A good example of a values-based FTA from an EU perspective is the *Trade Agreement between the European Union and its Member States, of the one part, and Colombia and Peru, of the other part* (EU-Peru-Colombia FTA). O.J. L 354/3, 21.12.2012, signed on 26.06.2012. Provisionally applied from 01.03.2013 between the EU and Peru; provisionally applied (with the exception of Articles 2, 202(1), 291 and 292) from 01.08. 2013 between the EU and Colombia (http://www.consilium.europa.eu/en/documents-publications/agreements-conventions/agreement/?aid=2011057).

[199] For further explanation regarding negotiations on trade and environment under the Doha Round, see *An Introduction to Trade and Environment in the WTO*, WTO, http://www.wto.org/english/tratop_e/envir_e/envt_intro_e.htm [https://perma.cc/EH57-K8TF].

[200] *Trade and Environment*, WTO, https://www.wto.org/english/tratop_e/envir_e/envir_e.htm [https://perma.cc/WCZ7-5C64].

[201] *See* Leal-Arcas (2013), ch. 2–3.

[202] Norberg (2016), p. 120.

[203] *See id.*

[204] This section was written by Alesandra Salaza.

humanity faces today, but international efforts aiming at coping with it evolve slowly, this section aims to make an impact on the international agenda by suggesting ways in which trade agreements may be used to help decarbonize the economy, contributing, as a result, to sustainable development. It will argue that we can use trade policy as a vehicle not only for fostering the transition from the vast use of fossil fuels to the use of sustainable energy resources, such as renewables, but also for tackling climate change and contributing to sustainable development. The adoption of an approach that uses trade agreements to contribute to a low-carbon economy could potentially be more efficient than the approach that has been adopted so far, based on international environmental/climate agreements. The reason for such a statement is that, in the past, some of the commitments announced at international conferences remained merely declarations without a proper implementation, let alone enforcement. This is consistent with the opinion of many authors that consider international environmental law a typical example of "soft law".[205] Trade agreements however, are legally binding, containing substantial provisions and relevant enforcement proceedings.

The following section analyses how trade policy has already contributed to sustainable development in the past through important achievements, such as contributing to poverty reduction and the protection of human rights. Building on these previous experiences, the section aims at questioning whether trade policy could achieve the same positive results in the transition towards a decarbonised economy, contributing, consequently, to sustainable development as well.

International trade and climate change have been often perceived as opposed issues and not many efforts have been made to create a link between the two communities. In 1992, the Earth Summit, which took place in Rio de Janeiro, focused specifically on issues of sustainable development and climate change for the first time at such wide international level. Following this important achievement, there have been at least two occasions where the international community missed the opportunity to address trade and environmental concerns jointly: the WTO Agreement of 1994 and the Paris Agreement on Climate Change of 2015.

This Section will also describe such missed opportunities: (1) a multilateral trade agreement, namely the WTO Agreement, that, despite being concluded only 2 years after the Rio Earth Summit, only mentions sustainable development in its preamble, and (2) an environmental/climate agreement, namely the Paris Agreement on Climate Change, that does not even mention the term "trade". As a consequence, the potential benefits of addressing the issues resulting from the interaction of trade and climate change were missed.

This Section will also explain the link between trade policy and sustainable development through the use of sustainable energy sources, the decarbonisation of the economy, and the consequent fight against climate change. Furthermore, it will describe different proposals and ideas that help trade agreements be potentially

---

[205]Mensah (2008), p. 51; Boyle (2014), pp. 118 ff.

efficient tools to enhance decarbonize the economy and contribute to sustainable development.

4.4.3.2 How Has Trade Policy Helped Achieve Sustainable Development?

This section argues that trade policy has been a very powerful instrument for change, through two examples: (1) poverty reduction; and (2) the protection of human rights. It also makes the claim that, similarly to such issues, trade could contribute to the transition to a low-carbon economy, consequently tackling climate change and contributing to sustainable development, as it will foster not only the economy, but also the use of sustainable energy resources, the creation of jobs related to renewables, energy security, the diffusion of the role of the "prosumers" and the related consequences in terms of energy poverty and democratization of the energy sector.

The intersection between all the above issues is well captured by the structure of the Sustainable Development Goals (SDGs).[206] This set of 17 goals and 169 proposed targets is based on the main characteristic of sustainable development: integration between economic, social and environmental aspects.[207] Tackling one of the issues often results in benefits in other sectors, which explains why the SDGs require a holistic approach for all the goals.[208]

An analysis of the issues of poverty reduction and human rights protection, where trade policy has reached such important and wide results, can inspire the application of the same principle for decarbonisation, sustainable energy, climate change mitigation, and, ultimately, sustainable development.

In the past and still today, trade has hugely contributed to improving the living conditions of many people worldwide in terms of poverty reduction and human rights protection, contributing therefore to acheving the SDGs, especially SDG 1: "End poverty in all its forms everywhere".

---

[206]Resolution adopted by the General Assembly on 25 September 2015, *Transforming our world: the 2030 Agenda for Sustainable Development*, A/RES/70/1, available at http://www.un.org/ga/search/view_doc.asp?symbol=A/RES/70/1&Lang=E.

[207]Cutter, A. *et al.*, "Sustainable Development Goals and Integration: Achieving a better balance between the economic, social and environmental dimensions," Stakeholder Forum, available at http://www.stakeholderforum.org/fileadmin/files/Balancing%20the%20dimensions%20in%20the%20SDGs%20FINAL.pdf.

[208]Sachs (2016), available at http://worldhappiness.report/ed/2016/.

4.4.3.2.1 Poverty Reduction

Poverty can be defined and measured in many different ways.[209] It is often measured in monetary terms, capturing by levels of income or consumption per capita or per household.[210] The most common example of an income-focused approach to poverty is the Millennium Development Goals.[211] In 2000, the international community made the commitment to eradicate absolute poverty by halving the number of people living on less than US$ 1.25 dollar a day.[212] This commitment was reached 5 years ahead of schedule, in 2010, rather than 2015. Based on this measure, there has been marked progress on reducing poverty over the past decades. In 2013, the World Bank estimated that 10.7% of the world's population lived on less than US$1.90 a day, which means a reduction of the 35% from 1990, or 1.1 billion people.[213]

Has trade been the cause of extreme poverty reduction? Although the academic literature is not unanimous to sustain that trade has had an important impact in reducing poverty,[214] the substantial reduction in extreme income poverty in absolute terms has been associated widely with the diffusion of free trade and international trade agreements.[215] Although the drivers of poverty are multi-dimensional, according to the World Bank, opening up to trade allows each country to use its resources more efficiently by specializing in the production of the goods and services

---

[209]The most popular unitary measures are the Human Poverty Indexes (HPIs) and the World Bank's poverty lines, which uses income-based measures of poverty. The Human Development Index (HDI) combines Gross Domestic Product (GDP) per capita purchasing power parity (PPP), literacy, (primary, secondary and tertiary) school enrollment rates, and life expectancy at birth. The HDI tries to capture the insight in Nussbaum and Sen's capability theories. Nussbaum (1997), p. 273; Sen (1999).

[210]United Nations. Department of Economic and Social Affairs (2010), pp. 13 ff.

[211]UN General Assembly, United Nations Millennium Declaration, 55/2, 18/09/2000. Available at: http://www.un.org/millennium/declaration/ares552e.pdf.

[212]World Bank Data available at: https://data.worldbank.org/indicator/SI.POV.DDAY?locations=ZJ-8S-Z4-Z7-ZQ-ZG&start=2013&end=2013&view=bar. The World Bank (2016), Available at https://openknowledge.worldbank.org/bitstream/handle/10986/25078/9781464809583.pdf.

[213]*Ibid.*

[214]N. Hassoun, "Free Trade, Poverty, and Inequality", Dietrich College of Humanities and Social Sciences at Research Showcase @ CMU, available at: http://repository.cmu.edu/cgi/viewcontent.cgi?article=1354&context=philosophy, p. 25. Castilho et al. (2012), pp. 821–835; Cornia (2004).

[215]M. Hayashikawa, OECD, "Trading Out of Poverty How Aid for Trade Can Help", 3–-4 November 2008 OECD Conference Centre, Paris, available at: https://www.oecd.org/site/tadpd/41231150.pdf. Bannister and Thugge (2001), available at: http://www.imf.org/external/pubs/ft/fandd/2001/12/banniste.htm.
OECD (2011), available at: http://www.oecd.org/publications/trade-for-growth-and-poverty-reduction-9789264098978-en.htm.
Upendra Das (2009), available at: http://ris.org.in/images/RIS_images/pdf/DIE%20Regional%20trade%20WP.pdf.

that it can produce more cheaply, while importing the others.[216] This drives to growth from an economic point of view.

Ultimately, however, trade also allows to reach long-term objectives since it helps the access to technologies available in other countries and facilitates the creation of new employment opportunities.[217] With particular regard to least-developed countries, which suffer the most the consequences of extreme poverty, malnutrition, unemployment and poor health conditions, alongside trade, growing flows of capital across countries have widely contributed to tackle such problems.[218] A recent example that shows how trade is still nowadays on the agenda of many developing countries in the path of fighting poverty is given by the result of the African Trade Week.[219] In December 2016, the participants agreed that "there is now a consensus among African countries and the international community to use trade as a tool for economic transformation and poverty eradication".[220]

In developing countries, "RTAs have the potential to promote higher standards in terms of labour, environment, transparency and other progressive reforms and non-economic policy objectives," ultimately demonstrating that trade agreements help determine national trade policy and consequently could amplify the impact of trade on development.[221]

The changes in poverty are almost entirely attributable to economic growth itself, not to changes in the equal and fair distribution of income. This means that, together with poverty, it is important to consider also other aspects directly related with it, such as inequality,[222] and other human rights that, as will be shown in the next section, have received benefits from trade policy.

Inequality is particularly interesting if related to trade, especially to free trade and integration. Free trade is considered one of the main factors that contribute to the

---

[216] World Bank Group and World Trade Organization (2015), available at: https://www.wto.org/english/res_e/booksp_e/worldbankandwto15_e.pdf. Ricardo (2004), first edited in 1817, when the author explained for the first time the theory of comparative advantage.

[217] World Bank Group and World Trade Organization (2015), available online: https://www.wto.org/english/res_e/booksp_e/worldbankandwto15_e.pdf, p. 7.

[218] A. U. Santos-Paulino, "Trade, Income Distribution and Poverty in Developing Countries: A Survey", UNCTAD Discussion Papers, No. 207, July 2012, available at: http://unctad.org/en/PublicationsLibrary/osgdp20121_en.pdf.

[219] African Trade Week 2016, "The Continental Free Trade Area and Trade Facilitation", November 28 – December 2, 2016, Information note, available at: https://au.int/web/sites/default/files/newsevents/conceptnotes/31409-cn-atw2016_information_note_sept30.pdf.

[220] African Trade Week 2016, "The Continental Free Trade Area and Trade Facilitation", November 28 – December 2, 2016, Information note, available at: https://au.int/web/sites/default/files/newsevents/conceptnotes/31409-cn-atw2016_information_note_sept30.pdf, p. 2.

[221] "Developing countries have become more active participants in regional trade agreements", A. DiCaprio, "Regional trade agreements, integration and development", UNCTAD Research Paper No. 1, July 2017, UNCTAD/SER.RP/2017/1, available at: http://unctad.org/en/pages/PublicationWebflyer.aspx?publicationid=1823, p. 5.

[222] The World Bank (2016), Available at: https://openknowledge.worldbank.org/bitstream/handle/10986/25078/9781464809583.pdf.

4.4 Capitalizing on RTAS

concentration of wealth and to the gap in social inequality, leading to social discontent. At the same time, "individuals express concern with rising inequality, broadly defined. In fact, their perceptions of increasing inequality—even though objective measures of inequality declined—have been argued to be one of the factors contributing to the Arab Spring"[223] and, most recently, to Brexit[224] and Donald Trump's victory in the presidential elections of the United States.[225] Yet, aware of the problems that inequality is still causing, especially in developing countries, another element to consider is the impact on poverty and inequality reduction of alternative markets known as Fair Trade markets. Fair Trade has recently gained the reputation of a strong critique of conventional global inequalities.[226] Fair Trade not only generates $1.5 billion per year, but also has become a valuable instrument for alternative globalization grounded in sustainable development, moving (sustainably) numerous commodities and involving thousands of producers, consumers, and distributors world-wide.[227]

4.4.3.2.2 Human Rights Protection

Although the WTO rules do not directly refer to human rights, there are unquestionably social and ethical dimensions within international trade.[228] Including a human rights dimension in trade agreements, especially at the regional level (RTAs),[229] is quite frequent,[230] with regional agreements that apply trade preferences according to human rights standards.[231] In this regard, for example, Clair Gammage refers to the principle of conditionality that the EU has applied for "giving trade preferences on the basis that the beneficiary country complies with certain political and social standards".[232] This demonstrates how trade has been used in the past to protect

---

[223] The World Bank (2016), Available at: https://openknowledge.worldbank.org/bitstream/handle/10986/25078/9781464809583.pdf, p. 69 ff at p. 70.

[224] D. Dorling, B. Stuart and J. Stubbs, "Brexit, inequality and the demographic divide", EUROPP – European Politics and Policy, London School of Economics, available at: http://blogs.lse.ac.uk/politicsandpolicy/brexit-inequality-and-the-demographic-divide/.

[225] World Economic Forum (2017), available at: http://www3.weforum.org/docs/GRR17_Report_web.pdf.

[226] Raynolds and Murray (2008). For a case study: Leigh Taylor (2002).

[227] *Ibid.*

[228] For an in-depth account of the links between trade and human rights, see Cottier et al. (2005) and Lang (2011).

[229] Human rights "are better addressed regionally than multilaterally", "RTAs tend to be highly political [and] the pressures put on one country by other members of the agreement could prove highly influential (p. 105), Barnekow and Kulkarni (2017), pp. 99–117.

[230] Gammage (2014), pp. 779–792.

[231] *Ibid.*

[232] *Ibid*, p. 781.

human rights, such as privacy, participation, due process, access to information and to affordable medicines, and many other issues.[233]

Labour rights have received a particularly positive impact due to the globalization of trade and the consequent application to developing countries to higher working conditions and standards for workers worldwide. Women, in particular, represent a very vulnerable category in the labour market, but have received some important benefits from international trade, especially if considering those markets targeted by small and medium enterprises.[234]

4.4.3.3 Missed Opportunities for Cooperation Between Trade and Climate Change

Having analysed how trade policy has contributed to poverty reduction and to the protection of human rights, this section aims at making the claim that trade (similarly to the positive impact it has had on poverty and human rights) can also be seen as a novel tool to mitigate climate change and decarbonize the economy.[235] Trade agreements should refer to climate change and, vice versa, international climate change agreements should take into account trade issues. Some of the EU trade agreements, for example, have already started to refer to the protection of the environment, sustainable development and, more recently, climate change.[236] This is a first step towards the adoption of an approach that addresses the two issues jointly.[237] Unfortunately, in the past, the international community has missed two crucial opportunities for cooperation between trade and climate change, namely the WTO Agreement and the Paris Agreement on climate change.

The 1990s have represented a crucial period in the process of building up awareness towards environmental issues and their impacts on the climate at the international level.[238] In 1992, the United Nations Conference on Environment and Development (UNCED), also known as the Earth Summit of Rio de Janeiro, put together 172 countries worldwide, as well as thousands of non-governmental organizations, people from the civil society and other stakeholders.[239] The aim of the conferences was to pursue economic development in ways that would protect the

---

[233] S. A. Aaronson and J. P. Chauffour, "The Wedding of Trade and Human Rights: Marriage of Convenience or Permanent Match?", WTO Research and Analysis, available at: https://www.wto.org/english/res_e/publications_e/wtr11_forum_e/wtr11_15feb11_e.htm.

[234] United Nations Inter-Agency Network on Women and Gender Equality (IANWGE), "Gender Equality & Trade Policy" Resource Paper, 2011, available at: http://www.un.org/womenwatch/feature/trade/gender_equality_and_trade_policy.pdf. For a different opinion: Dine (2005).

[235] Bartels (2013), pp. 297–314.

[236] *Ibid.*

[237] French (2006), p. 103. For an in-depth analysis of the links between trade and climate change, see Leal-Arcas (2013).

[238] Harris (2012), p. 44.

[239] Sax (2000), pp. 2375–2402.

4.4 Capitalizing on RTAS

Earth's environment. Among the official documents that came out as a result of the conferences, there is the United Nations Framework Convention on Climate Change (UNFCCC). Its ultimate objective was to stabilize greenhouse gases (GHGs) concentrations in the atmosphere at a level that had to prevent dangerous human interference with the climate system.[240] For the first time in history, climate change and its anthropocentric causes were discussed at the international level. The awareness of principles such as sustainable development, integration, intergenerational responsibilities, the international dimension of climate change, ocean acidification due to global warming and the need for cooperation between Northern and Southern countries started to gain increasing relevance within the international community, who decided to take action to jointly tackle those issues.[241]

Another crucial international event of the 1990s was the Agreement Establishing the World Trade Organization (WTO), in 1994. Regardless of the awareness raised in Rio on climate change and environmental issues just 2 years earlier, the WTO Agreement merely mentions "sustainable development" and "the environment" in its preamble:

> Recognizing that their relations in the field of trade and economic endeavour should be conducted with a view to raising standards of living, ensuring full employment and a large and steadily growing volume of real income and effective demand, and expanding the production of and trade in goods and services, while allowing for the optimal use of the world's resources in accordance with the objective of sustainable development, seeking both to protect and preserve the environment and to enhance the means for doing so in a manner consistent with their respective needs and concerns at different levels of economic development.[242]

The reasons why the WTO Agreement does not address climate change are many. At that point of history, the world's understanding of such an issue was not mature enough. After all, early human rights agreements did not contain any reference to environmental or climate change concerns either.[243] Also, in 1994, the holistic approach adopted by the recent SDGs was still far from being considered[244] and there were still insufficient evidences on the anthropogenic origin of climate change.

On 12 December 2015 the United Nations Framework Convention on Climate Change (UNFCCC) adopted by consensus the Paris Agreement, which does not

---

[240]United Nations Framework Convention on Climate Change, FCCC/INFORMAL/84 GE.05-62220 (E) 200705.

[241]Sands and Peel (2012).

[242]WTO, "Agreement Establishing the World Trade Organization", 1994, available at: https://www.wto.org/english/docs_e/legal_e/04-wto_e.htm.

[243]Weiss (2011), p. 15.

[244]A. Cutter, "Sustainable Development Goals and Integration: Achieving a better balance between the economic, social and environmental dimensions", Stakeholder Forum, available at: http://www.stakeholderforum.org/fileadmin/files/Balancing%20the%20dimensions%20in%20the%20SDGs%20FINAL.pdf. McEntire et al. (2002), pp. 267–281.

even mention the term 'trade'.[245] The Paris Agreement is certainly an example of a missed opportunity for cooperation between the trade and climate regimes. By then, the factors mentioned above for the WTO Agreement no longer hold true in the 2010s. First of all, the anthropogenic contribution to climate change has been largely demonstrated.[246] Secondly, the international community had already discussed, in September 2015, the Agenda 2030, with which it introduced the SDGs.[247] They consider holistically all the aspects of sustainable development, including trade (SDG 17) and other aspects related to trade, such as consumption, production and waste (SDG 12) and transport (SDG 9). Considering the holistic approach of the SDGs that was adopted just a few months before the Paris Agreement, it seems unclear why the international community in Paris decided not to mention international trade—which is a large contributor to GHG emissions and impacts on the anthropogenic causes of climate change.[248]

4.4.3.4 Balancing Trade Policy and Climate Change for the Achievement of Sustainable Development

This section links trade with climate change and energy security in the context of the decarbisation of the economy and of sustainable development. More than 80% of the global energy supply comes from fossil fuels, such as oil, coal and natural gas.[249] The use of fossil fuels is common everywhere, from rural to urban areas, from developed and developing countries to less developed countries, from home heating and electricity to fuel for private and mass transportation.[250] Although fossil fuels are largely used worldwide, they present several issues.

First of all, they emit greenhouse gases that contribute to global warming and, ultimately, to climate change.[251] Fossil fuels represent also a limited natural resource: their depletion means the end of the resource from the Planet.[252] As they represent the major source of supply of energy, their depletion could also lead to

---

[245]UN Paris Agreement, available at http://unfccc.int/files/essential_background/convention/application/pdf/english_paris_agreement.pdf.

[246]Rosenzweig et al. (2008), pp. 353–357; Houghton et al. (2001).

[247]*Transforming our world: the 2030 Agenda for Sustainable Development* (A/70/L.1) Resolution adopted by the General Assembly on 25 September 2015 http://www.un.org/ga/search/view_doc.asp?symbol=A/RES/70/1&Lang=E.

[248]From a WTO analysis on "The impact of trade opening on climate change": "The general presumption is that trade opening will increase economic activity and hence energy use. Everything else being equal, this increase in the scale of economic activity and energy use will lead to higher levels of greenhouse gas emissions," available: https://www.wto.org/english/tratop_e/envir_e/climate_impact_e.htm.

[249]World Energy Council, 2010 Survey of Energy Resources, 22nd edition, 2010, p. 3.

[250]*Ibid.*

[251]Intergovernmental Panel on Climate Change (IPCC) (2015).

[252]*Ibid.*

4.4 Capitalizing on RTAS

energy insecurity.[253] Energy security concerns the balance of supply and demand,[254] but, at the global level, the demand for energy is rapidly increasing,[255] while fossil fuels are rapidly decreasing.[256] This is why one of the main priorities of every country at the moment is to satisfy their national energy demand and to cooperate at the international level in the energy sector.[257]

These complex issues implies geopolitical decisions and international cooperation to avoid possible diplomacy and even military conflicts, which is the main concern behind energy insecurity. As climate change and energy insecurity are both partially caused by the massive use of non-renewable energy sources, the use and diffusion of renewables represents one of the solution to these increasing global concerns. Switching from fossil fuels to renewables represents one of the priority of the international agenda. Even though the need to reduce global GHGs emissions is not new, there is an increasing urgency for a "transition to a low-carbon, green and resource-efficient global economy to mitigate the risk of dangerous climate change."[258] A step towards this achievement is the Paris Agreement, which indeed aims at holding the increase in global average temperature "well below 2°C above pre-industrial levels and to pursue efforts to limit the temperature increase to 1.5°C above pre-industrial levels."[259]

Such a goal can only be achieved with a profound transition from fossil fuels to a low-carbon economy and trade policy can play a crucial role within this transition. The premise for such a statement is that trade agreements should contain measures to protect the environment, reduce GHGs emissions and tackling climate change, ultimately contributing to the achievement of sustainable development.

A change is indeed necessary because the traditional international instruments to tackle climate change present the following issues. First of all, international climate

---

[253]Cherp and Jewell (2014), pp. 415–421.

[254]R. Hoggett, N. Eyre and M. Keay, Energy demand and energy security, Presentation of the Oxford Institute for Energy Studies, University of Oxford, available at https://www.exeter.ac.uk/energysecurity/documents/ESMW_closing_conf/Energy-efficiency_and_energy_demand-Keay.pdf.

[255]Asif and Muneer (2007), pp. 1388–1413.

[256]Höök and Tang (2013), pp. 797–809.

[257]A. Silva-Calderón, Speech, "International cooperation in the oil & gas sector & the development of energy diplomacy," OPEC Secretary General to the seminar organized by the International Institute of Energy Policy & Diplomacy of MGIMO-University, Ministry of Foreign Affairs of the Russian Federation, 3rd Russian Oil Gas Week, Moscow, Russia, 5th November 2003, available at http://www.opec.org/opec_web/en/911.htm.

[258]Achim Steiner interviewed by Inch on "Debate: Can we Really Decarbonise the Economy?" Paris 2015, available at https://www.ineos.com/inch-magazine/articles/issue-9/debate-can-we-really-decarbonise-the-economy/.

[259]Paris Agreement, Article 2.1(a). Recent research shows that climate researchers have been underestimating the amount of carbon dioxide that is possible to emit to be compatible with the ambitions expressed in the Paris Agreement on Climate Change. In other words, the world may be in a position to emit significantly more CO2 in the next few decades than was previously announced and still be in compliance with the requirements of the Paris Agreement. See Millar et al. (2017).

agreements are legally non-binding. They only contain voluntary commitments, without any enforcement mechanisms. As Geoffrey Palmer noticed, soon after the Earth Summit in 1992, *"Rio did not produce enough binding new principles of international environmental law sufficient to protect the environment against known threats or secure its future."*[260] As a result of the lack of binding rules and legal instruments, climate change is still a very urgent problem nowadays.

Secondly, even when international climate agreements are ratified, a country can always withdraw without any legal or economic repercussion, if not only in terms of reputation at the international level and in front of the public opinion. Thirdly, more than 190 countries were involved in the Paris negotiations, making it difficult to reach consensus on some important issues. Achieving consensus during the negotiations for the adoption of the text of the Paris Agreement on Climate Change came at the expenses of setting more ambitious goals.[261] The differences in the countries' commitments have to be dealt with through other reviewing measures that have been adopted, such as the regular submission of the Intended Nationally Determined Contributions (INDCs). Prioritising universal participation led, for instance, to the consequence that the term "decarbonisation" was not included in the text of the Paris Agreement, because of the opposition of countries, such as Saudi Arabia.[262] Multilateralism in the international scenario is indeed a very difficult effort in terms of diplomacy and political decision-making.

The issues of climate change agreements mentioned above do not belong to trade agreements. Indeed, trade agreements are, first of all, legally binding and enforceable. Secondly, withdrawing from them would mean large economic repercussions for the country that takes such decision. Lastly, the regionalism that has been spreading in recent years in this sector[263] makes the content of the agreements much easier to be agreed upon by the parties simply because regional trade agreements are only signed between a limited number of parties. For all the above reasons, if trade agreements contain specific references to climate change issues, they could create binding provisions for countries to mitigate climate change and, eventually, they can be even more effective than international environmental agreements in the fight against climate change.

Regarding cumulative emissions,[264] almost half of global emissions of greenhouse gases comes from just four parties to the UN Framework Convention on Climate Change (China, the United States, the European Union, and India).[265]

---

[260] http://openscholarship.wustl.edu/cgi/viewcontent.cgi?article=1867&context=law_lawreview.

[261] S. Korwin, Analysis of the Paris Agreement, Climate and Law Policy, January 2016. Available at http://www.climatelawandpolicy.com/en/blog/29-blog-climate-change/102-analysis-paris.html.

[262] European Parliament, Implementing the Paris Agreement – Issues at Stake in View of the COP 22 Climate Change Conference in Marrakesh, IP/A/ENVI/2016-11, pp.73–74.

[263] T. Risse, The Diffusion of Regionalism, Regional Institutions, Regional Governance, Freie Universität Berlin, Paper presented at the EUSA 2015 Conference, Boston, USA, March 5–7, 2015. Leal-Arcas (2011a), pp. 597–629.

[264] Cumulative emissions describe a country's total historic emissions. See https://wri.org/blog/2014/11/6-graphs-explain-world%E2%80%99s-top-10-emitters.

[265] https://wri.org/blog/2014/11/6-graphs-explain-world%E2%80%99s-top-10-emitters.

Therefore, for example, if a regional trade agreement is signed by the major emitters of GHGs and contains legally binding and enforceable rules on the environment, climate change, and sustainable development, this agreement could help the decarbonization of the economy, lead to climate change mitigation, and to energy security.

Ultimately, economic development (from trade), environmental benefits (climate change mitigation and decarbonization), and social benefits (tackling unemployment, inequality, energy poverty) will be reached by including climate change considerations within trade agreements. Those three positive results in terms of environmental, economic, and social aspects are also the three components of sustainable development. Therefore, as depicted in Fig. 4.2, if trade agreements take into account environmental/climate change/sustainable development issues, rather than only focusing on trade itself, trade policy can contribute to the transition to decarbonization of the economy and, ultimately, to sustainable development.

Before moving to an analysis of recent FTAs and their sustainability-related provisions, Table 4.2 shows a list of some recently concluded or currently being negotiated EU FTAs that relate to sustainability. What is important is the implementation and enforceability of these provisions, which will be analysed in the following section.

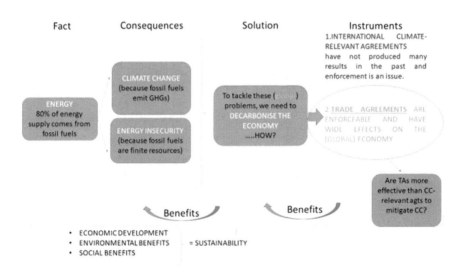

Fig. 4.2 Trade agreements as effective instruments to mitigate climate change (In Fig. 4.2, 'CC' stands for climate change; 'TAs' stands for trade agreements)

Table 4.2 Selected EU free-trade agreements and their sustainability-related provisions

FTA	Year	Topic	ART.	Excerpt
EU-Korea	2011	Environment	Art. 13.5	1. The Parties recognise the value of **international environmental governance** and agreements as a response of the international community to global or regional environmental problem
		Climate change	Art. 13.5	3. The Parties reaffirm their commitment to reaching the ultimate objective of the **United Nations Framework Convention on Climate Change and its Kyoto Protocol**
		Sustainable development	Art. 1.1	2. The objectives of this Agreement are: [...] (g) to commit [...] to the development of international trade in such a way as to **contribute to the objective of sustainable development**
		Energy	Art. 13.6	2. The Parties shall strive to facilitate and promote [...] environmental technologies, **sustainable renewable energy, energy efficient** products and services and eco-labelled goods
		Natural resources	Art. 7.50	2. Nothing in this Chapter shall be construed to prevent the adoption or enforcement by either Party of measures: [...] relating to the conservation of **exhaustible natural resources**
		Environmental goods	Art. 13.6	2. The Parties shall strive to facilitate and promote trade and foreign direct investment in **environmental goods and services**
EU-Japan (JEEPA)	2017	Environment	Art. 16.4	5. Nothing in this Agreement prevents a Party from adopting or maintaining measures to implement the **multilateral environmental agreements**
		Climate change	Art. 16.4	4. The Parties recognise the importance of achieving the ultimate objective of the **United Nations Framework Convention on Climate Change (UNFCCC).** The Parties commit to working together to **take actions to address climate change** towards achieving the purpose of **the Paris Agreement**
		Sustainable development	Art. 16.1	1. The Parties recognise the importance of promoting the development of international trade in a way that contributes to **sustainable**

(continued)

4.4 Capitalizing on RTAS   151

Table 4.2 (continued)

FTA	Year	Topic	ART.	Excerpt
				development, for the welfare of **present and future generations**
		Energy	Art. 16.5	(c) The Parties shall strive to facilitate trade and investment in goods and services of particular relevance for climate change mitigation, such as **sustainable renewable energy** and **energy efficient goods and services**
		Natural resources	Art. 16.12	(h) cooperate on trade-related aspects of the international climate change regime, including on means to **promote low-carbon technologies**
		Environmental goods	Art. 16.5	(b) The Parties shall strive to facilitate and promote trade and investment in **environmental goods and services**
EU-Colombia-Peru	2013	Environment	Art. 270	2. The Parties reaffirm their commitment to effectively **implement in their laws and practices the following multilateral environmental agreements:** [...]
		Climate change	Art. 270	2. ... the **Kyoto Protocol to the United Nations Framework Convention on Climate** ...
		Sustainable development	Art. 270	3. The Trade Committee may recommend the extension of the application of paragraph 2 to other multilateral environmental agreements following a proposal by the Sub-committee on Trade and **Sustainable Development**
		Energy	Art. 275	5. (b) promoting measures for **energy efficiency and renewable energy** that respond to environmental and economic needs and minimise technical obstacles to trade
		Natural resources	Art. 167	1. [...] nothing in this Title and Title V [...] shall be construed to prevent the adoption or enforcement by any Party of measures: [...] (c) relating to the **conservation of living and non-living exhaustible natural resources**
		Environmental goods	Art. 271	2. The Parties shall strive to facilitate and promote trade and foreign direct investment in **environmental goods and services**
EU-Singapore	2018	Environment	Art. 12.6	2. The Parties shall effectively implement in their respective laws, regulations or other measures and practices

(continued)

Table 4.2 (continued)

FTA	Year	Topic	ART.	Excerpt
				in their territories, **the multilateral environmental agreements** to which they are party
		Climate change	Art. 12.6	3. The Parties reaffirm their commitment to reaching the ultimate objective of the **UN Framework Convention on Climate Change** [...] and to support efforts to **develop a post-2020 international climate change** agreement under the UNFCCC
		Sustainable development	Art. 12.1	1. The Parties reaffirm their commitment to developing and promoting international trade and their bilateral trade and economic relationship in such a way as to **contribute to sustainable development**
		Energy	Art. 12.11	2. The Parties shall pay special attention to facilitating the removal of obstacles to trade or investment concerning climate-friendly goods and services, such as **sustainable renewable energy goods** and related services and **energy efficient products and services**
		Natural resources	–	–
		Environmental goods	Art.12.10	The Parties may initiate cooperative activities of mutual benefit in areas including but not limited to: [...] (j) exchange of views on the liberalisation of **environmental goods and services**
TTIP[a]	Under negotiations	Environment	Art. 10[a]	1. The Parties **recognise the value of global environmental governance and rules**, including Multilateral Environmental Agreements, to tackle environmental challenges
		Climate change	Point 1[b]	The Parties acknowledge that the urgent threat of **climate change** requires collective **action for low-emission and climate-resilient development**
		Sustainable development	Art. 1[a]	1. The Parties reaffirm their commitment to **pursue sustainable development**
		Energy	Point 4[b]	[...] the Parties shall: a) facilitate and promote trade and investment in

(continued)

4.4 Capitalizing on RTAS   153

Table 4.2 (continued)

FTA	Year	Topic	ART.	Excerpt
				environmental goods and services, such as **sustainable renewable energy goods** and related services, and **energy efficient goods** and related services
		Natural resources	Art. 11[a]	2. [...] the Parties shall: b) promote conservation and **sustainable use of natural resources** when undertaking trade activities
		Environmental goods	Point 4[b]	[...] the Parties shall: b) effectively implement the WTO Environmental Goods Agreement (EGA) and in this context cooperate to reduce or, as appropriate, eliminate non-tariff barriers related to environmental goods and services. See also above in the energy section

[a]European Union's initial proposal for legal text on "Trade and Sustainable Development" (negotiation round of the 19th–23rd October 2015)
[b]European Union's initial proposal for a legal text on climate that would be included in the "Trade and Sustainable Development" chapter (negotiation round of the 11th–15th July 2016)

4.4.4 Trade Law As the Enforcer of Climate Change Obligations[266]

This section will introduce the contribution that black-letter international trade law may make to the global decarbonization, beyond trade policy's suggested contribution (the greening effect on economy and enhancement of sustainable development previously described). International trade law may be used to help decarbonize the economy by bundling the fate of free trade agreements—hereinafter FTAs—and international climate-change obligations, i.e., the breach of the latter would give rise to legal consequences under the dispute settlement mechanisms of the former (or, in other words, no free-trade-benefits without effective fulfilment of international climate-change-related obligations). The feasibility of this idea will be assessed via the textual analysis of the current wording and content of environmental/climate-change-related elements already present in a sample of relatively recent and/or upcoming FTAs. However, as a complement to pure textual analysis, a brief reference will be made to similar academic endeavours by other authors, which have resorted to either resembling or alternative methodologies. The rather lengthy textual analysis and the brief academic reference will be completed by a synthesis and some conclusions.

---

[266]This section was largely written by Edoard Alvares.

This section offers a textual analysis of environmental or climate-change-related provisions in several recent FTAs and offers a case for bundling environmental and trade agreements.

4.4.4.1 Empirical Assessment: "Environmental" Language and Content of Sample FTAs

The first sub-section is a review of the relevant language and content of a selection of 14 FTAs[267] with climate-relevant wording based on two parameters (i.e., enforceability and substantive content (which can be weak, ambiguous or solid)). When analyzing their environmental and/or climate-change-relevant substantive content and its potential articulation with dispute settlement mechanisms, at least five different configurations can be identified therein. Let us assess them in sequence.

4.4.4.1.1 First Configuration—Enforceability, but Weak Substantive Content: The CPTPP and the ASEAN Agreements with China and India

The first possible configuration to be found within the sample selected is embodied in three different agreements within the sample: (1) the *"Comprehensive and Progressive Agreement for Trans-Pacific Partnership"* (CPTPP), a trade agreement originally signed by twelve Asia-Pacific nations on February the 4th 2016 under the name Trans-Pacific Partnership (TPP)[268]; (2) the *"Framework Agreement on Comprehensive Economic Co-operation between the Association of South East Asian Nations and the People's Republic of China"* (ASEAN-China Agreement), an international agreement which created a free trade area between the members of the ASEAN and China, and which entered into force on January the 1st 2010; and (3) the *"Framework Agreement on Comprehensive Economic Cooperation Between the Republic of India and the Association of Southeast Asian Nations"* (ASEAN-India Agreement); another international agreement which also created a

---

[267]The 14 FTAs are: the CPTPP, ASEAN trade agreements with China and India, the TTIP, US-Korea FTA, US-Peru FTA, US-Colombia FTA, US-Mexico-Canada Agreement (USMCA), EU-Korea FTA, EU-Colombia-Peru FTA, EU-Ukraine FTA, JEEPA, CETA, and EU-Singapore FTA. All these FTAs have been chosen either because they include major GHG emitters or have sustainable-development wording.

[268]As soon as he took office, Presidet Trump of the US withdrew from the Trans-Pacific Partnership (TPP). For further details on the TPP, see https://ustr.gov/tpp/. The Agreement was then re-configured without the US to become the Comprehensive and Progressive Agreement for Trans-Pacific Partnership (CPTPP). For further details about the CPTPP, see https://international.gc.ca/trade-commerce/trade-agreements-accords-commerciaux/agr-acc/cptpp-ptpgp/index.aspx?lang=eng. For the purpose of this section, reference to the TPP and CPTPP are made interchangeably since the relevant provisions in the TPP are identical in the CPTPP.

free trade area between the members of the ASEAN and India, and which equally entered into force on January the 1st 2010.

Interestingly for this chapter's purposes, the CPTPP contains a chapter specifically entitled "environment" (Chapter 20). This chapter envisages in Article 20.23 resort to Dispute Resolution if other means to solve controversies arising therefrom have failed. This includes, notably, resort to Consultations (Article 28.5) and possibly to the establishment of a Panel (Article 28.7.) Following Article 28.12, the function of a Panel *"is to make an objective assessment of the matter before it, [...] and to make the findings, determinations and recommendations as are called for in its terms of reference and necessary for the resolution of the dispute."* The Panel will prepare an initial report (Article 28.17), and, where appropriate, a final report (Article 28.18), whose implementation will be assessed (Article 28.19).

In the event of non-implementation, the non-complying party may either need to provide compensation, or may either face a suspension of the benefits granted under the CPTPP (or, under certain conditions, may have to pay a monetary assessment) (Article 28.20). This regime would, in principle, provide significant chances of international enforceability to the obligations contained in Chapter 20 of the CPTPP.[269] However, while the said chapter does contain an article entitled *"Transition to a Low Emissions and Resilient Economy"* (Article 20.15), whose content is relevant for the purposes of this chapter, the article's language is very weak, making it hard to substantiate any solid claim:

> 1. The Parties **acknowledge** that transition to a low emissions economy requires collective action.
> 2. The Parties **recognise** that each Party's actions to transition to a low emissions economy should reflect domestic circumstances and capabilities and, consistent with Article 20.12 (Cooperation Frameworks), Parties shall **cooperate** to address matters of joint or common interest. [...] Further, the Parties shall, as appropriate, **engage in cooperative and capacity-building activities** related to transitioning to a low emissions economy. [Emphasis added]

Such weak substantive content makes irrelevant the fact that Article 20.15 comes within the scope of the Dispute Settlement scheme provided for in the Treaty.

A similar, albeit slightly different description can be made of the ASEAN-China and the ASEAN-India Agreements, the second and third agreements mentioned above. Since their structure and content are almost identical, they may be dealt with simultaneously: while the respective contents of these agreements are meant to be enforceable under additional dispute-settlement agreements between the parties,[270] their actual environmentally-relevant contents are restrained to simply

---

[269] As pointed out by J. Hillman (See Hillman 2016, p. 223): *"All three options are intended to be temporary. They are applied as an incentive to encourage full compliance and applicable only until full compliance is achieved."*

[270] See, amongst others, Article 2 ("Scope and Coverage") and Article 13 ("Compensation and Suspension of Concessions or Benefits") of the Agreement on Dispute Settlement Mechanism of the Framework Agreement on Comprehensive Economic Co-operation between the People's Republic of China and the Association of Southeast Asian Nations. See also Article 2 ("*Coverage and*

providing for cooperation on environmental issues, as an area of *"economic co-operation"*.[271] This renders the availability of enforcement mechanisms and redress, again, irrelevant.

4.4.4.1.2 Second Configuration—Ambiguous Substantive Content and Non-enforceability: Selected EU FTAs

The second possible configuration that may be found within the sample of treaties is embodied in the following five EU-promoted trade agreements: the *"Free trade Agreement between the European Union and its Member States, of the one part, and the Republic of Korea, of the other part"* (EU-South Korea FTA)[272]; the *"Trade Agreement between the European Union and its Member States, of the one part, and Colombia and Peru, of the other part"* (EU- Colombia-Peru TA)[273]; the *"Association Agreement between the European Union and its Member States, of the one part, and Ukraine, of the other part"*[274]; the *"Comprehensive Economic and Trade Agreement (CETA) between Canada, of the one part, and the European Union and its Member States, of the other part"* (CETA)[275]; and the *"Japan-EU Economic Partnership Agreement"* (JEEPA).[276] Overall, the environmental language and content of the post-2008-financial-crisis EU FTAs seem to be relatively stable (with the notable exception, as it will be seen below, of the EU-Colombia-Peru TA), and, therefore, they will be assessed simultaneously. In terms of substantive content, generally speaking, these agreements tend to contain a chapter on *"Trade*

Application") and Article 16 (*"Compensation and the Suspension of Concessions or Benefits"*) of the Agreement on Dispute Settlement Mechanism under the Framework Agreement on Comprehensive Economic Cooperation between the Republic of India and the Association of Southeast Asian Nations.

[271]See Article 7 ("Other Areas of Economic Co-operation"), second paragraph of the ASEAN-China Agreement: *"2. Co-operation shall be extended to other areas, including, but not limited to, banking, finance, tourism, industrial co-operation, transport, telecommunications, intellectual property rights, small and medium enterprises (SMEs), environment, bio-technology, fishery, forestry and forestry products, mining, energy and sub-regional development."* The first paragraph, which refers to other sectors, merely states that *"1. The Parties agree to strengthen their co-operation in 5 priority sectors [...]."* See, along similar lines, Article 6 of the ASEAN-India Agreement.

[272]O.J. L 127/6, 14.5.2011, signed on 06.10.2010; in force as of 13.12.2015 (http://www.consilium.europa.eu/en/documents-publications/agreements-conventions/agreement/?aid=2010036).

[273]O.J. L 354/3, 21.12.2012, signed on 26.06.2012. Provisionally applied from 01.03.2013 between the EU and Peru; provisionally applied (with the exception of Articles 2, 202(1), 291 and 292) from 01.08. 2013 between the EU and Colombia (http://www.consilium.europa.eu/en/documents-publications/agreements-conventions/agreement/?aid=2011057).

[274]O.J. L 161/3, 29.5.2014, signed on 26.06.2014, in force as of 01.09.2017.

[275]CETA entered into force provisionally on 21 September 2017; it was signed on 30.10.2016.

[276]The Japan-EU Economic Partnership Agreement entered into force on 1 February 2019, creating an open trading area that covers over 600 million people and almost one third of the world's total GDP, making it the largest bilateral trade deal.

and sustainable development" or "*Trade and environment*", within which an article on "*Multilateral environmental agreements*" may be found.[277] The content of the said article has a constant basis, upon which other elements are pinned. The said basis is constituted by the following language:

> 1. The Parties **recognise the value** of international environmental governance and agreements as a response of the international community to global or regional environmental problems.
> 2. The Parties **reaffirm their commitments** to the effective **implementation** in their laws and practices of the multilateral environmental agreements to which they are party.[278]

This common basis is sometimes accompanied by further paragraphs which, despite depicting progressive intentions, are unable to lead to ambitious environmental outcomes. As brief examples, in the Agreements with South Korea and Japan, the parties mention the United Nations Framework Convention on Climate Change and the Kyoto Protocol (in the case of the JEEPA, the Paris Agreement), but only to state that they "*reaffirm their commitment to reaching* [the Agreements'] *ultimate objective*".[279] The Agreement with Ukraine establishes further means through which the parties "*shall cooperate*"[280]; as does CETA, which adds that the parties "*commit to consult*" and "*exchang*[e] *information*". Equally, the first paragraph above is sometimes accompanied by additional elements which remain within the same vague narrative (the parties "*commit to consulting and cooperating*", "*stress the need to enhance*", etc.)[281]

---

[277] Korea: Chapter Thirteen, "*Trade and sustainable development*", Article 13.5; Ukraine: Chapter 13, "*Trade and sustainable development*", Article 292; CETA: Chapter twenty-four, "*Trade and environment*", Article 24.4; JEEPA: Chapter twenty-four, "*Trade and sustainable development*", Article 4.

[278] This "constant basis" has specifically been taken from Article 13.5 of the EU-South Korea FTA.

[279] Compare the Korea and JEEPA wordings. Korea: "*3. The Parties* [...] **commit** *to cooperating on the development of the future international climate change framework in accordance with the Bali Action Plan (85)*".
JEEPA: "[...] *The Parties commit to work together to take actions to address climate change towards achieving the purpose of the Paris Agreement adopted by the Conference of the Parties to the UNFCCC at its 21st session.*" [Emphasis added, original footnotes omitted].

[280] 4. The Parties **shall ensure that** environmental **policy shall be based on** the precautionary principle and on the principles that preventive action should be taken, that environmental damage should as a priority be rectified at source and that the polluter should pay.
5. The Parties **shall cooperate** in order to promote the prudent and rational utilisation of natural resources in accordance with the objective of sustainable development with a view to strengthening the links between the Parties' trade and environmental policies and practices.
[Emphasis added]

[281] Further language in the various additions includes the following terms: "[...] *they* **commit to consulting and cooperating** *as appropriate* [...]"; "[...] *and* **stress the need to enhance** *the mutual supportiveness between trade and environment policies, rules, and measures.*" "*The Parties* **stress the importance** *of multilateral environmental agreements,* [...] *as a means of multilateral environmental governance* [...] *the Parties shall* **exchange views and information** *on trade-related environmental matters of mutual interest for a* [...]." [Emphasis added].

However, beyond these considerations, which would tend to sustain the conclusion that the substantive content of the concerned Agreements is purely and simply weak, a qualification may be made: the second paragraph is ambiguous in its content. The paragraph begins with a typically "weak" wording (*"The Parties reaffirm their commitments"*) but, instead of referring to weak policy objectives, it refers to the *"effective implementation in their laws and practices [...] of the multilateral environmental agreements to which they are party."*[282] It, therefore, could be argued that, despite its beginning, the paragraph actually links, in a solid manner, the trade agreement to the relevant multilateral environmental agreements (any relevant agreement to which they are party, without limitation): in other words, it could be argued that a lack of *"effective implementation in their laws and practices"* of the relevant environmental agreements entails a breach of paragraph 2.

This potential interpretation is further reinforced by the specific wording of the relevant provisions of the EU-Colombia-Peru FTA. The second, third and fourth paragraphs of Article 270 (*"Multilateral Environmental Standards and Agreements"*), included in Title IX (*"Trade and sustainable development"*) of the said FTA, read as follows[283]:

> 2. The Parties **reaffirm their commitment to effectively implement** in their laws and practices the following multilateral environmental agreements: the Montreal Protocol on Substances that Deplete the Ozone Layer adopted on 16 September of 1987, the Basel Convention on the Control of Transboundary Movements of Hazardous Wastes and their Disposal adopted on 22 March 1989, the Stockholm Convention on Persistent Organic Pollutants adopted on 22 May 2001, the Convention on International Trade in Endangered Species of Wild Fauna and Flora signed on 3 March 1973 (hereinafter referred to as "CITES"), the CBD, the Cartagena Protocol on Biosafety to the CBD adopted on 29 January 2000, the Kyoto Protocol to the United Nations Framework Convention on Climate Change adopted on 11 December 1997 (hereinafter referred to as "Kyoto Protocol") and the Rotterdam Convention on the Prior Informed Consent Procedure for Certain Hazardous Chemicals and Pesticides in International Trade adopted on 10 September 1998.
>
> 3. The Trade Committee **may recommend the extension of the application** of paragraph 2 to other multilateral environmental agreements [...].
>
> 4. Nothing in this Agreement shall limit the right of a Party to **adopt or maintain measures to implement** the agreements referred to in paragraph 2. [...][284] [Emphasis added, original footnotes omitted]

---

[282]There are, however, minor variations on the wording. For instance, JEEPA refers to *"related practices"*; and in CETA, each party implements multilateral environmental agreements *"in its whole territory"*.

[283]The wording of the first paragraph, which has been omitted, is very similar to the wording of the equivalent paragraph in the Agreement with Ukraine and in CETA, the relevant verbs used being: *"1. The Parties **recognise** the value [...] **stress the need to enhance** the mutual supportiveness [...and...] shall **dialogue and cooperate** as appropriate [...]."*

[284]Compare this wording of paragraph 4 of Article 270 of the EU-Colombia-Peru FTA with the wording of the fourth paragraph of Article 24.4 CETA: *"4. The Parties **acknowledge** their right to use Article 28.3 (General exceptions) in relation to environmental measures, including those taken pursuant to multilateral environmental agreements to which they are party."*

4.4 Capitalizing on RTAS

Paragraph 2 of the article contains a list with a limited number of precisely identified agreements. Notwithstanding the fact that, at its present state, this provision does not seem very prone to providing coverage to climate-change instruments,[285] it may be argued that it does not contain vague policy objectives and statements, but clear, precise and solid obligations. If this is the case, the above-referred parallel provisions in other EU FTAs may be interpreted accordingly, as being open-ended clauses (hence, not restrained to a limited list of treaties), containing the same kind of clear, precise and solid obligations.

However, in order for this ambiguous possibility to have any practical impact, the said provisions would need to be enforceable under the relevant *"Dispute Settlement"* provisions in the respective EU FTAs, which does not seem to be the case. While, in most cases, the dispute settlement chapters of the selected FTAs are assorted with provisions which state that their scope is general and covers the entirety of the said agreements, these provisions operate only *"unless otherwise provided"*.[286] The environmentally-relevant chapters, therefore, exclude resort to

---

[285] Only one of the instruments listed in paragraph 2 may be characterized as a climate-change instrument. Beyond this, in the EU-Colombia-Peru FTA, Climate change agreements are the object of a separate, notably weak provision, Article 275. Article 275 reads as follows: *"1. Bearing in mind the United Nations Framework Convention on Climate Change [...] and the Kyoto Protocol, the Parties **recognise** that climate change is an issue of common and global concern that calls for the widest possible cooperation by all countries and their participation in an effective and appropriate international response, [...]*

*2. The Parties are **resolved to enhance their efforts** regarding climate change, which are led by developed countries, including through the promotion of domestic policies and suitable international initiatives to mitigate and to adapt to climate change, on the basis of equity and in accordance with their common but differentiated responsibilities and respective capabilities and their social and economic conditions, and taking particularly into account the needs, circumstances, and high vulnerability to the adverse effects of climate change of those Parties which are developing countries.*

*3. The Parties also **recognise** that the effect of climate change can affect their current and further development, and therefore **highlight** the importance of increasing and supporting adaptation efforts, [...]*

*4. Considering the global objective of a rapid transition to low-carbon economies, the Parties will **promote the sustainable use of natural resources** and will promote trade and investment measures that promote and facilitate access, dissemination and use of best available technologies for clean energy production and use, and for mitigation of and adaptation to climate change.*

*5. The Parties **agree to consider** actions to contribute to achieving climate change mitigation and adaptation objectives through their trade and investment policies, inter alia by:*

*(a) **facilitating** the removal of trade and investment barriers to access to, innovation, development, and deployment of goods, services and technologies that can contribute to mitigation or adaptation, taking into account the circumstances of developing countries;*

*(b) **promoting** measures for energy efficiency and renewable energy that respond to environmental and economic needs and minimise technical obstacles to trade."* [Emphasis added]

[286] See, for instance, Article 29.2 (*"Scope"*) of Chapter twenty-nine (*"Dispute settlement"*) of CETA: *"Except as otherwise provided in this Agreement, this Chapter applies to any dispute concerning the interpretation or application of the provisions of this Agreement."*

dispute settlement under various formulations.[287] In general, they only provide for soft and conciliatory mechanisms, such as *"Government consultations"*, or the establishment of *"panels"* or *"groups"* of experts, who draft *"initial"* and, *"final"* reports, on how to solve issues.[288] The parties then *"shall take into account the final report"*, or *"inform the* [relevant entity] *of its intentions as regards the recommendations"*, or *"discuss actions or measures to resolve the matter in question, taking into account the [...] report and suggestions."*[289] The parties' courses of action are often *"monitor*[ed]*"*, but no consequence is provided for.

4.4.4.1.3 Third Configuration—Solid Substantive Content, but Non-enforceability: The EU-Singapore FTA

A third possible configuration is reflected in the text of the *"Free Trade Agreement between the European Union and the Republic of Singapore"* (EU-Singapore FTA.)[290] Chapter thirteen (*"Trade and sustainable development"*) contains in Section C (*"Trade and Sustainable Development – Environmental Aspects"*) an Article on *"Multilateral Environmental Standards and Agreements"* (Article 12.6) which bears a structure which greatly resembles the one described in the analysis of the previous configuration (i.e., *"ambiguous substantive content but weak enforceability"*). However, in this occasion there is no possible ambiguity:

> 2. The Parties shall **effectively implement** in their respective laws, regulations or other measures and practices in their territories, **the multilateral environmental agreements to which they are party**. [Emphasis added; original footnote omitted]

The wording of the second paragraph of this Article is clearly assertive, without any sort of resort to the typical vague language found in weak policy statements. It can hardly be argued that it does not contain clear, precise and solid obligations. *A priori*, its only fault, for the purposes of this chapter, is that it clearly does not cover climate-change obligations, for climate change is the object of the third paragraph of the Article, which is undoubtedly weak in its content[291]:

---

[287]EU-Korea FTA, Article 13.16 (*"Dispute settlement"*): *"For any matter arising under this Chapter, the Parties shall only have recourse to the procedures provided for in Articles 13.14 and 13.15."* (*See* similarly Article 300.7 of the Agreement with Ukraine, and the first paragraph of Article 24.16 of CETA).

[288]See, as examples, Article 13.15 (*"Panel of experts"*) of the EU-Korea FTA; Article 16 of the *"Trade and sustainable development"* chapter of JEEPA.

[289]Respectively, Article 24.15.11 of CETA, Article 285.4 of the EU-Colombia-Peru FTA, and Article 17.6 of the *"Trade and sustainable development"* chapter of JEEPA.

[290]Not yet in force; Not yet signed: See Opinion 2/15 [Opinion of the Court (Full Court) of 16 May 2017 ECLI:EU:C:2017:376]: *"The Free Trade Agreement between the European Union and the Republic of Singapore falls within the exclusive competence of the European Union, with the exception of the following provisions, which fall within a competence shared between the European Union and the Member States: [...]"*.

[291]This third paragraph resembles the third paragraph of the corresponding provisions of JEEPA, and of the EU-Korea FTA.

3. The Parties **reaffirm their commitment** to reaching the ultimate objective of the UN Framework Convention on Climate Change (hereinafter referred to as "UNFCCC"), and of its Kyoto Protocol in a manner consistent with the principles and provisions of the UNFCCC. They **commit to work** together to strengthen the multilateral, rules-based regime under the UNFCCC building on the UNFCCC's agreed decisions, and to support efforts to develop a post-2020 international climate change agreement under the UNFCCC applicable to all parties. [Emphasis added]

In any case, again, as was the case in the second model, for the second paragraph of this provision to have any practical impact, it would need to be enforceable under the relevant "*Dispute Settlement*" chapter. And, again, this does not seem to be the case.[292]

4.4.4.1.4 Fourth Configuration—Enforceability, but Ambiguous Substantive Content: The TTIP

The "*Transatlantic Trade Investment and Partnership*" (TTIP), that the European Union and the United States began negotiating between the 7th and the 12th of July 2013, is at the roots of the fourth configuration detected in the sample of Treaties. Notwithstanding the fact that a significant amount of uncertainty swirls around its future,[293] draft-documents on this envisaged agreement, reflecting different stages in the negotiations, will serve as the basis for the analysis. The first of such documents is the EU's "*initial proposal for legal text on "Trade and Sustainable Development*"[294] drafted to serve during the negotiating round with the United States that took place between the 19th and the 23rd October 2015. Within "*Section III*" (named "*Trade and Sustainable Development – Environmental aspects*"), Article 10 ("*Multilateral environmental governance and rules*") contains the following language:

1. The Parties **recognise** the value of global environmental governance and rules, including Multilateral Environmental Agreements, to tackle environmental challenges of common concern and stress the need to enhance the mutual supportiveness between trade and environment policies, rules and measures.

2. Each Party **reaffirms its commitment** to effectively **implement** in its domestic laws and practices the Multilateral Environmental Agreements to which it is a party.

3. The Parties should **continue to strive** towards further ratification of Multilateral Environmental Agreements and cooperate in this regard, including through **exchanging information** on advancement and **supporting each other**'s full participation in, or membership to, multilateral environmental agreements, international bodies, and processes.

---

[292]See Article 13.16 ("*Government Consultations*"): "*1. In case of disagreement on any matter arising under this Chapter, the Parties **shall only** have recourse to the procedures provided for in Article 13.16 (Government Consultations) and Article 13.17 (Panel of Experts). **Chapter Fifteen (Dispute Settlement)** and Chapter Sixteen (Mediation Mechanism) **do not apply to this Chapter**.*"

[293]Following a meeting report of the "*Transatlantic Trade & Investment Partnership Advisory Group*" dated March the 9th 2017, the negotiations are "*frozen*" (p. 3) [http://trade.ec.europa.eu/doclib/docs/2017/april/tradoc_155484.pdf, retrieved August the 16th 2017.]

[294]To see the textual proposal, visit: http://trade.ec.europa.eu/doclib/docs/2015/november/tradoc_153923.pdf.

4. The Parties **commit to consult and cooperate** with each other as appropriate in Multilateral Environmental Agreements and other global environmental fora, in particular trade-related environmental issues.

5. The Parties **acknowledge** that nothing in the Agreement should prevent either Party from adopting or maintaining measures to implement the Multilateral Environmental Agreements to which it is a party, [...]. [Emphasis added; original footnotes omitted]

In general terms, the wording of Article 10 roughly corresponds with that of equivalent substantive provisions in the EU FTAs that were classified as having an "*ambiguous substantive content*" but lacking enforceability (i.e., the second configuration above). However, what distinguishes this early TTIP document from the above-commented EU FTAs is precisely the fact that, following an EU's "*initial proposal for legal text on "Dispute Settlement (Government to Government)"* the TTIP's relevant provisions would be enforceable. Article 2 ("*Scope of application*") in Section 1 ("*Objective and Scope*") establishes that the dispute settlement chapter applies to "*any dispute concerning the interpretation and application of the provisions of this Agreement, except as otherwise provided.*"

This assessment, made on the basis of a rather early-stage document, may be altered if the content of a second document, stemming from a later round of the negotiations (11–15 July 2016), gets to be inserted into the relevant chapter of the TTIP. The EU's "*initial proposal for a legal text on climate which would be included* [sic] *to the "Trade and Sustainable Development" chapter in TTIP*"[295] contains lengthy language on "*Trade favouring low-emission and climate-resilient development*" and "*Protection of the Ozone Layer and Measures Related to Hydrofluorocarbons*". For the sake of brevity (especially taking into account the above-referred uncertainty surrounding the negotiations), suffice it to say that most of the content of the draft provisions can be easily qualified as "weak", for it reflects the same kind of language as other provisions transcribed in previous pages. However, each of the two unnumbered draft provisions features a paragraph bearing more solid content[296]:

---

[295]To access the text, see http://trade.ec.europa.eu/doclib/docs/2016/july/tradoc_154800.pdf.

[296]Further paragraphs in the draft provisions resorts to the following language:

Trade favouring low-emission and climate-resilient development

1. The Parties **acknowledge** that the urgent threat of climate change requires collective action for low-emission and climate-resilient development.

2. The Parties **recognise** the importance of international rules and agreements in the area of climate change [...]

3. The Parties **agree to promote** the positive contribution of trade to the transition to a low-carbon economy and to climate-resilient development.

4. To this end, the Parties shall:

a) **facilitate and promote** trade and investment in environmental goods and services, such as sustainable renewable energy goods and related services, and energy efficient goods and related services,

[...]

c) **cooperate as appropriate** in relevant international fora[...];

d) **cooperate, exchange information and share experience**, i.a. on:

4.4 Capitalizing on RTAS 163

4. [...] the Parties shall: [...] b) **effectively implement** the WTO Environmental Goods Agreement (EGA) and in this context cooperate to reduce or, as appropriate, eliminate non-tariff barriers related to environmental goods and services; [Emphasis added]

2. [...] each Party **shall take measures** to control the production and consumption of, and trade in, substances within the scope of the Montreal Protocol on Substances that Deplete the Ozone Layer, including any future amendments thereto. [Emphasis added]

Nevertheless, as none of these paragraphs with "solid" content refers to any of the core climate-change agreements, they may only be collaterally relevant, if anything, for the purposes of this section.

The potential insertion of this language into the TTIP would make it unlikely that climate change be covered by the above referred Article 10. This would mutate this fourth configuration, thus making it partially resembling to the first configuration described above (namely, "*strong enforceability, but weak content.*") This fourth configuration would, nevertheless, remain a distinct one, for despite the provision's general weakness, it does contain solid obligations as regards collaterally-relevant agreements, which is not the case regarding its counterpart in the CPTPP.

In any case, irrespective of the "ambiguity" in Article 10, and irrespective of the potential impact of the insertion of the unnumbered provisions just discussed, a major hindrance to the enforceability dimension of the TTIP, which is also shared by the CPTPP, would always remain: putting aside the actual architecture and specific features that a final TTIP may possibly come to display, what is really fundamental is the fact that the United States has not ratified the Kyoto Protocol and is possibly going to withdraw from the Paris Agreement on Climate Change.[297] Consequently,

---

- trade-related aspects of climate action, and means to promote mitigation and adaptation including through low-emission, energy-efficient and climate-resilient policies,

- the development of cost-effective, clean, safe, secure and sustainable low emission technologies, energy-efficient solutions, and renewable energy sources; sustainable transport and sustainable urban infrastructure development; addressing deforestation and forest degradation; emissions monitoring; market and non-market mechanisms.

5. Both Parties will **actively promote** the development of a sustainable and safe lowcarbon economy, such as investment in renewable energies and energy-efficient solutions. The Parties share the goal of progressively phasing out inefficient fossil fuel subsidies that encourage wasteful consumption. Such a phasing out may take into account economic aspects and security of supply considerations and be accompanied by measures to alleviate the social consequences associated with the phasing out.

Protection of the Ozone Layer and Measures Related to Hydrofluorocarbons

1. The Parties **recognise** that emissions of certain substances can significantly deplete and otherwise modify the ozone layer in a manner that is likely to result in adverse effects on human health and the environment. The Parties also **recognize** that hydrofluorocarbons (HFCs) are replacements for the ozone-depleting substances (ODS) and potent greenhouse gases whose emissions are likely to result in adverse effects on the environment and climate.

[...]

3. The Parties shall **cooperate** to address matters of mutual interest related to substances within the scope of the Montreal Protocol [...].

[297] New York Times, "Trump Will Withdraw U.S. From Paris Climate Agreement", June the 1st 2017 (https://www.nytimes.com/2017/06/01/climate/trump-paris-climate-agreement.html?mcubz=1).

there would be no relevant climate-change obligation that could potentially come to be enforced via this trade agreement.

As interesting for this chapter's purposes as they are, these developments on the TTIP negotiations, as it has been already highlighted, do not provide a sufficient degree of certainty. Therefore, discussion will stop at this point, and switch to the last treaty configuration detected in the sample.

4.4.4.1.5 Fifth Configuration—Solid Substantive Content and Enforceability, but Restrictive Scope: Selected US FTAs

The fifth and final configuration found in the sample of treaties chosen can be found in three US international economic agreements: the "*Free Trade Agreement between the United States of America and the Republic of Korea*" (KORUS),[298] the "*United States-Peru Trade Promotion Agreement*" (US-Peru TPA),[299] and the "*United States-Colombia Trade Promotion Agreement*" (US-Colombia TPA).[300] As was the case with the EU FTAs, these agreements bear a great resemblance in their wordings, for the purposes of this chapter. Therefore, the following commentary will focus on the US-Peru TPA, but all observations are applicable, *mutatis mutandis*, to the other two legal instruments. Chapter Eighteen of the US-Peru TPA ("*Environment*") contains an article on "*Environmental Agreements*" (Article 18.2) which reads as follows[301]:

> A Party **shall adopt, maintain, and implement** laws, regulations, and all other measures **to fulfill its obligations** under the multilateral environmental agreements listed in Annex 18.2 ("covered agreements"). [Emphasis added]

The restrictive list of "*covered agreements*" referred to (which is identical in all three treaties) is composed of seven international agreements, amongst which the "*Convention on International Trade in Endangered Species of Wild Fauna and Flora*" (CITES), the "*Montreal Protocol on Substances that Deplete the Ozone Layer*", and the "*Protocol of 1978 Relating to the International Convention for the Prevention of Pollution from Ships, 1973*" (MARPOL), as well as other less

---

[298] Free Trade Agreement between the United States of America and the Republic of Korea, entered into force on March 15, 2012.

[299] Signed on April 12, 2006, entered into force on February 1, 2009.

[300] Entered into force on May 15, 2012.

[301] In each of the treaties discussed, the article is assorted with two footnotes, the first of which will be dealt with below. The second one placed at the very end of the article establishes that: "*For purposes of Article 18.2: (1) "covered agreements" shall encompass those existing or future protocols, amendments, annexes, and adjustments under the relevant agreement to which both Parties are party; and (2) a Party's "obligations" shall be interpreted to reflect, inter alia, existing and future reservations, exemptions, and exceptions applicable to it under the relevant agreement.*"

significant instruments.[302] The relationship between, on the one hand, the US-Peru TPA, and any of the targeted multilateral environmental agreements is further shaped by Article 18.13 ("*Relationship to Environmental Agreements*")[303]:

1. The Parties **recognize** that multilateral environmental agreements to which they are all party, play an important role globally and domestically in protecting the environment and that their respective implementation of these agreements is critical to achieving the environmental objectives thereof. The Parties further **recognize** that this Chapter and the ECA can contribute to realizing the goals of those agreements. [...]

2. To this end, the Parties shall **consult**, as appropriate, with respect to negotiations on environmental issues of mutual interest.

3. Each Party **recognizes the importance** to it of the multilateral environmental agreements to which it is a party.

4. In the event of any inconsistency between a Party's obligations under this Agreement and a covered agreement, the Party shall **seek to balance** its obligations under both agreements, but this shall not preclude the Party from taking a particular measure to comply with its obligations under the covered agreement, [...]. [Emphasis added]

---

[302]In the Peruvian Treaty, Annex 18.2 establishes that, notwithstanding the Parties' prerogative to agree on additions to the list: "[...] *covered agreement means a multilateral environmental agreement listed below to which both Parties are party:*

(a) *the Convention on International Trade in Endangered Species of Wild Fauna and Flora, done at Washington, March 3, 1973, as amended;*
(b) *the Montreal Protocol on Substances that Deplete the Ozone Layer, done at Montreal, September 16, 1987, as adjusted and amended;*
(c) *the Protocol of 1978 Relating to the International Convention for the Prevention of Pollution from Ships, 1973, done at London, February 17, 1978, as amended;*
(d) *the Convention on Wetlands of International Importance Especially as Waterfowl Habitat, done at Ramsar, February 2, 1971, as amended;*
(e) *the Convention on the Conservation of Antarctic Marine Living Resources, done at Canberra, May 20, 1980;*
(f) *the International Convention for the Regulation of Whaling, done at Washington, December 2, 1946; and*
(g) *the Convention for the Establishment of an Inter-American Tropical Tuna Commission, done at Washington, May 31, 1949.*"

[303]Three comments need to be made. First, footnote 11 (placed at the end of paragraph 4 of Article 18.13, reads as follows: "*11 For greater certainty, paragraph 4 is without prejudice to multilateral environmental agreements other than covered agreements.*" Second, "ECA" stands for "*Environmental Cooperation Agreement*", in the case of the Republic of Korea, it is the "*Agreement between the Government of the United States of America and the Government of the Republic of Korea on Environmental Cooperation*", signed January the 23rd 2012. Third, the first paragraph of the Korean version of Article 18.13 ("*Article 20.10: Relation to Multilateral Environmental Agreements*") features some apparently minor changes which, however, could have a deep impact in the scope of the article: "*1. The Parties recognize that* **certain** *multilateral environmental agreements play an important role globally and domestically in protecting the environment. The Parties further recognize that this Chapter and the ECA can contribute to realizing the goals of such agreements. Accordingly, the Parties shall continue to seek means to enhance the mutual supportiveness of multilateral environmental agreements to which they are both party and trade agreements to which they are both party.*" (Emphasis added) (Nota Bene: This version does not feature the reference to "*agreements to which they are all party.*").

Putting aside the fact that Article 18.13 resorts to weak and vague language (possibly in order not to force the preeminence of environmental values over economic ones), the main article discussed, namely Article 18.2, is precise, unconditional and assertive, and neatly establishes clear obligations which, in this case, are actually enforceable under the relevant dispute settlement provisions. In this sense, Article 18.12 (*"Environmental Consultations and Panel Procedure"*) establishes, amongst other aspects,[304] in paragraph 6, that a complaining party may *"as provided in Chapter Twenty-One (Dispute Settlement), thereafter have recourse to the other provisions of that Chapter"*. Specifically, by establishing how *"a panel convened under* [the] *Dispute Settlement* [chapter]" is meant to make its *"findings"*, paragraph 8 of the same article provides guidance as regards the operation of the dispute settlement provisions in respect of Article 18.2 (*"Environmental Agreements"*). Within the provisions on dispute settlement of the US-Peru TPA, Article 21.16 (*"Non-Implementation – Suspension of Benefits"*) establishes that an infringing party may have to provide compensation or may, under certain conditions, see her rights under the Treaty suspended.[305]

However, despite the Agreement's clear enforceability (and solid substantive content), there are, at least, two restrictions in the scope of the relevant provisions, which render this fifth Treaty configuration unfit for the purposes of this chapter. The first one arises from the fact that, as already mentioned, the Treaties feature a restrictive list of *"covered agreements"* which does not contain any climate-change-relevant international instrument. The second one, which is just equally important, arises from a footnote placed immediately after the title of the *"Environmental Agreements"* provision in each of the Treaties. The one in Article 18.2 of the US-Peru TPA reads as follows:

> 1 To establish a violation of Article 18.2 a Party must demonstrate that the other Party has **failed** to adopt, maintain, or implement laws, regulations, or other measures **to fulfill an obligation** under a covered agreement **in a manner affecting trade or investment between the Parties**. [Emphasis added]

This entails that only a specific kind of breach of the *"covered agreements"* may potentially lead to legal consequences under the dispute settlement provisions of the relevant Treaties. As it may be understood, therefore, the potential of this fifth Treaty configuration is severely hindered. While it is still too early to say, it is likely that the recast of the North American Free Trade Agreement (NAFTA), namely the US-Mexico-Canada Agreement (USMCA),[306] whose renegotiation rounds begun during

---

[304]See paras 1-5 of Article 18.12.

[305]In the case of KORUS, Article 22.13 (*"Non-implementation"*); in the case of the US-Colombia TPA, Article 21.16 (*"Non-Implementation – Suspension of Benefits."*).

[306]https://ustr.gov/trade-agreements/free-trade-agreements/united-states-mexico-canada-agreement/agreement-between.

Summer 2017,[307] may end up belonging to this category. Following a *"Summary of Objectives for the NAFTA Renegotiation"* released by the office of the United Stated Trade Representative on July the 17th 2017,[308] the United States intend to insert provisions on environmental issues into the text of the NAFTA.[309] Specifically, the very little guidance provided by the said document possibly hints at the replication of the structural configuration described above, including its restrain to a limited list of *"covered agreements"*.[310]

Having concluded the description of the fifth configuration, let us briefly consider other similar academic accounts which have resorted to other methodologies.

4.4.4.2 Other Academic Accounts

The textual analysis in previous pages needs to be complemented by a brief reference to other academic assessments,[311] which have been performed along similar lines (resembling focuses, either under different or slightly varying methodologies) in recent years. These studies provide a wider context to the referred textual analysis and further enlighten the potential contribution of FTAs to international environmental and climate change-related issues.[312]

A first account to be mentioned, was performed by S. Jinnah & E. Morgera in 2013.[313] It comprised a sophisticated analysis, based on coding, of Environmental provisions in American and EU Free Trade Agreements. This interesting and timely work does not focus, however, on climate-change-relevant provisions, which are only very briefly and generically addressed.[314] Nevertheless, this piece is extremely valuable, inasmuch as it shows (although not explicitly stated) the potential that future climate-relevant provisions could come to bear, if, in policy terms, an

---

[307] See the July 2017 press release on the matter by the office of the United States Trade Representative (https://ustr.gov/about-us/policy-offices/press-office/press-releases/2017/july/ustr-releases-nafta-negotiating).

[308] https://ustr.gov/sites/default/files/files/Press/Releases/NAFTAObjectives.pdf.

[309] See the heading *"Environment"* in pages 13 and 14 of the said Summary of Objectives.

[310] See notably the second and third indents in the *"Environment"* section, which point to *"strong and enforceable environmental obligations"* but limiting the scheme to *"select"* agreements, including CITES.

[311] For scholarship on compensation for climate change loss, see Grossman (2003), p. 1; Hunter and Salzman (2007), p. 1741; Penalver (1998), p. 563; Kysar (2011); Kaminskaite-Salters (2011); Brunnee et al. (2011).

[312] For scholarship on climate change litigation, see Burns and Osofsky (2009); Lin (2012), p. 35; Richard Lord et al. (2011); Peel (2011), p. 15; Wilensky (2015), p. 131; Fisher (2013), p. 236; Peel and Osofsky (2015); Fisher et al. (2017), p. 173. Also, see Vanhala and Hilson (2013), p. 141; Fisher and Scotford (2016), p. 1.

[313] Jinnah and Morgera (2013), pp. 324–339.

[314] See, for instance, pp. 326, 332, 333, 335, and 337.

approach similar to the one taken in respect of generic environmental provisions was adopted.

A second account, which runs along similar lines, and deserves being highlighted is the very recent work performed by J.F. Morin & R. Gauthier Nadeau on what they label as *"environmental gems"* in trade agreements[315]: in their opinion, several very innovative environmentally-relevant provisions found in a rather limited number of trade agreements are good examples of *"best practices"* that should actually be amplified and replicated in future agreements. Most of these "legal one-hit-wonders" are not directly climate-change relevant.[316] Yet, again, they point to the huge potential that a similar approach in climate-change terms would have.

Finally, N.A.J. Croquet has performed an in-depth analysis of "The climate-change norms under the EU-Korea Free Trade Agreement"[317] which rightly points to the need of assessing, in every case, what is the actual value of the climate-change relevant provisions that may be found in trade agreements. When discussing this point as regards the EU-Korea FTA, Croquet, following assessments by A. Boyle, and by K. Abbott and D. Snidal, places this question within the academic soft law-hard law continuum.[318] It is under these coordinates that he asserts, following the referred authors, that *"given that the EU-Korea FTA constitutes hard law vis-à-vis the contracting parties"*, climate change provisions are to be considered from the standpoint of two additional elements: *"their degree of precision"* and *"their enforcement/dispute settlement mechanism"*.[319]

Consequently, Croquet goes on to consider: i) that what we have labelled as *"ambiguous provisions"* in previous pages are actually *"renvoi clauses"*, i.e. *"confirmatory clauses that urge the Contracting parties to live up to their pre-existing climate change treaty commitments"*; ii) that in the EU-Korea FTA case the renvoi clause is unambiguous and reflects an unequivocal obligation to *"effectively implement"* previously assumed climate change-relevant obligations; iii) *"that these preexisting treaty commitments now fall under the Chapter 13 dispute settlement mechanism."*[320] As relevant and enlightening as this account is, it is not possible to fully subscribe it, nevertheless: while it may be possible to agree with ii), as Croquet quotes documentation by the "WTO Committee on Regional Trade Agreements", it is not possible to fully agree with iii). What Croquet labels as "dispute settlement" is a set of provisions within Chapter 13 ("Trade and sustainable development") which is independent from the "Dispute settlement"

---

[315]Morin, J.F. and Gauthier Nadeau, R. "Environmental Gems in Trade Agreements: Little-known Clauses for Progressive Trade Agreements," *CIGI Papers* No. 148, 2017.

[316]There is only a brief reference to *"two Stabilisation and Association Agreements of the European Union* [calling] *for the ratification of the Kyoto Protocol on Climate Change"*). Ibidem, pp. 7–8.

[317]Croquet (2015), pp. 124–157.

[318]*Ibidem* pp. 128–131.

[319]*Ibidem*, p. 131.

[320]All quotations and information are taken from *ibidem*, pp. 131–132.

4.4 Capitalizing on RTAS   169

Chapter (Chapter 14) and which, ultimately, is not accompanied by any sort of remedy.

In the light of all of these considerations, let us synthetize the findings of previous pages.

4.4.4.3 Synthesis: The Case for Bundling Trade and Environmental Agreements

After a textual analysis of 14 FTAs with climate-relevant wording, Fig. 4.3 provides an overview of each of the 14 FTAs and where they stand vis-à-vis enforceability and substantive content.

Empirical research shows that, as the law stands nowadays, there are, at least, four sorts of obstacles to the hypothesis presented in this chapter:

1) Weak and/or vague language of the relevant "primary" obligation (for example, the CPTPP): whenever a given climate-change relevant provision is covered by the dispute settlement mechanism, the actual content of the relevant obligation is weak, and thus rather useless.
2) Lack of coverage by the relevant dispute-settlement provisions (for example, the EU-Singapore FTA): reverse problem; while the content/language of the relevant obligations is strong, the provision is not covered by the dispute settlement mechanism.
3) Severe limitations in the scope of the relevant provisions (for example, the KORUS): whenever both the content/language of the primary obligations is unequivocal and the enforceability dimension thereof is strong, the relevant

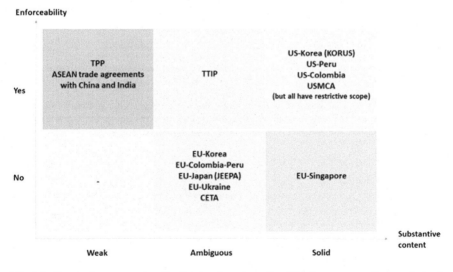

Fig. 4.3 Free-trade agreements with climate-relevant wording (For the purposes of our analysis of enforceability in Fig. 4.3, it makes no difference whether we refer to the TPP or CPTPP)

provision only covers a selected number of environmental agreements and/or is only triggered under certain narrow conditions.
4) Lack of environmental commitment by at least one of the parties (for example, the TTIP): No bundled climate-change enforcement would be possible because the relevant party has not ratified any meaningful international climate-change obligation.

Depending on the reading that may be made of the provisions labelled above as "ambiguous", a fifth point could be added: If the said provisions are to be considered as actually "ambiguous", an additional difficulty would arise from the fact that the precise content of the relevant obligation is not, in principle, easy to ascertain due to the combination of seemingly "weak" and "solid" elements in its wording. If, however, the said provisions are to be understood, following Croquet's analysis, as being actually clear, no addition needs to be made.

So far, there has been limited input of trade agreements to climate change mitigation. FTAs deliver when they are fully implemented and enforced. The enforcement of environmental provisions in EU FTAs seems to be an issue: they do not offer enforceability for climate change-relevant provisions, whereas recent US FTAs do.[321] However, when it comes to substantive content of climate-related provisions in FTAs, where there is solid content, there is either no enforceability (for instance, the EU-Singapore FTA) or the restrictive scope of the specific FTA blocks the potential of enforcement of such provisions (for instance, in US FTAs).[322]

Notwithstanding all of these facts, in normative terms (*de lege ferenda*), an effective way for the European Union to move forward regarding the enforcement of international climate-change-mitigation obligations would be to bundle the fate of climate-change-relevant obligations with that of trade agreements. Since the relevant elements for "bundling" are already disseminated in various FTAs (some feature solid climate-change-linkage language, others feature coverage by the relevant dispute-settlement mechanism), law-makers would need to simultaneously include both sorts of elements in future FTAs to obtain the desired result:

1) In substantive terms, possibly the clearest wording found in the analyzed agreements is the one featured in Article 12.6 of the EU-Singapore FTA:

> 2. The Parties shall **effectively implement** in their respective laws, regulations or other measures and practices in their territories, **the multilateral environmental agreements to which they are party**. [Emphasis added; original footnote omitted]

---

[321] See *Free Trade Agreement between the United States of America and the Republic of Korea* (KORUS; entered into force on March the 15th 2012), the *United States-Peru Trade Promotion Agreement* (US-Peru TPA; entered into force on February the 1st 2009), and the *United States-Colombia Trade Promotion Agreement* (US-Colombia TPA; entered into force on May the 15th 2012).

[322] In fact, in June 2018, "a federal judge in California ruled that climate change was a matter for Congress and diplomacy, not judges." The Economist, "Crude awakening," 9 February 2019, p. 9. This view shows the difficulty of enforcing climate change matters in courts.

It is not a closed or limited list of agreements, but a general clause, and leaves no doubt regarding the kind and extent of obligation being assumed.

2) In enforceability terms, it will suffice to follow the example of the CPTPP and the US FTAs: in other words, the relevant provision needs to be covered by the Dispute Settlement Chapter of the relevant agreement, which needs to ultimately provide access to remedies such as compensation and suspension of benefits in case of "non-implementation" of panel reports. In fact, empirically, EU FTAs have a higher number of climate change-related provisions than US FTAs. However, quantity does not correlate with quality of climate change-related provisions because these cannot be enforced.

One would think that environmental agreements, and not trade agreements, would be the solution to environmental protection. However, they lack enforceability. A case in point is the Paris Climate Agreement, which does not have a dispute settlement system, unlike the WTO. Moreover, US FTAs seem to be more innovative and effective legal instruments than EU FTAs when it comes to the enforceability of their environmental provisions. This phenomenon is due to the fact that the environmental-protection provisions in EU FTAs are not enforceable. A solution would be to make sure that the relevant environmental-protection provisions are covered by the dispute settlement chapter of a given EU FTA.

4.4.5 Towards the Drafting of Twenty-First-Century Trade Agreements[323]

4.4.5.1 Introduction

This section offers options for the drafting of future trade agreements with the aim to help mitigate climate change and enhance sustainable energy. It focuses on some legal and political-economy suggestions within the trade sector that can promote sustainable energy and mitigate climate change. Considering trade and climate-related issues jointly will be at the core of these proposals. Such proposals aim at moving forward from the twentieth-century trade agreements, suggesting new approaches that suit the evolving economy and society of the twenty-first-century. They can be used as a platform for the creation of "new generation" trade agreements, which are more aligned with the twenty-first-century challenges. Climate change-related issues cannot be ignored in this century, just as the protection of human rights and poverty reduction could not be ignored in the previous century. This fact is reflected in trade to such an extent that, in some cases, trade has also contributed to tackle issues related to human rights protection and poverty reduction.

---

[323]This section was written by Alesandra Salaza.

In this section, an analysis of the state of the art in the European context will show some of the instruments that have already been adopted to include climate-related issues in trade: from some of the early-stage trade agreements to some of the most recent EU actions. We offer a few proposals as possible tools to merge trade and climate-related concerns and to function as a basis for modern twenty-first-century trade agreements.

4.4.5.2 The EU Context: From Early-Stage Trade Agreements to Today's Actions

The European Union remains committed to promoting sustainability, decarbonizing the economy, and tackling climate change-related issues. The EU has the potential to be the leader in trade and climate change-related issues jointly.

Reference to climate change-related concerns can be found in the GATT 1947. Article XX(b) and (g) contains exceptions to the general application of the GATT rules. More specifically:

> [N]othing in this Agreement shall be construed to prevent the adoption or enforcement by any contracting party of measures:
> (b) necessary to protect human, animal or plant life or health;
> (g) relating to the conservation of exhaustible natural resources if such measures are made effective in conjunction with restrictions on domestic production or consumption.[324]

In the evolution of trade agreements concerning climate-related issues at the EU level, the introduction of chapters entirely dedicated to sustainable development was firstly adopted in 2008. The first agreement containing such a chapter was the 2008 EU-Cariforum Economic Partnership Agreement.[325] This agreement offers the possibility of establishing a Consultative Committee to "encompass all economic, social and environmental aspects of the relations between the [parties], as they arise in the context of the implementation of this Agreement" (Article 232(1)).

After that Agreement, the importance of considering trade and the environment jointly has been recognized in many documents. In 2010, the European Parliament Resolution on Human Rights and Social and Environmental Standards in International Trade Agreements stated that: "in the context of free trade agreements, conditional liberalisations, including shortening the timetable for abolishing restrictions or access to an additional market, could be envisaged when environmental and social standards are complied with".[326] Furthermore, it calls to consider trade not only as an end, but also as a tool for promoting environmental standards (paragraph 1) as well as other values. It also calls for a 'new-generation' of free trade agreements

---

[324]General Agreement on Tariffs and Trade (GATT), 1947, Article XX (b) and (g).

[325]EU-Cariforum Economic Partnership Agreement, 30.10.2008, L 289/II/1941.

[326]European Parliament resolution of 25 November 2010 on human rights and social and environmental standards in international trade agreements (2009/2219(INI), paragraph 18).

4.4 Capitalizing on RTAS   173

Table 4.3 Tools for action in the EU

Area of action	Description
Bilateral trade agreements	Trade agreements contribute to climate action by: – committing to the implementation of international climate change conventions – providing rules for trade and investment in environmental goods and services – removing of non-tariff barriers in the case of renewable energy
Environment	Studies on trade and environment in collaboration with the United Nations Environment Programme (UNEP)
Waste	Controls in trade in hazardous and non-hazardous waste are part of the EU waste policy
Fishing	Imports of fish from illegal, unreported and unregulated fishing is prevent into the borders of the EU countries, together with trade in such products
Transport	Working with the International Maritime Organization and International Civil Aviation Organization to address the issue of GHGs emission from shipping and air transport
Generalised scheme of Preferences+ (GSP+)	Developing countries can access the EU market by ratifying and implementing some multilateral environmental agreements
Forest Law Enforcement Governance (FLEG)	Bilateral trade agreements regarding the export to the EU of legally-harvested timber

(paragraph 12) and for a systematic inclusion of chapters on sustainable development in the application of EU trade agreements (paragraph 13).

More recently, an increased awareness of the importance of linking trade with climate change-related issues can be found at the European level. To summarise some of the steps that the EU has taken in this sector, a list of tools can be found in Table 4.3 below.

The following section will present some proposals to start considering trade and climate change jointly and to initiate the change towards a low-carbon economy.

4.4.5.3 Proposals for New-Generation Trade Agreements

The introduction of chapters in FTAs focusing on sustainable development is not enough a tool to actually consider climate change-relevant issues and trade matters jointly. It is crucial to start reflecting on possible ways to effectively bring trade and climate change together. Below are some proposals that may help reach the goal of considering trade and climate change-related issues jointly. The first example derives from human rights developments because, as has been demonstrated above, human rights have received some benefits from trade and have a longer history than climate change-relevant concerns. Some of the principles and measures used for human rights in trade could be applied to the environment and climate in trade agreements.

The first proposal is inspired by the Canada-Colombia Free Trade Agreement (CCOFTA, 2008).[327] This agreement features a separate side-agreement which was signed on the 27th of May 2010, namely the Agreement concerning Annual Reports on Human Rights and Free Trade between Canada and the Republic of Colombia.[328] This Agreement requires that both parties produce annual public reports on how the actions adopted under the CCOFTA impact on human rights in both countries. The side-agreement contains very basic rules and it is made of just five articles, but it "is particularly important because it represents the first attempt by governments to undertake this kind of analysis."[329] The Agreement has been developed on the basis of the Guiding Principles on Human Rights Impact Assessments of Trade and Investment Agreements, adopted under the auspices of the United Nations.[330] Such principles expressly distinguish the human rights impact assessments "from other impact assessments (such as social, environmental or sustainability impact assessments)."

These principles could be an inspiration for other sectors and similar guiding principles could be adopted to act as a model for climate-relevant side-agreements of future international trade agreements. This could be a measure to consider concerns about the environment, climate change, and sustainability in a trade agreement. Considerations about climate change are already starting to appear in the annual reports on human rights adopted, for example, by Canada in 2015: "Canada and Colombia share important values and priorities: fighting climate change."[331] If, similarly to the human rights sector, an annual report on climate-relevant concerns is required, it could achieve the joint consideration of climate concerns together with trade.

Another proposal is to harness effectively international trade approaches that encourage and support the path to a low-carbon economy. This is possible by incorporating trade elements into countries' climate change contributions, namely the INDCs. In the text of the Paris Agreement on Climate Change, there is no single reference to trade. Nonetheless, there is still the opportunity to include trade policy through the Parties' commitments. The communication of the INDCs is a

---

[327]For the full text of the Agreement, see http://international.gc.ca/trade-commerce/trade-agreements-accords-commerciaux/agr-acc/colombia-colombie/fta-ale/index.aspx?lang=eng.

[328]Agreement Concerning Annual Reports on Human Rights and Free Trade between Canada and the Republic of Colombia, E105278, 27th of May 2010, available at http://www.treaty-accord.gc.ca/text-texte.aspx?id=105278.

[329]James Harrison, cited by United Nations Human rights Office of the High Commissioner, Human rights in the trade arena, 25 October 2011, available at http://www.ohchr.org/EN/NewsEvents/Pages/HRInTheTradeArena.aspx.

[330]Report of the Special Rapporteur on the right to food, Olivier De Schutter, Guiding principles on human rights impact assessments of trade and investment agreements, A/HRC/19/59/Add.5, UN General Assembly, 19 December 2011, commentary point 2.1, p. 6.

[331]Annual Report Pursuant to the Agreement concerning Annual Reports on Human Rights and Free Trade between Canada and the Republic of Colombia for the Period January 1, 2015 to December 31, 2015.

requirement by Article 4 of the Paris Agreement[332] and they represent the commitments of the Parties to the Paris Agreement towards the achievement of the objective of the Paris Agreement reflected in Article 2 (namely to hold the increase in global average temperature to "well below 2°C above pre-industrial levels" and to pursue efforts "to limit the temperature increase to 1.5°C above pre-industrial levels").[333] Most Parties[334] have already submitted their INDCs, which became their nationally determined contributions (NDC) upon ratification of the Paris Agreement.

The Parties are expected to implement their NDCs, starting from 2020. These contributions are a crucial element of the Paris Agreement. The ambitions of the climate targets and the actions communicated in the NDCs, together with their implementation, decide whether the Parties to the UNFCCC will be able to achieve the goal of the Agreement. Although guidelines are available for countries on what the INDCs could address, the advantages of using trade opportunities should be stressed in greater detail.[335] As of late 2017, 45% of the INDCs that were submitted contained some reference to trade. Only around 22% include trade measures that specifically address climate change mitigation.[336] There are still integral opportunities to see this percentage grow in the future. In future, the Parties to the Paris Agreement will assess the guidelines for updated and more ambitious NDCs, which will be revised every 5 years. That will be a good opportunity to incorporate trade measures for climate change mitigation in the guidelines and consequently have new NDCs that expressly address such aspects jointly.

4.5 Conclusion

Based on empirical evidence, it seems that trade agreements are stricter on environmental protection (see, for instance, the Comprehensive and Progressive Agreement for Trans-Pacific Partnership's (CPTPP) chapter on environment in relation to a low-emissions economy) than climate change agreements such as the Paris Climate

---

[332] Article 4, paragraph 2 of the Paris Agreement: "Each Party shall prepare, communicate and maintain successive nationally determined contributions that it intends to achieve. Parties shall pursue domestic mitigation measures, with the aim of achieving the objectives of such contributions."

[333] Article 2.1(a) of the Paris Agreement.

[334] 181 Parties had submitted their first NDCs as of 24 December 2018. Data from the UNCFFF Interim NDC Registry, available at: http://www4.unfccc.int/ndcregistry/Pages/Home.aspx.

[335] C. Brandi, "Strengthening climate-friendly trade elements in future nationally determined contributions," Opinion, ICTSD, 10 May 2017, available at https://www.ictsd.org/opinion/strengthening-climate-friendly-trade-elements-in-future-nationally-determined.

[336] See the German Development Institute database "NDCs Explorer" available at http://klimalog.die-gdi.de/ndc/#NDCExplorer/worldMap?INDC??climatechangemitigation???cat14.

Agreement. That the CPTPP is enforceable on the reduction of GHG emissions,[337] whereas the Paris Climate Agreement is not,[338] enhances those stricter provisions.

Trade policy has never been under more scrutiny than it is today. It has become heavily politicized and values-based.[339] Without trade policy or trade agreements, one cannot bring forward values such as sustainability, which are increasingly becoming part of trade agreements. Today, more than ever, trade must be based on sustainable values.

There are anti-globalization movements rising in the West against large free-trade agreements. Issues regarding transparency in the trade context are important for two main reasons: (1) political accountability and (2) they show citizens the results of trade agreements. Policymakers should explain to civil society the benefits of trade. More inclusiveness will be necessary to avoid the current lack of trust among citizens in the context of trade. Experience has shown the importance of working with citizens, involving them in trade policymaking, and taking their advice to make sure that policymakers do not lower environmental or labor standards in the conclusion of trade agreements. For future FTAs, one could imagine high standards in major FTAs and then other countries could follow to make a global standard.

The aim of the international community should be to have open, predictable, and equitable trade relations among countries. Traditionally, in-depth free-trade agreements have been about free movement of goods, services, capital, and people. Although it is now starting to happen, one should add more assertively environmental protection to such agreements. One could propose environmental conditionality in FTAs, where the condition for countries to be Parties to such FTAs would be to be legally bound to complying with international environmental agreements. This is how the UK abolished slavery: if it wanted to be a Party to the Congress of Vienna of 1815, it had to put an end to slave trade. The trading system needs to do more to work for the environment, resilience, prosperity, and sustainability. Very little is still known on the interaction between trade agreements and climate action and environmental green goods in terms of production, consumption, and transport. Yet trade is increasingly becoming an important element in the intersection with climate action.

Sustainable development is a WTO objective since the creation of the WTO. There is a policy dialogue at the WTO between the trade and environment epistemic communities via its Committee on Trade and Environment. Similarly, there is an increasing interest in inserting climate-related provisions in the WTO context and in FTAs. Reference to climate change, the UNFCCC, the Kyoto Protocol, and the Paris Agreement in FTAs in increasing.

It is a fact that free-trade agreements typically take years to agree on. Nevertheless, in the early 1990s, both the US and the EU concluded innovative

---

[337]CPTPP, Art. 20.15, ¶ 2.

[338]Paris Agreement, Art. 6(4).

[339]Gone are the days when trade policy was solely technocratic. For instance, when the GATT 1947 was established. Some years ago, trade policy was about tariffs. Today trade is about standards and values. But since the Trump administration, it seems to be moving back to tariffs.

4.5 Conclusion

FTAs regarding environmental protection (and over the years both actors seem to be converging on a shared set of environmental norms[340]), whereas recently concluded or currently being negotiated FTAs (such as CETA or the CPTPP) are less innovative because they copy good ideas from previous FTAs. One also infers from the empirical analysis in this chapter that there is correlation between democracy and the inclusion of environmental-protection provisions in FTAs: democratic states include more environmental-protection provisions in their FTAs than non-democracies. The empirical analysis in this chapter has also shown that there are still limitations to how trade agreements can contribute to climate change mitigation. This denotes that there is still a major gap between the trade and climate communities to answer questions such as how can one reduce GHG emissions from energy-intensive traded goods without economic loss?

We are already witnessing a progressive trend towards using trade agreements as legal instruments to protect the environment: while CETA and JEEPA face non-enforceability and ambiguous content when it comes to environmental-protection provisions (see Fig. 4.3 above), the EU-Singapore FTA only suffers from non-enforceability, but offers a solid content in its environmental-protection provisions. This lack of enforceability of environmental law is very clear in multilateral environmental agreements, which are typically governed by mediation or conciliation, such as the Vienna Convention on the Protection of the Ozone Layer.

Being creative in legislative terms and accepting the fact that most international trade rules were written for a twentieth-century reality, with very limited mention of climate change because the science of climate change was weak, should be enough reasons to be optimistic that we may soon see trade agreements that address twenty-first-century challenges such as climate change and clean energy technologies. In fact, evidence of this trend is the fact that recent bilateral and regional trade agreements cover, albeit in a light manner, climate-related issues by inserting chapters in such agreements that refer to the links between trade and climate change or chapters on sustainable development and environmental protection. Therefore, as time progresses and the science of climate change becomes more robust, there is a great chance that future free trade agreements may incorporate strong chapters on climate-related issues. The EU is perfectly placed to exploit this nexus within the realm of its external relations.

Finally, a 'peace clause' for climate action in future (EU) FTAs could be a way forward in finding supportive avenues between trade and climate action. The trading system could also be more accommodating to climate objectives if countries reduced tariffs[341] for environmental goods and services and removed or reduced fossil fuel subsidies. Such actions would contribute greatly to GHG emissions reduction. In any event, environmental protection should not be lowered in the name of trade

---

[340]See Morin and Rochette (2017), pp. 621–658.

[341]It is curious to see that tariffs are back in the trade field. Tariffs were the norm during the Uruguay Round (1986–1994). The international trade regime then moved to sustainable development, but is now back to tariffs since 2017.

liberalization or investment promotion. This pressure could come from consumers from the bottom up.

References

Abbott K, Snidal D (2000) Hard and soft law in international governance. Int Organ 54(3):421–456
Anthony VanDuzer J (2016) Sustainable development provisions in international trade treaties: what lessons for international investment agreements? In: Hindelang S, Krajewski M (eds) Shifting paradigms in international investment law: more balanced, less isolated, increasingly diversified. Oxford University Press, Oxford
Asif M, Muneer T (2007) Energy supply, its demand and security issues for developed and emerging economies. Renew Sustain Energy Rev 11(7):1388–1413
Baier S et al (2008) Do economic integration agreements actually work? Issues in understanding the causes and consequences of the growth of regionalism. World Economy 31:461–497
Bannister GJ, Thugge K (2001) International trade and poverty alleviation. Finance Dev, IMF 38(4)
Barnekow SE, Kulkarni KG (2017) Why regionalism? A look at the costs and benefits of regional trade agreements in Africa. Global Bus Rev 18(1):99–117
Bartels L (2013) "Human rights and sustainable development obligations in EU free trade agreements", Legal issues of economic integration 40, no. 4. Kluwer Law International BV, Leiden, pp 297–314
Berger A et al (2017) "Towards "Greener" trade? Tracking environmental provisions in the preferential trade agreements of emerging countries," Discussion Paper 2/2017. Deutsches Institute fuer Entwicklungspolitik, Tulpenfeld
Bhagwati J (1992) Regionalism versus multilateralism. World Economy 15(5):535–556
Boyle A (2014) Soft law in international law-making. In: Evans M (ed) International Law, 4th edn. Oxford University Press, Oxford, p 118 ff
Brewer T (2015) "Arctic Black carbon from shipping: a club approach to climate and trade governance," Issue Paper No. 4, Global Economic Policy and Institutions Series. International Centre for Trade and Sustainable Development, Geneva
Brunnee J et al (2011) Overview of legal issues relevant to climate change. In: Richard Lord QC et al (eds) Climate change liability: transnational law and practice. Cambridge University Press, Cambridge
Buchanan J (1965) An economic theory of clubs. Economica 32(125):1–14
Burns WCG, Osofsky HM (eds) (2009) Adjudicating climate change. Cambridge University Press, Cambridge
Cassimatis A (2007) Human rights related trade measures under international law. Koninklijke Brill NV, Leiden
Castilho M, Menéndez M, Sztulman A (2012) Trade liberalization, inequality, and poverty in Brazilian States. World Dev 40(4):821–835
Cherp A, Jewell J (2014) The concept of energy security: beyond the four as. Energy Policy 75:415–421
Cornia G (ed) (2004) Inequality, growth, and poverty in an era of liberalization and globalization. Oxford University Press, Oxford
Cottier T, Pauwelyn J, Burgi E (eds) (2005) Human rights and international trade law. Oxford University Press, Oxford
Croquet NAJ (2015) The climate-change norms under the EU-Korea free trade agreement: between soft and hard law. In: Wouters J, Marx A, Geraerts D, Natens B (eds) Global Governance through trade EU policies and approaches (Leuven Global Governance series). Edward Elgar Publishing, Cheltenham, pp 124–157

Das R (2011) Regional trade-FDI-poverty alleviation linkages – some analytical and empirical explorations. In: Volz U (ed) Regional integration, economic development and global governance. Edward Elgar Publishing, Cheltenham

Dellink R et al (2017) "International trade consequences of climate change," OECD Trade and Environment Working Papers, 2017/01. OECD Publishing, Paris

Dine J (2005) "Companies, international trade, and human rights", Cambridge studies in corporate law. Cambridge University Press, Cambridge

Fisher E (2013) Climate change litigation, obsession and expertise: reflecting on the scholarly response to Massachusetts v. EPA. Law Policy 35:236

Fisher E, Scotford E (2016) Climate change adjudication: the need to foster legal capacity: an editorial comment. J Environ Law 28:1

Fisher E, Scotford E, Barritt E (2017) The legally disruptive nature of climate change. Modern Law Rev 80:173

French D (2006) Supporting the principle of integration in the furtherance of sustainable development: a sideways glance. Environ Law Manag 18(3):103

Friedrich J, Ge M, Pickens A (2017) This interactive chart explains World's Top 10 Emitters, and How They've Changed. World Resources Institute, Washington, D.C.

Fujiwara N, Egenhofer C (2007) Do regional integration approaches hold lessons for climate change regime formation? The case of differentiated integration in Europe'. In: Carraro C, Egenhofer C (eds) Climate and trade policy: bottom-up approaches towards global agreement. Edward Elgar, Cheltenham

Gallagher K (2017) Trade, investment, and climate policy: the need for coherence. In: Gallagher K (ed) Trade in balance: reconciling trade and climate policy. Boston University, Boston, p 7

Gammage C (2014) Protecting human rights in the context of free trade? The case of the SADC group economic partnership agreement. Eur Law J 20(6):779–792

Ge M et al (2014) 6 graphs explain the world's top 10 emitters. World Resources Institute, Washington, D.C.

Grossman DA (2003) Warming up to a not-so-radical idea: tort-based climate change litigation. Columbia J Environ Law 28:1

Grossman GM, Krueger AB (1993) Environmental impact of a North American free trade agreement. In: Garber PM (ed) Mexico-U.S. free trade agreement. MIT Press, Cambridge, MA

Harris F (2012) Global environmental issues. John Wiley & Sons, Oxford, p 44

Hildreth V (2014) Renewable energy subsidies and the GATT. Chicago J Int Law 14(2):702–730

Hillman J (2016) Dispute settlement mechanism. In: Cimino-Isaacs C, Schoot JJ (eds) Trans-pacific partnership: an assessment. Peterson Institute for International Economics, Washington, D.C., p 223

Höök M, Tang X (2013) Depletion of fossil fuels and anthropogenic climate change—A review. Energy Policy 52:797–809

Houghton JT et al (2001) Climate change 2001: the scientific basis. Cambridge University Press, Cambridge

Hovi J et al (2016) Climate change mitigation: a role for climate clubs? Palgrave Commun 2(1):16020

Hunter D, Salzman J (2007) Negligence in the air: the duty of care in climate change litigation. Univ Pa Law Rev 155:1741

Intergovernmental Panel on Climate Change (IPCC) (2015) Climate change 2014: mitigation of climate change. Cambridge University Press, Cambridge, p 26

Irwin D (2017) Clashing over commerce: a history of US trade policy. University of Chicago Press, Chicago

Jackson T (2011) Prosperity without growth: economics for a finite planet. Earthscan Publications, London

Jinnah S, Morgera E (2013) Environmental provisions in American and EU free trade agreements: a preliminary comparison and research agenda. Rev Eur Comp Int Environ Law 22(13):324–339

Kaminskaite-Salters G (2011) Climate change litigation in the UK: it's feasibility and prospects. In: Faure M, Peeters M (eds) Climate change liability. Edward Elgar Publishing Limited, Cheltenham

Kennedy P (2017) The rise and fall of the great powers: economic change and military conflict from 1500-2000. William Collins, London

Kysar D (2011) What climate change can do about tort law. Environ Law 41:1

Lang A (2011) World trade law after neoliberalism: reimagining the global economic order. Oxford University Press, Oxford

Leal W, Leal-Arcas R (eds) (2018) University initiatives in climate change mitigation and adaptation, chapter 21. Springer, Berlin

Leal-Arcas R (2007) Theories of supranationalism in the EU. J Law Soc 8:88

Leal-Arcas R (2008a) Theory and practice of EC external trade law and policy. Cameron May, London, pp 486–487

Leal-Arcas R (2008b) The GATS in the Doha Round: a European perspective. In: Kern A, Andenas M (eds) The World Trade Organization and trade in services. Brill, Leiden, p 28

Leal-Arcas R (2010) International trade and investment law: multilateral, regional and bilateral governance. Edward Elgar Publishing, Cheltenham, p 39

Leal-Arcas R (2011a) Proliferation of regional trade agreements: complementing or supplanting multilateralism? Chicago J Int Law 11:597

Leal-Arcas R (2011b) Top-down versus bottom-up approaches for climate change negotiations: an analysis. IUP J Governance Public Policy 6:7–52

Leal-Arcas R (2011c) Alternative architecture for climate change: major economies. Eur J Legal Stud 4(1):25–56

Leal-Arcas R (2013) Climate change and international trade. Edward Elgar, Cheltenham, pp 291–293

Leal-Arcas R (2015a) How governing international trade in energy can enhance EU energy security. Renew Energy Law Policy Rev 6(3):203

Leal-Arcas R (2015b) Mega-regionals and sustainable development: the transatlantic trade and investment partnership and the trans-Pacific partnership. Renew Energy Law Policy Rev 6 (4):248–264

Leal-Arcas R (2016) Energy security, trade and the EU: regional and international perspectives. Edward Elgar, Cheltenham

Leal-Arcas R (2017) Sustainability, common concern and public goods. George Washington Int Law Rev 49(4):801–877

Leal-Arcas R (2018) Small is beautiful: why a club approach is the way to go in climate change mitigation. In: Leal W, Surroop D (eds) The Nexus: energy, environment and climate change. Springer, Berlin, pp 305–317

Leal-Arcas R, Alvarez Armas E (2018) The climate-energy-trade nexus in EU external relations. In: Minas S, Ntousas V (eds) EU climate diplomacy: politics, law and negotiations. Routledge, Abingdon, pp 153–168

Leal-Arcas R, Filis A (2014) Legal aspects of the promotion of renewable energy within the EU and in relation to the EU's obligations in the WTO. Renew Energy Law Policy Rev 5:3–25

Leal-Arcas R, Filis A (2015) Certain legal aspects of the multilateral trade system and the promotion of renewable energy. In: Lim CL, Mercurio B (eds) International economic law after the global crisis: a tale of fragmented disciplines, vol 482. Cambridge University Press, Cambridge

Leal-Arcas R, Minas S (2016) Mapping the international and European governance of renewable energy. Oxford Yearb Eur Law 35(1):621–666

Leal-Arcas R, Filis A, Gosh ESA (2014) International energy governance: selected legal issues. Edward Elgar Publishing, Cheltenham, pp 489–494

Leal-Arcas R, Caruso V, Leupuscek R (2015) Renewables, preferential trade agreements and EU energy security. Laws 4:472

Lester S (2016) Is the Doha round over? The WTO's negotiating agenda for 2016 and beyond. Cato Institute, Washington, D.C.

Lin J (2012) Climate change and the courts. Legal Stud 32:35
Lindsey B, Teles S (2017) The captured economy: how the powerful enrich themselves, slow down growth, and increase inequality. Oxford University Press, Oxford
McEntire DA, Fuller C, Johnston CW, Weber R (2002) A comparison of disaster paradigms: the search for a holistic policy guide. Public Adm 62(3):267–281
Mensah TA (2008) Soft law: a fresh look at an old mechanism. Environ Policy Law 38(1/2):50–56
Millar R et al (2017) Emission budgets and pathways consistent with limiting warming to 1.5 °C. Nat Geosci 10(10):741
Morin J-F, Rochette M (2017) Transatlantic convergence of preferential trade agreements environmental clauses. Bus Polit 19(4):621–658
Munda G (2005) 'Measuring sustainability': a multi-criterion framework. Environ Dev Sustain 7:117–134
Norberg J (2016) Progress: ten reasons to look forward to the future. Oneworld Publications, London, p 120
Nordhaus W (2015) Climate clubs: overcoming free-riding in international climate policy. Am Econ Rev 105(4):1339–1370
Nussbaum MC (1997) Human rights and human capabilities. Fordham Law Rev 66:273
OECD (2011) Trade for growth and poverty reduction: how aid for trade can help. OECD Publishing, Paris
Oye K (1985) Explaining cooperation under anarchy. World Polit 38(1):21
Pakpahan B (2012) Deadlock in the WTO: what is next? World Trade Organization, Geneva
Pauwelyn J (2009) Legal avenues to multilateralizing regionalism: beyond Article XXIV. In: Baldwin R, Low P (eds) Multilateralizing regionalism: challenges for the global trading system. Cambridge University Press, Cambridge, p 368
Peel J (2011) Issues in climate change litigation. Carbon Climate Law Rev 5:15
Peel J, Osofsky HM (2015) Climate change litigation. Cambridge University Press, Cambridge
Penalver EM (1998) Acts of god or toxic torts - applying tort principles to the problem of climate change. Nat Resour J 38:563
Raynolds LT, Murray DL (2008) The fair trade future. Policy Innovations, The Carnegie Council, New York
Ricardo D (2004) On the principles of political economy and taxation. Courier Corporation, North Chelmsford
Richard Lord QC et al (eds) (2011) Climate change liability: transnational law and practice. Cambridge University Press, Cambridge
Rosenzweig C et al (2008) Attributing physical and biological impacts to anthropogenic climate change. Nature 453(7193):353–357
Sachs J (2016) Happiness and sustainable development: concepts and evidence. In: Helliwell J, Layard R, Sachs J (eds) World happiness report 2016 update, vol I. Sustainable Development Solutions Network, New York
Sands B, Peel J (2012) Principles of international environmental law, 3rd edn. Cambridge University Press, Cambridge
Sax JL (2000) Environmental law at the turn of the century: a reportorial fragment of contemporary history. Calif Law Rev 88(6):2375–2402
Schrag D (2007) Confronting the climate-energy challenge. Elements 3:171–178
Sen A (1999) Development as freedom. Anchor Books, New York
Stewart R et al (2013) Building a more effective global climate regime through a bottom-up approach. Theor Inquiries Law 14(1):273–305
Suttle O (2014) Equality in global commerce: towards a political theory of international economic law. Eur J Int Law 25(4):1043–1070
Suttle O (2017) Distributive justice and world trade law: a political theory of international trade regulation. Cambridge University Press, Cambridge

Taylor PL (2002) "Poverty alleviation through participation in fair trade coffee networks: synthesis of case study research question findings", Department Colorado State University, September, Community and Resource Development Program. The Ford Foundation, New York

The World Bank (2016) Taking on Inequality. International Bank for Reconstruction and Development/The World Bank, Washington

United Nations. Department of Economic and Social Affairs (2010) Rethinking poverty: report on the world social situation. United Nations Publications, New York, p 13 ff

Upendra Das R (2009) "Regional trade-fdi-poverty alleviation linkages. some analytical and empirical explorations", Deutsches Institut für Entwicklungspolitik, n. 18/2009. DIE, Bonn

Vanhala L, Hilson C (2013) Climate change litigation: symposium introduction. Law Policy 35:141

Victor D (2017) Three-dimensional climate clubs: implications for climate cooperation and the G20. International Centre for Trade and Sustainable Development, Geneva

Walton D (2007) Geopolitics and the great powers in the 21st Century: multipolarity and the revolution in strategic perspective. Routledge, Abingdon

Weiss EB (2011) The evolution of international environmental law. Jpn Yearb Int Law 54:1–27

Wilensky M (2015) Climate change in the courts: an assessment of non-U.S. climate litigation. Duke Environ Law Policy Forum 26:131

World Bank Group and World Trade Organization (2015) The role of trade in ending poverty. WTO Press, Geneva

World Economic Forum (2017) The global risks report, 12th edn. World Economic Forum, Cologny

Chapter 5
Trading Sustainable Energy

5.1 Introduction[1]

This chapter discusses energy sustainability through the trading system and proposes regional trade agreements as means to further enhance peace and energy security. The chapter makes the case that, to achieve security of energy supply, two main factors are important: diversification to minimize risk and regional cooperation. It is suggested that all regions of the world should reduce or eliminate the technical barriers to energy trade.

5.2 Energy Sustainability Through Trade

5.2.1 The Economic Burden of Energy Supply

The electricity and gas sectors in many countries rely heavily on government support through direct or indirect subsidies due to largely inefficient systems. Given that the energy sector in some countries is run through state corporations, the inefficiencies of these corporations, and the resultant lending activity, particularly in the electricity sector, to support the production and distribution of electricity, become public debts. In addition, retail tariffs for electricity and gas tend to be well below the cost of supply, resulting in the need for most governments to adopt some form of subsidization to ensure that energy remains affordable to consumers.[2] Not only do these

---

[1] Sections of this chapter have been co-written with Nelson Akondo.
[2] The World Bank, *Middle East and North Africa - Integration of Electricity Networks in the Arab World: Regional Market Structure and Design* (Report No: ACS7124, December 2013).

subsidies divert resources away from sectors where they are much needed, but they distort the economy and discourage further investment in the energy sector.[3]

A type of subsidization which primarily affects the liberalization of the energy market is regulated price control. Most governments in a bid to control the cost of electricity to the populace adopt under-pricing policies that prevent realistic pricing of energy. While the socio-political objective for such price control mechanisms may be well-intended, they tend to favor the wealthier sections of the society and not the poor. Consequently, private investment in the energy sector is discouraged because the attendant difficulty in recovering costs impinges on the economic justification for investment in the sector.

A progressive shift to a more market-oriented approach to pricing, which allows prices to be determined by the markets and not governments, would therefore attract private participation in the energy market. Not only would such liberalized markets result in the creation of employment through increased private sector participation, but they also promote competition. A more competitive energy market would promote competitive pricing for consumers and is more likely to enure to the benefit of the poorer sections of society than current subsidy regimes.

The growing demand for energy in many regions of the world has required governments to direct significant resources towards augmenting supply, consequently increasing the fiscal burden of the sector.[4] This burden has become even more pronounced as many countries are still recovering from the global effects of the 2008 financial crisis. Owing to the economic pressures faced by these countries, several infrastructural projects have been halted,[5] while other countries have had to revise and, in some cases, scrap subsidies in the sector.[6]

While the importance of improving production efficiency, cost recovery and their associated cost-reduction benefits to governments cannot be underestimated, it is important to take steps to reduce subsidies in the energy sector to strengthen its financial sustainability. Although subsidies are perceived as a positive social intervention strategy, they have created graver problems than the imbalances they are intended to address in many countries. Subsidy reform would free-up resources for productive public spending, which would further boost economic growth by reducing lower budget deficits and stimulating private sector investment.[7]

---

[3]International Monetary Fund, *Energy Subsidies in the Middle East and North Africa: Lessons for Reform* (March 2014).

[4]It is estimated that the Middle East and North Africa region would require infrastructure investment of US$ 450 billion by 2020 to meet 135 GW generation capacity. See The World Bank, *Middle East and North Africa - Integration of Electricity Networks in the Arab World: Regional Market Structure and Design* (Report No: ACS7124, December 2013) page 76.

[5]The World Bank, *Middle East and North Africa - Integration of Electricity Networks in the Arab World: Regional Market Structure and Design* (Report No: ACS7124, December 2013).

[6]International Monetary Fund, *Energy Subsidies in the Middle East and North Africa: Lessons for Reform* (March 2014).

[7]International Monetary Fund, *Energy Subsidies in the Middle East and North Africa: Lessons for Reform* (March 2014).

The natural response by most governments would be to protect reserves and adopt strategies to protect the primary source of budget funding. This has negative implications for regional energy integration. For instance, countries with a surplus of oil are more likely to sell the surplus to Europe and Asia to support their budgets, rather than entering into regional initiatives, which would require the sharing of such surplus to support energy supply. A change in economic paradigms is therefore required to influence policy that affects the energy sector. Such a change would involve a diversification of their economies to reduce the need to adopt policies targeted at protecting oil revenue.

Undoubtedly, there is room for the creation of a shared vision for economic diversification within energy suppliers. The OECD, through the Centre for Tax Policy and Administration and the MENA-OECD Investment Program, has already taken steps at engaging such countries on these issues. These discussions provide a useful building block for the development of a diversification program to reduce overall reliance on oil.

Less reliance on oil revenue would not only facilitate the adoption of a sustainable energy policy, but it could offer opportunities at directing investment for the energy sector. Diversification would also foster private participation in many sectors of the economy, thereby freeing states from the burden of service provision, and ultimately reduce the fiscal burden of the government.

5.2.2 Maximizing the Societal Gains Through Trade

The energy sector is quintessential to a country's development. Not only can it serve as the catalyst for economic development, but it is a necessary condition for infrastructural development. Beyond this, improving energy trade has the potential of addressing several of the social challenges that many countries are currently facing.

Primarily, the aim of an integrated energy market would be to secure energy provision within a given region. This would mean not only the augmentation of supply levels to meet demand, but also to ensure that energy is affordable to all segments of consumers. The social intervention policies have done very little to ensure that the poorer sections of the public can actually afford energy. The affordability of energy is largely impinged by high production costs and inefficiency, which energy suppliers naturally would pass on to consumers. Consequently, through regional integration, countries can optimize supply and reduce inefficiency, while adopting more dynamic tariff systems and conservation strategies. These will contribute to the lowering of prices for consumers, particularly for poorer sections of the public. This is, however, dependent on a transition from a state-controlled pricing policy to a market-based approach to pricing.

Further, the diversification of the energy market both in terms of production sources and private sector participation would result in the creation of jobs in the energy and other related sectors. This potential would further be augmented by regional integration, which would see these gains being replicated on a regional level through the creation of jobs internationally. Regional job-creation and its attendant potential for migration would also foster greater social interaction that could contribute towards promoting greater socio-political tolerance.[8] Although the development of local and regional job markets may prove to be a challenge in the short term due to the human-capital needs associated with the deployment of new technologies, it also presents an opportunity for the education sector, which would be required to contribute to filling the human-capital requirement gap.

5.2.3 Towards Clean Energy

Some countries have fossil fuels, others do not, but all of them have, to various degrees, potential for renewable energy.[9] Yet, many regions of the world have very low levels of investment in renewable energy. This means that renewable energy has phenomenal potential. Indeed, some parts of the world are blessed with unique solar irradiance and the wind energy prospects are also important. Research by the International Energy Agency (IEA) estimates that concentrated solar power technology alone (which is used when the sun does not shine) could amount to 100 times the electricity demand of the Middle East and North Africa (MENA) region and Europe combined.[10] Some countries such as Egypt, Jordan, Morocco and the UAE are leading the way on this front. Broadly speaking, the declining oil prices and the need for economic diversification are the major reasons to engage in renewable energy. For instance, 85% of Saudi Arabia's export earnings stem from oil and gas exports.[11]

There are still barriers to renewable energy promotion (which also applies to trade in electricity and gas):

1. Energy subsidies given to the fossil-fuel industry;
2. Legal and administrative complexities;
3. Weak regional interconnections; and
4. Awareness of affordability.

---

[8]This view resonates with Mahatma Gandhi's famous remark: "It is easier to bridge the oceans that lie between continents than it is to bridge the gap between individuals or the peoples." See https://www.azquotes.com/quote/877472.

[9]For further details on renewable energy governance, see Leal-Arcas and Minas (2016), pp. 621–666.

[10]International Energy Agency, "Technology Roadmap. Concentration Solar Power", p. 10, available at https://www.iea.org/publications/freepublications/publication/csp_roadmap.pdf.

[11]The Organisation of the Petroleum Exporting Countries, "Saudi Arabia facts and figures", available at http://www.opec.org/opec_web/en/about_us/169.htm. Saudi Arabia would need high oil prices to achieve a fiscal breakeven.

In some countries, electricity is heavily subsidized by the Government, which means that there is little incentive to make a transition to renewables. So a vision and policies are needed to implement renewable energy. Yet, it is encouraging to note that quite a few countries are part of the International Solar Alliance,[12] a group of sunshine countries based between the Tropic of Cancer and the Tropic of Capricorn,[13] where the potential of solar energy is phenomenal. Furthermore, a strong link between energy security and climate change mitigation is the use of renewable energy to diversify energy sources and, therefore, enhance energy security. The use of renewable energy, in turn, is a way to mitigate climate change.[14]

The desire to improve supply security has, however, led to a commitment to improving the situation, with many countries setting targets for the improvement of the renewable energy share of their energy mix. This shift is motivated by different factors—while for countries like Jordan and Morocco, it is to meet shortfalls in energy production, countries like the UAE, Qatar and Kuwait aim to preserve domestic oil for exportation.[15] In furtherance of this commitment, various countries have set renewable energy targets to be attained by the year 2030.[16] However, recognizable efforts can only be seen in Jordan, Morocco, Qatar, the UAE and Kuwait, where significant reform and investment appear to have been made in furtherance of the set targets. Countries like Algeria, Turkey, Egypt and Israel, however, have made some commendable progress with smaller scale projects.

Morocco is one of the most remarkable examples. This is so due to its spirited energy transition, but not least because of its role of the clean-energy gateway to Europe. Morocco committed itself to increase its share of renewable energy electricity to 52% by 2030. To that end, Morocco has taken several measures such as the creation and expansion of the Noor solar complex, located near its Southern desert town of Ouarzazate, one of the largest solar plants in the world.[17] In fact, although 100% energy from renewables may not yet be possible, 100% electricity from renewables is a possibility.

---

[12]http://www.intsolaralliance.org/. For a list of prospective countries of the International Solar Alliance, see http://www.intsolaralliance.org/countries_2.html.

[13]For a map of the so-called global sunbelt, see http://www.greenrhinoenergy.com/solar/radiation/empiricalevidence.php.

[14]For further analysis on the links between climate change, international trade and energy, see Leal-Arcas (2013).

[15]Squire Sanders, 'The Future For Renewable Energy in the MENA Region' (Clean Energy Pipeline Report).

[16]Rahmatallah Poudineh, Anupama Sen & Bassam Fattouh, "Advancing Renewable Energy in Resource-Rich Economies of the MENA" (*Oxford Institute for Energy Studies* (OIES PAPER: MEP 15), September 2016).

[17]The Guardian, "Morocco lights the way for Africa on renewable energy", available at https://www.theguardian.com/global-development/2016/nov/17/cop22-host-morocco-lights-way-africa-renewable-energy-2020.

There is currently only one electrical transmission interconnection between North Africa and Europe, namely the Morocco-Spain interconnector.[18] In that vein, the EU also seeks to diversify the supply of its energy sources. Thus, the European continent could constitute an attractive market for Morocco to export its surplus clean energy power. The Morocco-Spain interconnection enjoys a capacity of 1400 megawatts (MW) which should hopefully reach 5400 MW by 2019. Morocco could export more renewable energy electricity to Spain, and therefrom to the rest of the EU, without either party having to prominently invest in transmission infrastructure.[19]

Such ventures could be further ways to advance energy integration. If like-minded nations to Morocco such as Egypt, Jordan or the UAE were to cooperate, the rise of renewables could entail several benefits. First, these projects would create employment and economic stability in a politically unsettled region. Second, these schemes would deliver energy that is clean, reliable and sustainable. Indeed, renewable energy constitutes the cleanest and, in the long term, safest alternative to diversify the energy mix that heavily relies on fossil fuels in some countries. Finally, further trans-continental transmission interconnections would reinforce inter-regional ties between energy exporters and eager energy importers by exporting surplus lower-cost renewable energy.

A question for which critical consideration is required, is whether investment in renewable energy can be justified, as opposed to continuous reliance on fossil fuels, which come at a lesser cost to many energy-rich countries. Indeed, the high costs associated with the development and deployment of renewable energy cannot be ignored.[20] However, given the vast availability of renewable energy sources (solar and wind) in some regions of the world, it is believed that specific regions present viable options for a transition to renewable-energy-based power system.[21] Further, given the global consensus on benefits of renewable energy development, there exist various funding options which lessen the burden on states. In addition to the various funding mechanisms that exist for renewable energy development, there exists a strong case for concerted national action, as well as regional integration if the economic barrier to the development of renewable energy is to be addressed.

---

[18]In the EU context, there is the so-called projects of common interest, which are "key cross border infrastructure projects that link the energy systems of EU countries. They are intended to help the EU achieve its energy policy and climate objectives: affordable, secure and sustainable energy for all citizens, and the long-term decarbonisation of the economy." See https://ec.europa.eu/energy/en/topics/infrastructure/projects-common-interest.

[19]Mobarek, S. "Renewable energy export-import: a win-win for the EU and North Africa", World Bank Blog, available at http://blogs.worldbank.org/energy/renewable-energy-export-import-win-win-eu-and-north-africa.

[20]It is worth noting that the cost of renewable energy deployment continues to decrease. The cost of large-scale renewable energy sources such as offshore wind, concentrated solar power (CSP) and photovoltaic (PV) have fallen over the years. See: "Levelised Cost of New Generation Technologies", Institute for Energy Research (May 2009) - http://www.instituteforenergyresearch.org/2009/05/12/levelized-cost-of-new-generating-technologies/.

[21]Dii (Desertec Industrial Initiative), *Desert Power: Getting Started, The Manual for Renewable Electricity in MENA* (2013), Policy Report, Dii.

A successful implementation of the EU's model would therefore see the setting targets for the production of energy from renewable sources, greenhouse gas emissions, and energy efficiency gains in other regions of the world. Given that the responsibility for the development of renewable energy is largely at the national level, several national actions such as the following should be considered:

1. The re-evaluation of energy pricing and subsidy policies, as current levels of subsidization would render energy from renewable sources uncompetitive.[22]
2. Regulatory reform, aimed both at providing fiscal incentives for renewable energy, and the creation of competitive markets to promote private sector participation. This could be done through the institution of tax benefits and reward schemes like feed-in-tariffs, and the introduction of competitive procurement procedures.
3. The establishment of institutions with requisite capacity to support, promote, and facilitate the development of renewable energy.
4. The integration of environmental and climate-change concerns into broader national strategies. This will ensure that complementary policies are adopted in other sectors to prevent the erosion of the gains made by reforms in the energy sector.

These national efforts could then be augmented and supported by a regional integration program. For instance, the integration of electricity markets would ensure that the negative effects of the unstable and unpredictable nature of renewable energy supply in individual states are mitigated through capacity-sharing and load-distribution. Regional integration could also provide an opportunity for joint investment in renewable energy research, development, and deployment. Doing so would ensure that countries that are unable to harness their renewable energy potential are assisted, to the benefit of the entire region. Further, an integrated market also facilitates information sharing which, in turn, would support research and development. Finally, a region-wide institutional arrangement modelled on the EU's system would go a long way to facilitate the attainment of shared goals.

Another component of energy sector administration that affects the environment is supply and demand management. Developmental planning in many countries fails to address this issue. As a result, energy supply is usually incongruent with both the economic reality and energy needs of the country. This has proven to be the case in many countries. While adopting policies to address the problem of production efficiency, it is prudent to consider a region-wide, conservation and consumption efficiency policy.

Given that efficient energy management schemes require considerations for sustainable development of the energy sector, this would not only ensure adequate resource management, but would go a long way to promote the development of

---

[22]Laura El-Katiri, 'A Road Map for Renewable Energy in the Middle East and North Africa' (*Oxford Institute for Energy Studies* (OIES PAPER: MEP 6), January 2014).

renewable energy. In other words, development is not possible without energy and sustainable development is not possible without sustainable energy.

While efforts are being taken to increase energy generation from renewable energy sources, it is also important to consider the role of trade in natural gas in the transition from reliance on fossil fuel to 'greener' sources. Owing to its environmental friendliness, natural gas is the fossil fuel of choice. Therefore promoting gas-fueled generation is worth considering as an interim measure. Desertec industrial initiatives (Dii) projections indicate that once the full potential of renewable energy is realized in MENA, renewable energy will account for 98% of the regions energy needs while 2% would be through gas generation. This means current infrastructure, trade, and cooperation, which are largely centered on oil, would become redundant.[23]

Although it may be argued that these scenarios are based on 30–40 years scenarios and are therefore of little relevance to today's policies, the introduction of trade in gas in current energy trade discussions would go a long way to ensure a smooth transition towards the realization of the envisaged renewable energy scenarios. Furthermore, it is estimated that, although the initial cost of installing gas-fired production units is high (compared to nuclear energy, which is the proposed alternative), the overall cost of a nuclear plant is much higher, including the risk of nuclear disasters.[24] Finally, in the aspiration towards clean energy, a roadmap should be put in place:

1. Long-term target: leaving fossil fuels behind;
2. Mid-term target: making fossil fuels as environmentally friendly as possible. For instance, via carbon capture and storage;
3. Short-term target: trying to reach quick solutions.

5.3 Regional Trade Agreements As a Means to Promote Peace and Energy Security

5.3.1 Fundamentals for Effective Regional Trade

Beyond the development of infrastructure and the entry into force of cooperation agreements, the success of regional trade initiatives is dependent, as depicted in Fig. 5.1, on the complex interplay of norm creation, the existence of robust institutional frameworks, as well as strong political will to pursue the objectives of the initiative. The complexity of the relationship between these complementary factors lies in the inherent conflict of the primary objective of each factor. For instance,

---

[23]Dii (Desertec Industrial Initiative), *Desert Power: Getting Started, The Manual for Renewable Electricity in MENA* (2013), Policy Report, Dii page 74.

[24]David R Jalilvand, 'Renewable Energy for the Middle East and North Africa: Policies for a Successful Transition' (Friedrich Ebert Stiftung Study, February 2012).

5.3 Regional Trade Agreements As a Means to Promote Peace and Energy Security

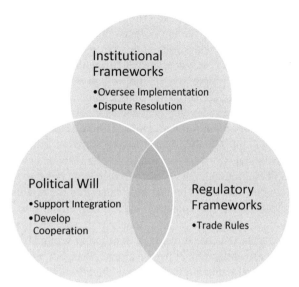

Fig. 5.1 Requirements for effective development of regional trade

while institutional frameworks for regional initiatives are developed with neutral political sensitivity, and usually with an eye on the "bigger picture", governments that participate in these initiatives often do so in line with parochial political motivations. However, it is often the case that opposing considerations make a case for fostering the development of the initiative, rather than frustrating it. Therefore, achieving synergies between these factors ensures the sustainability and progress of regional trade initiatives.[25]

The creation of a regional institution or structural framework is critical to the growth and development of any regional initiative. This would usually take the form of an administrative or quasi-administrative body with relevant expertise and representation from all member countries, whose main objective would be to oversee the implementation of the goals of the initiative. The effectiveness of such a regional institution could be augmented by giving it dispute resolution functions. The nature of such institutional frameworks could range from a formally established entity to standing or *ad hoc* committees that meet regularly to undertake their duties. This would largely depend on the amount of financial resources countries would be willing to contribute towards the running of the institution. Regardless of what nature it takes, it is critical that the composition and remits of such an institutional framework be fully delineated if the regional initiative is to succeed.

A properly functioning trade regime would also require the existence of a robust and dynamic set of rules to regulate, amongst others, market access; market standards and operations; pricing; taxation; and competition within the market. The existence of these provides the degree of certainty that is required for the sustenance

---

[25]For an in-depth analysis of sustainability in the context of common concern, see Leal-Arcas (2017).

and growth of any market. The trend in developed regional markets such as the EU points to a multilevel approach to regulation, where minimum standards are set at the regional level, with different degrees of implementation at the national level. While this may be a practical approach during the embryonic stages of the market, the goal should be a progressive harmonization of rules across Member States. This would make for greater certainty, transparency, and predictability within the market and ensure uniform development across the market.

As important as the afore-discussed elements are, they are largely dependent on the will of Member States to see them to fruition. It is axiomatic to say that governments of Member States are the main actors in any regional trade initiative. Without the willingness of governments to develop and participate in the proposed initiative, they will remain but plans. It is important that governments be made to understand that, despite their individual national developmental agenda, cooperation through trade serves as a catalyst to socio-economic development. Therefore, resources that are directed at promoting regional integration and cooperation do not necessarily detract from national plans, but rather promote, in the long term, the individual national goals. Such a "bigger picture" outlook also goes a long way to help displace other divisive considerations such as culture, language, and religion.

The fact that establishing the requisite balance is desired by many actors in proposed regional initiatives is not in doubt. However, seeing the proposals through to completion often proves quite challenging. In this regard, established multinational organizations such as the World Bank and regional multilateral financial institutions may prove valuable in providing the required technical and experiential assistance.

5.3.2 Past Experiences

Regional Trade Agreements (RTAs) have experienced a striking evolution, both in terms of number and breadth, since the 1990s. This development has been so prominent that, as of June 2016 with the notification of the RTA between Japan and Mongolia, all WTO members have an RTA in force.[26] The basic premise of an RTA is the reduction of tariff barriers between its parties. Liberalization is the best approach to instigate regional trade.[27] In turn, regional trade entails benefits that can go well beyond the economic growth and competitiveness of the parties involved.

Several RTAs contain provisions which directly address the energy sector. Examples of such agreements include the North American Free Trade Agreement

---

[26]World Trade Organization, "Regional Trade Agreements", available at https://www.wto.org/english/tratop_e/region_e/region_e.htm.

[27]World Bank, "Regional Trade Agreements: Effects on Trade" *Global Economic Prospects*, p. 57, 2005.

(NAFTA) or the Energy Charter Treaty (ECT).[28] RTAs are perceived by several countries as providing a better suited legal framework to multilateralism when it comes to energy trade regulation. This is so as they typically include more detailed provisions that tackle the specific energy concerns of the parties to the RTA.[29]

Furthermore, RTAs have the potential to be conducive to peace. Broadly speaking, regional trade integration fosters interdependence between countries, creates economic incentives for peace, and nurtures diplomatic avenues to solve disputes.[30] The European Union (EU), the European Atomic Energy Community (EURATOM), the Association of Southeast Asian Nations (ASEAN), and the Southern Common Market (MERCOSUR) are success stories that illustrate how regional trade integration can promote peace. The inception of these trading blocs was pivotal to allay the fears of further armed conflicts between Member States in the region. As such, they constitute valuable examples that may spur other regions of the world to emulate a similar effort. What constitutes these regions of the world to emulate the establishment of a trading bloc is the fact that, as a result, they could reach peace in the region, generate wealth and jobs, solve disputes, and avoid future wars between themselves, as we have seen in the case of the EU.

In short, this section emphasizes the role of RTAs as peacemakers as well as avenues to promote energy security.

5.3.2.1 The EU and EURATOM[31]

The creation of the EU ensued from the Treaty of Paris establishing the European Coal and Steel Community, concluded in 1951 between its six founding Member States (Belgium, France, West Germany, Italy, the Netherlands and Luxembourg). In the wake of World War II, these countries agreed to cooperate to ensure peace, but also to prevent competition between European countries over energy resources. Ever since, the EU has progressively expanded to 28 Member States which have arguably achieved prosperity through trade liberalization. The national energy markets of EU Member States have become further integrated over the course of the years.

The EU has encouraged the development of energy network in a number of ways:

- by financing, at the European level, the construction of trans-European energy networks;
- by adopting measures to foster the liberalization and opening-up of electricity and gas markets; and

---

[28]For an analysis of the ECT, see Leal-Arcas (2018).

[29]Shih, W. "Energy Security, GATT/WTO and Regional Agreements" *Society of International Economic Law Working Paper No.* 10/08, p. 41, 2008.

[30]Brown et al. (2005).

[31]This sub-section draws largely from Leal-Arcas, R. and Alemany Ríos, J. "How can the EU diversify its energy supply to improve its energy security?" *Queen Mary University of London School of Law Legal Studies Research Paper No.* 190/2015, pp. 3–5, 2015.

- by creating specialized European bodies to facilitate cooperation between transmission system operators (TSOs) and energy regulators, such as the European Agency for the Cooperation of Energy Regulators (ACER).

Instead of pressing for a top-down common European blueprint, the latest developments suggest that a bottom-up regional approach seems to be more attuned to effectively integrate energy markets. For instance, the success of the Nordic power market (Nord Pool) supports this statement. Indeed, this bottom-up initiative resulted in the effective integration of the electricity markets of several EU Member States (Denmark, Estonia, Finland, Germany, Latvia, Lithuania, Sweden and the UK) as well as Norway. This is so as nations usually prefer to cooperate with their neighbors with whom they share similar challenges and interests, rather than at the EU level where a common solution tends to be more difficult to reach. This trend has been acknowledged and is strongly advocated by the European Commission itself. Thus, the progressive fusion of ventures led by clumps of Member States within the EU is gradually paving the way to complete the Internal Energy Market as a whole.

The launching of a very ambitious initiative for the establishment of the European Energy Union[32] in February 2015 represented a milestone in terms of EU energy policy.[33] This initiative was undertaken in an effort to ensure resilience and facilitation of regional energy trade. The Energy Union is already delivering tangible results (*e.g.*, the inauguration of the Santa Llogaia–Baixàs power line, which doubles the interconnection capacity between France and Spain).[34] Furthermore, an array of projects are lined up to foster energy security as well as cross-border energy trade within the EU. This regional energy venture is set to streamline shared welfare while warranting affordable, secure, and sustainable energy through energy trade. The European Energy Union revolves around five pillars:

1. Security, solidarity and trust;
2. Finalization of the internal energy market;
3. Moderation of demand through energy efficiency;
4. Decarbonization of the European energy mix; and
5. Technologies, research and innovation.[35]

As for the European Atomic Energy Community (EURATOM), the Treaty establishing the EURATOM was signed in 1957. The EURATOM is a separate

---

[32] See Communication from the Commission to the European Parliament, the Council, the European Economic and Social Committee, the Committee of the Regions and the European Investment Bank, "A Framework Strategy for a Resilient Energy Union with a Forward-Looking Climate Change Policy" COM(2015) 80 final (25 February 2015). For an analysis of the European Energy Union, see Leal-Arcas (2016).

[33] For a comprehensive analysis of EU energy law and policy, see Leal-Arcas and Wouters (2017).

[34] European Commission, "Building the Energy Union: Key electricity interconnection between France and Spain completed" press release 20 February 2015, available at http://europa.eu/rapid/press-release_IP-15-4463_en.htm.

[35] For a comprehensive analysis of the European Energy Union see Leal-Arcas (2016).

international organization to the EU but has close ties with the latter (*e.g.* the EURATOM has the same membership as the EU). The rationale behind the creation of this entity was that nuclear energy represents an essential resource for industrialization and the advancement of peace. The community aims to create the conditions necessary for the development of a powerful nuclear industry which will provide extensive energy resources, lead to the modernization of technical processes, and contribute, through its many other applications, to the prosperity of its peoples.[36] More remarkably, the EURATOM enjoys exclusive jurisdiction over nuclear supply both in internal and external relations.[37] In the same vein as the EU with the ACER and TSOs, the EURATOM also counts with *ad hoc* institutions to regulate energy trade such as the EURATOM Supply Agency (ESA), for instance.

5.3.2.2 ASEAN

The ASEAN was established through the Bangkok Declaration in 1967 by its founding fathers (Indonesia, Malaysia, Philippines, Singapore and Thailand). The declaration was signed in the aftermath of a period fraught with tension and disputes in the region. The thorniest episode arguably being the Indonesian-Malaysian confrontation or *Konfrontasi* (1963-1966) which was a violent conflict where Indonesia initially opposed the creation of Malaysia. All these incidents prompted the gathering of these five nations under the ASEAN to prevent future disputes. To that end, the Bangkok Declaration spelled out cooperation in the economic, social, cultural, technical, educational and other fields; as well as the promotion of regional peace.[38] The ASEAN's membership gradually increased to its current ten Member States (Brunei, Cambodia, Laos, Myanmar and Vietnam joined at a later stage). The ASEAN has successfully achieved to avert major conflicts among its Member States. The bloc has done so by implementing two key principles: respect for sovereignty and peaceful settlement of disputes.[39]

The ASEAN energy policy is geared towards setting a regionally harmonized framework of energy supply as a means to encourage industrialization and economic growth.[40] The ASEAN created the ASEAN Centre for Energy (ACE) in 1999 to accelerate the integration of energy strategies of its Member States. More recently, the ASEAN Energy Market Integration (AEMI) initiative was launched in 2013. Connectivity and energy market integration are the prime concerns outlined by the

---

[36]Preamble of the Treaty establishing the EURATOM.

[37]Articles 52-76 of the Treaty establishing the EURATOM. Furthermore, Article 86 asserts that all special fissile materials automatically become the property of the community.

[38]The Association of Southeast Asian Nations, "History. The Founding of ASEAN", available at http://asean.org/asean/about-asean/history/.

[39]Heng, P. "The ASEAN Way and Regional Security Cooperation in the South China Sea", *Robert Schuman Centre for Advanced Studies Research Paper No.* 2014/121, p. 1, 2014.

[40]Sovacool (2009), p. 2357.

ASEAN Plan of Action for Energy Cooperation (APAEC) 2016–2025.[41] To that end, the plan delineates several key strategies (*e.g.* to initiate multilateral electricity trade in at least one sub-region by 2018; to enhance connectivity for energy security and accessibility via pipelines and regasification terminals; to increase the component of renewable energy to 23% by 2025 in the ASEAN energy mix; to reduce energy intensity by 20% in 2020 based on 2005 level).[42]

5.3.2.3 MERCOSUR

The Treaty of Asunción signed in 1991 by Argentina, Brazil, Paraguay and Uruguay established MERCOSUR. This regional grouping was initially founded to reduce the tensions between Argentina and Brazil. In addition, MERCOSUR also contributed to thwart a possible coup in Paraguay through the affirmation by the presidents of its Member States that democracy was a necessary requisite to join.[43] MERCOSUR is the world's fourth largest integrated economic bloc. It is the most advanced trade integration venture in the developing world.[44] Venezuela joined MERCOSUR in 2012 while Bolivia is currently in the process of becoming a full member. This initiative has enabled its Members States to overcome their historic mutual distrust and consolidated peaceful democracy in South America's Southern Cone.[45] Furthermore, MERCOSUR can be regarded as a Latin America initiative to counter the influence of other regional trading blocs (*i.e.* NAFTA). In addition, MERCOSUR was also envisioned as a platform to enable its Member States to tackle common security concerns such as drug trafficking.[46]

In general terms, the energy situation in Latin America is rather promising. South America boasts 10% of the world's oil resources as well as 5% of its gas reserves. Nevertheless, the distribution of these resources renders energy policy truly challenging. Indeed, most of this energy wealth is harbored in politically unstable countries such as Venezuela.[47] *Prima facie*, the latter's accession to MERCOSUR reinforced the belief that the bloc had the potential to become an energy juggernaut. Nevertheless, Venezuela's failure to meet human rights and trade standards motivated MERCOSUR's decision to suspend its membership in December 2016.[48]

---

[41] ASEAN Energy Market Integration, "AEMI initiative developing ASEAN solution to fight energy poverty", available at http://www.asean-aemi.org/.

[42] ASEAN Centre for Energy, "ASEAN Plan of Action for Energy Cooperation (APAEC) 2016-2025. Phase I 2016-2020", p. 3, 2015, available at https://app.box.com/s/g6b4ynph5wwiuvtxgol3s2rwjwcmrtbj.

[43] Khan et al. (2009), p. 30.

[44] *Ibid.*, p. 41

[45] O'Keefe (2004), p. 204.

[46] Brown et al. (2005), p. 6.

[47] Franca Filho et al. (2010).

[48] BBC, "MERCOSUR suspends Venezuela over trade and human rights", available at http://www.bbc.co.uk/news/world-latin-america-38181198.

Even though the Treaty of Asunción fails to address energy, MERCOSUR has taken several steps towards the integration of energy markets in the Southern Cone. Memorandums of Understanding were adopted to promote electricity and gas integration in 1998 and 1999, respectively. These measures endeavored to guarantee free market conditions thereby streamlining the free movement of electricity and gas among Member States.[49] Moreover, a Framework Agreement on Regional Energy Cooperation between MERCOSUR Member States and its Associated States (Chile, Colombia, Ecuador, Guyana and Surinam) was signed in 2005.

References

Brown O et al (2005) Regional trade agreements: promoting conflict or building peace? International Institute for Sustainable Development, Winnipeg, p 6
Franca Filho M et al (2010) The Law of MERCOSUR. Hart Publishing, Oxford, p 387
Khan S et al (2009) Regional trade integration and conflict resolution. Routledge, Abingdon, p 30
Leal-Arcas R (2013) Climate change and international trade. Edward Elgar Publishing Ltd, Cheltenham
Leal-Arcas R (2016) The European Energy Union: the quest for secure, affordable and sustainable energy. Claeys & Casteels Publishing, Deventer
Leal-Arcas R (2017) Sustainability, common concern and public goods. George Washington Int Law Rev 49(4):801
Leal-Arcas R (ed) (2018) Commentary on the Energy Charter Treaty. Edward Elgar Publishing, Cheltenham
Leal-Arcas R, Minas S (2016) Mapping the international and European governance of renewable energy. Oxford Yearb Eur Law 35(1):621–666
Leal-Arcas R, Wouters J (eds) (2017) Research handbook on EU energy law and policy. Edward Elgar Publishing Ltd, Cheltenham
O'Keefe T (2004) Economic integration as a means for promoting regional political stability: lessons from the European Union and MERCOSUR. Chicago Kent Law Rev 80(1):204
Roggenkamp M et al (2012) Energy networks and the law. Oxford University Press, Oxford, p 68
Sovacool B (2009) Energy policy and cooperation in Southeast Asia: The history, challenges and implications of the trans-ASEAN gas pipeline (TAGP) network. Energy Policy 37:2357

---

[49]Roggenkamp et al. (2012), p. 68.

Part II
From the Bottom Up

Chapter 6
Decentralization and Empowering the Citizen

6.1 Introduction[1]

The increasing role of new actors in law-making has received attention since the 1990s.[2] Developments in climate change and environmental law in this era have catalyzed innovative governance approaches by non-State actors and international organizations. These developments have created new legal challenges, both public and private, in a global multilevel governance context. New actors are not solely involved in contributing to thematic law and policy agenda setting, developing solutions, and providing oversight capacity; they are also becoming important players in delivering services. Opportunities to deliver services are growing as the global economy reconfigures around advancing information and communications technologies illustrated by the rapidly emerging 'gig' economy.[3]

In this new setting, ample space is created for the emergence of new energy actors, a principal one being prosumers, namely consumers who are also producers of (renewable) energy and who use energy in a smarter and more efficient manner. Energy prosumers is an umbrella term referring to self-generating energy providers, whether households or energy communities. Individuals contribute to the energy supply in their vicinity via their community-owned own-installed renewable energy

---

[1] Several ideas in this chapter are a development from Leal-Arcas (2018a), pp. 83–132.
[2] See for instance Clapham (2009), pp. 200–212; Halliday (2001), pp. 21–40; Josselin and Wallace (2001), pp. 1–20.
[3] 'A gig economy is an environment in which temporary positions are common and organizations contract with independent workers for short-term engagements.' (See WhatIs.com (2016) Definition: Gig Economy, updated May 2016, http://whatis.techtarget.com/definition/gig-economy).

capacity, more often than not solar roofing, wind energy, or combined heat and power.[4] Doing so will empower and enable a more sustainable future.

This chapter critically analyses the new challenges and opportunities that prosumers bring to achieving energy security goals in the European Union (EU). The chapter places decentralization as preferable to central government policymaking. The EU, along with the United States (US),[5] is a pioneer in engineering a hybrid electricity market model, where traditional power plants will be supplemented by virtual power plants, a plethora of small, individual energy producers and a corresponding new set of mechanisms to cater for the new market. That said, the adoption and customisation of (elements of) this new energy architecture by other countries will hinge upon the degree of its success within European soil. This chapter contributes to the literature in two specific ways. First of all, it critically discusses an emerging new actor in the EU's energy security that we refer to as prosumers. Second, it illustrates in broad terms the ways in which this new actor will cooperate with other actors in the EU energy market and contribute to the European Union's energy goals.

In this context, side by side with traditional threats and challenges, new risks, but also opportunities, arise for ensuring energy security.[6] The energy sector is undergoing a large-scale low-carbon transition. What is under-emphasized in this transition is that it involves a major paradigm shift from a supply-driven to a demand-side energy policy. Driven by a mix of geopolitical, economic, climate, and technological considerations, the energy sector is moving towards a new architecture,[7] the principal pillars of which are progressive electrification, a cleaner energy mix, renewable indigenous energy production, increased energy efficiency, and the development of

---

[4]J Roberts (2016) Prosumer Rights: Options for a legal framework post-2020, ClientEarth, May 2016, http://www.greenpeace.org/eu-unit/Global/eu-unit/reports-briefings/2016/ClientEarth%20Prosumer%20Rights%20-%20options%20for%20a%20legal%20framework%20-%20FINAL%2003062016.pdf.

[5]In the case of the US, new legislation plans (known as the Affordable Clean Energy (ACE) rule) try to move away from coal-fired generation, which is driven by fundamental economics. See Crooks, E. "Energy plan dashes Trump hopes for coal," *Financial Times*, 21 August 2018. Under the Affordable Clean Energy rule (which replaces the Obama administration's Clean Power Plan (CPP), whose aim was to set federal guidelines for states to reduce their $CO2$ emitted by power generators), states will be able to set less strict standards, even having no $CO2$-emissions reductions targets at all. Coal can be included as part of the energy mix. The CPP, however, sought to reduce $CO2$ emissions from power plants by 32% from their levels in 2005 by 2030. With the ACE rule, the level of $CO2$-emissions reduction is expected to be minimal.

[6]Leal-Arcas (2015), pp. 291–336, Oxford University Press.

[7]All these issues could be placed together under the concept of *ecological economics*, which addresses the relationships between ecosystems and economic systems in the broadest sense. See generally Costanza (1991). The main aims of ecological economics are:

- Establishing a historical perspective on social-natural interactions;
- Finding a common language and a set of concepts for the analysis of economies and ecosystems;
- Studying the intersection between natural science and social science.

See Faber et al. (1998).

new markets to produce, transmit, and, crucially, manage energy.[8] The key to this overhaul is the slow, but already underway, development of prosumer markets.

The chapter analyzes one of the mega-trends of the twenty-first century, namely a paradigm shift in the governance of sustainability from the bottom up. The chapter concludes with a future research agenda to fill the knowledge gap on the links between four major global concerns: trade, energy, climate change, and sustainability.

6.2 Megatrends of the Twenty-First Century[9]

The scientific community is by now in almost unanimous agreement that the greenhouse gas (GHG) effect is real,[10] and the level of GHG emissions in the atmosphere continues to increase.[11] There are clear policy actions to tackle climate change: mitigation,[12] adaptation,[13] and geoengineering.[14] As a result of the Paris Agreement, and prior to the signing of the Paris Agreement, new avenues to tackle climate change more effectively have emerged, such as the involvement of mayors,[15] governors,[16] and CEOs.[17] From this perspective, the Paris Agreement

---

[8] For an overview of the current legal and policy situation in EU energy, see Leal-Arcas and Wouters (2017).

[9] See generally Esty (2017), pp. 1–80.

[10] *The 97% Consensus on Global Warming*, Skeptical Science, https://www.skepticalscience.com/global-warming-scientific-consensus-intermediate.htm [https://perma.cc/LH6P-EMBT].

[11] *Global greenhouse gas emissions data*, U.S. Envtl. Prot. Agency, https://www.epa.gov/ghgemissions/global-greenhouse-gas-emissions-data [https://perma.cc/J8UG-TWSM].

[12] For example, by reducing the emissions of GHGs in the atmosphere with the promotion of electric cars or making use of the circular economy. See Global Compass, *The Earth Circle*, The Economist (Jan. 31, 2017), http://films.economist.com/globalcompass [https://perma.cc/PZ9U-W5CE].

[13] For example, by minimizing the damage caused by the effects of climate change; a case in point is using scarce water resources more efficiently. See Adaptation to Climate Change, European Comm'n, https://ec.europa.eu/clima/policies/adaptation_en#tab-0-0 [https://perma.cc/64DV-EU7S].

[14] For example, by enhancing surface brightness, such as painting roofs white. See *Geoengineering to Combat Global Warming*, U.N. Env't Programme (May 2011), https://na.unep.net/geas/getUNEPPageWithArticleIDScript.php?article_id=52 [https://perma.cc/9FKP-LQ56].

[15] C40 Cities, http://www.c40.org/ [https://perma.cc/J53Q-WNQ9]. Several cities throughout the world have agreed to make new buildings carbon-neutral from 2030 and to retrofit others to the same standards by 2050. The mayor of London Sadiq Khan has promised to make London zero-carbon by 2050. See The Economist, "Local government v global warming," 15 September 2018, pp. 67–68, at 67.

[16] Regions of Climate Action, http://regions20.org/ [https://perma.cc/8EPB-X2QN]. In September 2018, California's governor, Jerry Brown, issued an executive order to make California carbon-neutral by 2045. See The Economist, "The king of green," 15 September 2018, p. 7. Moreover, as of September 2018, Utah was considering a carbon tax. See The Economist, "Local government v global warming," 15 September 2018, pp. 67–68, at 67.

[17] *See, e.g., CEOs from Leading Companies Worth More Than $2 Trillion Ask COP21 to Secure a Prosperous World*, The Climate Group (Nov. 23, 2015), https://www.theclimategroup.org/news/

combines the action of both state and non-state actors during the negotiating phase and in its implementation.

Cities should take climate action because today the majority of the world's population lives in cities,[18] and this trend to urban migration is on the rise[19]; because they are the main polluters and the main implementers of legislation[20]; and because mayors of cities are pragmatic with global issues such as climate change, poverty, and terrorism.[21] Such issues are also too big for nation-states, and cities arguably offer better governance on these matters.[22] Furthermore, some of the greatest environmental and social challenges come from cities: food, water, waste, infrastructure, transport. Moreover, mayors tend to come from the cities they govern[23] and therefore have a much higher level of trust than politicians at the national level.[24] All of this means that using cities to mitigate climate change is a promising initiative, so educating citizens and raising awareness become crucial.

Global issue governance at city and local levels is on the rise. Some of these initiatives even go beyond climate action and have collateral effects such as job creation and prevention of premature pollution-related deaths. Examples of such bottom-up structures are: the C40 Mayors Summits[25]; the Compact of Mayors[26]; the Covenant of Mayors for Climate and Energy[27]; the Global Covenant of Mayors for

---

ceos-leading-companies-worth-more-2-trillion-ask-cop21-secure-prosperous-world [https://perma.cc/4S5U-QM2X].

[18] *See World's Population Increasingly Urban with More than Half Living in Urban Areas*, U.N. Dep't of Econ. & Soc. Affairs (July 10, 2014), http://www.un.org/en/development/desa/news/population/world-urbanization-prospects-2014.html [https://perma.cc/2WLK-6V3M].

[19] *Id.* By 2050, 70% of the world's population is expected to live in cities. See Mark Wilson, *By 2050, 70% of the World's Population Will Be Urban. Is That a Good Thing?*, Co.Design (Mar. 12, 2012), https://www.fastcodesign.com/1669244/by-2050-70-of-the-worlds-population-will-be-urban-is-that-a-good-thing [https://perma.cc/85TT-ASQV].

[20] *See* Regions of Climate Action, http://regions20.org/.

[21] *Mayors Get Things Done. Should They Run the World?*, The Globe & Mail (Mar. 11, 2014), http://www.theglobeandmail.com/opinion/ideas-lab/should-mayors-lead-the-world/article17275044/ [https://perma.cc/T2H4-FYX5].

[22] For further details on the potential of cities to solve global problems locally, see Benjamin R. Barber, *If Mayors Ruled the World: Dysfunctional Nations, Rising Cities* 5–9, 342 (2013) (arguing that local executives exhibit a nonpartisan and pragmatic style of governance that is lacking in national and international halls of power).

[23] *Id.* at 98.

[24] *Id.* at 226–27.

[25] *See About Us*, C40 Cities: Mayors Summit, https://mayorssummit2016.c40.org/ [https://perma.cc/7S7Z-VCF3].

[26] *See About*, Compact of Mayors, https://www.compactofmayors.org/history [https://perma.cc/6YW2-HEWT].

[27] *See About*, Covenant of Mayors for Climate & Energy, http://www.covenantofmayors.eu/about/covenant-of-mayors_en.html [https://perma.cc/VGQ7-L7GE]. The idea behind the Covenant is to support local authorities in the implementation of local sustainable energy policies.

Climate and Energy[28]; RESURBE[29]; the "100 resilient cities" scheme pioneered by the Rockefeller Foundation[30]; United Cities and Local Governments[31]; International Council of Local Environmental Initiatives[32]; CityNet[33]; City Protocol[34]; the United States Conference of Mayors; Habitat III[35]; and the Making Cities Resilient campaign[36] in the framework of the U.N. Office for Disaster Risk Reduction.[37] All of these examples show that, until recently, there has been a legal and policy vacuum at the city level regarding climate action and that city networks for climate deliberation are on the rise. It also means that there is a lot that cities can do even when national governments refuse to act on climate change or other global issues. This could even lead to the creation of a "[L]eague of [C]ities," to quote the American political theorist Benjamin Barber.[38]

Mayors' and governors' plans of action for climate change mitigation and adaptation could be emulated in other cities and regions of the world with similar concerns. For instance, the mayor of Rio de Janeiro, Brazil, may have a plan to mitigate climate change that is opportune for Manila, Philippines. To make sure that intercity networks remain coordinated, there have been proposals for the creation of a Global Parliament of Mayors[39] to enable cities to have a stronger voice on global issues and address global priorities more democratically and directly by citizens.[40] The purpose is to democratize globalization or to globalize democracy.[41]

---

[28] *See About*, Global Covenant of Mayors for Climate & Energy, https://www.globalcovenantofmayors.org/about/ [https://perma.cc/VJ79-4UU8].

[29] *See RESURBE*, Cátedro UNESCO, http://www.unescosost.org/en/project/resurbe/ [https://perma.cc/V96T-TKX2].

[30] *See About Us*, 100 Resilient Cities, http://www.100resilientcities.org/about-us#/-_/ [https://perma.cc/QY8L-XKJB].

[31] *See About Us*, United Cities and Local Governments, https://www.uclg.org/en/organisation/about [https://perma.cc/LB8M-BBKM].

[32] *See About*, International Council of Local Environmental Initiatives, http://www.iclei.org/about/who-is-iclei.html [https://perma.cc/3KAR-VSL3].

[33] *See About Us*, Citynet, http://citynet-ap.org/category/about/ [https://perma.cc/WT2C-Q4DY].

[34] *See What's City Protocol*, City Protocol, http://cityprotocol.org/whats-city-protocol/our-mission/ [https://perma.cc/FZ2C-UAC5].

[35] *See About Habitat III*, Habitat III, https://habitat3.org/about [https://perma.cc/6LU7-5QTU].

[36] *See Making Cities Resilient*, United Nations Office for Disaster Risk Reduction, https://www.unisdr.org/we/campaign/cities [https://perma.cc/NG4P-2FXH].

[37] *Our Mandate*, United Nations Office for Disaster Risk Reduction, https://www.unisdr.org/who-we-are/mandate [https://perma.cc/MC5P-5YBG].

[38] Barber (2013), p. 22.

[39] *See id.* at 22–23, 336–59.

[40] *See id.*

[41] *See id.*

Moving forward, the international community may also consider putting a price on harm-causing.[42] Addressing climate change will require such top-down, centralized guidance from intergovernmental decisions and bottom-up, decentralized implementation of climate change goals through companies and citizens' participation. Both approaches are necessary to succeed. Although national action is not a prerequisite for local intervention on climate change, it certainly helps get things done more efficiently. For the implementation of any policy, good legislation is key. Incomplete policy is non-implementable policy.

Expanding clean energy choices is also an increasingly popular issue because clean energy is an effective way to decarbonize the economy and it is therefore necessary to find a way to finance it.[43] As a result of clean energy's popularity, there is an innovation race across the world.[44] It is necessary to create a policy framework for innovators to be willing to accept failure and not be afraid of making mistakes to encourage continued development.

All of these trends raise the interesting question of how to manage globalization in a sustainability era. Table 6.1 below offers the main trends of the twenty-first century in a sustainability context.

In January 2017, the U.S. National Intelligence Council (NIC) published its public Global Trends Report titled Global Trends: The Paradox of Progress.[45] Through 2035, the NIC noted that the global trends of climate change, the environment, and public health issues "will demand attention."[46] The next chapter discusses in more detail how increased citizen participation can help achieve that required attention.

---

[42]For an example of a recent domestic approach proposed in the United States by a group of senior Republican (among them, two former secretaries of state and of the treasury, namely James Baker and George Shultz), *see* Juliet Eilperin & Chris Mooney, *Senior Republican Statesmen Propose Replacing Obama's Climate Policies with a Carbon Tax*, Wash. Post (Feb. 8, 2017), https://www.washingtonpost.com/news/energy-environment/wp/2017/02/07/senior-republican-leaders-propose-replacing-obamas-climate-plans-with-a-carbon-tax/?utm_term=.1ceadf0fe007 [https://perma.cc/3C24-ZTDF].

[43]*Renewable Energy Proves Increasingly Popular*, Economist Intelligence Unit: United Arab Emirates (Dec. 30, 2015), http://country.eiu.com/article.aspx?articleid=1203814104# [https://perma.cc/773J-CAJ8].

[44]*See* Andrew Grant & Gaia Grant, The Innovation Race: How to Change a Culture to Change the Game (2016); *see also* Guevara-Stone (May 25, 2017), https://www.rmi.org/news/35-years-bold-steps-clean-energy-race-part-1/ [https://perma.cc/2B36-67HS].

[45]Nat'l Intelligence Council, Global Trends: Paradox of Progress (Jan. 2017), https://www.dni.gov/files/images/globalTrends/documents/GT-Full-Report.pdf [https://perma.cc/4KX4-TD94].

[46]*Id.* at 6.

Table 6.1 Megatrends of the twenty-first century[a]

Twentieth century	Twenty-first century
Focus of attention was government	Focus of attention should be business
Environmental information silos; little attention to economics	Since *vox populi* is that economics will always prevail over the environment, it is necessary to have an integrated approach between the environment, energy, and the economy. The international trading system unites the three sectors
Top-down approach to climate change mitigation through participation of presidents and prime ministers of countries	Bottom-up approach to climate change mitigation through participation of citizens, mayors, governors, CEOs, and billionaires
Command and control approach; "polluter pays" principle	Market mechanisms; economic incentives not to pollute
Prohibitions	Problem-solving
Good consumers were not rewarded	Reward individuals who solve problems
Gurus gave prescriptions on how to move forward	Big data[b] usage for better analysis to inform decisions
Success was based on money expenditure	Success is based on outcomes and implementation
Environmental protection as a moral good	Price-based approach to punish environmental harm
Innovation in technology	Innovation in government and finance
Limited infrastructure	Technological revolution: using technology to help with infrastructure

[a]This list is based on a "Decalogue" developed by Daniel Esty of Yale University, First Yale Sustainability Leadership Forum, September 2016 at Yale University

[b]*See generally Enter the Data Economy: EU Policies for a Thriving Data Ecosystem*, Eur. Pol. Strategy Ctr.: EPSC Strategic Notes (Jan. 11, 2017), https://ec.europa.eu/epsc/publications/strategic-notes/enter-data-economy_en (providing insight into the emerging role of "big data" in driving economic decisions) [https://perma.cc/XKZ4-FF89]. The current situation regarding barriers with data collection tends to be as follows: (1) unavailable data; or (2) available data, but owner are unwilling to share; or (3) available data, but we need greater expertise on how to make better use of big data for better policymaking

6.3 EU Energy Law Reform: An Example of Decentralization

To understand the increasing role of prosumers in a decentralized energy system and to make access to energy more democratic,[47] it is relevant to say a few words about EU energy law reform. A new energy ecosystem, one that increasingly harnesses renewable energies via decentralized providers, is evolving in EU Member States. Driving this change is EU regulation, which needs to be clear for investment to be

---

[47]In terms broader than energy access, a quote attributed to Thomas Jefferson seems pertinent here: 'That government is best which governs the least.'

predictable. The linking of energy with climate-related issues is relatively new within the EU. Energy issues have always been at the heart of European integration, but energy-related topics (such as climate change policy,[48] renewable energies,[49] energy planning, and security of energy supply[50]) have only gained importance in the EU's policy and regulation agenda following an environmentalism of energy law under the auspices of sustainability.[51] Arguably, today energy and environmental regulation are two sides of the same coin.[52]

The European Union has prioritized the goal of securing affordable and low-carbon energy as the basis for a green economy. It has set aspirational goals and associated targets to launch an Energy Union beyond 2020.[53] To encourage this transition to a more secure, affordable, and decarbonized energy system,[54] the EU adopted climate and energy targets for 2020 and 2030 along with a long-term goal to reduce EU-wide greenhouse gas emissions by 80–95% below 1990 levels by 2050.[55] In 2014, the EU set the target to reduce greenhouse gas emissions by at least 40% by 2030 from 1990 levels.[56]

Under Directive 2009/72/EC of the European Parliament and of the Council and Directive 2009/73/EC of the European Parliament and of the Council, EU Member States are required to ensure the implementation of smart metering systems that assist the active participation of consumers in the electricity and gas supply markets.[57] This is part of a broader goal to increase energy efficiency through developing the demand response market.[58] The European Commission explicitly acknowledged in its Energy Union strategy that citizens should be 'at its core, where [they] take ownership of the energy transition, benefit from new technologies to reduce their bills, participate actively in the market, and where vulnerable

---

[48]See, generally, Leal-Arcas (2013).

[49]Leal-Arcas and Minas (2016), pp. 621–666.

[50]Leal-Arcas et al. (2015), pp. 291–336, Oxford University Press.

[51]Solorio et al. (2013) cited in Galera (2017), p. 13.

[52]Orlando (2014), p. 74.

[53]Fujiwara (2016), pp. 605–609.

[54]Leal-Arcas, R. "The transition towards decarbonization: A legal and policy exploration of the European Union," *Queen Mary School of Law Legal Studies Research Paper 222/2016*, pp. 1–31.

[55]European Commission, Communication from the Commission to the European Parliament, the Council, the European Economic and Social Committee and the Committee of the Regions: Energy Roadmap 2050, COM(2011) 885 final, 15 December 2011, at 2.

[56]European Council, Conclusions on 2030 Climate and Energy Policy Framework, SN 79/14, 23 October 2014, para 2.

[57]European Commission, Commission Recommendation of 10 October 2014 on the Data Protection Impact Assessment Template for Smart Grid and Smart Metering Systems, 2014/724/EU, 10 October 2014, para (2).

[58]Directive 2012/27/EU of the European Parliament and of the Council of 25 October 2012 on energy efficiency, amending Directives 2009/125/EC and 2010/30/EU and repealing Directives 2004/8/EC and 2006/32/EC, OJ L 315/1 ('Energy Efficiency Directive').

consumers are protected'.[59] One way to increase participation is through the promotion of demand response measures amongst consumers. Demand response 'is a tariff or programme established to incentivise changes in electric consumption patterns by end-use consumers'[60] 'from their normal consumption patterns in response to changes in the price of electricity over time, or to incentivise payments designed to induce lower electricity use at times of high market prices or when system reliability is jeopardized.'[61]

The 2012 Energy Efficiency Directive constitutes a major step towards the development of demand response in Europe.[62] The Directive requires EU Member States to promote participation in and access to Demand Response.[63] It also requires them to define technical modalities for participation in these markets.[64] In this regard, there remains much work to be done by EU Member States.

In terms of barriers to demand response, there are barriers to implicit demand response and barriers to explicit demand response. Regarding implicit demand response, where consumers react to price signals that they receive, barriers can be the lack of access to a dynamic pricing contract (where a consumer receives as close to real-time signals as possible) or the lack of a smart meter with the correct functionalities to transmit signals and record the consumer's reactions.

As for explicit demand response, the barriers can be divided into legal barriers and logistical barriers.[65] In terms of legal barriers, in some countries, demand response is not allowed to participate in certain markets or at all. In other countries, for example Spain, aggregation is illegal, which is a major barrier for demand response. More generally, in other countries, there is not enough clarity on legislation, which poses a legal barrier. Regarding logistical barriers, these are where technical demand response and aggregation are legal, but market requirements, such as product definitions and minimum bid sizes, are very high (for instance in Sweden), or where there are limitations on aggregation.

---

[59]European Commission, Energy Union Package: Communication from the Commission to the European Parliament, the Council, the European Economic and Social Committee, the Committee of the Regions and the European Investment Bank: A Framework Strategy for a Resilient Energy Union with a Forward-Looking Climate Change Policy, COM (2015)080 final, 25 February 2015, at 2.

[60]Smart Energy Demand Coalition (2015) Mapping Demand Response in Europe Today 2015, http://www.smartenergydemand.eu/wp-content/uploads/2015/09/Mapping-Demand-Response-in-Europe-Today-2015.pdf, at 20.

[61]Balijepalli VSKM, Pradhan V, Khaparde SA, Shereef RM (2011) Review of Demand Response under Smart Grid Paradigm. 2011 IEEE PES Innovative Smart Grid Technologies – India. http://desismartgrid.com/wp-content/uploads/2012/07/review_of_demand_response_vskmurthy.pdf, at 1.

[62]Energy Efficiency Directive.

[63]Ibid., Article 15(8).

[64]Ibid.

[65]For further details on the barriers to explicit demand response, see Smart Energy Demand Coalition (2017) Explicit Demand Response in Europe: Mapping the Markets 2017, http://www.smartenergydemand.eu/wp-content/uploads/2017/04/SEDC-Explicit-Demand-Response-in-Europe-Mapping-the-Markets-2017.pdf.

At a technical level, there is a critical need for standardized regulation at the European level, including clarified roles and responsibilities, to help to realize expanded demand response provision within a decentralized energy system.[66] Information and communication technologies, especially new digital applications for smart girds, play a central role in enabling new renewable energy providers to monitor and process data generated across distributed infrastructure and create opportunities to meet the various EU energy policy goals including efficiency, security, and sustainability.[67] These opportunities are enabling the emergence of so-called 'energy prosumers', i.e., new, flexible market actors—whether individual households or cooperatives—that generate and sell energy to the grid via Distribution System Operators (DSOs). The opportunities across the EU for consumers to become prosumers are, however, mixed, due to Member States being slow at introducing relevant regulation. There are exceptions to this situation, though, as explained below.

In terms of smart metering, Italy represents a forerunner in the EU.[68] Smart metering implementation has been completed, covering 99% of electronic metering points. The DSO is the owner and responsible party for implementing the smart grid and for guaranteeing power quality.[69] Given that the low voltage remote control meters that were first rolled out in 2001 have a lifespan of 15 years, the first replacement campaign was launched in 2016.[70]

In 2016, the first generation (1G) meters reached their end-of-life and some companies have started installing 2G meters. Italian law laid down functional specifications for 2G meters and identifies some crucial criteria: 2G meters, once installed, shall remain in operation, presumably, for another 15 years; over this period, they must be able to support every electric system transformation, such as the new distributed production paradigm and the changes of the electricity market.[71] Other countries, such as Spain, have not developed an implementation plan for smart grids. Yet, the roll-out of smart meters is ongoing.

The EU's legal reforms are thus leading to providing new opportunities for both existing actors and new ones, such as prosumers. The following section offers reflections on the opportunities emerging for all actors in the energy system, in

---

[66]Smart Energy Demand Coalition (2015) Mapping Demand Response in Europe Today 2015.

[67]See Moreno-Munoz et al. (2016), pp. 1611–1616.

[68]Reuters (2013) Europe to Follow Italy's Lead on Smart Meters, 30 May 2013, http://www.reuters.com/article/energy-efficiency-smartmeters-italy-idUSL5N0EA3HL20130530.

[69]The metering activity in Italy is regulated by the Regulation ARG/elt 199/11 (TIT).

[70]Enel S.p.A., "Enel Presents Enel Open Meter, The New Electronic Meter," Enel S.p.A., 27 June 2016. [Online]. Available: https://www.enel.com/en/media/press/d201606-enel-presents-enel-open-meter-the-new-electronic-meter.html.

[71]See Italian Legislative Decree No. 102 of 2014, 4 July 2014; Italian Regulatory Authority for Electricity Gas and Water, Second Generation Smart Metering Systems for Low Voltage Electricity Measurement: Guidelines for Determining Functional Specifications in the Implementation of Article 9, Paragraph 3 of Legislative Decree 102/2014, 416/2015/R/EFL, 6 August 2015.

climate action, and in international trade. It shifts the current paradigm in the governance of sustainability by empowering the citizen.

6.4 A Paradigm Shift in the Governance of Sustainable Development: Citizens' Empowerment

This section provides a timely and forward-thinking analysis regarding the transition to clean energy, climate action, and international trade. In the case of energy transition, it does so by offering a behavioral-economics analysis of prosumer-market factors. We argue that transitioning to clean energy cannot be achieved solely through top-down or bottom-up methods; rather, a symbiotic relationship between government or businesses creating opportunities and individual prosumers is key. This section puts an emphasis on the effectiveness of bottom-up factors like smart cities, non-governmental organizations (NGOs), and ordinary citizens.

6.4.1 From Top-Down to Bottom-Up Governance

Multilateralism does not seem to be doing well these days, at least not at the WTO).[72] There are multiple reasons for the troubles at the WTO. For one, it has not kept pace with economic and geopolitical changes, with multilateral negotiations under its auspices appearing to go nowhere and its dispute settlement system stagnating. The failure of the Doha Round is attributed by some to the deadlock between the European Union and the United States regarding agricultural subsidies, the resulting disregard of the developing countries' interests in access to global markets for agriculture, and the fundamental lack of a shared social purpose among the major trading powers.[73]

Another possible reason is the fundamental lack of trust among citizens that their interests are being sufficiently considered by those negotiating behind closed doors. Current WTO procedures, which mainly contemplate actions by member states'

---

[72] The US has been withdrawing from a number of multilateral fora since President Trump came to office. As of June 2018, the most recent example was the withdrawal from the UN Human Rights Council. See Gardiner Harris, "Trump Administration Withdraws U.S. from U.N. Human Rights Council," *The New York Times*, 19 June 2018, available at https://www.nytimes.com/2018/06/19/us/politics/trump-israel-palestinians-human-rights.html. In September 2018, it was announced that the Trump administration was keen to reform the International Postal Union, which regulates the international postal system. See The Economist, "The Universal Postal Union: Stamping on the competition," 8 September 2018, pp. 60–61.

[73] "Deadlocked in Doha," *The Economist*, March 27, 2003, https://www.economist.com/leaders/2003/03/27/deadlocked-in-doha; Drache (2006), https://www.cigionline.org/sites/default/files/trade_development_and_the_doha_round.pdf; Muzaka and Bishop (2015), p. 386.

governments, are perceived as too inflexible, and thus unable, to pay attention to the concerns of various non-state actors. Some proposals for citizens' empowerment, such as the involvement of civil society in the Committee on Trade and Environment and their participation as stakeholders during the negotiation process of future trade agreements, are difficult to incorporate into these state-to-state procedures and processes at the WTO.

Arguably, sometimes one needs unilateralism to improve multilateralism. The US intends to withdraw from the Paris Agreement on Climate Change[74] and President Trump questions the validity of the US contribution to the UN; multilateral trade negotiations at the WTO seem to go nowhere and the WTO's dispute settlement system is stagnated.[75] It seems as if the WTO has not been up to par with economic change. State-centricity seems to be making people unhappy. There seems to be a fundamental lack of trust in current governance structures.

All of this puts into question the hegemonic stability theory that predicates that the international system is most likely to be stable when a single state is the dominant power in the world.[76] Based on the view that one should never waste a crisis to reach reform, would it be the right time to think of alternative ways of governance? It is often the case that what citizens think is overlooked by policymakers. Would greater involvement of citizens make a difference for a better and more effective global economic governance? Big crises can lead to big reforms and positive developments.

A top-down guidance to sustainable development will come from inter-governmental decisions[77] (i.e., high level of abstraction),[78] whereas a bottom-up approach means that action/implementation will happen from consumers'/citizens' participation (i.e., low level of abstraction).[79] National governments are essential,

---

[74]See Leal-Arcas and Morelli (2018); Stavins, Robert "Trump's Paris Withdrawal: The Nail in the Coffin of U.S. Global Leadership?" June 6, 2017, Blog Post at An Economic View of the Environment. In accordance with Article 29 of the Paris Agreement, it takes four years to withdraw from the Agreement. Interestingly, Mr Trump's intention to quit the Paris Agreement happens to be one day after he faces re-election in 2020.

[75]US Trade Representative Robert Lighthizer has repeatedly made the point that the WTO needs to be reformed and that US trade policy has gone in the wrong direction since the creation of the WTO. See S. Donnan, "We need to talk about the Lighthizer Doctrine," *Financial Times*, 12 February 2018, available at https://www.ft.com/content/7335e48c-0fe7-11e8-8cb6-b9ccc4c4dbbb?desktop=true&segmentId=7c8f09b9-9b61-4fbb-9430-9208a9e233c8#myft:notification:daily-email:content.

[76]Goldstein (2005), p. 107.

[77]On theories of decision, see Roy and Vanderpooten (1996), pp. 22–38; Hinloopen and Nijkamp (1990), pp. 37–56.

[78]For academic literature on top-down approaches, whether in the field of energy or climate change, see Leal-Arcas and Wouters (2017), Leal-Arcas (2018b), Cameron (2010), Bodansky et al. (2017), Carlane (2010), Freestone and Streck (2009), Helm (2007), Hofmann (2012), McAdam (2012), Klein et al. (2017), Zillman et al. (2008), Roggenkamp et al. (2016) and Viñuales (2012).

[79]For a previous analysis, see Leal-Arcas (2011a), pp. 1–54; Leal and Leal-Arcas (2018); Weston and Bollier (2013).

but are no longer the only key actors. This raises the question whether cities[80] can make effective change if national governments do not deliver. At what point should businesses have to step up if politicians fall short? Cities around the world are demonstrating innovative strategies for advancing solutions to climate change. Via this bottom-up approach to governance, citizens can ask states for reform.

In the case of international trade, during the WTO Ministerial Conference in Seattle in 1999 there were large crowds of people angrily demonstrating on the streets, asking trade technocrats to be transparent and share the outcome of multilateral trade negotiations that were happening behind closed doors. Those were the days when multilateral trade was sexy. More recently, with the rise of mega-regional trade agreements (as examples of plurilateralism, which seems to be the way forward in international trade)[81] such as the Comprehensive and Progressive Agreement for Trans-Pacific Partnership (CPTPP),[82] there have been large demonstrations on the streets of the US against the Trans-Pacific Partnership and, when it comes to demonstrations against the Trans-Atlantic Trade and Investment Partnership [TTIP], we have seen the same on the streets of the UK,[83] Germany, and Austria. All of this shows an increasing interest among citizens in international trade negotiations, who are concerned that the outcome of such negotiations may affect their daily life negatively as a result of "openness to investment from other members, the protection of patents, and environmental safeguards."[84]

---

[80]There is a vast body of literature on sustainable cities. See Button (2002), pp. 217–233; Lopez Moreno et al. (2008); Wiek and Binder (2005), pp. 589–608. See also United 4 Smart Sustainable Cities, available at https://www.itu.int/en/ITU-T/ssc/united/Pages/default.aspx. A sustainable city should meet the needs of the present without sacrificing the ability of future generations to meet their own needs.

[81]The following is evidence that plurilateralism, as opposed to multilateralism, seems to be the way forward in international trade negotiations: In December 2017, during the WTO Ministerial Conference in Buenos Aires, some, but not all, WTO Members (therefore, making this procedure an example of plurilateralism) issued joint statements that were signed by subgroups of WTO Members. The aim of these plurilateral statements was to deal with specific topics, including informal work programs for Micro, Small and Medium Enterprises (WT/MIN(17)/58/Rev.1), investment facilitation (WT/MIN(17)/59), electronic commerce (WT/MIN(17)/60), fossil fuel subsidies (WT/MIN(17)/54)), as well as on services domestic regulation (WT/MIN(17)/61) within the WTO Working Party on Domestic Regulation. For an analysis of plurilateral governance in climate change, see Leal-Arcas (2011b), pp. 22–56.

[82]After the US decided to withdraw from the Trans-Pacific Partnership, which never entered into force, it was agreed in January 2018 that negotiations would start on a new trade agreement called the Comprehensive and Progressive Agreement for Trans-Pacific Partnership. To see the newly agreed text, visit https://www.mfat.govt.nz/en/trade/free-trade-agreements/free-trade-agreements-concluded-but-not-in-force/cptpp/comprehensive-and-progressive-agreement-for-trans-pacific-partnership-text/#chapters. Crucial side letters were not yet available as of February 2018.

[83]Anecdotally, it is interesting to note that more people signed an anti-TTIP campaign in the UK—which is known as a free-trade country—than in France—which is known as a protectionist nation. See The Economist, "The politics of trade deals: Not so global Britain," 10 February 2018, pp. 27–28, at 27.

[84]The Economist, "Banyan: Trading places," 27 January 2018, p. 47.

In light of the abovementioned decline in popularity or political viability of the multilateral trade system, states are entering into more regional trade agreements (RTAs), a number of which also contain innovative provisions furthering the energy transition by promoting the use and development of clean or renewable energy technologies.[85] Energy-related provisions in regional trade agreements can take various forms: including those allowing exceptions from "normal" trade rules for sustainable energy mechanisms[86]; extending subsidies for the low-carbon economies[87]; establishing climate finance instruments and capacity-building activities to develop carbon markets[88]; and enhancing trade in environmental (climate-friendly) goods and services.[89] Some RTAs additionally contain provisions that encourage investment in the energy sector with the specific objective of expanding and diversifying the energy mix and reducing dependency on fossil fuels.[90]

Yet, while there are indications of its waning status, the multilateral trading regime represented by the WTO continues to be relevant because most countries will remain predominantly energy-dependent, and therefore, markets for climate-friendly goods and services need to be opened up.

6.4.2 A Local, Bottom-Up Perspective

This section focuses on the potential role of the citizen in promoting climate-friendly approaches to energy and trade. One of the mega-trends of this century is a paradigm shift from the twentieth-century's top-down approach to climate governance (e.g., the Kyoto Protocol) to a greater emphasis on bottom-up leadership. Not only does the Paris Agreement allow States to design their own mitigation contributions, but non-State actors, including citizens, nongovernmental organizations (NGOs), cities, local leaders (mayors and governors), and businesses, are playing a major role in implementing the Agreement's climate goals. In energy governance, we observe a similar push for energy democratization as control over energy security shifts and new energy actors emerge, namely, prosumers and renewable energy cooperatives. This section argues that, particularly in this age of declining multilateralism, such bottom-up approaches could help to expedite the changes in global energy patterns required to mitigate climate change.

The rest of the section proposes a bottom-up approach to governance that the implementation of the Paris Agreement on Climate Change offers.

---

[85]Gehring et al. (2013), p. 27.
[86]Gehring et al. (2013), pp. 12–15.
[87]Gehring et al. (2013), p. 23.
[88]Gehring et al. (2013), pp. 17–18.
[89]Gehring et al. (2013), pp. 21–22.
[90]Gehring et al. (2013), p. 26.

6.4.2.1 A Different Model: Climate Change

Unlike the WTO, the Paris Agreement takes a bottom-up approach to commitments. Rather than being negotiated, emissions targets are designed by each party according to its national circumstance. The idea is to attract wide participation, especially by all the major emitters of greenhouse gases, and—although it may sound counter-intuitive—to promote greater ambition by removing the fear of sanctions for non-compliance. Another distinction from the WTO is that the targets are not legally binding and are subject to reporting and review, rather than being enforceable through binding dispute settlement. As part of these reviews, the Paris Agreement encourages updating of targets because it was recognized that the initial set of targets from over 180 countries was not sufficient to meet the global temperature goal.

Beyond the largely bottom-up approach to targets, the Paris outcome created platforms and other opportunities for non-state actors to take on commitments for emissions reduction and participate in the multiple processes.[91] This strategy is expected to augment or supplement national governments' goals and commitments. This approach, also called "hybrid multilateralism" by some authors, is characterized by an intricate entanglement of public and private authority and involves a "more integrated role [for non-state actors] in multilateral processes through ... monitoring of national action and experimentation with local, regional and transnational mitigation and adaptation strategies."[92]

Indeed, a huge amount of activity on the part of both businesses and non-national governments is occurring worldwide. Among many examples, the Non-State Actor Zone for Climate Action (NAZCA) tracks their voluntary climate action commitments.[93] In the case of the United States, businesses and state/local governments were already taking significant climate action before President Trump's announcement of his intent to withdraw the United States from the Paris Agreement; however, that announcement triggered a multitude of additional initiatives (e.g., We Are Still In, the U.S. Climate Alliance).

Global issues, such as climate change, poverty, or terrorism,[94] are too big for nation-states, but are (somewhat counter-intuitively) more suitable for cities to tackle.[95] Multiple human activities today are concentrated in cities. They are

---

[91] Galvanizing the Groundswell of Climate Actions, *How Can Funders Accelerate Climate Action to 2018-2020?: Building A Catalytic "Ecosystem" for Subnational and Non-State Actors* (Memorandum), March 2017, https://www.cisl.cam.ac.uk/publications/publication-pdfs/ggca-memorandum-to-funders-on-sub-non-state-climate-action-mar-2017-1.pdf.

[92] See Bäckstrand et al. (2017), pp. 561–579.

[93] Global Climate Action, "NAZCA 2018," Global Climate Action, available at http://climateaction.unfccc.int/.

[94] "Mayors Get Things Done. Should They Run the World?," *The Globe & Mail*, March 11, 2014, http://www.theglobeandmail.com/opinion/ideas-lab/should-mayors-lead-theworld/article17275044/.

[95] See generally Friends of Europe, *Cities: The New Policy Shapers in the Energy Transition*, November 2017, https://www.friendsofeurope.org/sites/default/files/2017-11/Cities_web_2.pdf.

where the majority of the world population lives[96] (and this trend is still rising)[97]; where 50% of global waste is produced[98]; where 80% of global economic activity (as measured by gross domestic product) takes place[99]; and from which between 60% and 80% of greenhouse gas (GHG) emissions originate.[100] City mayors tend towards a more pragmatic approach that arguably offers better governance on these matters. Moreover, mayors tend to come from the cities they govern and therefore garner a much higher level of trust than politicians at the national level. Accordingly, as the main polluters and the main implementers of legislation,[101] cities (and therefore citizens) can, and should, take climate action. Indeed, cities around the world are demonstrating innovative strategies for advancing solutions to climate change.[102] Specifically, city-level climate action that includes a much greater participation of citizens is very promising.[103] The so-called "all hands on deck" approach means that climate action includes citizens' seemingly mundane daily choices, such as commuting to the workplace and upgrading the lighting system in buildings.[104]

Rapid decarbonization efforts have the greatest potential in cities, particularly in developing countries and in sectors such as construction, transport, energy, water, and waste. In these areas, private sector investments are most crucial.[105] While the private sector is not lacking in initiative to contribute to climate change mitigation

---

[96] See "World's Population Increasingly Urban with More than Half Living in Urban Areas," U.N. Department of Economic & Social Affairs, July 10, 2014, http://www.un.org/en/development/desa/news/population/world-urbanization-prospects-2014.html.

[97] Mark Wilson, "By 2050, 70% of the World's Population Will Be Urban. Is That a Good Thing?," Co.Design, March 12, 2012, https://www.fastcodesign.com/1669244/by-2050-70-of-theworlds-population-will-be-urban-is-that-a-good-thing.

[98] United Nations Environment Programme, "Resource efficiency as key issue in the new urban agenda: Advancing sustainable consumption and production in cities," 1, available at http://sdg.iisd.org/news/unep-international-panel-calls-for-improved-resource-efficiency-for-sustainable-urbanization/.

[99] Dobbs et al. (2011) https://www.mckinsey.com/~/media/McKinsey/Featured%20Insights/Urbanization/Urban%20world/MGI_urban_world_mapping_economic_power_of_cities_full_report.ashx.

[100] UNEP-DTIE Sustainable Consumption and Production Branch, *Cities And Buildings: UNEP Initiatives and Projects* (Paris: United Nations Environment Programme, undated) 5, http://www.oas.org/en/sedi/dsd/Biodiversity/Sustainable_Cities/Sustainable_Communities/Events/SC%20Course%20Trinidad%202014/ModuleVI/2.%20Cities%20and%20Buildings%20%E2%80%93%20UNEP%20DTIE%20Initiatives%20and%20projects_hd.pdf.

[101] "The 100 Climates Solutions Projects Campaign of the R20." *R20 - Regions of Climate Action* (blog). Available at https://regions20.org/our-projects/100-climate-solutions-projects-campaign/.

[102] Carbon Neutral Cities Alliance, *Framework for Long-Term Deep Carbon Reduction Planning*, (December 2015), https://www.usdn.org/uploads/cms/documents/cnca-framework-12-2-15.pdf?source=http%3a%2f%2fusdn.org%2fuploads%2fcms%2fdocuments%2fcnca-framework-12-2-15.pdf.

[103] Monstadt (2007), pp. 326–343.

[104] Carbon Neutral Cities Alliance, *Long-Term Deep Carbon Reduction*, 28.

[105] Echeverri (2018), pp. 42–51.

efforts, citizens can additionally put pressure on businesses to undertake environmentally responsible and beneficial activities.[106] Parenthetically, when politicians fall short, businesses may have a role to play in helping decarbonize the economy. While politicians are susceptible to short-termism (for obvious electoral reasons) and may be too risk-averse, entrepreneurs tend to be "riskophiles" and persistent, with longer-term visions and commitments.

Private sector-led climate action and initiatives matter on two main fronts: first, on actual contributions to the decrease in global GHG emissions; and second, through private investors financing the mitigation and adaptation efforts of other businesses or of capital-deficient developing countries (and the cities therein). Companies that have set, or committed to set, science-based climate targets include Honda Motor Company, The Port Authority of New York and New Jersey, AstraZeneca, IKEA, Tesco, and Dell Technologies.[107] Science-based targets refer to "the level of decarbonization required to keep global temperature increase below 2 °C compared to pre-industrial temperatures, as described in the Fifth Assessment Report of the Intergovernmental Panel on Climate Change."[108] The Portfolio Decarbonization Coalition, for example, comprises asset owners and managers that collectively endeavor to re-channel "capital from particularly carbon-intensive companies, projects and technologies in each sector" to "particularly carbon-efficient" ones and "will commit to a concrete decarbonization plan."[109]

A question remains, however, whether and how these types of bottom-up responses to climate change can be specifically imported into the energy transition sphere. As the world reduces its oil dependence, production and export of green technology and reliance on clean energy would be of considerable value. Two ingredients may help move forward the energy transition: international collaboration and energy decentralization. Potential international collaboration can be achieved in the field of technology, for which international trade will certainly play a major role. As for energy decentralization, the emergence of micro-/mini-grids dealing with locally produced wind and solar energy as well as electric-vehicle batteries is the way forward. All of these innovations will not only help in providing better access to energy, but it will also decentralize economies.

Softer, informal tools of governance, rather than treaties, seem to be central to the current crisis/transformation of multilateral governance. In the field of energy governance, regulatory alignment, technology alignment, and building common institutions might all help enhance sustainable energy.[110] New actors are emerging. One of them is the citizens.

---

[106]See, for instance, the initiative 'The consumer goods forum,' at https://www.theconsumergoodsforum.com/.

[107]https://sciencebasedtargets.org/companies-taking-action/ See also: https://www.bsr.org/en/about.

[108]Science Based Targets, "What Is a Science-Based Target?" Science Based Targets. Available at https://sciencebasedtargets.org/what-is-a-science-based-target/.

[109]UNEPFI, "Portfolio Decarbonization Coalition - About," UNEPFI, Available at http://unepfi.org/pdc/about/.

[110]See generally Leal-Arcas et al. (2014).

6.4.3 New Concept: Citizen Empowerment

Citizens' empowerment is a relatively new concept in global governance.[111] Consumers' and citizens' participation in the determination and implementation of solutions to global problems significantly deviates from the traditional approach under which inter-governmental decisions dictate commitments and actions from the top down. One way that private individuals are participating in climate action is through the Carbon Rationing Action Group (CRAG) initiative, wherein a voluntarily formed group tracks each member's personal CO2 emissions (air travel, household heating, car use, household electricity consumption, and/or other public transport use).[112] The CRAG initiative has a notable enforcement mechanism in the form of financial penalties and exclusion from the scheme.[113] Additionally, individuals are exercising their power as consumers by demanding "green products and services"; in this regard, eco-labeling aids in sending signals between consumers and producers and facilitating transactions.[114]

Empowering citizens has implications for societal change as it provides a human element to governance.[115] More direct participation by citizens is increasingly necessary to reach good governance. In the field of energy governance, one of the aims of this section is to explore how to effectively place citizens at the center of the transformation of the grid by allowing greater citizen participation and access to information. Citizen participation will bring stability, facilitate citizens' wellbeing, provide better access to energy, it will put pressure on companies to do the right thing,[116] and provide better management of climate change and environmental issues. By doing so, we are moving away from energy poverty towards a transition to energy democracy,[117] energy citizenship,[118] decentralized energy,[119] sustainable

---

[111] See generally Sassen (2003), pp. 5–28; Cullen and Morrow (2001), pp. 7–40.

[112] Carbonday, "CRAG : Carbonday." Carbonday, Accessed November 11, 2018. http://carbonday.com/get-involved/crag/.

[113] Dellinger (2013), pp. 621–623.

[114] Orts (2011), pp. 231–233.

[115] Leal-Arcas (2018c), pp. 1–37.

[116] See for instance the initiative 'The consumer goods forum,' at https://www.theconsumergoodsforum.com/. See also Carroll (1974), pp. 75–88; Carroll (1979), pp. 497–505; Carroll (1999), pp. 268–295; Chen and Bouvain (2009), pp. 299–317; Freeman and Hasnaoui (2010); Idowu and Towler (2004), pp. 420–437; O'Connor and Spangenberg (2008), pp. 1399–1415; Welford et al. (2007), pp. 52–62.

[117] Morris and Jungjohann (2016). The concept of *Energiewende* describes Germany's efforts to move away from fossil fuels and nuclear power by promoting renewable energy instead, whilst remaining a major industrial power. See Buchan (2012), p. 1.

[118] See for instance Devine-Wright (2007).

[119] Orehounig et al. (2015), pp. 277–289.

6.4 A Paradigm Shift in the Governance of Sustainable Development:...

energy enhancement,[120] more effective climate change mitigation and greater presence of citizens in trade policy/diplomacy.

Since more prosumers are entering the market, all of this, in turn, will lead to the creation of scalable micro-grids for prosumers[121] and utility companies, new policies and regulatory frameworks for smart grids, as well as a better grid management.[122] It will also encourage prosumers towards a more energy-efficient behavior. Further, it will change citizens' attitudes from being passive to active consumers by presenting a variety of local engagement opportunities. Local renewable energy communities are at the grassroots of the movement to change the current energy-security system. For instance, how can legal technical barriers to energy technology[123] be reduced or eliminated for smart grids to take off in different jurisdictions?[124] How could the legal environment be developed to benefit technology and create, say, a single smart grid in supranational structures like that of the EU?[125] Such a system would make energy security cheaper and consumers would be able to control and manage their energy bills.

The use of behavioral economics in public policy has been increasingly on the agenda. In energy policy, "it has become clear that efforts to steer people towards "better"—that is, more energy efficient—choices and behaviours are much needed."[126] As suggested by Lucia Reisch, there is increasing evidence that the right incentives do spur behavioral change.[127] This has certainly been the case in Nordic countries, where the so-called Nordic model has failed in top-down policies (such as the creation of common defense policy, a single currency), but has been

---

[120]See for instance Omar (2018), available at http://www.file:///C:/Users/lcw197/Downloads/FETWorkshoponFutureBatteryTechnologiesforEnergyStorageA4webpdf.pdf.

[121]For further details on prosumers, see Leal-Arcas et al. (2017); Leal-Arcas (2018d), pp. 4–20.

[122]A UK National Infrastructure Commission report states that 'smart power' could save consumers up to £8bn per year by 2030 via smart grid demand management, energy storage, and interconnectors. See National Infrastructure Commission, "Smart Power," available at https://assets.publishing.service.gov.uk/government/uploads/system/uploads/attachment_data/file/505218/IC_Energy_Report_web.pdf.

[123]On new energy technologies, see Afgan and Carvalho (2002), pp. 739–755; Pehnt (2006), pp. 55–71; Madlener and Stagl (2005), pp. 147–167.

[124]According to Stanford University researchers, 'utilities around the world can rely on multiple methods to stabilize their electricity grids in a shift to 100% wind, solar, and hydroelectricity.' See T. Kubota, "Jacobson study shows multiple paths to grid stability in 100% renewable future," *The Energy Mix*, 14 February 2018, available at http://theenergymix.com/2018/02/14/jacobson-study-shows-multiple-paths-to-grid-stability-in-100-renewable-future/.

[125]For an initiative in this direction towards energy cooperation between the North Seas countries, see The North Seas Countries' Offshore Grid Initiative, available at http://www.benelux.int/nl/kernthemas/holder/energie/nscogi-2012-report/. Similar thinking is taking place for the creation of a single, shared 5G wireless network. See The Economist, "Telecoms: Next-generation thinking," 10 February 2018, pp. 11–12.

[126]L. Reisch, "Nudging Europe's Energy Transformation," *The Globalist*, 20 August 2012, available at https://www.theglobalist.com/nudging-europes-energy-transformation/.

[127]Ibid.

very successful in the design of bottom-up approaches to policies with the right incentives and market integration.[128]

This shift in the governance of sustainable development implies putting citizens at the center of this process. The phenomenon of what we describe as a 'bottom-up approach' to the *democratic*[129] implementation of climate change mitigation plans is one of the mega-trends of the twenty-first century.[130] Since the majority of the world population lives in cities[131] (and this trend is on the rise),[132] since 50% of global waste is produced in cities,[133] since 80% of global economic activity takes place in cities,[134] since the urban heat island effect is a fact,[135] since cities consume about

---

[128]Hans-Arild Bredesen, Terje Nilsen, Elizabeth S. Lingjærde, *Power to the People: The first 20 years of Nordic power-market integration*, 2013.

[129]In the true sense of the term, namely that power remains with the citizens. For analyses of democracy, see Deneen (2018); Frum (2018); Levitsky and Ziblatt (2018).

[130]A creation of the Paris Agreement, which has become the locomotive of climate action.

[131]*See World's Population Increasingly Urban with More than Half Living in Urban Areas*, U.N. DEP'T OF ECON. & SOC. AFFAIRS (July 10, 2014), http://www.un.org/en/development/desa/news/population/world-urbanization-prospects-2014.html.

[132]By 2050, 70% of the world's population is expected to live in cities. *See* Mark Wilson, *By 2050, 70% of the World's Population Will Be Urban. Is That a Good Thing?*, CO.DESIGN (Mar. 12, 2012), https://www.fastcodesign.com/1669244/by-2050-70-of-theworlds-population-will-be-urban-is-that-a-good-thing.

[133]UNEP, "Resource efficiency as key issue in the new urban agenda: Advancing sustainable consumption and production in cities," available at http://web.unep.org/ietc/sites/unep.org.ietc/files/Key%20messages%20RE%20Habitat%20III_en.pdf, p. 1.

[134]As measured by global Gross Domestic Product (GDP). Richard Dobbs et al., McKinsey Global Institute, Urban world: Mapping the economic power of cities 1 (2011). For various analyses of cities as centers of human interaction, see Geddes (1915), Weber (1921) and Batty (2013).

[135]The urban heat island effect explains why urban areas are significantly warmer than rural areas due to anthropogenic activity. On the links between climate change and anthropogenic activity, see Schrag (2012), pp. 425–436. In 120 years, world population has increased fivefold: from 1.5 to 7.5 billion. By 2030, around 60% of the world population may be living in cities. Urban growth is a multiplier of human impacts. Cities typically increase living standards as well as impacts. The so-called 'I = PAT' equation is the mathematical notation of the formula $I = P \times A \times T$ that describes the impact of human activity on the environment, where 'I' represents the environmental impact, 'P' population, 'A' affluence, and 'T' technology. Society has changed over the centuries, from living in the agropolis (i.e., a city of around 20,000 people, surrounded by concentric circles of markets, forest, cultivation, and pasture) to the petropolis (i.e., the modern city that depends on fossil fuel inputs) to eventually the ecopolis (i.e., a city based on the four laws of ecology, namely (1) everything is connected to everything else; (2) everything must go somewhere; (3) nature knows best; and (4) nothing comes from nothing). See The Next Wave, 2 March 2015, available at https://thenextwavefutures.wordpress.com/2015/03/02/from-urban-paradox-to-ecopolis-22/. The idea of ecopolis is adapted from the work by Commoner (1971). In this sense, the Chinese concept of 'ecological civilization' is most welcome. President Xi of China famously said: "Promoting eco-civilization is an important part of China's overall plan to develop its economy, politics, culture, social progress and ecology. It is also the internal need of China's modernization construction. The people's well-being also relies on a beautiful environment. The concept of green development should be embedded in every step of social construction." See "China on its way to promote ecological civilization," 11 August 2017, available at http://www.china.org.cn/china/

6.4 A Paradigm Shift in the Governance of Sustainable Development:...

75% of global primary energy,[136] and since between 60% and 80% of GHG emissions comes from cities,[137] this new mega-trend of climate action at the city level with a much greater participation of citizens is very promising.[138] However, when trying to get to 100% green energy, many cities may have to import power from rural and offshore areas, where space is less of an issue than in cities.[139]

So why should cities (and therefore citizens) take climate action? Because cities are the main polluters and the main implementers of legislation[140]; and because mayors of cities are pragmatic with global issues such as climate change, poverty or terrorism.[141] Also because such issues are too big for nation-states and because cities arguably offer better governance on these matters. Moreover, mayors tend to come from the cities they govern and therefore have a much higher level of trust than politicians at the national level.

What should be the role of citizens in the shift towards a circular economy (i.e., recycling and reusing products) and in trade diplomacy? What should be the role of the emerging environmental goods and services sector? In the specific case of international trade, one could imagine as citizens' empowerment the involvement of civil society, as stakeholders of trade agreements, in committees on trade and environment via their participation during the negotiation process of future trade agreements. Moreover, with the rise of e-commerce, one could think of the increasing participation of micro, small and medium enterprises via apps on their smartphones. How can trade policy have more contact with private companies that are involved in international trade? Regarding the process of negotiation of trade agreements, however, there are technical barriers to bringing participation to the grassroots level. Potential areas for improvement and participation at the grassroots

---

2017-08/11/content_41390769.htm; H. Wang-Kaeding, "What does Xi Jingping's new phrase 'Ecological Civilization' mean?" The Diplomat, 6 March 2018, available at https://thediplomat.com/2018/03/what-does-xi-jinpings-new-phrase-ecological-civilization-mean/.

[136]See UN Habitat, available at https://unhabitat.org/urban-themes/energy/.

[137]U.N. Env't Program, "Cities and buildings: UNEP initiatives and projects," at 5, http://www.oas.org/en/sedi/dsd/Biodiversity/Sustainable_Cities/Sustainable_Communities/Events/SC%20Course%20Trinidad%202014/ModuleVI/2.%20Cities%20and%20Buildings%20%E2%80%93%20UNEP%20DTIE%20Initiatives%20and%20projects_hd.pdf [https://perma.cc/QZC9-V8TR].

[138]Monstad (2007), pp. 326–343.

[139]Herbert Girardet speaks of two types of cities: (1) a regenerative city, which is powered, heated, cooled, and driven by renewable energy, and which restores degraded ecosystems; and (2) a resource-wasting city, which emits large amounts of CO2 without ensuring re-absorption, uses resources without concern for their origins or destination of their wastes, and consumes large amounts of meat produced mainly with imported feed. Moreover, cities tend to be centers of cultural excellence, places where ideas are shared, and where people's talents are magnified by stimulating interaction and innovation. See generally Girardet (2014).

[140]Regions of climate action, http://regions20.org/.

[141]*Mayors Get Things Done. Should They Run the World?*, The Globe & Mail (Mar. 11, 2014), http://www.theglobeandmail.com/opinion/ideas-lab/should-mayors-lead-theworld/article17275044/.

Fig. 6.1 Conceptualizing the empowerment of citizens in trade, energy and climate

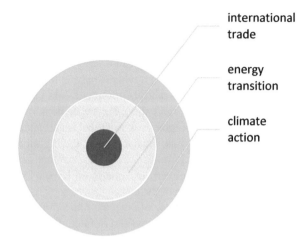

level are transparency,[142] NGO involvement, the implementation of trade agreements, information asymmetry, and due process, among others.

Following the so-called Thünen's model of agricultural land,[143] one can think of the following graphic representation of concentric circles to describe citizens' priorities when it comes to their empowerment in trade, energy transition, and climate action:

Figure 6.1 explains that, in the priorities of empowering citizens, trade comes first because it is a daily need with the widest and most tangible impact, unlike access to energy (which is desirable, but not essential for survival) or being a victim of the consequences of climate change (which is the least tangible and most abstract of the three concepts).

The rest of this section proceeds as follows: First, it elucidates the citizens' involvement in trade institutions and agreements. It then explains the concept of citizen empowerment by presenting the energy transition goals and proposes bottom-up approaches to the energy transition and climate action. Let us now, in turn, deal with each one of the three concepts (i.e., international trade, energy, and climate action) in the context of empowering citizens.

---

[142] See for instance the Green Paper of 3 May 2006 on European transparency initiative, COM (2006) 194 final, Official Journal C 151 of 29 June 2006.

[143] The Thünen's model of agricultural land, named after Johann Heinrich von Thünen, is the first serious treatment of spatial economics and economic geography, connecting it with the theory of rent. The model made the following assumptions: The city is located centrally within an "isolated State;" the isolated State is surrounded by wilderness; the land is completely flat and has no rivers or mountains; soil quality and climate are consistent; farmers in the isolated State transport their own goods to market via oxcart, across land, directly to the central city. There are no roads; and finally farmers behave rationally to maximize profits. See Wikipedia, "Johann Heinrich von Thünen," available at https://en.wikipedia.org/wiki/Johann_Heinrich_von_Th%C3%BCnen#Th%C3%BCnen's_model_of_agricultural_land.

6.4.3.1 International Trade

International trade can aid in responding to today's sustainability challenges, specifically those pertaining to climate change and energy transition, by offering means for various actors to cooperate in enhancing sustainable energy. International collaboration is important in making clean energy cheaper and encouraging its wider use, and international trade in clean energy technologies can "mitigate greenhouse gas emissions in addition to the economic gains that can be expected from an optimization of supply chains."[144]

Accordingly, the continued relevance of the multilateral trade system can no longer mean that only states would be involved in the development of trade agreements that relate to, or affect, the energy sector. Due consideration must be given to the interests of non-state stakeholders who should be allowed to participate in multilateral trade processes either directly, by giving them seats at negotiating tables or committees, or indirectly through procedures facilitating transparency and greater consultation.

With regard to existing trade remedies, for instance, various stakeholders should be consulted because their inputs are relevant in ascertaining whether an antidumping or countervailing duty is in the public interest.[145] At the WTO, there already exist promising efforts to engage with civil society: at the Eleventh Ministerial Conference in Buenos Aires, 251 NGOs from 52 countries were accredited.[146]

Moreover, in December 2017, the EU Commission announced the creation of a new advisory group on EU trade agreements.[147] The aim of the group is to increase transparency and inclusiveness in EU trade policy. The EU Commission is committed to this cause.[148] The perspective of this wide group of stakeholders[149] (consumer groups, trade unions, and other non-governmental organizations) on EU trade policy will certainly help towards better trade policymaking in the future. The EU Commission has also acknowledged elsewhere EU citizens' expectation that EU trade

---

[144]Meléndez-Ortiz and Sugathan (2017), p. 934.

[145]Meléndez-Ortiz and Sugathan (2017), p. 949.

[146]https://www.wto.org/english/thewto_e/minist_e/mc11_e/ngo_e.htm.

[147]European Commission, "Commission decision of 13.9.2017 setting up the Group of Experts on EU Trade Agreements," C(2017) 6113 final, available at http://ec.europa.eu/transparency/regexpert/index.cfm?do=groupDetail.groupDetailDoc&id=34613&no=1.

[148]See speech by European Commission President Jean-Claude Juncker on the State of the Union 2017, 13 September 2017, available at http://europa.eu/rapid/press-release_SPEECH-17-3165_en.htm; see also European Commission, Communication from the Commission to the European Parliament, the Council, the European Economic and Social Committee and the Committee of the Regions, "A balanced and progressive trade policy to harness globalisation," COM(2017) 492 final, 13.9.2017, available at https://ec.europa.eu/transparency/regdoc/rep/1/2017/EN/COM-2017-492-F1-EN-MAIN-PART-1.PDF.

[149]To access the list of members in the expert group on EU trade agreements, see http://trade.ec.europa.eu/doclib/docs/2017/december/tradoc_156487.pdf.

agreements should support sustainable-development objectives such as climate action.[150]

The role of citizens and micro, small and medium-sized enterprises (MSMEs) in international trade governance is another example of a bottom-up approach to sustainable development governance that would shift the current paradigm. A report authored by the WTO Secretariat states how the current trade governance system can support MSMEs in their participation in the international trading system:

1. By helping them meet sustainability standards and conforming with other international regulations to take advantage of the opportunities resulting from global supply chains;
2. By ensuring that MSMEs can trade their goods and services in a timely and competitive manner, which will result in greater consumer confidence; and
3. By making sure that trade finance is available. Doing so will contribute to gender equality (which is the solution to achieving many of the SDGs, not the challenge), increasing economic growth, fostering innovation, and increasing participation in international trade.[151]

The WTO[152] has focused on trade-policy related factors that affect MSMEs' limited but gradually growing participation in international trade, and tackled how e-commerce and other information and communications technology (ICT)-oriented services encourage their access to world markets and global value chains.[153] Participation in international trade helps MSMEs grow and become more profitable, and cooperation among states helps lessen obstacles to such participation.[154] Specifically, trade agreements can, among others, reduce variable and fixed costs of trade that are particularly burdensome for SMEs, and can ease the information deficiency or asymmetry relating to non-tariff measures (i.e., standards and regulations) that SMEs have to deal with.[155] Trade cooperation likewise enables states to provide preferential treatment (and access) to SMEs and assist their technological development.[156] The report also highlighted the MSMEs' contribution to the creation of more inclusive employment which potentially brings more citizens closer to the global marketplace.[157]

---

[150]Non-paper of the Commission services, "Trade and Sustainable Development (TSD) chapters in EU Free Trade Agreements (FTAs), 11 July 2017, available at http://trade.ec.europa.eu/doclib/docs/2017/july/tradoc_155686.pdf.

[151]World Trade Organization, "Mainstream trade to attain the sustainable development goals," p. 64, available at https://www.wto.org/english/res_e/booksp_e/sdg_e.pdf.

[152]WTO Secretariat (2016).

[153]WTO Secretariat (2016), pp. 20–21.

[154]WTO Secretariat (2016), p. 130.

[155]WTO Secretariat (2016), pp. 131–134.

[156]WTO Secretariat (2016), pp. 130–131.

[157]WTO Secretariat (2016), p. 18.

Trading is not possible without trust. Trust is based on incentives. Citizens need to have the necessary framework that enables them the required trust to believe in a trading system where they can be participants. For instance, green consumer behavior in trade (such as gradually getting rid of using fossil fuels) will help towards the mitigation of climate change. The more harmonized the market, greater economic incentives will derive from it. A key ingredient to improving trade (in energy) is better and more efficient connection between markets. All of this can be achieved if markets work towards "zero tariffs, zero non-tariff barriers and zero subsidies."[158]

6.4.3.2 Energy Transition

6.4.3.2.1 The Role of Citizens

Promoting the use of renewable energy is one of the most pressing concerns for climate change and long-term sustainability at a global level. The long-term goal should be 100% energy use from wind, solar, and hydropower sources. At present, the EU has set a binding target of 20% final energy consumption from renewable sources by 2020,[159] and the Council has endorsed achieving at least 27% renewable energy consumption in 2030.[160] While the transition is happening at a slow pace,[161] it is promising that the energy mix is changing to low carbon[162] and is getting cheaper.[163]

Apart from the power sector, heating, cooling, and transport are sectors where fossil fuels need to be gradually replaced with renewables.[164] Sector coupling—either among sub-units within the energy sector or between energy and other sectors—might be a way to make this replacement possible.

---

[158]The Economist, "Trade and tariffs: Sealed with a kiss," 28 July 2018, p. 27.

[159]Renewable Energy Directive: https://ec.europa.eu/energy/en/topics/renewable-energy/renewable-energy-directive.

[160]"European Council agrees climate and energy goals for 2030," October 23, 2014, https://ec.europa.eu/energy/en/news/european-council-agrees-climate-and-energy-goals-2030. See also: http://www.consilium.europa.eu/en/policies/climate-change/2030-climate-and-energy-framework/.

[161]Florence Schulz, "Energy transition goals lack ambition, local projects lead the way, study suggests," *EURACTIVE Germany*, April 12, 2018, https://www.euractiv.com/section/energy/news/energy-transition-goals-lack-ambition-local-projects-lead-the-way-study-suggests/.

[162]Agora Energiewende and Sandbag, *The European Power Sector in 2017: State of Affairs and Review of Current Developments* (January 2018) 5-7, https://sandbag.org.uk/wp-content/uploads/2018/01/EU-power-sector-report-2017.pdf; Michael Holder, "Renewables push low carbon sources to almost 55 per cent of UK power mix," *BusinessGreen*, 8 February 2018, https://www.businessgreen.com/bg/news-analysis/3026287/renewables-push-low-carbon-sources-to-almost-55-per-cent-of-uk-electricity.

[163]Frankfurt School-UNEP Centre/BNEF (2018), pp. 16–17, http://fs-unep-centre.org/sites/default/files/publications/gtr2018v2.pdf.

[164]de Llano-Paz et al. (2015), pp. 54–55.

However, the achievement of energy transition goals cannot rely on reduced energy demand alone as it is not realistic to expect such a considerable reduction. Instead, the focus should be directed to a smart policy design for energy demand, with "smartness" measured along the dimensions of sustainability, collaboration, and intelligence.[165] As researchers explain:

> [A] holistic approach is required to effectively deal with the current challenges posed by the sustainable development principles: consumers (and their communities at large) should be explicitly encouraged to be directly engaged through a more participative and collaborative behavior, factually realizing a collaborative consumption strategy, that is a technology-enabled sharing of goods and services between consumers that requires enhanced forms of collaborations.[166]

Smart energy demand policy needs to be complemented with technological and institutional improvements on the supply side. The "smart grid" concept and the emphasis on collaborations among producers and consumers capture this suggestion.[167] As will be elaborated below, making energy access cheaper and more secure requires the reduction or elimination of legal technical barriers to energy technology so that smart grids can take off in different jurisdictions.[168] If we succeed at a more efficient and sustainable energy system, energy imports and energy dependency will gradually fall, costs will be cut, and GHG emissions will be reduced.[169]

The energy transition, which is happening at a slow pace,[170] is an opportunity to protect the planet, as is also to create jobs and provide economic growth.[171] Currently, we have central generation of energy, one-way flow of energy, and passive consumers. With the energy transition, we could have distributed generation of energy and a two-way power flow.[172] The long-term goal in the energy field is 100% energy use from wind, solar, and hydropower.

---

[165] Zaheer Tariq, Sergio Cavalieri, and Roberto Pinto, "Determinants of Smart Energy Demand Management: An Exploratory Analysis," 550-51.

[166] Tariq, Cavalieri, and Pinto, "Determinants of Smart Energy Demand Management," 548-49.

[167] For an initiative in this direction towards energy cooperation between the North Seas countries, see The North Seas Countries' Offshore Grid Initiative, http://www.benelux.int/nl/kernthemas/holder/energie/nscogi-2012-report/. Similar thinking is taking place for the creation of a single, shared 5G wireless network. See "Telecoms: Next-generation thinking," *The Economist,* February 10, 2018, 11-12.

[168] According to Stanford University researchers, "utilities around the world can rely on multiple methods to stabilize their electricity grids in a shift to 100% wind, solar, and hydroelectricity." See Taylor Kubota, "Jacobson study shows multiple paths to grid stability in 100% renewable future," *The Energy Mix,* February 14, 2018, http://theenergymix.com/2018/02/14/jacobson-study-shows-multiple-paths-to-grid-stability-in-100-renewable-future/.

[169] In 2018, global energy-related CO_2 emissions grew by 1.7%. This was due to adverse weather, which increased the demand for cooling and heating. See The Economist, 30 March 2019, p. 10.

[170] This is largely due to the fact that the financial returns from oil are higher than those from renewables.

[171] On this point, see Meléndez-Ortiz (2016).

[172] This situation raises the question of how to price power produced at home.

6.4 A Paradigm Shift in the Governance of Sustainable Development:...

The EU has called for a paradigm shift in the energy area, encouraging a shift towards a more prosumer-oriented market, placing citizens at the heart of energy security by promoting self-consumption and smaller local energy communities, where prosumers can produce their own energy. The idea is to give more control to consumers. It is believed that smart grids have enormous potential to make this happen. To enable the successful implementation of smart grids, states must have a framework in place that will enable that transition to happen. To successfully implement smart grids, states must take action to ensure the democratization, decarbonization, digitalization, and decentralization of the economy and the energy market.

Since the energy sector and the economy go hand in hand, the future of the energy transition and the future of countries' economies will inevitably go hand in hand. There are several factors to take into account in the energy transition: circularity/cradle-to-grave principle (recycling over and over again),[173] consumer's engagement, decarbonization, long-term thinking, minimizing social impact on consumers, multilevel governance (at local, regional, national, supranational, international level), simplicity, speed (namely making sure that the energy transition happens within a reasonable timeframe), affordability, and transparency with data.

But what are the main drivers of the energy transition in the energy market? Several factors seem to come to mind: increased access to information and communication; energy decentralization which, as a result, leads to energy democratization[174] via a multilevel governance system; citizens' empowerment[175] aiming at a state of autarky (in as much as this is possible) in a customer-centered system that enables them to exploit market opportunities; new business models; innovation; stronger and smarter grids; better and smarter regulation aiming at reducing or eliminating technical barriers[176]; and electrification, which drives the deployment of renewable energy.[177]

What is the role of the market in securing a successful energy transition? It is, among other things, to set price signals, to provide regulatory adjustments to new situations, to influence the drivers that will make the energy transition a reality, to provide a level playing field, to act as an enabler for business models, to drive competition, to provide further economic liberalization, to drive consumer behavior (and vice versa, i.e., consumer behavior will drive the market), and to enable innovation.

---

[173] Geissdoerfer et al. (2017), pp. 757–768.

[174] By energy democratization, we mean a situation where regions and consumers gradually become more self-sufficient in their access to energy.

[175] Leal-Arcas (2018c), pp. 1–37.

[176] Leal-Arcas et al. (2018).

[177] On renewable energy in the context of energy transition, see IRENA, IEA, and REN21, "Renewable energy policies in a time of transition," 2018, available at http://www.irena.org/-/media/Files/IRENA/Agency/Publication/2018/Apr/IRENA_IEA_REN21_Policies_2018.pdf.

The implementation of the energy transition will inevitably vary from country to country, based on access to technology and economic conditions.[178] It will require the convergence of centralized with decentralized energy systems. For instance, in the case of the EU, it will require solar and wind energy integration for the implementation of the energy transition. Greater flexibility will be necessary for cross-border energy trade and for local/regional smart grids.

The energy mix is changing to low carbon and is getting cheaper. Moreover, in addition to the power sector, heating, cooling, and transport are sectors where fossil fuels need to be gradually replaced with renewables. Sector coupling (i.e., interconnecting the energy-consuming sectors with the power-producing sector) may be a way to make this possible within the energy sector and between the energy sector and other sectors.[179] In addition, reducing energy demand may not be an option in the future, given our life style in the West, which is increasingly replicated in the rest of the world. Instead, what is needed is a smart policy design for energy demand, which needs to be complemented with technological and institutional improvements on the supply side. If we succeed at a more efficient and sustainable energy system, energy imports and energy dependency will gradually fall, costs will be cut and GHG emissions reduced. One can also provide incentives for CO2 emissions reduction.[180]

How can we get there? By empowering citizens in access to energy.[181] Gordon Walker has identified four types of community-owned means of renewable-energy production in the UK: (1) cooperatives, (2) community charities, (3) development trusts, and (4) renewable-energy projects with shares owned by a local community organization.[182] In addition, there are examples of cooperative models for wind turbine companies in several EU countries (namely Austria, Germany, Denmark, The Netherlands), which are illustrations of innovative models of citizens' participation and community involvement in energy production.[183] What citizens want from the grid is security of supply, lower bills, protecting the environment, and smartness.

---

[178]Think for instance of the polymer problem, where having proper waste-management systems makes a difference to solve it. See The Economist, "Plastic Pollution: Too much of a good thing," 3rd March 2018, pp. 50–52. More generally, see Nakamura (1999), pp. 133–145; Nakamura and Kondo (2006), pp. 494–506.

[179]See for instance European Commission, Communication from the Commission to the European Parliament, the Council, the European Economic and Social Committee, the Committee of the Regions and the European Investment Bank, "Clean energy for all Europeans," COM(2016) 860 final, 30 November 2016, available at http://eur-lex.europa.eu/resource.html?uri=cellar: fa6ea15b-b7b0-11e6-9e3c-01aa75ed71a1.0001.02/DOC_1&format=PDF; European Commission, "A Roadmap for moving to a competitive low carbon economy in 2050," COM(2011) 112 final, 8 March 2011. These two policy papers aim at the convergence of liberalization with climate action.

[180]California is considering the possibility of subsidies to remove CO2. See The Economist, "The power of negative thinking," 9 June 2018, p. 78.

[181]See for instance Gladwell (2002).

[182]Walker (2008), pp. 4401–4405.

[183]Bauwens et al. (2016), pp. 136–147.

6.4.3.2.2 Bottom-Up Approaches to the Energy Transition

Pursuing energy transition through citizen empowerment takes various forms. Undertakings can be conceptually categorized into those directed at consumer information dissemination and raising public awareness; access to cheaper, smarter, and more secure energy; and cooperation among prosumers and collaboration between producers and consumers. Let's analyze each one in turn.

6.4.3.2.2.1 Information Dissemination and Awareness

In light of evidence that the younger generation wants to consume in a sustainable manner,[184] use of social media (Twitter, Facebook, YouTube) could be leveraged to further educate the youth (the segment of society that makes most use of it) about the links among trade, climate change, and energy consumption, and more broadly, to involve them in policymaking and public life.

General public awareness and empowerment can additionally be enhanced through wider use of eco-labeling,[185] which enables informed decision-making about consumption and "encourage[s] the behavioral change of producers and consumers towards long-term sustainability".[186] Among the better-known examples of government-supported eco-labels are ENERGY STAR[187] (US) and Blue Angel (Germany).[188] To deliver cost-saving energy efficiency solutions, the former partners with the U.S. Environmental Protection Agency (EPA),[189] which, in turn, enters into partnership agreements with foreign governments—Canada, Japan, Switzerland, and Taiwan—to promote specific Energy Star qualified products in their markets.[190]

---

[184] Nielsen, "Green Generation: Millennials Say Sustainability is A Shopping Priority," November 5, 2015, http://www.nielsen.com/us/en/insights/news/2015/green-generation-millennials-say-sustainability-is-a-shopping-priority.html; Sarah Landrum, "Millennials Driving Brands To Practice Socially Responsible Marketing," *Forbes,* March 17, 2017, https://www.forbes.com/sites/sarahlandrum/2017/03/17/millennials-driving-brands-to-practice-socially-responsible-marketing/#5b07bce4990b.

[185] Abhijit Banerjee and Barry D. Solomon, "Eco-labeling for energy efficiency and sustainability: a meta-evaluation of US programs (https://econpapers.repec.org/article/eeeenepol/v_3a31_3ay_3a2003_3ai_3a2_3ap_3a109-123.htm).

[186] UNEP, "Eco-labelling," UNEP, Available at https://www.unenvironment.org/explore-topics/resource-efficiency/what-we-do/responsible-industry/eco-labelling.

[187] Energy Star, "ENERGY STAR Overview." Energy Star. Available at https://www.energystar.gov/about.

[188] The Blue Angel. "The Blue Angel." Available at https://www.blauer-engel.de/en.

[189] Energy Star, "About ENERGY STAR – 2017." Available at https://www.energystar.gov/sites/default/files/asset/document/Energy%20Star_factsheets_About%20EnergyStar_508_1.pdf.

[190] Energy Star. "ENERGY STAR International Partners," Available at https://www.energystar.gov/index.cfm?c=partners.intl_implementation.

6.4.3.2.2.2 Improvement of Access to Cheaper, Smarter, and More Secure Energy

Energy transition through decentralized power generation is linked to enabling individuals, households, schools, and small businesses, among other entities, to become "energy citizens" or "prosumers" who produce and, in certain circumstances, sell their own renewable electricity.[191] Prosumers can additionally further the goals of environmental protection and market competition by providing essential grid services, like energy storage, and by committing to reduce or shift their consumption patterns as part of efficiency and demand response programs.[192]

New policies and regulatory frameworks, as well as the reduction or elimination of any legal barriers,[193] are critical to encourage the creation of scalable micro-grids for prosumers and utility companies, achieve better grid management, and harmonize these developments with other policy objectives. Foremost of these barriers is the lack of definition of the concept of prosumers and the concomitant lack of recognition of their rights and obligations under existing laws and regulations.[194]

6.4.3.2.2.3 Cooperation and Collaboration

Citizen empowerment also involves helping prosumers help themselves. In the wind energy sector of several EU countries (i.e., Austria, Germany, Sweden, The Netherlands), the formation of cooperatives—pioneered by Denmark—enables individuals and small-scale entities to invest their pooled resources in infrastructure to generate and distribute the energy that they themselves are primarily using.[195] Notably, this model has also been gaining attention in North America.[196]

Private actor-led initiatives occurring at the transnational level are likewise emerging. One example is Breakthrough Energy Coalition,[197] which comprises

---

[191] Frédéric Simon, "Study maps potential of 'energy citizens' in push for renewable power," EURACTIV.com, Oct. 5, 2016, https://www.euractiv.com/section/energy/news/study-maps-potential-of-energy-citizens-in-push-for-renewable-power/; GREENPEACE briefing, "Putting energy citizens at the heart of the Energy Union," September 2016, https://www.greenpeace.org/sweden/PageFiles/448269/GreenpeaceEnergyCitizensbriefingSeptember2016Sweden.pdf.

[192] Jacobs (2016), pp. 527–528.

[193] See Lavrijssen and Parra (2017), p. 1207.

[194] Josh Roberts, *Prosumer Rights: Options for an EU legal framework post-2020* (ClientEarth, 2016), https://www.documents.clientearth.org/wp-content/uploads/library/2016-06-03-prosumer-rights-options-for-an-eu-legal-framework-post-2020-coll-en.pdf; Josh Roberts, "What does the future hold for energy 'prosumers'?," *Environment Journal*, July 14, 2016, https://environmentjournal.online/articles/future-hold-energy-prosumers/.

[195] See International Labour Office, *Providing clean energy and energy access through cooperatives* (2013); Mark Bolinger, "Community Wind Power Ownership Schemes in Europe and their Relevance to the United States," May 2001, http://citeseerx.ist.psu.edu/viewdoc/download?doi=10.1.1.204.794&rep=rep1&type=pdf.

[196] Tildy Bayar, "Community Wind Arrives Stateside," *Renewable Energy World*, July 5, 2012, https://www.renewableenergyworld.com/articles/print/volume-15/issue-3/wind-power/community-wind-arrives-stateside.html.

[197] Breakthrough Energy "Breakthrough Energy Coalition." Available at http://www.b-t.energy/coalition/.

6.4 A Paradigm Shift in the Governance of Sustainable Development:...

patient and tolerant visionary billionaires with diverse backgrounds, as well as "global corporations that produce or consume energy in vast quantities, and financial institutions with the capital necessary to finance the world's largest infrastructure projects," who are collectively determined to provide reliable, affordable, and carbon-neutral energy by investing in and building innovative technologies. In its so-called "Landscape of Innovation," the Coalition aims to address emissions in five key areas: electricity, transportation, agriculture, manufacturing, and buildings.[198] Another initiative called Mission Innovation[199] brings together a group of 23 countries and the EU,[200] aiming to reinvigorate and accelerate clean energy innovation throughout the world to make clean energy affordable for all. Various multinational companies, including 40% of the Fortune 500 companies, are collectively undertaking to contribute to energy transition by aiming to procure 100% of their energy needs from renewable sources.[201]

6.4.3.3 Climate Action

An influential climate-change thinker, William Nordhaus, recommended in the 1990s gradual, modest reductions of GHG emissions in his book on the economics of climate change.[202] Another thinker of climate change, Nicholas Stern, demanded a few years later immediate and dramatic efforts to mitigate climate change, including spending 1–2% of GDP in advanced economies.[203] For both approaches, it is necessary to have cooperation from the bottom up.[204] This is possible without compromising on economic growth.[205]

Predictions are that there will be a 7-degree rise in global temperatures by 2100, which would make life very difficult in various part of the world, especially those

---

[198] ibid.

[199] *Mission Innovation: Accelerating the Clean Energy Revolution*, http://mission-innovation.net/.

[200] *Mission Innovation: Member Participation*, http://mission-innovation.net/countries/.

[201] Stephan Jungcurt, "Energy Update: Non-state Actors, Regional Initiatives Show Progress Towards Energy Transition," November 22, 2016, http://sdg.iisd.org/news/energy-update-non-state-actors-regional-initiatives-show-progress-towards-energy-transition/; TeamIRENA, "How Corporates are Taking the Lead in Renewable Energy," November 15, 2016, https://irenanewsroom.org/2016/11/15/how-corporates-are-taking-the-lead-in-renewable-energy/.

[202] Nordhaus (1994).

[203] Stern (2007).

[204] The issue of climate change mitigation has even reached democratic levels as close to citizens as teenagers suing the US federal government as part of efforts to force action to request climate action. See M. Nijhuis, "The teen-agers suing over climate change," The New Yorker, 6 December 2016. See also N. Geiling, "In landmark case, Dutch citizens sue their government over failure to act on climate change," Think Progress, 14 April 2015, available at https://thinkprogress.org/in-landmark-case-dutch-citizens-sue-their-government-over-failure-to-act-on-climate-change-e01ebb9c3af7/.

[205] Cowen (2018) (which makes a moral case for economic growth and is a source of inspiration and optimism about our future possibilities). See also Leal-Arcas (2018d), pp. 4–20.

near the Equator.[206] So international cooperation is crucial for climate change mitigation. A promising way forward is bringing together environmental NGOs and businesses for greater and close cooperation on issues of climate action.[207] A case in point that became a surprising fact is the very well organized social movement in the US to implement the Paris Climate Agreement as soon as President Trump announced his intention to withdraw from that Agreement. Cities, states and businesses gathered together for climate action. Outside the main conference building of the 2017 UN climate summit, a coalition of people gathered under the heading 'We are still in.'[208] Many cities in the US today are requiring tougher energy-efficient standards on buildings and electrifying their public buses, as signs of commitment to the Paris Agreement.[209]

Since cement-making produces around 6% of the world's CO2 emissions and steel around 8% (half of it goes into buildings), the most logical building material to use would be wood because it is renewable energy in that a mature tree is cut down, but a new tree can be planted and it will capture CO2.[210] Staying way from cement in construction is a serious challenge for China, which "used more cement between 2011 and 2013 than the U.S. used in the entire twentieth Century."[211] A further incentive for using wood for construction purposes is that it is much lighter than steel, brick or concrete. From January 2019, all new public-sector buildings in the EU need to be built to nearly zero-energy standards.[212]

Equally, joint actions between countries could have a 'trickle-down effect' from governments to citizens and businesses for the promotion of business opportunities in clean energy, especially for small and medium enterprises (SMEs), the facilitation of trade and investment in environmentally friendly goods and services such as

---

[206]J. Eilperin, D. Dennis, and C. Mooney, "Trump administration sees a 7-degree rise in global temperatures by 2100," *The Washington Post*, 28 September 2018, available at https://www.washingtonpost.com/national/health-science/trump-administration-sees-a-7-degree-rise-in-global-temperatures-by-2100/2018/09/27/b9c6fada-bb45-11e8-bdc0-90f81cc58c5d_story.html?noredirect=on&utm_term=.8fe8267d2343.

[207]See Leal and Leal-Arcas (2018). See also the views of Andrew Guzman on the real-world consequences of climate change in *Overheated: The human cost of climate change*, Oxford University Press, 2014.

[208]O. Milman and J. Watts, "One nation, two tribes: opposing visions of US climate role on show in Bonn," The Guardian, 9 November 2017, available at https://www.theguardian.com/environment/2017/nov/09/bonn-climate-change-talks-us-two-tribes.

[209]The Economist, "Local government v global warming," 15 September 2018, pp. 67–68, at 67.

[210]The Economist, "The house made of wood," 5 January 2019, p. 12.

[211]A. Swanson, "How China used more cement in 3 years than the U.S. did in the entire 20th Century," *The Washington Post*, 24 March 2015, available at https://www.washingtonpost.com/news/wonk/wp/2015/03/24/how-china-used-more-cement-in-3-years-than-the-u-s-did-in-the-entire-20th-century/?noredirect=on&utm_term=.4113c0ceda3f.

[212]The Economist, "The house made of wood," 5 January 2019, p. 12.

energy efficient goods and services, and cooperation on trade-related aspects of climate change mitigation.[213]

6.4.3.4 Power to the Citizens

One very promising development in the twenty-first century is the empowerment of citizens on issues of common concern such as climate change, sustainable energy, and international trade. Citizens' empowerment means that civil society can play an important role in the new challenges of trade diplomacy, such as the integration of noneconomic aspects of trade in trade policy and the inclusion of trade policies in the democratic debate.[214] This approach makes the system of decision making closer to the citizens and therefore less technocratic (see Fig. 6.2 below).

This novel idea of greater citizen participation, engendered by citizens' empowerment, is a promising way of providing better management of environmental issues and helping achieve the Sustainable Development Goals (SDGs).[215] Moving forward, citizens must contribute to finding more effective ways to obtain sustainable energy, mitigate climate change, and develop a more democratic and transparent trade policy-making process. Figure 6.2 represents several specific means by which citizens can ostensibly help enhance sustainable energy initiatives, mitigate climate change, and make citizens richer through free and open environmental trade.

Citizens' empowerment can be achieved by allowing for more participation in the process of decision making. More broadly, regression analyses show that when society allows free choice, it has a considerable impact on happiness.[216] Since the beginning of the 1980s, democratization, economic development, and increasing social tolerance have all increased citizens' perception that they have free choice, and consequently increased citizen happiness.[217]

---

[213]Leal-Arcas (2012), pp. 875–927; Non-paper of the European Commission services, "Feedback and way forward on improving the implementation and enforcement of Trade and Sustainable Development chapters in EU Free Trade Agreements," 26 February 2018, available at http://trade.ec.europa.eu/doclib/docs/2018/february/tradoc_156618.pdf.

[214]For analyses of democracy, see Deneen (2018) (asserting that an increasingly oppressive liberal world order will likely give way to authoritarian illiberalism); Frum (2018) (exploring how Donald Trump was elected and asserting that his governance has harmed American democracy); Levitsky and Ziblatt (2018) (examining historical threats to democracies and drawing lessons from other nations to offer strategies to follow in the current American context).

[215]*See* G.A. Res. 70/1, Transforming Our World: The 2030 Agenda for Sustainable Development 14–35 (Sept. 25, 2015).

[216]Inglehart et al. (2008), pp. 264, 270. One wonders whether lack of freedom, high levels of pollution or social inequality may explain why, say, China ranked so poorly in the 2018 World Happiness Report. See http://worldhappiness.report/.

[217]*Id.* at 279–80.

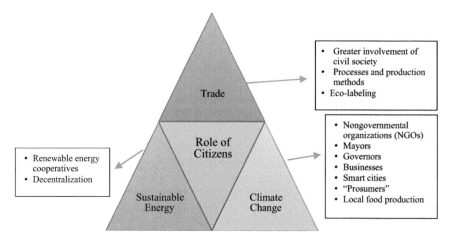

Fig. 6.2 Citizens' empowerment and sustainable development goals

6.4.3.5 Citizens and Trade (and Climate Change)

Traditionally, governments discuss trade measures and their links with climate change without allowing for citizens' participation.[218] This rather technocratic exercise of mitigating climate change and its links to trade policy has the potential to become more democratic.

Trade will need to be substantially reconceptualized to empower individuals within the international trade framework. If global society wants to emancipate people around the world and benefit from the wealth of transnational insights, perceptions, and resources, society should aim at facilitating access to global knowledge via international trade. Moreover, trade agreements should emphasize and encourage the trade of technological equipment, smart appliances, and applications that serve to reduce energy consumption and GHG emissions. Furthermore, trade subsidization distorts markets and leads to more GHG emissions than would otherwise result.[219]

Trade places a spotlight on the dynamic shifts that are taking place and will take place globally in the so-called processes and production methods (PPMs) of goods. Consumers increasingly seek information on how the PPMs of the products they buy affect the environment and request eco-labeling as well as labeling and traceability

---

[218] *See* Leal-Arcas (2008), pp. 425–439.

[219] *See* Leal-Arcas et al. (2014), pp. 430–431. One could make the case that some World Trade Organization (WTO) rules need clarification, especially in the field of subsidies, and ask the question whether trade subsidies should exist if they are for a good purpose, such as a public good like climate change mitigation. *Id.* at 136–137.

6.4 A Paradigm Shift in the Governance of Sustainable Development:...

regarding genetically modified organisms.[220] This change in consumer demand will transform the geographies of trade, both spatially and temporally. The importance of new technologies in PPMs is a crucial aspect of this advancement.

International trade agreements could have provisions that empower citizens as consumers to better scrutinize trade agreements. This addition would make trade governance closer to citizens. Close scrutiny is necessary to examine the rules of international trade that need to be amended to reduce the impact of global trade on the environment.[221] In broad terms, trade rules are not guided towards environmental protection as much as they could be.[222]

The ease of proliferation of news and information through the Internet—which provides more transparency and access to information than ever before—has allowed people to become more aware of trade negotiations and their effects. This increased awareness has resulted in demonstrations against what many citizens consider unfair and detrimental trade agreements that are supposed to benefit ordinary citizens but in reality only benefit a few.[223] Classic examples are the massive demonstrations against the Trans-Pacific Partnership (TPP) in the United States[224] and against the Trans-Atlantic Trade and Investment Partnership (TTIP) in Germany, Austria, and Sweden.[225] These demonstrations occur because citizens widely consider trade to be designed by and for the interest of large transnational

---

[220]*See* OECD, *Processes and Production Methods (PPMs): Conceptual Framework and Considerations on Use of PPM-Based Trade Measures*, 7 (1997), http://www.oecd.org/officialdocuments/publicdisplaydocumentpdf/?cote=OCDE/GD(97)137&docLanguage=En [https://perma.cc/MN2L-XHN6].

[221]Esty (1994), p. 225.

[222]See for instance the Preamble to the Agreement Establishing the World Trade Organization, which states:

> *Recognizing* that their relations in the field of trade and economic endeavour should be conducted with a view to raising standards of living, ensuring full employment and a large and steadily growing volume of real income and effective demand, and expanding the production of and trade in goods and services, while allowing for the optimal use of the world's resources in accordance with the objective of *sustainable development*, seeking both to protect and preserve the environment and to enhance the means for doing so in a manner consistent with their respective needs and concerns at different levels of economic development.... (emphasis added)

Marrakesh Agreement Establishing the World Trade Organization, Apr. 15, 1994, 1867 U.N.T.S. 154 (1994).

[223]*See Anti-Globalists: Why They're Wrong*, The Economist (Oct. 1, 2016), http://www.economist.com/news/leaders/21707926-globalisations-critics-say-it-benefits-only-elite-fact-less-open-world-would-hurt [https://perma.cc/95KB-VUFC].

[224]*See TPP Signing Sparks Dozens of Protests across US over Biggest Trade Pact*, RT AM. (Feb. 5, 2016), https://www.rt.com/usa/331356-tpp-signing-protests-usa/ [https://perma.cc/DZD2-TVL3].

[225]Michael Nienaber, *Tens of Thousands Protest in Europe Against Atlantic Free Trade Deals*, Reuters (UK) (Sept. 17, 2016), http://www.reuters.com/article/us-eu-usa-ttip-idUSKCN11N0H6 [https://perma.cc/ZP4Q-KQNC].

corporations, rather than for the needs of the general population.[226] So, reshuffling political procedures by drawing citizens into these processes is necessary, and arguably indispensable. It is, therefore, worth exploring how local and regional governments,[227] such as those of cities or municipalities represented by their mayors, can better represent the interests of their people.

Accountability, efficiency (via more rapid feedback), and transparency are strongest at the governing level closest to citizens. In a post-Westphalian world, neomedievalism[228] may prevail, but the role of the city can be preponderant. The involvement of citizens can be encouraged in different intellectual and cultural ways, such as within civil society's role in liberal Western democracies, within the Asian values[229] context in China,[230] or citizens' empowerment in theocracies. Politically, the principles of subsidiarity, devolution, federal systems, regional schemes, and closer ties between specific cities—not least within the European Union—form the background for a rising role for the cities of the world to come together. All these innovative options of governance make decision-making easier and more impactful and aim at a decentralized system of governance.

Lastly, given that citizens' roles in trade are primarily as consumers, for their activities to have an impact on climate change mitigation efforts, consumer activity (i.e., purchases) must be significantly valued within the broader economic dynamic of a country. Table 6.2 assesses the consumer habits in eight major GHG-emitting states that are also parties to three mega-regional trade agreements (RTAs) (the CPTPP, TTIP, and the Regional Comprehensive Economic Partnership (RCEP)) to ascertain whether consumer spending is of significant importance such that a change in consumer habits could influence trade patterns in these jurisdictions. Table 6.2 indicates consumer spending as a percentage of gross domestic product (GDP). The figures are based on the market value of all goods and services, including durable products (such as cars, washing machines, and home computers), purchased by households.

Consumer spending contributes significantly to the GDPs of the countries in Table 6.2, with the exception of China. So, empowering citizens to be more climate change-conscious in their purchasing habits could spur the growth of "greener"

---

[226] For an example of this perspective, see Robert Reich, *Trade Deals Like the TPP Only Benefit the One Percent*, MOYERS & CO. (Feb. 21, 2015), http://billmoyers.com/2015/02/21/trade-deals-boost-top-1-bust-rest/ [https://perma.cc/42FF-2NZ3].

[227] Local governments today are bearing the costs of climate-change disasters. Conversely, if CO2 is removed from the air by planting trees in a given part of the world, who benefits from this GHG emission sequestration?

[228] "Neomedievalism" is a term often used as a political theory about modern international relations. *See* Kobrin (1998), p. 361.

[229] Values are desirable, tans-situational goals, varying in importance, that serves as guiding principles in people's lives. See Schwartz (1992), pp. 1–65.

[230] The "Asian values context" refers to the notion of collectivism, rather than individualistic approaches to society that are more prevalent in Western societies. *See generally* Wee (1999), p. 332.

6.4 A Paradigm Shift in the Governance of Sustainable Development:... 237

Table 6.2 Household final consumption expenditure[a]

Country	Consumer spending as percentage of GDP
China	37.1
United States	68.1
European Union	56.3
India	59.2
Russia	52.1
Indonesia	55.4
Brazil	63.8
Japan	56.6
Canada	57.5
Mexico	67.1

[a]*Household Final Consumption Expenditure, Etc. (% of GDP)*, The World Bank, http://data.worldbank.org/indicator/NE.CON.PETC.ZS?end=2015&name_desc=false&start=1967&view=chart [https://perma.cc/WE3N-Z8JP]

markets in the jurisdictions that are parties to the three mega-RTAs mentioned above by creating high demand for greener goods.

6.4.3.6 Citizens, Climate Change, and Sustainable Energy (and Trade)

The empowerment of citizens is a promising tool for climate change mitigation, but it depends upon support from NGOs, mayors and governors representing citizens, smart cities,[231] digitalization of data, prosumers, and local food production.[232] The same is true with the enhancement of sustainable energy via renewable energy cooperatives[233] and energy decentralization. The decentralization and localization

---

[231] By smart city, we mean a city that is a self-sufficient unit and has digital technology embedded across all its functions to enhance performance, well-being and communication, and also to reduce costs and resource consumption.

[232] "Smart cities" refers to an urban development vision to integrate information and communication technology and Internet-of-things technology in a secure fashion to manage a city's assets. See Matt Hamblen, Just What is a Smart City? Computerworld (Oct. 1, 2015), http://www.computerworld.com/article/2986403/internet-of-things/just-what-is-a-smart-city.html [https://perma.cc/ZX5F-67QF]. "Prosumers" refers to consumers who are also producers of (renewable) energy and who use energy in a smarter and more efficient manner. *See Electricity Prosumers*, European Parliament Think Thank, http://www.europarl.europa.eu/thinktank/en/document.html?reference=EPRS_BRI(2016)593518 [https://perma.cc/W6YM-EXZB]. And "local food" refers to a movement that aims to connect food producers and food consumers in the same geographic region to develop more self-reliant and resilient food networks; improve local economies; or have an impact on the health, environment, community, or society of a particular place. *See Local Food*, Wikipedia, https://en.wikipedia.org/wiki/Local_food [https://perma.cc/4A6L-DWT5].

[233] Renewable energy at cooperative level is big enough to be technologically and economically efficient, and small enough to be locally owned or controlled. In addition, local projects can help in local economic and social regeneration. The benefits of 'localisation' have to be set against the technical and economic advantages of larger scale systems.

of energy dependency could potentially lead to a change in the relationship between energy producers and governance institutions, including municipal administrations and city mayors. The Paris Agreement can be characterized as a hybrid global agreement that facilitates these changes within a multipolar world. The global stock-take (Article 14 of the Paris Agreement) will foster new ways of valuing, seeing, and comparing sectors, communities (rich and poor, urban and rural), countries, and regions. This data will inform other agreements as well as policy on resource management (such as eco-labeling and PPMs).

The opportunities ahead are partially the result of technology enabling a decentralization of production and processing of goods—for instance, 3-D printing as opposed to Fordist-style manufacturing—and a dynamic hybridization of services—for instance, the gig economy—away from old hierarchical and linear models towards multilevel and circular ones. The form these will take depends upon how the power dynamics will manifest, including backlash by citizens, corporations, and countries with the most to lose within the existing globalized trade system. This hybridization indicates a recognition that there is no inevitable, single pathway or outcome; rather, that the political economy within, and between, regional contexts will influence the potential opportunities and outcomes for citizens' engagement.

6.5 Conclusion, Recommendations and a Future Research Agenda

This chapter sheds light on the emergence of a new actor, namely the prosumer, in the EU's energy security arena. The chapter has shown that prosumers can share extra energy production with others and serve as a shelter for micro-grids. To reach scalability within sustainability in the case of micro/mini-grids, they would need to be subsidized.

Questions would need to be addressed about how accountability would work in the context of citizens' empowerment and what the role of private citizens in this bottom-up design is. If there is a shift towards a prosumer focus, there needs to be a corresponding recognition of the importance of scrutiny and accountability. Democratic governance must take full account of the role of decision-makers at every level. Decentralization plays an important role, but so does accountability. NGOs and all forms of local governance should be co-ordinated with consultation processes that engage in a transparent manner.

Sustainable energy is rapidly becoming an EU special brand, like the protection of human rights, in the quest for looking after the environment. Achieving sustainable energy encompasses the following points: decarbonising the economy (by using less energy based on fossil fuels and making greater use of clean energy), democratising access to energy (namely everyone has the right to participate), digitalization, diversification of energy supply, and disrupting traditional energy cycles. Leadership is shifting from national politics to local politics and, therefore,

6.5 Conclusion, Recommendations and a Future Research Agenda

power is being decentralised. For instance, when there is a disaster in a given neighborhood, citizens do not contact the head of State or government of the nation, but the mayor of the city. A clear example of this trend towards local politics is the Local Governments for Sustainability platform.[234] These decentralized solutions are the way forward because they enable democratization of energy throughout the world, which empowers citizens, shifting the current paradigm.

As part of the global energy transition, two factors are crucial: (1) continued global economic growth, which is a must, and (2) the protection of the environment. Therefore, it is necessary to have a strategy that is realistic, fair, and pragmatic. Realistic because the transition will take time and cannot be done overnight. Fair because energy poverty deserves attention. And pragmatic because GHG emissions must be addressed via more efficient technologies. Access to energy will depend on the link between consumers and producers. Having in place the right regulations will help speed up the process of energy transition.

Following trends in the EU towards decentralisation and the emergence of a 'gig' economy, the energy sector is currently undergoing a large-scale transition. One of its core aspects is the progressive top-down diffusion of the potential, competences, and leverage from EU institutions, States, and corporate actors across the energy value chain towards prosumers, who need to be at the centre of the energy transition for it to happen democratically in a bottom-up manner. This phenomenon can be conceptualised as energy democratisation, namely moving away from a few energy companies monopolising access to energy towards energy owned mainly by consumers, making consumers of the utmost importance. Therefore, energy transition can only happen if there is citizen participation.

All of this is achievable by shifting the current paradigm to one that is more human-centric, by linking projects to people, and more collaborative in how it tackles various obstacles, whether legal or behavioural. Think of the analogy of organic food: it is more expensive, but for many, its benefits outweigh the costs. Moreover, consumers have the power to choose either organic or non-organic. By the same token, many citizens are interested in climate-friendly products even if they are more expensive. This means that we need to look at the whole production process, not just the end product, if we are serious about consumer empowerment. To get there, legislation must remove barriers to participation and protect and promote consumers to enable them to produce, store, sell, and consume their own energy.

Moreover, energy choices are governed by the price of fossil fuels, their convenience, and availability. Making policy choices at a local level, while desirable, requires co-ordination to be strategic. It is important to stress the link between local and central government in such a strategy. When designing policies from the bottom

---

[234]'Local Governments for Sustainability is an international association of local and metropolitan governments dedicated to sustainable development.' (See UNEP Climate Initiatives Platform (2017) ICLEI – Local Governments for Sustainability, http://climateinitiativesplatform.org/index.php/ICLEI_-_Local_Governments_for_Sustainability).

up, a distinction needs to be made between federal and non-federal systems of government, and one should address the role of local communities and local governments. What might work under one system may not work under another. For the design of policies, central governments have many policy options and economic instruments, including renewable energy subsidies, carbon price caps, and environmental taxation.

While all of the above creates ample potential for facilitating and improving the EU's security of supply as well as fulfilling its climate targets, several caveats exist. These not only are confined within energy security prerogatives, but also extend to the critical management of digital security, which the digitalisation of energy services brings to the fore. So, for consumers to become prosumers and engage in the energy transition, it will be crucial to make the process interesting and simple and to inform them much more, given the current level of energy consumer dissatisfaction.

Here is where cities can play a major role at educating citizens on energy transition and climate change mitigation, not least because cities consume three quarters of the world's energy,[235] and because they are smaller entities than countries or regions, so it is easier to get things done. Even more impactful would be to educate companies and policymakers on sustainable development, since there are fewer of them than there are citizens. Doing so will shift the paradigm from a system that is producer-centric to one that will be consumer-centric. With development have come environmental and social problems such as climate change, poverty, wars, unhappiness, depletion of resources, and environmental pollution that need to be rectified. This paradigm shift is crucial because development is not possible without energy and sustainable development is not possible without sustainable energy.

Society can minimize the level of suffering resulting from climate change by doing a lot of climate change mitigation and adaptation. As John Holdren says, "we need enough mitigation to avoid the unmanageable and enough adaptation to manage the unavoidable."[236] In an ever-shifting context, demand management emerges as a key issue. The provision of adequate and precise information to prosumers—so that they can optimise their use of smart grids—as well as the transition to targeted, flexible contracts to adjust to the needs of prosumers need to be embedded in well-articulated broader policy and market regulatory frameworks. Moreover, private and public finance should be effectively attracted and directed to indispensable infrastructure schemes that will enable the transition from the traditional centralized power network to the decentralized nexus of smart grids. And it is well known that, where finance flows, action happens. Currently, both centralized and decentralized energy are interdependent and help each other.

Renewable energy can supply all our energy forever at costs comparable with existing energy sources and without their major environmental and social impacts.

---

[235] ARUP, Energy in Cities, at 2, available at https://www.arup.com/-/media/Arup/Files/Publications/A/A2-Issue-7.pdf.

[236] Lecture given by John Holdren at Harvard University, 2 October 2018.

6.5 Conclusion, Recommendations and a Future Research Agenda

The technologies that will be prioritised in terms of energy generation to back renewable-energy generation will play a crucial role in facilitating the role of prosumers in the new market in-the-making. Since renewable energy is becoming more competitive, more green jobs will be created in the future and the trend towards a clean energy revolution is ever closer. This energy transition into renewable energy, in turn, will help both enhance energy security and mitigate climate change. So rather than investing large amounts of money into building liquefied natural gas terminals and gas pipelines,[237] the EU should make a greater effort to invest in renewable energy.

The emerging establishment of prosumer markets is an invaluable development that will enable the transition from supply-driven to demand-side EU energy policy. This cannot but have far-reaching ramifications for the amply politicised and securitised gas trade with Russia as well as for furthering the internal EU market architecture. It is expected that it will decrease flows of energy as well as dependence on Russian gas in the medium term while at the same time acting as a stimulus for further market integration in the energy, climate, and digital economy realms.

Giving civil society a greater voice is imperative for the energy transition to happen. Below are some of the necessary actions:

1. Speeding up action on the ground and localising global agendas;
2. More alliances between countries and donors in the decarbonisation process;
3. Greater collaboration between civil society, governments, and NGOs to include all layers of governance;
4. Bringing together different camps of governments;
5. Scaling up the capacity of local governments;
6. Webbing[238] will be necessary: we need to look at issues and challenges, not sectors; temporal linkages are required, namely using time as an indicator, given its importance in the context of decarbonisation, and there needs to be policy coherence.

Moreover, in the future, energy will be consumed near where it is produced. How will this impact international trade (in energy), especially in an era of trade restrictions? Furthermore, the protectionist concept of 'buy local' seems to be going global. This policy is suggested, among other things, to reduce greenhouse gas emissions from transportation, which will benefit climate change. But, in the international trade domain, there are no winners with protectionism.

What implications will 'buy local' have for international trade at a time when world trade is slowing? Unless there is more innovation in transportation, there is a

---

[237] See for instance the case of Poland, whose main gas company signed a long-term contract in November 2018 to receive deliveries of liquefied natural gas from the United States as part of a larger effort to reduce its energy dependence on Russia. See Monika Scislowska, "Poland signs deal for long-term deliveries of US gas," 8 November 2018, *AP News*, available at https://apnews.com/72568e5c41bd49d0a3515926311ebc7b.

[238] By webbing, we are referring to connecting different issues in a broader policy approach, rather than approaching them in silos.

chance that this policy will result in less demand for international trade. New actors and modes of governance are changing the traditional global trading system, or at least are contributing to the transformation from inter-State dealings to completely different forms of governance in which non-State actors (including individuals) play a role. The EU has been a social laboratory to test hypotheses of multi-level governance in the past, which are pertinent for the case of energy transition.

At a time when multilateral and exclusively state-to-state action is quite unpopular, while global challenges are simultaneously mounting and increasingly intertwined, the capabilities of citizens, cities, and other non-State actors need to be tapped and harnessed. Decarbonization is one global challenge that lies at the intersection of the trade and climate change regimes. A key ingredient to facilitating trade in climate-friendly energy is participation by a greater number of diverse stakeholders, who are more attuned to local environmental conditions and to energy demand and supply.

The reform of the WTO is urgent. That requires cooperation among WTO members. One country alone cannot do so or even establish a new global trade order. WTO members should safeguard the WTO rules and oppose any type of trade protectionism.

In the absence of a well-functioning WTO, and drawing on the precedent of climate action and the Paris Agreement innovations, attention should focus on the ways in which non-state actors can be empowered to play a greater role in accelerating the energy transition. The rise of the so-called energy prosumers—consumers *cum* producers—is an important development that holds great potential and requires further support from states, such as through the establishment of a legal framework recognizing prosumers' rights and obligations. The small scale in which these non-state entities typically operate to improve energy access and security makes collaboration among them worth encouraging. Various co-operatives in the European renewable energy sector can be taken as models.

Current discussions regarding WTO reforms should also include how these new entities can access global markets and how WTO rules should be applied to the energy sector. Member States should not be constrained in implementing measures like trade-related concessions and local content requirements to encourage these entities' participation, particularly in the production and diffusion of clean energy. In this regard, proposals for a General Agreement on Trade in Energy and/or a Sustainable Energy Trade Agreement,[239] which covers liberalization of trade in environmental goods and services, deserve further study. No solution to big challenges, such as climate change and trade liberalization, would be possible without social mobilization and cooperation among governments, companies, and researchers, whose role is to provide good information to create good policy.

---

[239]International Centre for Trade and Sustainable Development (ICTSD) (2011). Available at https://www.ictsd.org/sites/default/files/downloads/2011/12/fostering-low-carbon-growth-the-case-for-a-sustainable-energy-trade-agreement.pdf.

The issues raised above are all very relevant to a future research agenda in the broad field of international economic law and governance.

References

Afgan N, Carvalho M (2002) Multi-criteria assessment of new and renewable energy power plants. Energy 27:739–755
Bäckstrand K et al (2017) Non-state actors in global climate governance: from Copenhagen to Paris and beyond. Environ Politics 26(4):561–579
Barber BR (2013) If mayors ruled the world: dysfunctional nations, rising cities. Yale University Press, New Haven, p 22
Batty M (2013) The new science of cities. MIT Press, Cambridge
Bauwens T, Gotchev B, Holstenkamp L (2016) What drives the development of community energy in Europe? The case of wind power cooperatives. Energy Res Soc Sci 13:136–147
Bodansky D et al (2017) International climate change law. Oxford University Press, Oxford
Buchan D (2012) The Energiewende – Germany's gamble. The Oxford Institute for Energy Studies, Oxford, p 1
Button K (2002) City management and urban environmental indicators. Ecol Econ 40:217–233
Cameron P (2010) International energy investment law. Oxford University Press, Oxford
Carlane C (2010) Climate change law and policy, EU and US approaches. Oxford University Press, Oxford
Carroll A (1974) Corporate social responsibility: its managerial impact and implications. J Bus Res 2(1):75–88
Carroll A (1979) A three-dimensional conceptual model of corporate performance. Acad Manag Rev 4(4):497–505
Carroll A (1999) Corporate social responsibility: evolution of a definitional construct. Bus Soc 38(3):268–295
Chen S, Bouvain P (2009) Is corporate social responsibility converging? A comparison of corporate social responsibility reporting in the USA, UK, Australia, and Germany. J Bus Ethics 87:299–317
Clapham A (2009) Non-state actors. In: Chetail V (ed) Post-conflict peacebuilding: a Lexicon. Oxford University Press, New York, pp 200–212
Commoner B (1971) The closing circle: nature, man, and technology. Random House Inc., New York
Costanza R (ed) (1991) Ecological economics: the science and management of sustainability. Columbia University Press, New York
Cowen T (2018) Stubborn attachments: a vision for a society of free, prosperous, and responsible individuals. Stripe Press, San Francisco
Cullen H, Morrow K (2001) International Civil Society in International Law: The Growth of NGO Participation. Non-State Actors Int Law 1(1):7–40
de Llano-Paz F et al (2015) The European low-carbon mix for 2030: the role of renewable energy sources in an environmentally and socially efficient approach. Renew Sustain Energy Rev 48:54–55
Dellinger M (2013) Localizing climate change action. Minn J Law Sci Technol 14:621–623
Deneen P (2018) Why liberalism failed? Yale University Press, New Haven
Devine-Wright P (2007) Energy citizenship: psychological aspects of evolution in sustainable energy technologies. In: Murphy J (ed) Governing technology for sustainability. Earthscan, London
Dobbs R et al (2011) Urban world: mapping the economic power of cities. McKinsey Global Institute, New York. https://www.mckinsey.com/~/media/McKinsey/Featured%20Insights/

Urbanization/Urban%20world/MGI_urban_world_mapping_economic_power_of_cities_full_report.ashx

Drache D (2006) "Trade, development and the Doha Round: a sure bet or a Train Wreck?," (working paper). Centre for International Governance Innovation, Waterloo. https://www.cigionline.org/sites/default/files/trade_development_and_the_doha_round.pdf

Echeverri LG (2018) Investing for rapid decarbonization in cities. Curr Opinion Environ Sustain 30:42–51

Esty DC (1994) Greening the GATT: trade, environment, and the future. Peterson Institute, Washington, D.C, p 225

Esty D (2017) Red lights to green lights: From 20th century environmental regulation to 21st century sustainability. Environ Law 47(1):1–80

Faber M, Manstetten R, Proops J (1998) Ecological economics: concepts and methods. Edward Elgar, Cheltenham

Frankfurt School-UNEP Centre/BNEF (2018) Global trends in renewable energy investment. Frankfurt School of Finance & Management gGmBH, Frankfurt am Main, pp 16–17. http://fs-unep-centre.org/sites/default/files/publications/gtr2018v2.pdf

Freeman I, Hasnaoui A (2010) The meaning of corporate social responsibility: The vision of four nations. J Bus Ethics 100:419–443

Freestone D, Streck C (eds) (2009) Legal aspects of carbon trading: Kyoto, Copenhagen and beyond. Oxford University Press, Oxford

Frum D (2018) Trumpocracy: the corruption of the American Republic. Harper, New York

Fujiwara N (2016) Overview of the EU climate policy based on the 2030 framework. In: Heffron RJ, Little GFM (eds) Delivering energy law and policy in the EU and the US: a reader. Edinburgh University Press, Edinburgh, pp 605–609

Galera MDS (2017) The integration of energy and environment under the paradigm of sustainability threatened by the hurdles of the internal energy market. Eur Energy Environ Law Rev 26:13–25

Geddes P (1915) Cities in Evolution: An introduction to the town planning movement and to the study of civics. Williams & Norgate, London

Gehring MW et al (2013) "Climate change and sustainable energy measures in Regional Trade Agreements (RTAs): an overview," Issue Paper No. 3. International Centre for Trade and Sustainable Development, Geneva, p 27

Geissdoerfer M, Savaget P, Bocken NMP, Hultink EJ (2017) The circular economy – a new sustainability paradigm? J Cleaner Production 143:757–768

Girardet H (2014) Creating regenerative cities. Routledge, Abingdon

Gladwell M (2002) The tipping point: how little things can make a big difference. Black Bay Books, New York

Goldstein JS (2005) International relations. Pearson-Longman, New York, p 107

Guevara-Stone L (2017) 35 years of bold steps in the clean energy race: part 1. Rocky Mountain Institute, Colorado

Halliday F (2001) The romance of non-state actors. In: Josselin D, Wallace W (eds) Non-state actors in world politics. Palgrave, New York, pp 21–40

Helm D (2007) The new energy paradigm. Oxford University Press, Oxford

Hinloopen E, Nijkamp P (1990) Qualitative multiple criteria choice analysis. Qual Quantity 24:37–56

Hofmann M (2012) Climate governance at the crossroads: experimenting with a global response after Kyoto. Oxford University Press, Oxford

Idowu S, Towler B (2004) A comparative study of the contents of corporate social responsibility reports of UK companies. Manag Environ Quality 15(4):420–437

Inglehart R et al (2008) Development, freedom, and rising happiness: a global perspective (1981–2007). Perspect Psychol Sci 3:264–285

International Centre for Trade and Sustainable Development (ICTSD) (2011) Fostering low carbon growth: the case for a sustainable energy trade agreement. ICTSD, Geneva. Available at https://

References

www.ictsd.org/sites/default/files/downloads/2011/12/fostering-low-carbon-growth-the-case-for-a-sustainable-energy-trade-agreement.pdf

Jacobs SB (2016) The energy prosumer. Ecol Law Q 43:527–528

Josselin D, Wallace W (2001) Non-state actors in world politics: a framework. In: Josselin D, Wallace W (eds) Non-state actors in world politics. Palgrave, New York, pp 1–20

Klein D et al (2017) The Paris agreement on climate change, commentary and analysis. Oxford University Press, Oxford

Kobrin SJ (1998) Back to the future: neomedivalism and the postmodern digital world economy. J Int Aff 51:361

Lavrijssen S, Parra AC (2017) Radical prosumer innovations in the electricity sector and the impact on prosumer regulation. Sustainability 9:1207

Leal W, Leal-Arcas R (eds) (2018) University initiatives in climate change mitigation and adaptation. Springer, Berlin

Leal-Arcas R (2008) Theory and practice of EC external trade law and policy. Cameron May, London, pp 425–439

Leal-Arcas R (2011a) A bottom-up approach for climate change: the trade experience. Asian J Law Econ 2(4):1–54

Leal-Arcas R (2011b) Alternative architecture for climate change – major economies. Eur J Legal Stud 4(1):22–56

Leal-Arcas R (2012) Unilateral trade-related climate change measures. J World Invest Trade 13 (6):875–927

Leal-Arcas R (2013) Climate change and international trade. Edward Elgar Publishing, Cheltenham

Leal-Arcas R (2018a) New frontiers of international economic law: the quest for sustainable development. Univ Pa J Int Law 40(1):83–132

Leal-Arcas R (ed) (2018b) Commentary on the energy charter treaty. Edward Elgar, Cheltenham

Leal-Arcas R (2018c) Empowering citizens for common concerns: sustainable energy, trade and climate change. GSTF J Law Soc Sci 6(1):1–37

Leal-Arcas R (2018d) Re-thinking global climate change: A local, bottom-up perspective. Seton Hall J Diplomacy Int Relations XX(1):4–20

Leal-Arcas R, Minas S (2016) Mapping the international and European governance of renewable energy. Oxford Yearb Eur Law 35(1):621–666

Leal-Arcas R, Morelli A (2018) The resilience of the Paris Climate Agreement: negotiating and implementing the climate regime. Georgetown Environ Law Rev 31(1):1–64

Leal-Arcas R, Wouters J (eds) (2017) Research handbook on EU energy law and policy. Edward Elgar Publishing Ltd, Cheltenham

Leal-Arcas R, Filis A, Abu Gosh ES (2014) International energy governance: selected legal issues. Edward Elgar Publishing, Cheltenham, pp 430–431

Leal-Arcas R et al (2015) The European Union and its energy security challenges. J World Energy Law Bus 8(4):291–336

Leal-Arcas R et al (2017) Prosumers: New actors in EU energy security. Netherlands Yearb Int Law 48:139–172

Leal-Arcas R et al (2018) Smart grids in the European Union: Assessing energy security, regulation & social and ethical considerations. Columbia J Eur Law 24:291

Levitsky S, Ziblatt D (2018) How democracies die: what history tells us about our future. Crown, New York

Lopez Moreno E et al (2008) State of the world's cities 2008/2009. Earthscan, London

Madlener R, Stagl S (2005) Sustainability-guided promotion of renewable electricity generation. Ecol Econ 53:147–167

McAdam J (2012) Climate change, forced migration and international law. Oxford University Press, Oxford

Meléndez-Ortiz R (2016) Enabling the energy transition and scale-up of clean energy technologies: options for the global trade system. International Centre for Trade and Sustainable Development, Geneva

Meléndez-Ortiz R, Sugathan M (2017) Enabling the energy transition and scale-up of clean energy technologies: options for the global trade system – synthesis of the policy options. J World Trade 51(6):934

Monstad J (2007) Urban governance and the transition of energy systems: institutional change and shifting energy and climate policies in Berlin. Int J Urban Reg Res 31(2):326–343

Monstadt J (2007) Urban governance and the transition of energy systems: institutional change and shifting energy and climate policies in Berlin. Int J Urban Reg Res 31(2):326–343

Moreno-Munoz A, Bellido-Outeirinoa FJ, Sianob P, Gomez-Nietoc MA (2016) Mobile social media for smart grids customer engagement: Emerging trends and challenges. Renewable Sustain Energy Rev 53:1611–1616

Morris C, Jungjohann A (2016) Energy democracy: Germany's Energiewende to renewables. Palgrave Macmillan, Basingstoke

Muzaka V, Bishop ML (2015) Doha stalemate: the end of trade multilateralism? Rev Int Stud 41:386

Nakamura S (1999) An interindustry approach to analysing economic and environmental effects of the recycling of waste. Ecol Econ 28:133–145

Nakamura S, Kondo Y (2006) A waste input-output life-cycle cost analysis of the recycling of end-of-life electrical home appliances. Ecol Econ 57:494–506

Nordhaus W (1994) Managing the global commons: the economics of climate change. MIT Press, Cambridge

O'Connor M, Spangenberg J (2008) A methodology for CSR reporting: assuring a representative diversity of indicators across stakeholders, scales, sites and performance issues. J Cleaner Production 16:1399–1415

Omar N (2018) Future and emerging technologies: workshop on future battery technologies for energy storage. Publication office of the European Union, Luxembourg. Available at http://www.file:///C:/Users/lcw197/Downloads/FETWorkshoponFutureBatteryTechnologiesforEnergyStorageA4webpdf.pdf

Orehounig K, Evins R, Dorer V (2015) Integration of decentralized energy systems in neighbourhoods using the energy hub approach. Appl Energy 154:277–289

Orlando E (2014) The evolution of EU policy and law in the environmental field: achievements and current challenges. In: Bakker C, Francioni F (eds) The EU, The US and Global Climate Governance. Routledge, Abingdon, p 74

Orts EW (2011) Climate contracts. Virginia Environ Law J 29:231–233

Pehnt M (2006) Dynamic life cycle assessment (LCA) of renewable energy technologies. Renewable Energy 31:55–71

Roggenkamp M et al (eds) (2016) Energy law in Europe, National, EU and international regulation, 3rd edn. Oxford University Press, Oxford

Roy B, Vanderpooten D (1996) The European School of MCDA: emergence, basic features and current works. J Multi-Criteria Decis Anal 5:22–38

Sassen S (2003) The participation of states and citizens in global governance. Indiana J Global Legal Stud 10(5):5–28

Schrag D (2012) Geobiology of the anthropocene. In: Knoll A, Canfield D, Konhauser K (eds) Fundamentals of geobiology. Blackwell Publishing, Hoboken, pp 425–436

Schwartz S (1992) Universals in the content and structure of values: theory and empirical tests in 20 countries. In: Zanna M (ed) Advances in experimental social psychology, vol 25. Academic Press, New York, pp 1–65

Solorio I, Bechberger M, Popartan L (2013) The European energy policy and its 'Green Dimension': discursive hegemony and policy variations in the greening of energy policy. In: Barnes P, Hoerber T (eds) Sustainable development and Governance in Europe. Routledge, Abingdon

Stern N (2007) The economics of climate change: the stern review. Cambridge University Press, Cambridge

Viñuales J (2012) Foreign investment and environment in international law. Cambridge University Press, Cambridge

Walker G (2008) What are the barriers and incentives for community-owned means of energy production and use? Energy Policy 36:4401–4405

Weber M (1921) The City. Free Press, Glencoe

Wee CJWL (1999) "Asian Values," Singapore, and the Third Way: re-working individualism and collectivism. SOJURN J Sci Issues Southeast Asia 14:332

Welford R, Chan C, Man M (2007) Priorities for corporate social responsibility: A survey of businesses and their stakeholders. Corporate Soc Responsibility Environ Manag 15:52–62

Weston B, Bollier D (2013) Green governance: ecological survival, human rights and the commons. Cambridge University Press, Cambridge

Wiek A, Binder C (2005) Solution spaces for decision-making—a sustainability assessment tool or city-regions. Environ Impact Assess Rev 25:589–608

WTO Secretariat (2016) World trade report 2016: levelling the trading field for SMEs. WTO, Geneva

Zillman D et al (eds) (2008) Beyond the carbon economy. Oxford University Press, Oxford

Chapter 7
Smart Grids and Empowering the Citizen

7.1 Introduction[1]

"Smart grids" can be defined in a variety of ways. The following definition is proposed by the European Regulators Group for Electricity and Gas (ERGEG), and used also by the Council of European Energy Regulators (CEER) and the Commission: "A smart grid is an electricity network that can cost-efficiently integrate the behaviour and actions of all users connected to it – generators, consumers and those that do both – in order to ensure economically efficient, sustainable power systems with low losses and high levels of quality and security of supply and safety".[2] It might be interesting to note that this definition does not define smart grids by the kind of technology used. The term describes the complex connection between electricity generation, transmission, distribution, utilization, and information communication platforms via a system of sensors and other equipment across various levels of the electricity market.[3] One major purpose of smart grids is to target future behaviour of the most important grid user, namely the consumer, with the view to finding more means to use energy when and where necessary, and under more convenient conditions.

---

[1] Many ideas in this chapter draw from Leal-Arcas et al. (2018), pp. 311–410.
[2] European Regulators Group for Electricity & Gas (ERGEG), "Position Paper on Smart Grids: An ERGEG Public Consultation Paper," ERGEG, E09-EQS-30-04, 2009, p. 12; European Regulators Group for Electricity & Gas (ERGEG), "Position Paper on Smart Grids; An ERGEG Conclusion Paper," ERGEG, E10-EQS-38-05, 2010; European Commission, Communication from the Commission to the European Parliament, the Council, the European Economic and Social Committee and the Committee of the Regions "Smart Grids: from innovation to deployment", COM (2011) 202 final, Brussels, 12 April 2011; Council of European Energy Regulators (CEER), "CEER Status Review on European Regulatory Approaches Enabling Smart Grids Solutions ("Smart Regulation")," CEER, C13-EQS-57-04, Brussels, 2014; see also Swora (2010), pp. 465–480.
[3] Xiufeng (2016), p. 154.

Smart metering issues are of course related to smart grid issues. Yet, while smart meters are enablers for smart grids, they are merely one of many components of a smart grid. The ERGEG suggests that it is technically possible to develop smart grids and to roll out smart meters independently of each other.[4] Indeed, smart grids represent an amalgam of existing energy infrastructure and new information technology. Consequently, smart grid regulation transcends energy law and policy; it represents a balance between promoting the development of new technologies aimed at promoting the development of renewable energy and the need to protect consumers and consumer interests. Moreover, it is about access to energy without compromising the environment.

In this chapter, the development of smart grids[5] will be analyzed regarding its implications for social and ethical matters. This chapter draws primarily on the EU context, although it has its conceptual background in international law and policy. Indeed, the ethical framework is founded on international human rights law as incorporated into the Treaty on the Functioning of the European Union (TFEU). The primary aim of considering the ethical framework when dealing with the development of smart grids is to ensure that smart grids contribute to the further realization of economic and social rights within a period of transition to a low-carbon society. Society needs to be engaged and should benefit from the technological transformations occurring in energy generation and consumption. This chapter highlights opportunities and potential downsides of the path towards the achievement of such goals.

The key issue to be considered is how, by whom, and in what ways energy is governed at the EU level. Energy governance can be defined as multi-level management and regulation of energy supply, calling for variable degrees of coordination and cooperation between several actors.[6] In the words of Florini and Sovacool, energy governance refers to "collective action efforts undertaken to manage and distribute energy resources and provide energy services", and can hence serve as "a meaningful and useful framework for assessing energy-related challenges."[7]

With regard to EU energy governance, a definite dualism is at play. On the one hand, EU Member States implement energy policies at the national level. On the other, the European Commission sets the energy blueprint at the EU level. In particular, Member States retain their sovereignty in the energy sector on the grounds that energy is a strategic good. Consequently, decisions on the domestic energy mix should lie solely with national authorities.[8] Since the Lisbon Treaty,

---

[4]European Regulators Group for Electricity & Gas (ERGEG), "Position Paper on Smart Grids; An ERGEG Conclusion Paper," ERGEG, E10-EQS-38-05, 2010.

[5]Smart grids are called 'smart' because they allow some energy loads to be switched off for a while at peak demand times.

[6]See generally Leal-Arcas et al. (2014b).

[7]Florini and Sovacool (2009), pp. 5239–5248.

[8]Maltby (2013), pp. 435–444.

7.1 Introduction

energy has come under the shared competences of the EU and its Member States.[9] National energy measures have to be designed in conformity with a number of EU policies. Examples of such strategies include the 2020 climate and energy package[10] and the 2030 climate and energy framework,[11] for instance. The Commission has thus pioneered an ambitious climate-change mitigation agenda that is bound to impact on the Union's energy policy.[12]

EU energy policy is driven both by Member States' governments and supranational institutions. It is within this institutional framework that the European Commission is currently fostering research on ground-breaking technologies, the elaboration of forward-looking regulation, the transformation of the traditional energy market towards low-carbon systems, and the establishment of prosumers' markets. Such schemes are deeply rooted in the EU's vision to revitalize its energy security.

While Sect. 7.2 deals with smart grids as a multivalent instrument and Sect. 7.3 with the gig economy in the context of new technologies, in Sect. 7.4, the chapter focuses on how smart grids can contribute to a broader economic transformation. It considers the economic transition occurring globally towards collaborative economics and how the EU aims to incorporate new market exchange models into smart grids energy systems. The section considers the potential social and environmental benefits in addition to the challenges that lie ahead in realizing policy goals about the future.

Section 7.5 explores how the EU is working towards fostering more flexible, open, transparent, and dynamic policies within the energy sector. To achieve a low-carbon sustainable society that is fair and equitable for all, the new model also has to reduce the use of resources and to use them efficiently.

Section 7.6 provides an analysis of prosumers as the new actors in energy security. It outlines the key drivers of change behind energy service provision, identified as energy security, climate change, and sustainability. It then explores in greater detail the changing roles for old actors and the emerging roles for new actors within a decentralized energy system that incorporates new digital technologies in the form of smart grids. Section 7.7 concludes the chapter.

---

[9]Energy, in its wide sense, is expressly referred to as a matter of shared competence between the EU and its Member States. See Article 4 TFEU.

[10]European Commission, "2020 climate and energy package," European Commission, 26 July 2017. [Online]. Available: https://ec.europa.eu/clima/policies/strategies/2020_en.

[11]Conclusions of the European Council of 23 October 2014, available at http://www.consilium.europa.eu/uedocs/cms_data/docs/pressdata/en/ec/145397.pdf.

[12]Maltby (2013), pp. 435–444.

7.2 Smart Grids: A Multivalent Instrument

Smart grids, together with the promotion and integration of renewable energy generation in the electricity network, bear significant potential for achieving: low-carbon energy security; protection from the vagaries of international energy markets; affordable energy costs; enhanced access to energy; existent and future climate goals; empowerment of citizens; and enhanced competitiveness for the European economy.

As the International Energy Agency (IEA) underlines, the sweeping renewable energy generation revolution has propelled a profound debate over the design of the evolving power market and electricity security.[13] What makes the ongoing energy transition different to previous ones is the parallel change in both the energy and digital technology sectors. The contemporary energy transition is characterized by common changes in integrated systems. As such, the scope and scale of this transformation is ubiquitously potent and unprecedented.

This transition basically concerns the electricity sector. This industry expands exponentially at the cost of other sectors, and is projected to account from 25% in the last 25 years to close to 40% of the anticipated growth in energy consumption by 2040.[14] The electricity industry fosters crucial spill-overs to other sectors as well. The transportation sector, with the use of Electric Vehicles (EVs) as an inherent part of the grid, is an indicative example. Verbong, Beemsterboer and Sengers highlight the differences between the old and the emerging energy system as follows: "[it] will be more hybrid, in terms of the location and type of generation; lower carbon because of a larger contribution of renewable energy sources (RES); more complex and vulnerable; and less hierarchical".[15]

These changes are bound to profoundly impact on society at large and energy users in particular.[16] Indeed, smart grids can serve a multitude of goals, such as: spearheading economically optimal performance; fostering energy market competition; managing energy consumption and efficiency; achieving maximum possible carbon emissions reductions; maximizing the network efficiency; fomenting system and technology safety, security and resilience; altering and cleaning the energy mix; creating storage capacity and new technologies in the storage sector; expanding to the transportation sector through electric, plug-in vehicles; democratizing the energy systems; and empowering citizens/customers.

Smart grids are not being deployed only in the EU, but also in several other countries, most prominently in China, Japan, South Korea and the United States

---

[13]The International Energy Agency (2016), p. 1, available at https://www.iea.org/publications/freepublications/publication/WorldEnergyOutlook2016ExecutiveSummaryEnglish.pdf.
[14]*Ibid.*, p. 3.
[15]Verbong et al. (2013), pp. 117–125.
[16]*Ibid.*

(US).[17] It is important to stress that there are different drives for the roll-out of smart grids in each case. The frequent outages in the US electricity system, usually caused by ageing infrastructure, have motivated the substitution of the conventional grid with smart grids. China's main preoccupation has been with air quality and pollution. Smart grids have been part of the answer to this environmental question.

The EU is set to proceed with the large-scale roll-out of smart grids to fight climate change and improve energy efficiency in order to hit climate and energy goals set for the next decades.[18] In this context, smart grids are not per se climate policy instruments but speak to a wider set of goals.[19] The way power markets evolve depends on "the innovators' and designers' imagination producing market designs and outcomes better aligned with their political and value preferences".[20]

7.3 The 'gig' Economy and New Technologies

The emergence of the gig economy implies the introduction of new actors. The increase in service provisions such as AirBnB, contracting, free-lancing, self-employment and on-demand web-based platforms such as Uber are challenging traditionally regulated economic relations. It is not possible to isolate the gig economy per se.

To economists, the gig economy represents what is termed as a "disruptive innovation."[21] The gig economy offers opportunities for existing and new market players to engage in new forms of economic exchanges. However, it also has negative impacts on current relations between market participants, policymakers, and regulatory authorities. It remains to be seen how this new economy alters the mechanisms of traditional economic schemes.

The growth of the gig economy is intrinsically linked with new technologies. Many innovations depend on access to data at reduced costs. Arguably, there is a chance to maximize economic growth if we have more openly shared data under

---

[17]In February 2019, congresswoman Alexandria Ocasio-Cortez injected new energy into the climate-change mitigation debate with her proposed Green New Deal, available at https://www.congress.gov/bill/116th-congress/house-resolution/109/text. The aim of the Green New Deal is to make the economy both green and more equitable, thereby targeting climate change and social inequality. It intends to reach a zero net emissions by the middle of the twenty-first century. This goal can be reached via extensive support for green industries that eventually will make the US a leading exporter of clean technologies.

[18]C. Eid, R. Hakvoort and M. de Jong, *Global trends in the political economy of smart grids: A tailored perspective on 'smart' for grids in transition,* UNU-WIDER Working Paper 22/2016.

[19]*Ibid.*

[20]Bressand (2013), p. 25.

[21]Hatzopoulos and Roma (2017), pp. 81–127.

proper ethical structures, instead of competing data silos.[22] Data become "most valuable when open and shared."[23] In the EU, for instance, economic security and growth is associated with the provision of cloud computing. Special Rapporteur Hans Graux claims:

> allowing easy on-demand access to information technology services, cloud computing can significantly reduce capital expenditure, as cloud users only pay for what they actually use. ... [and that] this will foster innovative business models and services across all industries, generating new advantages for customers and companies alike.... Small businesses (SMEs) in particular can benefit from the cloud, as they can get access to high-performance IT solutions, which will help them to adapt quickly to new market developments and to innovate and grow their businesses faster.[24]

Given this perspective, the cloud has an enormous part to play in decentralised energy provision in the EU energy generation, as it will open up opportunities for new small- and medium-scale actors. To achieve this, Graux envisages a sharing economy that is not held back by regulation and barriers to market access.[25]

In the context of a decentralized energy system—a system that places the consumer at the centre of action, empowers the consumer and therefore democratizes the energy system[26]—it is important to talk about smart grids. Smart metering systems and smart grids foreshadow the impending 'Internet of Things,' and the potential risks associated with the collection of detailed consumption data are likely to increase in the future when combined with data from other sources. Across the EU, decentralization, as well as the development of smart grids and smart metering installation, is occurring at differing rates, with the purpose of reducing greenhouse gas emissions. Enabling the necessary regulatory reforms across the spectrum of issues will require innovative approaches to law and regulatory design. Gig-based economic activity often raises issues with regard to the application of existing legal frameworks and blurs established lines between consumer and provider, employee and self-employed, or the professional and non-professional provision of services.

All of this can result in uncertainty over applicable rules, especially when combined with regulatory fragmentation stemming from divergent regulatory approaches at the national or local level. When it comes to energy, fragmentation of policies, regulations and cooperation platforms are a constant problem in utilization and trade, creating barriers to communication and effective solutions.[27] It remains to be seen whether the impact of such disintegration may slow down the spread of smart grids or, on the contrary, whether it will have the potential to boost it

---

[22]See the views expressed by Mark Parsons, secretary-general of Research Data Alliance, in *The Economist*, 27th May 2017, p. 20.

[23]Idem.

[24]Hans Graux, Special Rapporteur, Establishing a Trusted Cloud for Europe, (European Commission's European Cloud Partnership Steering Board, 2014).

[25]Ibid.

[26]See 7th Citizens' Energy Forum Conclusions, 12–13 March 2015, available at https://ec.europa.eu/energy/sites/ener/files/documents/2015_03_13_LF_conclusions.pdf.

[27]Leal-Arcas and Filis (2013), pp. 348–405.

in the future. The EU Commission has noted that there is a risk that regulatory grey zones are being exploited to circumvent rules designed to preserve the public interest.[28] In reality, the gig economy is merely adding to a pattern of decentring governance and the emergence of a "post-regulatory" world.[29]

7.4 Smart Grids: Contributing to the EU Collaborative Economy

The introduction of smart grids into the EU energy grid heralds a crucial transformation. The EU is in the process of investing in radical reform of the economic foundations upon which it depends. The strategic decisions that the EU adopts are driven by many interconnected factors and the main difficulties seem to be found not much within the technical aspects, but more within the policy-related, social, or regulatory issues.[30]

The approach to the transition to a low-carbon economy that the European Commission has embraced is based on new, flexible, dynamic, digital, and resource-efficient economic models.[31] This will increase the reuse of materials to add value to each product's life-cycle and reduce dependency on sourcing natural resources externally. Such a moment of transition could be a substantial opportunity to overcome existing inequalities throughout the EU Member States while the EU economy continues to recover from the 2008 economic crisis.[32]

This section highlights the interlinkages between the different policy and governance approaches to sustainable development[33] and resource efficiency within a collaborative economy. It considers such approaches to emphasize the role that smart grids could play towards achieving the EU's policy goals.

An introduction to the concept of the collaborative economy will be provided in Sect. 7.4.1. Section 7.4.2 focuses on the EU context, while Sect. 7.4.3 specifically

---

[28]Communication from The Commission to the European Parliament, The Council, The European Economic and Social Committee and The Committee of the Regions A European Agenda for the Collaborative Economy (2016) 184 final - Brussels, 2.6.2016 COM(2016) 356 final.

[29]Black (2001).

[30]Vincenzo Giordano et al., Smart Grid Projects in Europe: Lessons Learned and Current Developments 9 (2013).

[31]Press Release IP/10/225, European Comm'n, On Europe 2020: Commission Proposes New Economic Strategy in Europe; *Commission Communication for a Roadmap for Moving to a Competitive Low Carbon Economy in 2050*, COM (2011) 112 final.

[32]Andreas A. Papandreou, *The Great Recession and the Transition to a Low-Carbon Economy* (FESSUD Working Paper Series no. 88, 2015) http://fessud.eu/wp-content/uploads/2015/01/The-Great-Recession-and-the-transition-to-a-low-carbon-economy-Working-paper-88.pdf.

[33]Sustainable development has been one of the main objectives of the European Union since it was included in the Treaty of Amsterdam (signed in 1997) as an overarching principle that inspires all the other EU policies objective of EU policies.

links the potential of a collaborative economy with a smart grids energy system. The final section focuses specifically on energy poverty, as an example of the social benefits that the collaborative economy can provide.

7.4.1 The Collaborative Economy: A "Disruptive Innovation"

The collaborative economy has become a major phenomenon in recent years due to increased business opportunities made possible by advances in digital Information and Communications Technologies (ICT).[34] The digital economy has opened up new innovative ways for people to engage in the market exchange of goods and services that circumvent existing institutional economic structures.[35] The collaborative economy provides the opportunity for individuals and/or communities to offer their assets, time, and skills within the digital market place.[36] This is particularly relevant to those looking to develop market mechanisms to tap into low-carbon energy generation and distribution from decentralized energy communities.[37]

The collaborative economy is a phenomenon that can profoundly change the way consumers buy or rent goods and services. It can also allow consumers to enter the market to provide goods, services, time, or skills themselves and become prosumers. Within such business models, the traditional business-to-consumer relationship is no longer the norm. A trilateral relationship is created instead: the consumer, the provider of a service or good, and the intermediary platform, with anyone being one or more of these actors.[38] The collaborative economy business models, unlike traditional markets, are based on relationships of trust, reputation, and reviews systems.

The advent of the collaborative economy, also referred to as the sharing economy, is what economists call a "disruptive innovation" while some even talk of it being, alongside the digital economy, "the fourth industrial revolution."[39] The concept of sharing goods and services is not without historical precedence. What differentiates traditional collaborative economic activities with the proper collaborative economy

---

[34]Various terms are used, mostly interchangeably, such as collaborative economy, sharing economy, peer-to-peer (P2P) economy, access economy, collaborative consumption and demand economy—among others—to describe the new economic phenomena. *See* Hatzopoulos and Roma (2017), pp. 81–127. This chapter will use the preferred term by the European Commission "collaborative economy."

[35]*Commission Communication Regarding Entrepreneurship 2020 Action Plan, Reigniting the Entrepreneurial Spirit in Europe*, COM (2012) 795 final.

[36]Lougher and Kalmanowicz (2016), p. 87.

[37]Ciocoiu (2011), pp. 33–43.

[38]Lougher and Kalmanowicz (2016), p. 87.

[39]Klaus Schwab, *The Fourth Industrial Revolution: What It Means and How to Respond*, Foreign Aff., (Dec. 12, 2015) https://www.foreignaffairs.com/articles/2015–12–12/fourth-industrial-revolution.

is that the sharing/collaborative model "has progressed from a community practice into a profitable business model."[40] The concept has a certain dynamism that fits within the advent of artificial intelligence (which is the brain and locomotive of the industrial revolution), big data, and 3D printing.[41] The collaborative economy represents a big change from traditional markets by bringing operators to modernize their offer and business models. This competition is generally good for consumers.[42] It can indeed make consumer markets more efficient, as it brings down transaction costs and is able to offer cheaper products and services.

As the phenomenon penetrates more into people's everyday lives, it is important that appropriate regulatory frameworks are adopted in order to provide essential services, such as energy. This must be done in such a way that the dynamism and flexibility of the exchanges between new small-scale enterprises providing services is not undermined. The collaborative economy offers many benefits to consumers and prosumers. But it also presents risks. Advantages and disadvantages of the collective economy will be analyzed in the following sub-section, which focuses on the European context.

7.4.2 The EU and the Collaborative Economy

Assisting consumers, businesses, and public authorities to participate and contribute to the success of a collaborative economy is central to the future economic strategy of the EU and the EU sees the collaborative economy as a new opportunity.[43] Commission Vice-President Jyrki Katainen even stated that "Europe's next unicorn could stem from the collaborative economy," stressing the innovative potential that might be revealed through the collaborative economy in the area of products or services.[44] When considering such a new business model, the EU is also aware of the

---

[40]Marco Böckmann, *The Shared Economy: It Is Time to Start Caring About Sharing; Value Creating Factors in the Shared Economy*, (2013) https://static1.squarespace.com/static/58d6cd33f5e231abb448d827/t/58ea595e1b10e3a416e8ab5b/1491753311257/bockmann-shared-economy.pdf.

[41]Klaus Schwab, *The Fourth Industrial Revolution: What It Means and How to Respond*, Foreign Aff., (Dec. 12, 2015) https://www.foreignaffairs.com/articles/2015-12-12/fourth-industrial-revolution.

[42]Guillermo Beltra, *The Consumer-Policy Nuts and Bolts of the Sharing Economy*, BEUC: The European Consumer Org., (Oct. 26, 2011) http://www.beuc.eu/blog/the-consumer-policy-nuts-and-bolts-of-the-sharing-economy/.

[43]European Commission Press Release, On a European Agenda for the Collaborative Economy, June 2, 2016, http://europa.eu/rapid/press-release_IP-16-2001_en.htm.

[44]Guillermo Beltra, *The Consumer-Policy Nuts and Bolts of the Sharing Economy*, BEUC: The European Consumer Org., (Oct. 26, 2011) http://www.beuc.eu/blog/the-consumer-policy-nuts-and-bolts-of-the-sharing-economy/.

scale of challenges faced by the delivery of such benefits.[45] The new economic model should happen without undermining existing consumer and employment rights, alongside other regulations on health, safety, and the environment. The European Commission cautions that a "fragmented approach to new business models creates uncertainty for traditional operators, new services providers, and consumers alike and may hamper innovation, job creation, and growth."[46]

The implications of the sharing economy for law, regulation, and policy-making are only beginning to be considered.[47] The European Commission, national competition authorities, and consumer protection regulators in Europe are currently in the process of formulating their regulatory approach to address some idiosyncratic issues raised by the sharing economy. When adopting the Single Market Strategy in 2015, the European Commission announced that it "will develop a European agenda for the collaborative economy, including guidance on how existing EU law applies to collaborative economy business models."[48] Currently, the non-regulatory approach followed by the EU relies on many pre-existing legal concepts. These concepts are often ill-adapted to this new model of doing business, thus bearing the risk of extreme fragmentation along national lines.[49] This will frustrate efforts to incorporate the collaborative economy into the updated Single Market Strategy,[50] including the European Energy Union.[51]

---

[45] European Commission Press Release, On a European Agenda for the Collaborative Economy, June 2, 2016 http://europa.eu/rapid/press-release_IP-16-2001_en.htm.

[46] *Id.*

[47] *See* Daniel E. Rauch & David Schleicher, *Like Uber, but for Local Governmental Policy: The Future of Local Regulation of the "Sharing Economy"*, (George Mason University Law & Economics Research Paper, no. 15–01, 1–61, 2015); Koopman et al. (2014), pp. 529; Katz (2015), pp. 1068–1126.

[48] *Commission Staff Working Document Accompanying the Document Communication from the Commission to the European Parliament, the Council, the European Economic and Social Committee and the Committee of the Regions - A European Agenda for the Collaborative Economy - Supporting Analysis*, SWD (2016) 184 final (June 2, 2016).

[49] Hatzopoulos and Roma (2017), pp. 81–127.

[50] The Single Market Strategy aims at enabling people, services, goods and capital to move more freely, offering opportunities for businesses and lowering prices for consumers. It also makes possible for citizens to travel, live, work or study wherever they prefer. In 2015, the European Commission presented a new Single Market Strategy to deliver a deeper and fairer Single Market, that takes into account new concepts and other strategies, such as the European Energy Union and the Digital Single Market Strategy. *Communication from the Commission, Upgrading the Single Market: More Opportunities for People and Business*, COM (2015) 550 final (Oct. 28, 2015).

[51] The European Energy Union was launched in February 2015 by the Commission, and it aims at ensuring that consumers and businesses have access to secure, affordable and climate-friendly energy and making the internal energy market a reality across the EU. The last report on the state of the Energy Union has been released in February 2017. *Commission Staff Working Document: Monitoring Progress Towards the Energy Union Objectives – Key Indicators*, SWD (2017) 32 final (Feb. 1, 2017).

The collaborative economy's expansion and success is intrinsically linked with new technologies. Cloud computing[52] facilities are considered integral by the European Commission for creating new opportunities to foster innovative business models, including the collective economy, because many new innovations depend on access to data at reduced costs.[53] Special Rapporteur Hans Graux notes that "small businesses [] in particular can benefit from the cloud, as they can gain access to high-performance IT solutions, which will help them to adapt quickly to new market developments and to innovate and grow their businesses faster."[54] Given this perspective, the cloud has an enormous role to play in delivering decentralized energy provisions in the EU energy generation. It will open up opportunities for new small- and medium-scale actors to manage data from wireless and internet applications that increasingly constitute smart grids.

7.4.3 Smart Grids: A Platform for the Collaborative Economy

Smart grids are "an integrated system that includes technologies, information (availability, accessibility, utility), human and social influences, organizational and managerial supporting arrangements, and political (policy) constraints as well as facilitation considerations."[55] Smart metering systems are one stepping stone towards smart grids, empowering consumers to actively participate in the energy market. Under Directive 2009/72/EC and Directive 2009/73/EC of the European Parliament and of the Council, EU Member States are required to "ensure the implementation of intelligent metering systems to assist the active participation of consumers in the electricity and gas supply markets."[56] It is also an initiative to increase the number of energy providers within the European Energy Union Strategy.[57]

The European Commission explicitly acknowledged its Energy Union as a strategy "with citizens at its core, where they take ownership of the energy transition,

---

[52]Cloud computing makes possible for users to access scalable and shareable pool of remote computing resources (such as networks, servers, storage, applications and services). This, consequently, means that investing in their own IT infrastructure is not necessary and that they can better share that IT infrastructure. The current policy on cloud computing is set within the Digital Single Market Strategy for Europe. Shawish and Salama (2014).

[53]*Communication from the Commission, Unleashing the Potential of Cloud Computing in Europe*, COM (2012) 529 final (Sept. 27, 2012).

[54]European Cloud Partnership Steering Board, Establishing a Trusted Cloud Europe, 8 (2014) https://www.asktheeu.org/en/request/3990/response/12724/attach/11/20140318%20ECP%20Vision%20Document%20FINAL%20v4.pdf.

[55]Katina et al. (2016), pp. 1–20.

[56]Parliament and Council Directive 2012/27/EU, On Energy Efficiency, 2012 O.J. (L 315) 1, ¶ 31.

[57]Miguel Arias Cañete, Commissioner for Climate Action & Energy, Speech at the European Commission, Smart Grids for a Smart Energy Union, (March 31, 2015).

benefit from new technologies to reduce their bills, participate actively in the market, and where vulnerable consumers are protected."[58] Local energy consumers are crucial to delivering a new power market design that enables consumers to participate in the market through demand-side response, auto-production, smart metering, and storage. In the Winter Package proposed by the European Commission in 2016, EU Member States are required to provide an enabling regulatory framework for local energy communities and users.[59]

With the appropriate regulatory and legal frameworks to incentivize the participation of consumers, the energy economy has the potential to switch from a traditional supply-side driven system controlled by energy cartels into a demand-led decentralized model that fosters competition from localized providers.[60] This potentially opens economic and societal space for the emergence of the energy prosumer at a level that is truly transformative. Political priority will need to support decentralization, countering decades of investment of political capital—and the requisite legal infrastructure—for large-scale energy business, including national companies. This demonstrates that decentralization can deliver secure, affordable, and sustainable energy supplies and could potentially provide the necessary persuasion to governments and citizens alike to embrace new energy systems.

7.4.4 Delivering Social Benefits in a Collaborative Economy

The relationship between new technologies and social change is at the core of the energy/climate debate.[61] There is an overwhelming belief that informed individuals will make rational choices that will benefit society and the environment. Nonetheless, the embedding of new technologies within society can have unforeseen consequences. It is very interesting to consider the unplanned consequences, and perhaps even the distorted incentives, that the upscaled adoption of new technologies into the

---

[58] *Commission Communication to the European Parliament, the Council, the European Economic and Social Committee, the Committee of the Regions and the European Investment Bank: A Framework Strategy for a Resilient Energy Union with a Forward-Looking Climate Change Policy*, at 2, COM (2015) 080 final (Feb. 25, 2015).

[59] *Commission Proposal for a Directive of the European Parliament and of the Council on Common Rules for the Internal Market in Electricity*, at 68, COM (2016) 864 final/2 (Feb. 23, 2017). The proposal defines the concept of local energy community as "an association, a cooperative, a partnership, a non-profit organisation or other legal entity which is effectively controlled by local shareholders or members, generally value rather than profit-driven, involved in distributed generation and in performing activities of a distribution system operator, supplier or aggregator at local level, including across borders." *Id.* at 52.

[60] Clastres (2011), pp. 5399–5408.

[61] Bickerstaff et al. (2016), pp. 2006–2025.

very structure of society and our economy can have. There is a need to question the "smart utopia" being offered.[62]

One goal underpinning energy reforms is to address energy poverty across Europe. On average, 11%—over 54 million—of EU citizens experienced some form of energy poverty (being unable to keep homes at ambient temperatures, having difficulty with bill payments and/or living with inadequate energy infrastructure services).[63] The situation is especially pervasive in Central Eastern and Southern European Member States.[64]

In addition to the cost in economic terms, the negative social and environmental impacts of energy poverty severely curtail the quality of life of vulnerable individuals and communities. Despite this, only a few EU countries have adopted legal definitions recognizing energy poverty.[65] The causes of energy poverty are multiple. A key issue is the structure of energy markets, which impacts energy pricing and determines, to some level, incentives for more efficient energy use. Investments in upgrading and incorporating modern digital ICT into the energy system need to tackle energy poverty at the forefront of their ambitions.

The potential of smart grids to contribute to addressing energy poverty in the EU will be determined by key policy and regulatory decisions. Policy design needs to take account of the interconnections with other related strategies being pursued by the EU. The Digital Single Market Strategy is central to smart grids' achieving economic value. Such a strategy focuses on maximizing the growth of Digital Economy potential by boosting competitiveness.[66] It is clear that ICT is already leading to new business models—as part of the new collaborative economy—and there is great speculation that, with the appropriate regulation, such new models could facilitate a more social just and equitable economy within Europe, and globally.[67] Nonetheless, whether these models can actually play a role in tackling some of the energy poverty issues remains to be seen.

To determine how best to ensure energy poverty is addressed, a distinction needs to be made between traditional consumers and those who are active service providers in the collaborative economy. The demographic affected by energy poverty and new service providers within the collaborative economy are by no means aligned. Energy poverty occurs largely in marginalized, vulnerable, and poorer communities, often in

---

[62] Strengers (2013).

[63] INSIGHT_E, Policy Report on Energy Poverty and Vulnerable Consumers in the Energy Sector Across the EU: Analysis of Policies and Measures 1 (May 2015).

[64] Bouzarovski and Petrova (2015), pp. 129–144.

[65] They are the UK, Ireland, France and Cyprus. INSIGHT_E, Policy Report on Energy Poverty and Vulnerable Consumers in the Energy Sector Across the EU: Analysis of Policies and Measures, at v (May 2015).

[66] *Commission Communication to the European Parliament, the Council, the European Economic and Social Committee and the Committee of the Regions: A Digital Single Market Strategy for Europe*, at 15, COM (2015) 192 final (May 6, 2015).

[67] Halff et al. (2014).

rural areas and small towns.[68] The actors driving the collaborative economy tend to be from urban and affluent communities.[69] Individually, the profiles also differ from that those who are active in forming and benefitting from the opportunities of the collaborative economy come from well-educated, younger, and technologically literate cohorts of the population.[70] However, it is argued that the collaborative economy opens up opportunities to young marginalized communities, who can enter the business sector without the need to meet professional cultural standards.[71] There are also concerns that transnational corporate players within the collaborative economy could appropriate emergent micro-entrepreneurs. Such companies have actively sought to lobby the law-making process within the EU. In a 2016 open letter to the Netherlands Presidency of the Council of the European Union, 47 commercial sharing platforms, including Uber and Airbnb, urged the EU Member States to "ensure that local and national laws do not unnecessarily limit the development of the collaborative economy to the detriment of Europeans" by citing the benefits stemming from sharing services.[72] It is integral that "benefits" are understood to be social ones and not just "commercial" benefits. For the collaborative economy to be socially sustainable, these benefits need to be available not just to those who can become market providers, but also to service users.[73]

The collaborative economy as a fluid, flexible organizing market, will not per se result in affordable energy pricing targeting those most in need.[74] However, it can deliver opportunities in terms of efficiency and affordability to consumers. Such potential depends on the structure of the energy market. Decentralization to increase competition, although part of the EU energy reform packages, has resulted in limiting competition even amongst large-scale providers. The goal under EU energy strategies to increase energy cooperatives that can deliver energy locally with the greatest efficiencies requires clear policy incentives. This will need government intervention to ensure that social opportunities are realized. Delivering social and environmental benefits to all must be at the core of the pathways to achieve a low-carbon energy transition. The next section considers how the EU is approaching the challenges.

---

[68]INSIGHT_E, Policy Report on Energy Poverty and Vulnerable Consumers in the Energy Sector Across the EU: Analysis of Policies and Measures 1 (May 2015).

[69]JCB Science for Policy Report, The Passions and the Interests: Unpacking the "Sharing Economy," (2016).

[70]Tawanna R. Dillahunt & Amelia R. Malone, *The Promise of the Sharing Economy Among Disadvantaged Communities*, (CHI '15 Proceedings of the 33rd Annual ACM Conference on Human Factors in Computing Systems, no. 23, at 2285–94, 2015).

[71]JCB Science for Policy Report, The Passions and the Interests: Unpacking the "Sharing Economy," (2016).

[72]Gemma Newlands *et al.*, Report from the EU H2020 Research Project Ps2Share: Participation, Privacy, and Power in the Sharing Economy: Power in the Sharing Economy (2017).

[73]Nica and Potcoravu (2015), pp. 69–75.

[74]Bauwens and Kostakis (2014), pp. 356–361.

7.5 Low-Carbon Transition Pathways and Smart Grids

The adoption of smart grids can have a vast positive impact on EU policy on energy and climate. The 2015 Paris Agreement has provided a significant boost to deliver the policies agreed by the EU countries on energy and climate.[75] The Agreement is a global driver of investment in technology, law, and policy to achieve a low-carbon world. The potential pathways to achieve this energy transition are many but principles of justice, equity, and fairness should inspire the whole approach to the change.

The United Nations (UN) Paris Agreement's stated goal for the maximum increase of the global average temperature is between 2 °C and 1.5 °C above pre-industrial levels.[76] A warming of 2 °C will result in a new climate regime, particularly in tropical regions, whilst 1.5 °C of warming will bring the Earth to a climate at the outer edge of historical experience for human civilization.[77] The risks associated with the rising global temperature are driving action that will have political, economic, environmental, and social impacts.[78] Either temperature outcome under the Paris Agreement will have impacts on existing energy systems, especially the infrastructure for generation and distribution.[79] Both the 2 °C and 1.5 °C targets are likely to be missed. Maintaining security and resilience requires engineers, policy-makers, and regulators to create climate-proofed energy systems as part of the process towards a low-carbon new model.

The EU has recognized the scale of the task. The EU's Sixth Environmental Action Programme (EAP) identified climate change as the "outstanding challenge of the next 10 years and beyond."[80] It has deliberately interlinked climate change policy with energy policy to develop pathways towards a low-carbon economy.[81]

---

[75] Miguel Arias Cañete, Speech on EU's Climate and Energy Policies After COP 21, (Feb. 8, 2016), (transcript available at http://bruegel.org/2016/02/speech-by-miguel-arias-canete-on-eus-climate-and-energy-policies-after-cop21/).

[76] Paris Agreement, art. 2.1(a).

[77] Schleussner et al. (2016), pp. 327–351.

[78] Burke et al. (2015), pp. 235–239.

[79] Rogelj et al. (2016), pp. 631–639.

[80] *Commission Communication to the European Parliament, the Council, the European Economic and Social Committee and the Committee of the Regions on the Mid-term review of the Sixth Community Environment Action Programme*, COM (2007) 225 final (Apr. 30, 2007). However, recent research shows that climate researchers have been under-estimating the amount of carbon dioxide that is possible to emit to be compatible with the ambitions expressed in the Paris Agreement on Climate Change. In other words, the world may be in a position to emit significantly more CO2 in the next few decades than was previously announced and still be in compliance with the requirements of the Paris Agreement. *See* Millar et al. (2017), pp. 741–747.

[81] The issue of climate change mitigation has even reached democratic levels as close to citizens as teenagers suing the US federal government as part of efforts to force action to request climate action. *See* Michelle Nijhuis, *The Teen-agers Suing over Climate Change*, The New Yorker (Dec. 6, 2016). *See also* Natasha Geiling, *In Landmark Case, Dutch Citizens Sue Their Government over Failure to Act on Climate Change*, Think Progress, (Apr. 14, 2015) https://thinkprogress.org/in-

To encourage the transition to a more secure, affordable, and decarbonized energy system,[82] the EU adopted climate and energy targets to be achieved in the coming decades. In 2007, the "Europe 2020 Strategy" set three key targets: 20% cut in GHG emissions (from 1990 levels), 20% of EU energy from renewables, and 20% improvement in energy efficiency.[83] In 2014, the EU set the target to reduce GHG emissions by at least 40% by 2030 from 1990 levels.[84] The EU also adopted a long-term goal aiming at reducing EU greenhouse gas emissions by 80–95% below 1990 levels by 2050.[85] In February 2015, the Energy Union Strategy was launched, with the goal of leading to a sustainable, low-carbon, and environmentally friendly economy.[86]

Despite such ambitious targets, the link between energy and climate-related issues is relatively new within the EU. Although energy issues have always been at the heart of European integration, energy-related topics (such as climate change policy, renewable energy, energy planning, and energy security of supply) have only gained in importance to the EU's policy and regulation agenda since the concept of sustainability increased in importance at the European and international[87] level.[88] Such a different approach has resulted in considering the three dimensions of sustainability (economic, environmental, and social) within any EU policy. It is encouraging that energy and environmental regulation are now clearly understood to be two sides of the same coin, whereas previously they were perceived as separate competences.[89] Developing strategies to achieve both climate and energy targets will require effective institutional management and good multilevel governance

---

landmark-case-dutch-citizens-sue-their-government-over-failure-to-act-on-climate-change-e01ebb9c3af7/.

[82] Rafael Leal-Arcas, *The Transition Towards Decarbonization: A Legal and Policy Examination of the European Union*, (Queen Mary School of Law Legal Studies Research Paper No. 222/2016, 2016).

[83] *Commission Communication, Europe 2020: A strategy for Smart, Sustainable and Inclusive Growth*, COM (2010) 2020 final (Mar. 3, 2010).

[84] *European Council, Conclusions on 2030 Climate and Energy Policy Framework*, SN 79/14 (Oct. 23, 2014).

[85] *Commission Communication to the European Commission, Parliament, the Council, the European Economic and Social Committee and the Committee of the Regions, Energy Roadmap 2050*, COM (2011) 885 final (Dec. 15, 2011). Fujiwara (2016), pp. 605–609.

[86] *Commission Communication on Unleashing the Potential of Cloud Computing in Europe*, COM (2012) 529 final (Sept. 27, 2012).

[87] Indeed, at the international level, a relatively new initiative called the International Solar Alliance, launched by India's Prime Minister Modi and France's President Francoise Hollande, is very promising as a mechanism to mitigate climate change. It is expected to channel $300 billion in 10 years for the promotion of renewable energy projects. *See* Twesh Mishra, *Sun Shines on $300-Billion Global Fund for Clean Energy*, Hindu Bus. Line (May 1, 2017), http://www.thehindubusinessline.com/economy/sun-shines-on-300billion-global-fund-for-clean-energy/article9675599.ece.

[88] Solorio et al. (2013).

[89] Orlando (2014), p. 74.

involving existing and new actors. A new transitional approach will help to achieve such a goal from an institutional point of view.

Until quite recently, the concept of transitional justice has been associated only with post-conflict truth and reconciliation processes.[90] However, an increasing number of justice scholars are seeing the value of applying the concept to other political and legal developments related to human rights, including natural resources management and climate change law.[91] A multidisciplinary approach to exploring the discourse and practice of transitional strategies within EU climate and energy policy can offer a conceptual foundation for understanding the justice dimension of the dynamic normative transition within other jurisdictions and contexts. A transitional justice approach to the transformation from a carbon-dominant energy system to one based on smart grids and renewables could offer the EU a methodological pathway that will help address pressing social issues such as energy poverty. This approach already exists in varying degrees in all European countries, as discussed in the previous section.

It is evident that the EU is seeking to undertake a transformation towards a low-carbon economy that can meet these challenges. The EU is increasingly seeking to include such principles within those laws and policies that aim at achieving resilient economic, social, and environmental systems.[92] The intersection of social, economic, environmental, and political rights across all communities of energy users, including marginalized and vulnerable groups, needs to be explored as part of a more interconnected examination of each of the EU's actions, especially considering its leading position of addressing environmental issues adopting a more inclusive, holistic and integrated approach.[93]

The Fifth EAP (1993) was a reaction to the perceived failure of regulatory measures to achieve environmental goals. The Fifth EAP abandoned the traditional "command-and-control" approach in favor of innovative regulatory models that implied "shared responsibility between various actors: government, industry, and the public."[94] The EU welcomed the principle of sustainable development, combining economic, social, and environmental aspects in 1997 when EU Member States adopted the Amsterdam Treaty.[95] This is now incorporated in Article 3(3) of the Treaty on European Union (TEU) and it can be considered a "constitutional objective" of the EU.[96] In 2001, the European Council adopted the EU Sustainable

---

[90]Roht-Arriaza and Mariezcurrena (2006).

[91]Teitel (2014); Franzki and Olarte (2014), pp. 201–218.

[92]ClientEarth, Identifying Opportunities for Sustainable Public Procurement Briefing Series, Briefing No. 1: Sustainable Development as a Key Policy Objective of the European Union (2011).

[93]*Id.*

[94]*Towards Sustainability: A European Community Programme of Policy and Action in Relation to the Environment and Sustainable Development,* 1993 O.J. (C 138) 5, (May 17, 1993).

[95]Treaty of Amsterdam Amending the Treaty on European Union, the Treaties Establishing the European Communities and Certain Related Acts, Oct. 2, 1997, 1997 O.J. (C 340) 1.

[96]The objective of sustainable development can be found in the Constitutions of other jurisdictions (such as South Africa), but the European Union as a supranational region is the only one that refers

Development Strategy, "a long-term strategy dovetailing policies for economically, socially, and ecologically sustainable development."[97] After this important step, the Sixth EAP (2002) advocated "a more inclusive approach including more specific targets and an increased use of market-based measures."[98] This aims at strengthening the integration of environmental concerns into other policies, in an attempt to foster greater engagement and implementation by EU Member States.[99] The most recent EAP, the Seventh EAP (2013),[100] emphasizes decoupling economic growth from carbon emissions and establishing a circular economy.[101] To achieve its goals, the Seventh EAP commits to a better integration of environmental concerns into other policy areas and ensures coherence when creating new policy. Strategic initiatives feeding into the Seventh EAP include the Roadmap to a Resource Efficient Europe[102] and the Roadmap for a low carbon economy by 2050.[103]

The EU Climate and Energy Package focuses on the fact that some contradictions can arise between the instruments to reduce GHG emissions and the protection of the environment. Although the EU is still not sure whether the package succeeds in balancing climate change mitigation with other environmental protection goals, it succeeds in supporting climate change mainstreaming.[104] The EU's climate policy and leadership on sustainability governance contrasts with the complexities of the internal energy market. Sustainability governance is still rather underdeveloped,[105] despite the overuse of the term "sustainability" in a significant number of legal

---

to such objective for more than one country. Article 3(3) of the Treaty on European Union (TEU) provides that "The Union shall establish an internal market. It shall work for the sustainable development of Europe based on balanced economic growth and price stability, a highly competitive social market economy, aiming at full employment and social progress, and a high level of protection and improvement of the quality of the environment. It shall promote scientific and technological advance." Also, according to Article 3(5) TEU, the EU shall contribute to "the sustainable development of the Earth, solidarity and mutual respect among peoples, free and fair trade, eradication of poverty and the protection of human rights." Treaty on European Union, art. 3, Oct. 26, 2012, 2012 O.J. (C 326) 1.

[97]*Commission Communication, A Sustainable Europe for a Better World: A European Union Strategy for Sustainable Development*, at 1, COM (2001) 264 Final (May 15, 2001).

[98]European Comm'n, Environment 2010: Our Future, Our Choice. 6th EU Environment Action Programme (2001), http://ec.europa.eu/environment/air/pdf/6eapbooklet_en.pdf.

[99]*Id.*

[100]Parliament and Council Decision 1386/2013/EU, On a General Union Environment Action Programme to 2020 "Living Well, Within the Limits of Our Planet," 2013 O.J. (L354) 171.

[101]*Environment Action Programme to 2020*, European Comm'n, (last updated June 8, 2016) http://ec.europa.eu/environment/action-programme/.

[102]*Commission Communication to the European Parliament, the Council, the European Economic and Social Committee and the Committee of the Regions, Roadmap to a Resource Efficient Europe*, COM (2011) 571 final (Sept. 20, 2011).

[103]*Commission Communication to the European Parliament, the Council, the European Economic and Social Committee and the Committee of the Regions, A Roadmap for Moving to a Competitive Low Carbon Economy in 2050*, COM (2011) 112 final (Aug. 3, 2011).

[104]Montini and Orlando (2012), p. 165.

[105]For an analysis, see Leal-Arcas (2017), p. 801.

7.5 Low-Carbon Transition Pathways and Smart Grids

instruments advocating for it.[106] Meeting renewable energy demands in a low-carbon economy will need to be done in a manner that does not result in negative impacts on the environment.[107]

The EU, as a governance body, continues to invest in advancing innovative approaches to policy-making in its pursuit of realizing sustainable development.[108] In the 1990s, Collier observed that environmental policy integration is necessary for "achieving sustainable development and preventing environmental damage; removing contradictions between policies as well as within policies, and realizing mutual benefits and the goal of making policies mutually supportive."[109] Given today's challenges of energy security of supply, climate change, biodiversity conservation, and the need for an equitable allocation of resources, sustainable development is perceived as a new constitutional paradigm, and is now even more essential to the EU's regulatory frameworks than when the concept was coined in 1987.[110] The adoption of the Sustainable Development Goals[111] by the international community at the UN General Assembly in September 2015 provided the EU with an opportunity to push forward the key principles of the TFEU and incorporate them into the very fabric of policy-making, both substantively and procedurally.[112]

As part of the 2030 Agenda for Sustainable Development,[113] the EU is keen to reform its policy-making approach to ensure that it considers long-term impacts. In measuring progress towards sustainable transitions and human well-being within the physical limits of the planet, it is necessary to assess environmental sustainability. The so-called "planetary boundaries"[114] for carbon emissions, water use, and land use are being modelled to determine the ecological space available for sustainable development. "Growing scientific evidence for the indispensable role of environmental sustainability in sustainable development calls for appropriate frameworks

---

[106] Galera (2017), p. 13.

[107] Hastik et al. (2016), p. 131.

[108] European Comm'n, *State of the Union 2016: Strengthening European Investments for Jobs and Growth*, (Sept. 14, 2016), http://europa.eu/rapid/press-release_IP-16–3002_en.htm [https://perma.cc/RW4A-PSY3].

[109] Ute Collier, Energy and Environment in the European Union 36 (1994).

[110] Galera (2017), p. 13.

[111] The 17 Sustainable Development Goals (SDGs) are part of the 2030 Agenda for Sustainable Development. They call for action by all countries, poor, rich and middle-income, to promote prosperity while protecting the planet. End of poverty must be achieved together with economic growth, considering both social needs, climate change and environmental protection. The SDGs are not legally binding, but governments are expected to take ownership and establish national policy strategies for their achievement. G.A. Res. 70/1, Transforming Our World: The 2030 Agenda for Sustainable Development (Sept. 25, 2015).

[112] European Comm'n, *Sustainable Development: EU Sets out Its Priorities* (Nov. 22, 2016).

[113] *Commission Communication to the European Parliament, The Council, The European Economic and Social Committee and the Committee of the Regions Next Steps for a Sustainable European Future European Action for Sustainability*, at 2–3, COM (2016) 739 final (Nov. 22, 2016).

[114] Rockström et al. (2009), p. 32. For a 2015 update, see Steffen et al. (2015), p. 736.

and indicators for environmental sustainability assessment."[115] Most decision-support systems and recommendations developed to analyze trade-offs between low-carbon energy generation and other interests have focused on single energy sources such as biomass, wind energy, and hydropower. A way to represent the pressure that humanity exerts on the Earth's ecosystems is to measure humanity's environmental footprint.[116] Recently, a growing list of such footprints has been created such as the ecological footprint, the carbon footprint, and the water footprint.[117] The anthropogenic impact on the planet needs to be taken up by policymakers, economists, and lawyers when designing long-term strategies for pathways to a low-carbon world, including those working on smart grids energy systems.

The concept of building resilience into the system has increasingly complemented the debate on sustainability[118] and has focused on long-term solutions. The European Environmental Agency has called for:

> increased use of foresight methods, such as horizon scanning, scenario development and visioning [which] could strengthen long-term decision-making by bringing together different perspectives and disciplines, and developing systemic understanding. Impact assessments of the European Commission and EU Member States, for example, could be enhanced if they were systematically required to consider the long-term global context.[119]

Technologies can either undermine or enhance the resilience of systems.[120] The energy/climate debate is one infused with a faith in the positive relationship between the introduction of new technologies and social change.[121]

It is not only the technological system, but also the social-ecological systems that need to be resilient to reduce the chances of exposure to shocks. "Social-ecological systems and socio-technical systems are understood to display complex, dynamic, multiscale, and adaptive properties; recommendations for their sustainable governance emphasize learning, experimentation, and iteration."[122] The transition phase is one where multiple pathways are being pursued and the social-ecological ecosystem is at its most dynamic and vulnerable stage.[123]

---

[115] Fang et al. (2015), p. 11285.

[116] The notion of measuring the carbon footprint as part of a sustainable world is even vivid in the "Clean Label" movement, which aims to provide honest information to the consumer and food professionals on questions such as what there is in our food, who made it, what is the carbon footprint and related issues. See *About*, Go Clean Label, (2017), https://gocleanlabel.com/about/.

[117] Sustainable Europe Research Institute (SERI), How to Measure Europe's Resource Use 30 (June 2009), http://www.foeeurope.org/sites/default/files/publications/foee_seri_measuring_europes_resource_use_0609.pdf.

[118] European Comm'n, European Political Strategy Centre, Sustainability Now! A European Vision for Sustainability 18 (2016).

[119] European Env't Agency, The European Environment State and Outlook 2015: Assessment of Global Megatrends 16 (2015).

[120] Smith and Stirling (2010), Article 11.

[121] Bickerstaff et al. (2016), p. 2006.

[122] Smith and Stirling (2010), Article 11.

[123] Chaffin et al. (2014), p. 56.

Research into the slow uptake of smart grids has emphasized the importance of developing a diverse approach and establishing multiple pathways for transformation amongst all stakeholders to build resilience within the system.[124] There is a need for flexible, responsive regulatory frameworks that are fit for a transformational social-economic system. This requires lawyers and policy-makers to recognize uncertainties within systems—in this case smart grid-based energy systems—and adopt a more adaptive approach to governance, which takes our incomplete knowledge of social-ecological systems into account.

The transition to a low-carbon world will need the EU Member States and others to carefully balance the new opportunities arising from ICT alongside societal and environmental needs in a just, fair, and equitable manner. The Member States must focus on delivering integrated sustainable outcomes across all sectors. One area where this is most necessary is the use and disposal of resources.

7.6 Prosumers As New Actors in Energy Security[125]

7.6.1 Drivers of Change

The global governance landscape is rapidly transforming. This is particularly evident in the energy sector. The energy sector's global transformation is being driven, directly and indirectly, by a changing geopolitical landscape, the urgent need to pursue low-carbon development to abate climate change, commitments to achieve integrated sustainable development,[126] innovative new digital information and communication technology, and transformations in economics. This section highlights the fact that these drivers are creating new spaces for both old and new actors to contribute to the transformation to a low-carbon world.

7.6.1.1 Energy Security Challenges

Driven by energy security concerns, energy actors, both state (governments, regulatory authorities and public-owned or controlled utilities) and non-state (private energy companies), are adapting the sourcing, use, and distribution of energy.[127] Diversifying energy supplies has opened opportunities for new actors to become involved in energy markets. The establishment of the International Renewable Energy Agency in 2009 to help countries transition to a sustainable energy future is an excellent example. The EU is faced with traditional energy challenges, the most

---

[124]Muench (2014), p. 80; Tuballa and Abundo (2016), p. 710.

[125]This section draws from Leal-Arcas et al. (2017), Chapter 5, pp. 139–172.

[126]See generally Leal-Arcas (2017).

[127]See for instance Leal-Arcas et al. (2016).

important of which revolves around dependence on single suppliers with high market shares and problematic relations (namely Russia).[128] Other challenges include dependence on increasing oil imports and fluctuating energy prices. The designated European Energy Union[129] reflects these concerns and endeavours to place the Union on a low-carbon trajectory.[130]

A number of forces are creating a shifting terrain, including the globalization of gas markets, the advent of liquefied natural gas, the development of fracking technology, the decarbonizing of natural gas and shale oil and gas exploration, and further market integration. A new energy architecture is emerging as a result, involving demand-side policies, new business models—which go hand-in-hand with digitalization—and new actors in the energy sector. In particular, disentangling the EU from overt dependence on energy exports and fossil energy, while increasing energy efficiency and renewable energy generation, will fundamentally reshape the energy scene in the dual direction of demand-side policies and will create a plethora of new, dispersed, small-scale energy producers. Increasing domestic energy generation through renewable sources is integral to achieving greater energy security for Europe's member states. The diversification of energy generation required will result in new energy actors, such as large-scale investors in renewable energy generation, prosumers, and energy communities. This is a situation mirrored in other countries adopting the same energy diversification strategies, for example the US, China and South Africa. In this landscape, the role of prosumers seems central.

7.6.1.2 Climate Change and Sustainable Development

Increasingly, a range of new actors that we refer to here as prosumers are actively contributing to initiatives to drive a transition to a low-carbon global economy.[131] The expanding environmental crisis[132] faced globally has resulted in new opportunities in governance. This has increasingly been recognised at the international level, where outcomes have recognised the value of non-state actors to achieving

---

[128]Leal-Arcas (2009), pp. 337–366.

[129]The European Energy Union could well be the flagship of a new outset towards a more prosperous, energy-secure and unified Europe, bearing in mind that EU Member States wish to guard their sovereignty over national energy systems. The aim is to make it easier to trade energy inside the EU. In the past, there have been divisions between EU Member States when trying to draft a unified energy policy. The European Energy Union tries to rectify this deficiency. For an analysis of the European Energy Union, see Leal-Arcas (2016), p. 12.

[130]Szulecki et al. (2016), p. 1.

[131]See generally Leal-Arcas et al. (2014a); "Better growth, better climate," The New Climate Economy, available at http://newclimateeconomy.report/2014/.

[132]One shocking environmental fact is that, in 2017, it rained in the Antarctica. See Potenza, A. "Unusual weather in Antarctica leads to rain and a Texas-sized melt," *The Verge*, 15 June 2017, available at https://www.theverge.com/2017/6/15/15811214/west-antarctica-ross-ice-shelf-melting-rain-el-nino.

transformation and resolving the crisis. International multilateral conferences, including the 2002 Johannesburg UN Conference on Environment and Development, 2012 Rio+20 Conference, the UN General Assembly adoption of the 2015 Sustainable Development Goals (SDGs), and the UN climate change negotiations in 2015, increasingly emphasise the role that new actors beyond the state can play in making the transformation take place.[133] As a consequence, many new fora are emerging to tackle entrenched problems within a range of sectors such as water, forests, and the urban environment, creating new spaces and fostering new participatory approaches. Increasingly, the interconnections between different sectors are being recognised when developing law and policy interventions to produce co-benefits.[134]

This is particularly evident in relation to the international climate change regime, which has actively engaged non-state actors to contribute towards delivering mitigation and adaptation projects and investments. In fact, since the COP 21 in Paris in 2015, the joining of international negotiations and non-state actors is shaping new global climate change governance, which means that there is a connection between global and local processes. Moreover, the global climate change governance regime is constituted by a broad array of public, private, and hybrid associations and mechanisms that have developed within and alongside the UNFCCC itself.[135] Alongside the UN Framework Convention on Climate Change (UNFCCC) Conference of the Parties (COP) 20 as part of the Lima-Paris Action Agenda in 2014, a Non-State Actor Zone for Climate Action (NAZCA) was launched by the Peruvian Presidency to track initiatives that increase climate change mitigation and adaptation. Commitments on the NAZCA platform illustrate the range of actors being involved across sectors, especially energy.[136] Initiatives undertaken to abate climate change now need to be integrated with all the goals outlined in the UN SDGs. This extra dimension provides greater opportunities for innovative solutions from multiple actors.

Old and new actors alike are seeking new alliances to achieve integrated outcomes. This is visible for instance with multinational enterprises in their efforts to design sustainability commodity supply chains standards, such as the Roundtable on Sustainable Palm Oil, which includes different stakeholders, including the Forest People's Programme, a non-governmental organisation run by indigenous people.[137] The effectiveness and impacts, especially socially and environmentally, of the

---

[133] See for instance Sustainable Development Goal number 7, available at http://www.un.org/sustainabledevelopment/energy/.

[134] See for instance the Sustainable Development Goals Knowledge Platform, available at https://sustainabledevelopment.un.org/sdgs. See also Northrop et al. (2016), Hufbauer and Suominen (2010), Victor (2011); Sachs (2015, 2017), Matsushita and Schoenbaum (2016) and Hufbauer et al. (2016).

[135] Van Asselt (2014) and Elliott (2013).

[136] See UNFCCC Nazca Platform at http://climateaction.unfccc.int/.

[137] Colchester (2016), pp. 150–165.

outsourcing of governance is much debated.[138] Despite concerns, however, the trend continues apace. It is a trend that provides opportunities that policymakers and regulators, as well as entrepreneurs at all levels, are looking to exploit for a variety of reasons. New technologies are providing platforms for innovative approaches by all actors seeking to contribute to the low-carbon global energy transition.

7.6.2 Energy Actors: Old and New

The transition to a low-carbon energy supply drawing on widely dispersed generation and supply can only result from new patterns of engagement.[139] In particular, a new market design is very much in demand, both differentiating the role of incumbents, as well as providing opportunities for the emergence of new actors.[140] It is important to note that differing needs and interests between old and new actors will influence the design of the regulatory landscape going forward in a rapidly changing economic context.

7.6.2.1 Old Actors

The traditional energy market landscape is under large-scale transition. Traditionally, and since the application of the unbundling regulation as part of the process of the establishment of a fully liberalized single energy market, Transmission System Operators (TSOs) have been in charge of balancing the load at high voltage levels and transmitting it from large generation plants to DSOs. The changing nature of electricity generation, increasingly moving from large power plants to an exponentially growing number of decentralized generation premises, casts doubt on the exact role of TSOs in the coming years and decades. DSOs, to the contrary, find themselves at the heart of the energy transition. The European Commission's proposed internal electricity market directive suggests an upgraded role for DSOs, not least in managing and coordinating all the new decentralized sources. DSOs are expected to integrate different forms of power generation into their grids, and distribute electricity to households, offices and all establishments in general in a secure and efficient way. Their function, at the same time, is expected to enormously gain from digitalization.[141]

---

[138]O'Rourke (2003); Brunsson and Jacobsson (2000).

[139]Upham (2016), p. 549.

[140]Parag and Sovacool (2016), Article number: 16032.

[141]European Commission, Proposal for a Regulation of the European Parliament and of The Council on the Internal Market For Electricity (recast), Brussels, 30.11.2016 COM(2016) 861 final 2016/0379 (COD) 4-5; Pavol Szalai, 'Power grid operators expect their "Uber moment"', EU Observer, 29 September 2016, https://www.euractiv.com/section/energy/news/power-grid-operators-expect-their-uber-moment/.

It is for these reasons, however, that DSOs also find themselves amidst competing actors for market share. For one, TSOs' marginalization may push them to ask for a redefined role in the changing energy market landscape. The precise division of labour between TSOs and DSOs remains murky for the time being. The rules of the game that will ensure their efficient cooperation within and across countries are yet to be defined.[142]

7.6.2.2 New Actors

The emergence of new actors on the energy market will influence the feasibility of fulfilling goals related to a more effective energy utilization. Prosumers have the potential of increasing energy efficiency and securing stable energy supplies for a wider range of consumers, including themselves. New opportunities arise for a new type of economic activity, that of energy aggregators, in what seems to be a much more variable business energy landscape. This role can be fulfilled by incumbent market players, as well as by new companies that will focus on encouraging their customers' efficient use of energy and contracting the surplus capacity, which they can sell in a "flexibility package" to the distributors and utilities. Small storage providers can also emerge in an evolving market that needs back-up capacity and last resort solutions to respond to energy supply and demand variability.

A high premium will be paid for such flexibility services, so the corporate rationale is evidently present. Importantly, there are strong grounds for such economic activity to take place at the community (or even at the district/neighbourhood) level, with co-operatives appearing as a potent form of entrepreneurial type of organization.[143] The energy market increasingly calls for integrated energy services companies which will optimize both digital technology and electricity distribution by means of trading flexibility services.[144] There are reasons to be optimistic about the affordability of technology advancement for future prosumers: when smartphones came out, they were unaffordable; today, around 80% of phones in the US are smartphones.[145] By analogy, the same effect should happen in energy technology.

There are many ways for citizens, small businesses and communities to contribute to the energy transition, actively participating in different aspects of the energy

---

[142]Pavol Szalai, 'Power grid operators expect their "Uber moment"', EU Observer, 29 September 2016, https://www.euractiv.com/section/energy/news/power-grid-operators-expect-their-uber-moment/.

[143]Ye Cai, Tao Huang, Ettore Bompard, Yijia Cao, Yong Li, "Self-Sustainable Community of Electricity Prosumers in the Emerging Distribution System," 2016 IEEE Transactions on Smart Grid (Volume: PP, Issue: 99).

[144]Louis Boscán and Rahmatallah Poudineh, 'Flexibility-Enabling Contracts in Electricity Markets' (2016) *Oxford Energy Comment, The Oxford Institute for Energy Studies* 2.

[145]Pew Research Center, "Mobile Fact Sheet," available at http://www.pewinternet.org/fact-sheet/mobile/.

market to become true "energy citizens." Citizens are no longer resigned to the role of passive consumers, but have the potential to be energy producers, or "prosumers," particularly through self-generation of renewable energy, storage, energy conservation and participation in demand response.[146] From a legal point of view, prosumers are still considered individuals, rather than commercial actors. Their likely coming together though, in energy communities, will necessitate a more commercial legal status. In general, the future legal status of prosumers is one of the issues that remains unsettled and will have to be determined by the upcoming EU energy regulation.

Crucially, the new energy market creates ample opportunities for individuals and households to become energy traders.[147] Either directly *vis-à-vis* established utilities or indirectly through aggregators, prosumers are empowered to trade the energy they have conserved or produced, thus killing two birds with one stone by facilitating flexibility and network optimization, as well as raising extra revenues for themselves. Indeed, the emphasis of the undertaken energy overhaul lies in distributed energy resources (DER), which enhance local generation and flows into the network.[148]

7.6.2.2.1 Supply Security

The development of prosumer markets fits into the EU's traditional diversification agenda regarding fuels, sources of supply, and routes of supply diversification.[149] This is because it adds to traditionally imported sources of energy such as oil and gas, and indigenous energy sources such as solar and wind energy via different pathways. On the one hand, prosumer markets can lead to improved security of supply and market resilience through increased indigenous energy generation, substantial rationalization of energy use, and increased energy conservation and efficiency. On the other hand, they can decrease needs and quantities of imports, thus also de-securitizing their importance for the smooth operation of the EU, in particular for national markets that are less interconnected and more dependent and vulnerable.

The exact shape prosumer markets will take is significantly contextualized and contingent upon both local conditions and national regulation. A one-size-fits-all approach hence is hardly feasible. In Greece, for example, geography plays a critical role, with the plethora of islands calling for different treatment than that of the

---

[146]Roberts (2016).

[147]This approach is quite in contrast to the typical top-down way of trading energy. See for instance Leal-Arcas (2015), pp. 202–219; Leal-Arcas et al. (2015), pp. 38–87.

[148]European Commission, 'Proposal for a DIRECTIVE OF THE EUROPEAN PARLIAMENT AND OF THE COUNCIL on the promotion of the use of energy from renewable sources (recast)', Brussels, 23.2.2017. COM(2016) 767 final/22016/0382(COD) CORRIGENDUM.

[149]Proedrou (2016).

mainland, as is reflected in the institutional energy structure and the associated regulatory provisions.[150] The existence of a big number of small islands in the Aegean Sea creates a strong rationale for autonomous energy generation, since connection to the main grid is rather costly. Utilizing the rich potential of energy generation through strong winds and ample sunshine could substantially boost indigenous energy generation. A smart grids architecture that would interconnect the grids of several adjacent islands, and then create interconnection points for these different groups of islands (most probably on the basis of existing administrative divisions), could also provide the appropriate scale, as well as interconnectivity options, necessary to ensure resilient security of supply.

With regard to affordability, the development of prosumer markets can bring a plethora of positive results. The first result is that it can lower electricity bills in two ways: self-consumption and enhanced demand management. The second is that is can add to prosumers' income via the financial gains they make by selling the (excess) energy they produce.[151] On top of these, prosumers become less sensitive to abrupt international energy price increases and/or supply disruptions. One can hardly overestimate the potential of prosumer markets in the long term if one thinks of the downward trajectory of renewable-energy investments costs and prices.[152] On the other hand, international energy prices are underpinned by consistent fluctuations, with occasionally high prices being a prerequisite for new rounds of investments[153] and juxtaposing them to the structurally cyclical nature of international energy prices and their ensuing boom-and bust-cycles. While high fossil energy prices are at times indispensable in order to finance a new wave of investments, this is not the case with renewable energy.[154]

While the potential is immense, the development of prosumer markets faces significant critiques as well as several hurdles. Starting with the former, it remains unclear how and whether prosumer markets will be able to offset the projected increase of energy demand. The ubiquitous electrification of society, seen in all sectors, from electric mobility (electric or hybrid cars, electrified public transportation), to heating (replacing oil, gas and biomass heating) and an exponentially growing number of electric appliances, evidently feeds electricity demand.[155] In other words, while smart grids target energy efficiency, there is no clear pathway to counter the well-known rebound effect, or Jevon's paradox. The more energy use becomes more efficient, the more energy use grows, with the end result being more amenities, but hardly any tangible benefits in the indispensable energy savings needed if even the contemporary mediocre climate targets (in the sense that they

---

[150]HEDNO, 'Regulatory Framework', http://www.deddie.gr/en/i-etaireia/ruthmistiko-plaisio.

[151]Proedrou (2017a).

[152]OECD/IEA, Renewable Energy Medium-Term Market Report. Market Analysis and Forecasts to 2020. Executive Summary, Paris (2015) 5.

[153]On investment law and policy, see generally Leal-Arcas (2010).

[154]Proedrou (2017b), p. 194.

[155]Verbong et al. (2013), p. 123.

do not bring us to the maximum acceptable level of a 2-degree Celsius increase in global temperature above pre-industrial levels) are to be met.

With respect to the hurdles, prosumer markets and EU regulation in this field lag significantly behind the pace of innovation and proliferation of renewable energy production. Some solar- rich countries such as Greece and Spain, for example, have only very recently passed laws (Spain in 2015[156] and Greece in 2016[157]) that allow for energy self-generation. This regulatory gap has been part of a broader emphasis on large corporate players, and the associated fiscal stimuli, to invest in large wind parks and photovoltaic panels, rather than in individuals and small-scale installations.[158] Such inertia has blocked the immense bottom-up potential of renewable energy generation at a massive scale, as has been the case in much less sunlit countries such as Belgium and Germany. The new law has come under severe scrutiny in Spain for two main reasons:

1. Firstly, individual investors are expected to pay a "tax on the sun;" and
2. Secondly, they are not remunerated for any energy quantities they ship to the grid.[159]

In countries where self-generation has been actively endorsed and subsidized, on the other hand, such as Belgium, the charging of a fee for individual energy producers to be granted the right to supply the grid provides a significant disincentive for cascading renewable generation schemes.[160] All these tax and regulatory impediments counter the very rationale and philosophy underpinning the transition towards the development of prosumer markets. Instead of compensating the DSOs for handling further supply, as is currently the case, a fairer and more stimulating approach seems to be that the prosumer has to be incentivized to increase his/her energy savings and shipments to the grids, rather than be charged for this right.

The operation of the electricity markets, moreover, calls for continuous balancing. In some cases, the incapacity of the grid to receive and utilize renewable energy meant that wind turbines had to be switched off, and that solar power was not brought into use.[161] On the other hand, in case demand exceeds supply what results is higher prices, or even load-shedding. Such problems are to be tackled in the

---

[156]Government of Spain, Ministry of Energy Tourism and Digital Agenda, 'Royal Decree 900/2015 on Self-Consumption' (2015), 9 October.

[157]Republic of Greece, Ministry of Environment and Energy, 'Decree 4414/2016' (2016) 9 August.

[158]Gaëtan Masson, Jose Ignacio Briano and Maria Jesus Baez, 'Review and Analysis of PV Self-Consumption Policies', *International Energy Agency, Photovoltaic Power Systems Program, Report IEA-PVPS T1-28:2016*, 13.

[159]Mira Galanova, 'Spain's Sunshine Toll: Row over Proposed Solar Tax', *BBC News*, 7 October 2013, Barcelona, Spain.

[160]European Environment Agency, 'Country Profile- Belgium'. Energy Support 2005-2012.

[161]Matthias Wissner, 'The Smart Grid – A Saucerful of Secrets?' (2011) *Applied Energy*; Eid, Rudi Hakvoort and Martin de Jong, 'Global Trends in the Political Economy of Smart Grids: A Tailored Perspective on "Smart" for Grids in Transition' (2016) *UNU-WIDER Research Paper wp2016-022. World Institute for Development Economic Research*.

underway prosumer markets by means of demand response management, real-time pricing, decentralized control automation, intra-day markets and flexible targeted contracts.[162]

7.6.2.2.2 Sustainability

Actors can contribute to integrated sustainability only if the energy they use and the technologies which enable their participation in the distribution are, throughout the entire life cycle, able to reduce environmental and climate change impacts. The EU energy targets promote energy efficiency, renewable energy as well as decentralisation. But these also need to fit within the broader EU sustainable development agenda post-SDGs[163] and its Circular Economy Program to increase resource efficiency and decrease waste.[164] In fact, one could divide by half the energy we use because much of it is wasted due to outdated technology, which means that saving energy is the most profitable thing we can do. The creation of a carbon tax can help here.[165] Others have come up with ecological solutions that bridge the gap between ecology and the economy.[166]

For prosumers, delivering on sustainability requirements will depend on obligations placed directly on them as commercial actors, rather than consumers, as opposed to manufacturers of technologies and physical infrastructure like solar panels.[167] Finding a balance between these is essential to secure sustainability and improve the decentralised energy ecosystem resilience. Moreover, we are living on credit because we are, on a daily basis, using large amounts of fossil fuels, which is not sustainable, and fossil fuels are being consumed without replacement.

The emergence of prosumers requires that they are empowered to themselves generate renewable energy. This, along with increased energy efficiency and conservation, should lead to a cleaner energy mix, all else being equal. This is because

---

[162] Clastres (2011).

[163] This is articulated in the Communication from the Commission to the European Parliament, The Council, The European Economic and Social Committee and the Committee of the Regions Next Steps for a Sustainable European Future European Action for Sustainability, Strasbourg, 22.11.2016 COM(2016) 739 final.

[164] Communication from the Commission to the European Parliament, Closing the Loop – An EU action plan for the Circular Economy, COM (2015) 614, Brussels 2.12.2015.

[165] See the proposal of senior Republican statesmen regarding a carbon tax in the United States. Mooney, C. and Eilperin, J. "Senior Republican statesmen propose replacing Obama's climate policies with a carbon tax," *The Washington Post*, 8 February 2017, available at https://www.washingtonpost.com/news/energy-environment/wp/2017/02/07/senior-republican-leaders-propose-replacing-obamas-climate-plans-with-a-carbon-tax/?utm_term=.1ceadf0fe007.

[166] See World Alliance for Efficient Solutions, available at http://alliance.solarimpulse.com/.

[167] Various countries have schemes that offer subsidies and loans for rooftop solar panels and energy-efficiency actions. For instance, in the US there is a program called Property Assessed Clean Energy. See https://www.energy.gov/eere/slsc/property-assessed-clean-energy-programs. In the UK, there is the Green Deal, at https://www.gov.uk/green-deal-energy-saving-measures.

self-consumption minimizes the losses involved in energy transmission. The end result is, hence, overall reduced energy consumption. Moreover, the development of prosumer markets enables distributed generation (micro-generation), giving rise to local energy communities. Belgium features here as an excellent case in point. Enhanced incentives for the installation of solar panels has rendered individuals and households producers of their energy,[168] both releasing pressure from the grid, as well as supplying the grid with renewable energy.[169]

Self-generation and self-consumption, moreover, will lead to cleaner energy systems, in that clean fuels will decrease the need for oil and gas imports.[170] This represents a potentially far-reaching game-changer. A recent study has convincingly shown that

> utilizing existing infrastructure such as existing building roofs and shade structures does significantly reduce the embodied energy requirements (by 20–40%) and in turn the EPBT [energy pay-back time] of flat-plate PV systems due to the avoidance of energy-intensive balance of systems (BOS) components like foundations ... [while] a greater life-cycle energy return and carbon offset per unit land area is yielded by locally-integrated non-concentrating systems, despite their lower efficiency per unit module area.[171]

A potential extension of prosumer markets involves bringing clean energy to the transportation sector via charging infrastructure, which holds high promise for further sustainability gains in terms of lower emissions.[172]

7.7 Conclusion

The overhaul of the energy systems through the implementation of smart grids is crucial to drive the EU's low-carbon transition. Smart grids enable consumers to use less energy, which is beneficial for climate change. While the smart grids' benefits

---

[168] Anecdotally, President Jimmy Carter was very keen to promote alternative energy after the Arab oil embargo of 1973. He, therefore, installed solar panels on the White House in 1979, which were subsequently removed by his successor, President Ronald Reagan. See D. Biello, "Where did the Carter White House's solar panels go?" *Scientific American*, 6 August 2010, available at https://www.scientificamerican.com/article/carter-white-house-solar-panel-array/. Today, there are still pending issues regarding the potential of solar energy: regulatory uncertainty and unreliable business models.

[169] Gaëtan Masson, Jose Ignacio Briano and Maria Jesus Baez, 'Review and Analysis of PV Self-Consumption Policies', *International Energy Agency, Photovoltaic Power Systems Program, Report IEA-PVPS T1-28:2016*, 13.

[170] Lehmann and Gawel (2013), p. 603.

[171] Halasah et al. (2013), p. 462.

[172] See generally the proposal of the EU Commission, "Clean energy for all Europeans," COM (2016) 860 final, 30 November 2016. One should add that global CO2 emissions from energy use remained flat in 2016 according to a report of BP. See BP Statistical Review of World Energy June 2016, https://www.bp.com/content/dam/bp/pdf/energy-economics/statistical-review-2016/bp-statistical-review-of-world-energy-2016-full-report.pdf.

make large-scale deployment compelling across the sustainability, security of supply, and affordability fronts, caveats remain and call for caution by policy-makers.

In conclusion, smart grids entail several benefits as they create the conditions for the proliferation of renewable energy generation; allow for self-consumption; boost energy efficiency via demand response; alleviate energy poverty; lead to decreases in fossil fuel imports; decrease dependence on unreliable oil and gas suppliers, and volatile prices; promote low-carbon energy security; and boost aggregate demand.

On the negative side, smart grids require high upfront investments costs; call for large-scale citizens' engagement, incentivization and education; presuppose functional markets; and require high attention on cybersecurity issues. The transition to the new energy architecture may also generate supplementary adverse results, such as higher prices, abuse of market power, and increase in overall energy consumption. These possibilities create the need to communicate these likely outcomes to European citizens in a timely and efficient manner, devise relevant policy tools, and engage with the emerging prosumers.

As the deployment of smart grids and the energy transition constitute uncharted waters, there is a voluminous regulatory vacuum. For instance, the role of both TSOs and DSOs remains unclear in the new energy setting. The emergence of integrated energy services companies, aggregators, and energy co-operatives is also going to be determined to a great extent by future regulation. How cross-border markets will develop is another unresolved issue. The roll-out of smart meters also raises critical questions regarding data privacy that go to the root of human rights issues. Finally, the policy tools that will incentivize renewable energy generation and pave the way for a cleaner future are of central importance. Feed-in tariffs, feed-in premiums, and a carbon tax can all provide stimuli to the cause.

References

Bauwens M, Kostakis V (2014) From the communism of capital to capital for the commons: towards an open co-operativism. Triple C – J for a Global Sustainable Info Soc 12:356–361

Bickerstaff K et al (2016) Decarbonisation at home: the contingent politics of experimental domestic energy technologies. Environ Plan A 48:2006–2025

Black J (2001) Decentring regulation: understanding the role of regulation and self-regulation in a "post-regulatory" world. Curr Legal Probl 54:103–146

Bouzarovski S, Petrova S (2015) The EU energy poverty and vulnerability agenda: an emergent domain of transnational action. In: Tosun J et al (eds) Energy policy making in the EU: building the agenda. Springer, Berlin, pp 129–144

Bressand A (2013) The role of markets and investment in global energy. In: Goldthau A (ed) The handbook of global energy policy. John Wiley & Sons, West Sussex, pp 15–29

Brunsson N, Jacobsson B (eds) (2000) A World of standards. Oxford University Press, Oxford

Burke M et al (2015) Global non-linear effect of temperature on economic production. Nature 527:235–239

Chaffin B et al (2014) A decade of adaptive governance scholarship: synthesis and future directions. Ecol Soc 19:56

Ciocoiu CN (2011) Integrating digital economy and green economy: opportunities for sustainable development. Theor Empirical Res Urban Manag 6:33–43

Clastres C (2011) Smart grids: another step towards competition, energy security and climate change objectives. Energy Policy 39:5399–5408

Colchester M (2016) Do commodity certification systems uphold indigenous peoples' rights? Lessons from the roundtable on sustainable palm oil and forest stewardship council. Policy Matt 21:150–165

Elliott L (2013) Climate diplomacy. In: Cooper AF, Heine J, Thakur R (eds) The Oxford handbook of modern diplomacy. Oxford University Press, Oxford

Fang K et al (2015) The environmental sustainability of nations: benchmarking the carbon, water and land footprints against allocated planetary boundaries. Sustainability 7:11285

Florini A, Sovacool BK (2009) Who governs energy? The challenges facing global energy governance. Energy Policy 37(12):5239–5248

Franzki H, Olarte MC (2014) Understanding the political economy of transitional justice: a critical theory perspective. In: Buckley-Ziste S et al (eds) Transitional justice theories. Routledge, Abingdon, pp 201–218

Fujiwara N (2016) Overview of the EU climate policy based on the 2030 framework. In: Heffron R, Little GFM (eds) Delivering energy law and policy in the EU and the US. Edinburgh University Press, Edinburgh, pp 605–609

Galera MDS (2017) The integration of energy environment under the paradigm of sustainability threatened by the hurdles of the internal energy market. Eur Energy Environ Law Rev 26:13

Halasah SA, Pearlmutter D, Feuermann D (2013) Field installation versus local integration of photovoltaic systems and their effect on energy evaluation metrics. Energy Policy 52:462

Halff A, Sovacool BK, Rozhon J (2014) Energy poverty: global challenges and local solutions. Oxford University Press, Oxford

Hastik R et al (2016) Using the "Footprint" approach to examine the potentials and impacts of renewable energy sources in the European Alps. Mountain Res Dev 36:130

Hatzopoulos V, Roma S (2017) Caring for sharing? The collaborative economy under EU law. Common Market Law Rev 54(1):81–127

Hufbauer G, Suominen K (2010) Globalization at risk: challenges to finance and trade. Yale University Press, New Haven

Hufbauer G, Melendez-Ortiz R, Samans R (2016) The law and economics of a sustainable energy trade agreement. Cambridge University Press, Cambridge

Katina PF et al (2016) A criticality-based approach for the analysis of smart grids. Technol Econ Smart Grids Sustain Energy 1(1):14

Katz V (2015) Regulating the sharing economy. Berkeley Technol Law J 30:1067

Koopman C et al (2014) The sharing economy and consumer protection regulation: the case for policy change. J Bus Entrepreneurship Law 8:530–545

Leal-Arcas R (2009) The EU and Russia as energy trading partners: friends or foes? Eur Foreign Aff Rev 14(3):337–366

Leal-Arcas R (2010) International trade and investment law: multilateral, regional and bilateral governance. Edward Elgar Publishing, Cheltenham

Leal-Arcas R (2015) How governing international trade in energy can enhance EU energy security. Renew Energy Law Policy Rev 6(3):202–219

Leal-Arcas R (2016) The European Energy Union: the quest for secure, affordable and sustainable energy. Claeys & Casteels Publishing, Deventer, p 12

Leal-Arcas R (2017) Sustainability, common concern and public goods. George Washington Int Law Rev 49(4):801

Leal-Arcas R, Filis A (2013) The fragmented governance of the global energy economy: a legal-institutional analysis. J World Energy Law Bus 6(4):348–405

Leal-Arcas R, Filis A, Abu Gosh E (2014a) International energy governance: selected legal issues. Edward Elgar Publishing, Cheltenham

Leal-Arcas R et al (2014b) International energy governance: selected legal issues. Edward Elgar Publishing, Cheltenham

Leal-Arcas R et al (2015) Multilateral, regional and bilateral energy trade governance. Renew Energy Law Policy Rev 6(1):38–87
Leal-Arcas R et al (2016) Energy security, trade and the EU: regional and international perspectives. Edward Elgar Publishing, Cheltenham
Leal-Arcas R, Lasniewska F, Proedrou F (2017) Prosumers: new actors in EU energy security. Netherlands Yearb Int Law 48:139–172
Leal-Arcas R et al (2018) Smart grids in the European Union: assessing energy security, regulation & social and ethical considerations. Columbia J Eur Law 24(2):311–410
Lehmann P, Gawel E (2013) Why should support schemes for renewable electricity complement the EU emissions trading scheme? Energy Policy 52:603
Lougher G, Kalmanowicz S (2016) EU competition law in the sharing economy. J Eur Competition Law Pract 7:87
Maltby T (2013) European Union energy policy integration: A case of European Commission policy entrepreneurship and increasing supranationalism. Energy Policy 55:435–444
Matsushita M, Schoenbaum T (eds) (2016) Emerging Issues in Sustainable Development: International trade law and policy relating to natural resources, energy, and the environment. Springer, Tokyo
Millar RJ et al (2017) Emission budgets and pathways consistent with limiting warming to 1.5 °C. Nat Geosci 10:741–747
Montini M, Orlando E (2012) Balancing climate change mitigation and environmental protection interests in the EU directive on carbon capture and storage. Climate Law 3:165
Muench S (2014) What Hampers energy system transformations? The case of smart grids. Energy Policy 73:80
Nica E, Potcoravu A (2015) The social sustainability of the sharing economy. Econ Manag Financ Markets 10:69–75
Northrop E et al (2016) Examining the alignment between the intended nationally determined contributions and sustainable development goals. World Resources Institute, Washington, D.C.
O'Rourke D (2003) Outsourcing regulation: analyzing nongovernmental systems of labor standards and monitoring. Policy Stud J 31:1–29
Orlando E (2014) The evolution of EU policy and law in the environmental field: achievements and current challenges. In: Bakker C, Francioni F (eds) The EU, the US and global climate governance. Routledge, Abingdon, p 74
Parag Y, Sovacool BK (2016) Electricity market design for the prosumer era. Nat Energy 1:16032
Proedrou F (2016) EU energy security beyond Ukraine: towards holistic diversification. Eur Foreign Aff Rev 21:57–73
Proedrou F (2017a) Are smart grids the key to EU energy security? In: Leal-Arcas R, Wouters J (eds) Research handbook on EU energy law and policy. Edward Elgar, Cheltenham
Proedrou F (2017b) A new framework for EU energy security: putting sustainability first. Eur Politics Soc 18:182–198
Roberts J (2016) Prosumer rights: options for a legal framework post-2020. ClientEarth, London
Rockström J et al (2009) Planetary boundaries: exploring the safe operating space for humanity. Ecol Soc 14:32
Rogelj J et al (2016) Paris agreement climate proposals need a boost to keep warming well below 2 °C. Nature 534:631–639
Roht-Arriaza N, Mariezcurrena J (2006) Transitional justice in the twenty-first century: beyond truth versus justice. Cambridge University Press, Cambridge
Sachs J (2015) The age of sustainable development. Columbia University Press, New York
Sachs J (2017) Building the New American economy: smart, fair, and sustainable. Columbia University Press, New York
Schleussner C-F et al (2016) Differential climate impacts for policy-relevant limits to global warming: the case of 1.5 °C and 2 °C. Earth Sys Dynamics 7:327–351

Shawish A, Salama M (2014) Cloud computing: paradigms and technologies. In: Xhafa F, Bessis N (eds) Inter-cooperative collective intelligence: techniques and applications, studies in computational intelligence. Springer, Berlin, p 495

Smith A, Stirling A (2010) The politics of social-ecological resilience and sustainable sociotechnical transitions. Ecol Soc 15

Solorio I et al (2013) The European energy policy and its green dimension – discursive hegemony and policy variations in the greening of energy policy. In: Barnes PC, Hoerber TC (eds) Sustainable development and governance in Europe. Routledge, Abingdon

Steffen W et al (2015) Planetary boundaries: guiding human development on a changing planet. Science 347:736

Strengers Y (2013) Smart energy technologies in everyday life: smart Utopia? Palgrave Macmillan, Basingstoke

Swora M (2010) Intelligent grid: unfinished regulation in the third EU energy package. J Energy Natl Resour Law 28:465–480

Szulecki K, Fischer S, Gullberg AT, Sartor O (2016) Shaping the "Energy Union": between national positions and governance innovation in EU energy and climate policy. Climate Policy 16:1

Teitel RG (2014) Globalizing transitional justice: contemporary essays. Oxford University Press, New York

The International Energy Agency (2016) World energy outlook. OECD/IEA, Paris, p 1

Tuballa ML, Abundo ML (2016) A review of the development of smart grid technologies. Renew Sustain Energy Rev 59:710

Upham P (2016) Public engagement and low carbon energy transitions: rationales and challenges. In: Heffron R, Little G (eds) Delivering energy law and policy in the EU and the US; a reader. Edinburgh University Press, Edinburgh, p 549

Van Asselt H (2014) Alongside the UNFCCC: complementary venues for climate action. Centre for Climate and Energy Solutions, Arlington

Verbong GP, Beemsterboer S, Sengers F (2013) Smart grids or smart users?: involving users in developing a low carbon electricity economy. Energy Policy 52:117–125

Victor D (2011) Global warming gridlock: creating more effective strategies for protecting the planet. Cambridge University Press, Cambridge

Xiufeng F (2016) Smart grids in China: industry regulation and foreign direct investment. Energy Law J 37(1):135–176

Chapter 8
Practical Applications of Decentralized Energy in the EU

8.1 Introduction[1]

The global energy market is still monopolized to a great extent by the production, trade and consumption of oil and gas.[2] The EU is no exception to this rule with a high import ratio of both oil and gas. Unreliable oil producers, geopolitical instability in many oil-rich countries, economic and resource nationalism—which are a threat to sustainable development—transportation-related hazards, and the high volatility of international oil prices are constraining importers to face significant risks.[3]

In the gas sector, the EU is confronting a practically oligopolistic external market with Russia, Algeria and Norway, supplying most of the imported gas.[4] Azerbaijan and more distant Liquefied Natural Gas (LNG) suppliers also contribute to the EU's import portfolio, without however changing the EU's dependence on a few exporters.[5] In particular, relations with the most important gas supplier, Russia, have become overtly problematic. This state of play must be borne in mind insofar as politics and international relations have a crucial influence in energy policies and international trade relations.

---

[1]Parts of this chapter draw from ideas developed in Leal-Arcas et al. (2019); Leal-Arcas et al. (2018), pp. 311–410. In addition, the sections on Germany and the Netherlands have been written by Alesandra Salaza; the one on France by Edoard Alvares.
[2]The International Energy Agency (2016), p. 5.
[3]Yergin (2011).
[4]Eurostat, "Main origin of primary energy imports, EU-28, 2005–2015 (% of extra EU-28 imports)," Eurostat, [Online]. Available: http://ec.europa.eu/eurostat/statistics-explained/index.php/File:Main_origin_of_primary_energy_imports,_EU-28,_2005-2015_%28%25_of_extra_EU-28_imports%29_YB17.png.
[5]Proedrou (2012).

Diversification of sources, routes and suppliers has been high on the EU's agenda. The Southern Gas Corridor[6] and a few LNG initiatives are the only tangible steps towards this direction. Nevertheless, these efforts have not produced sea changes in Russia's pivotal market role.[7] The rationale of liberalization and competition is in accordance with the logic of diversification. This is so as both premises aim to create a level playing field for external actors in a market well-shielded from monopolistic structures and practices.[8] While the application of the Third Energy Package[9] has blocked some of Russia's future investment moves, it cannot by itself substantially alter the EU's import portfolio.[10]

This is mainly due to the fact that Member States and their energy companies are responsible for negotiating and signing supply contracts. Indeed, Gazprom traditionally retains strategic alliances with a number of European oil and gas companies[11] (such as Italy's ENI, Austria's OMV, France's Gaz de France, and Germany's EON Ruhrgas and Wintershall).[12] Indeed, Russo-German relations have been remarkably cordial over the last decades with energy cooperation being at the center of this partnership. Interestingly, the recent fallout between Russia and Ukraine, and Russia's actions (invasion of Crimea and hybrid war in Eastern Ukraine) that evidently go against fundamental international law principles enshrined in several international treaties, have not resulted in any interruption of Russia-EU gas trade.[13]

As discussed throughout the book, an important goal for Europe at this time is achieving decentralized energy. This would mean shifting from large, centralized power stations, to small, local grids that better capitalize on renewable energy

---

[6]The Southern Gas Corridor is a term used to describe planned infrastructure projects aimed at improving EU energy security by bringing natural gas from the Caspian region to Europe. See Trans Adriatic Pipeline, "Southern Gas Corridor," Trans Adriatic Pipeline, 2017. [Online]. Available: https://www.tap-ag.com/the-pipeline/the-big-picture/southern-gas-corridor. The Southern Gas Corridor is also known as the Fourth Corridor (the other three corridors running from North Africa, Norway and Russia). See Leal-Arcas et al. (2015a), p. 19.

[7]Sidi (2017), pp. 51–66.

[8]Proedrou (2016), pp. 57–73.

[9]The EU's Third Energy Package is a legislative package for an internal gas and electricity market with the purpose of further opening up these markets in the European Union. It consists of two directives and three regulations: Directive 2009/72/EC, concerning common rules for the internal market in electricity; Directive 2009/73/EC, concerning common rules for the internal market in natural gas; Regulation (EC) No 714/2009, on conditions for access to the network for cross-border exchanges in electricity; Regulation (EC) No 715/2009, on conditions for access to the natural gas transmission networks; and Regulation (EC) No 713/2009 of the European Parliament and of the Council of 13 July 2009 establishing an Agency for the Cooperation of Energy Regulators.

[10]Goldthau and Sitter (2015), pp. 941–965; Goldthau (2016).

[11]It is interesting to note that, as of 2013, 90 companies caused two-thirds of anthropogenic greenhouse gas emissions. See Goldenberg, S. "Just 90 companies cause two-thirds of man-made global warming emissions," The Guardian, 20 November 2013, available at https://www.theguardian.com/environment/2013/nov/20/90-companies-man-made-global-warming-emissions-climate-change.

[12]Aissaoui et al. (1999).

[13]Casier (2016), pp. 763–778.

8.1 Introduction

sources (RES). It would mean a more flexible model, where consumers have control of their energy use and can avail of the option of becoming prosumers. A number of factors are driving the shift towards decentralized energy in Europe. These factors include:

- The need to address climate change by reducing GHGs;
- The need to increase renewable energy use;
- The need to increase energy efficiency;
- The need to increase energy security by decreasing imports and relying more on renewable energy;
- Growing electricity demand all over Europe, and
- The liberalization of Europe's energy markets.[14]

Achieving decentralization is a mammoth task for the EU, as it will require innovation and technical upgrades as well as coordination across legislation, policy, and a range of relevant actors. Indeed, regulation poses one of the most significant barriers at this time, as empowering individuals with prosumership and smarter demand response systems requires a host of new regulations. Such regulation must address matters such as pricing, monitoring, consumer protection, data protection, and subsidies and incentives, to name but a few.

In spite of the daunting nature of the task, efforts towards decentralization are already underway across the EU. EU governments are well aware of the need to achieve decentralization, driven by the factors listed above, and also to comply with the EU's 2020 Strategy goals.[15] To this end, many interesting developments abound. For example, researchers are developing new ways to make appliances more intelligent and energy efficient.[16] A pan-European project called WiseGRID (WG) that I have been involved with is working on how to effectively place citizens at the center of the transformation of the electricity grid by allowing greater citizen participation and, by doing so, moving towards energy democracy.[17] Throughout this chapter, I refer to WiseGRID-related applications and projects that have implications for decentralization, more empowered citizens, and progress on decarbonization.

This chapter examines progress on decentralization in several EU jurisdictions. It focuses on specific outcomes of decentralization, including deployment of smart grids and smart meters, promoting demand response, promotion of electric vehicles, storage, data protection issues, and greater interconnection with neighboring countries. It examines each country's progress on these fronts as well as any existing barriers. The chapter also examines regulation that has helped towards the goal of

---

[14] EU Directorate-General for Internal Policies, "Decentralized Energy Systems," 2010, available at: www.europarl.europa.eu/document/activities/cont/201106/20110629ATT22897/20110629ATT22897EN.pdf.

[15] European Commission, "2020 Energy Strategy," available at https://ec.europa.eu/energy/en/topics/energy-strategy-and-energy-union/2020-energy-strategy.

[16] About WiseGRID, WiseGRID, http://www.wisegrid.eu/ [https://perma.cc/CGM8-F2WK].

[17] Idem.

decentralization in each country, which could perhaps be adopted in other countries seeking to transition to more liberalized energy markets. For the purposes of this chapter, we focus mainly on electricity markets.

8.2 Progress on Decentralization

8.2.1 *The Operation of Prosumer Markets*

From the 1990s onwards, the EU electricity sector underwent a transition from vertically organized electricity companies that controlled production, transmission, distribution, and supply activities, to the unbundling of these services.[18] Transmission System Operators (TSOs)[19] were responsible only for the balancing of the load and its transmission from large electricity production plants at high voltage levels. From there, Distribution System Operators (DSOs)[20] distributed electricity to every corner. As we move to an electricity sector comprised of multiple large and small producers, Virtual Power Plants (VPPs), and decentralized energy production, the role, rationale for, and competences of the TSOs remain mired in uncertainty. DSOs, on the other hand, seem well-placed in the new energy setting. Indeed, according to the European Commission's proposed internal electricity market directive, their role will be significantly enhanced, principally when it comes to coordinating and managing the energy produced by the new decentralized energy producers.[21] DSOs are anticipated to absorb the energy thus produced, manage the load, and efficiently distribute electricity to households and corporate premises.[22] The

---

[18] *Understanding Electricity Markets in the EU*, European Parliament (Nov. 2016), http://www.europarl.europa.eu/RegData/etudes/BRIE/2016/593519/EPRS_BRI%282016%29593519_EN.pdf.

[19] A Transmission System Operator (TSO) can be defined as a natural or legal person responsible for operating, ensuring the maintenance of and, if necessary, developing the transmission system in a given area and, where applicable, its interconnections with other systems, and for ensuring the long-term ability of the system to meet reasonable demands for the transmission of electricity. *See* Parliament and Council Directive 2009/73/EC, Concerning Common Rules for the Internal Market in Natural Gas, art. 2(4), 2009 O.J. (L 211) 94, and Parliament and Council Directive 2009/72/EC, Concerning Common Rules for the Internal Market in Electricity, art. 2(4), 2009 O.J. (L 211) 55.

[20] A Distribution System Operator (DSO) can be defined as a natural or legal person responsible for operating, ensuring the maintenance of and, if necessary, developing the distribution system in a given area and, where applicable, its interconnections with other systems and for ensuring the long-term ability of the system to meet reasonable demands for the distribution of electricity or gas. *See* Parliament and Council Directive 2009/73/EC, Concerning Common Rules for the Internal Market in Natural Gas, art. 2(6), 2009 O.J. (L 211) 94, and Parliament and Council Directive 2009/72/EC, Concerning Common Rules for the Internal Market in Electricity, art. 2(6), 2009 O.J. (L 211) 55.

[21] *Commission Proposal for a Directive of the European Parliament and of the Council on Common Rules for the Internal Market in Electricity*, p. 68, COM (2016) 864 final (Feb. 23, 2017).

[22] *See id.*

digitalization of services through advanced metering infrastructure (AMI) will massively facilitate their upgraded role.[23]

This being the case, one could anticipate the TSOs' reaction and their pledge for a place in the sun. This potential friction raises questions as to how the competences of the new actors are going to be divided in the new energy landscape.[24]

Energy policy goals and correspondingly relevant national jurisdictions will play a pivotal role in moving the transition forward. Top-down, bottom-up, and hybrid (both top-down and bottom-up)[25] energy policy blueprints mandate variable leeway for different actors across the energy chain. Some aspects can be legally binding and perhaps commissioned to specific market players (e.g., smart meter roll-outs). Another energy policy goal would be allowing utilities, DSOs, and consumers to decide the ways, and pace at which, they move forward. For now, a hybrid model seems to be emerging. In this architecture, climate goals have been set at the higher governance level but the smart grid transition is carried out at the lower governance level. For example, environmental targets are set out by supranational instruments such as the 2020 Climate and Energy Package,[26] whereas the deployment of smart meters is effectively carried out on a national basis. Thus, certain EU Member States such as Spain are already well on their way to hit a 100% smart meter roll-out.[27] Conversely, other EU Member States such as the Czech Republic and Portugal have foregone replacing conventional meters with smart metering systems due to economic reasons.[28] Such stances are in accordance with the EU law principle of subsidiarity, according to which Member States are given the discretion to decide

---

[23]*Commission Proposal for a Regulation of the European Parliament and of the Council on the Internal Market for Electricity*, COM (2016) 861 final (Nov. 30, 2016).

[24]*See id.*

[25]A top-down approach to a problem is a situation that begins at the highest conceptual level and works down to the details. An example of such an approach would be where targets are set out at the international level and must be attained through national policies and measures. A bottom-up approach to a problem is one that begins with details and works up to the highest conceptual level. An example of such an approach would be where action starts at the national level based on each country's circumstances through a patchwork of national policies and measures (which are not necessarily binding) until they develop into unified policies at the international plane.

[26]These environmental targets aim to (1) reduce greenhouse gas (GHG) emissions by 20%; (2) reach 20% of renewable energy in the total energy consumption in the EU; and (3) increase energy efficiency to save 20% of EU energy consumption, all by 2020. See *2020 Climate and Energy Package*, European Comm'n (Sept. 9, 2017), https://ec.europa.eu/clima/policies/strategies/2020_en.

[27]Comisión Nacional de los Mercados y la Competencia, *El 62% de los Contadores Analógicos ya han sido Sustituidos por Contadores Inteligentes, Nota de Prensa* 1 (2017).

[28]*Commission Report on Benchmarking Smart Metering Deployment in the EU-27 with a Focus on Electricity*, p. 4, COM (2014) 356 final (June 17, 2014).

for themselves how they are going to reach the goals mutually agreed upon at the top EU political level.[29]

The previous reform of the electricity markets carries its important heritage to today's transition. Unbundling[30] has taken place in different ways in the various Member States.[31] In cases where legal unbundling took place, corporate links between the generation and distribution network companies, although they constitute two different legal entities, may well be maintained. This will create benefits to actors in the retail market. This is not the case in ownership unbundling, where the generation and network companies are fully separated. A level-playing field is indispensable if we are to avoid privileging certain actors vis-à-vis others.[32]

The specific market conditions also impact the pace and scale of investments. For example, market players with dominant market shares naturally prioritize retaining their central position, rather than investing in new network infrastructure and smart grid roll-outs, as the benefits that will accrue are unlikely to match the costs of reduced revenues resulting from a lessened market share.[33] On the other hand, investments are very pertinent not only in consideration of existing legislation, but also for tackling and anticipating market competition. In this context, DSOs are keen to invest in AMI.[34] Private investors can find a niche investing in control boxes downstream from the meter. A significant caveat is that private investment can render customers captive in light of the long contractual lead times that are imposed so that costs are recovered.[35] This in itself obstructs competition. Such issues must be seriously considered when designating the new regulatory framework for smart grid deployment. Waiting games are also typical corporate tactics that should be anticipated and treated appropriately, since existing market power determines future over- or under-investment plans.[36]

In this new energy landscape, opportunities are opening for new energy actors as well. One such type is energy aggregators. The rationale for their emergence is to

---

[29]Cherrelle Eid, Rudi Hakvoort & Martin de Jong, *Global Trends in the Political Economy of Smart Grids: A Tailored Perspective on 'Smart' for Grids in Transition*, 1–19, p. 10 (United Nations University World Institute for Development Economics Research Working Paper 2016/22, 2016).

[30]Ownership unbundling is the "process by which a large company with several different lines of business retains one or more core businesses and sells off the remaining assets, product/service lines, divisions or subsidiaries. Unbundling is done for a variety of reasons, but the goal is always to create a better performing company or companies." See *Unbundling*, Investopedia, http://www.investopedia.com/terms/u/unbundling.asp.

[31]*Understanding Electricity Markets in the EU*, European Parliament (Nov. 2016), http://www.europarl.europa.eu/RegData/etudes/BRIE/2016/593519/EPRS_BRI%282016%29593519_EN.pdf.

[32]*Id.* at 9.

[33]Donoso (2015), p. 37.

[34]EDSO, European Distributed System Operator for Smart Grids (2014).

[35]Clastres (2011), p. 5399.

[36]*See id.*

provide flexibility and join the Balancing Responsible Parties (BRPs)[37] in what will be a much more variable corporate electricity landscape. Such a role can also be taken up by incumbents. In the new market, however, flexibility services and packages will be crucial, and hence there seems to be much space for new corporate actors, services, and associated innovation. These services revolve around collecting decentralized prosumers' savings and energy generation and selling it back to utilities and BRPs in the form of "flexibility packages."[38]

Yet, another type of actors to emerge may be small storage providers. These can store the energy they have produced (in batteries or EVs, for instance)[39] and resell it for a high premium in a market in dire need of flexibility, back-up capacity, and last resort solutions. Such services can be developed at the community, district, or neighborhood level. In this case, the emergence of energy co-operatives may take shape. Integrated energy services companies are the key to the new electricity market.[40]

At an even lower level, individuals, households, and energy cooperatives can become energy actors themselves. They can sell the energy they produce or conserve to utilities and/or aggregators. Both flexibility and network optimization are achieved in this way. Distributed energy resources and storage facilities are central to the energy transition.[41]

Whether storage capacity will be incorporated successfully in smart grids will be critical to their eventual performance. Leaving aside the contested debate over the likelihood of success, storage capacity will tackle peak consumption, reduce system-wide generation costs, and minimize network congestions, thereby optimizing the operation of the electricity network.[42] EVs are a storage capacity option that is also highly contested.[43] Charging infrastructure costs, logistics, and issues regarding charging time and efficiency both for the vehicle and for the grid must still be resolved. Nevertheless, EVs have the potential to decarbonize the transport sector.

---

[37]Balance Responsible Party (BRP) can be defined as a market participant or its chosen representative responsible for its imbalances in the electricity market. *See Proposal for a Regulation of the European Parliament and of the Council on the Internal Market for Electricity*, p. 38, COM (2016) 861 final (Nov. 30, 2016).

[38]For further details on prosumers, see Leal-Arcas et al. (2017), p. 139.

[39]These producers will be able to store in batteries the electricity they generate, instead of selling it back to utilities. Doing so will encourage more customers to invest in batteries.

[40]Boscan and Poudineh (2016), p. 2.

[41]*Proposal for a Directive of the European Parliament and of the Council on the Promotion of the Use of Energy from Renewable Sources,* COM (2016) 767 final (Feb. 23, 2017).

[42]Cherrelle Eid, Rudi Hakvoort & Martin de Jong, *Global Trends in the Political Economy of Smart Grids: A Tailored Perspective on 'Smart' for Grids in Transition,* 1–19, p. 3 (United Nations University World Institute for Development Economics Research Working Paper 2016/22, 2016).

[43]Zachary Shahan, *Tesla CTO JB Straubel On Why EVs Selling Electricity To The Grid Is Not As Swell As It Sounds*, Clean Technica (Aug. 22, 2016), https://cleantechnica.com/2016/08/22/vehicle-to-grid-used-ev-batteries-grid-storage/.

This would represent a huge leap forward in meeting the EU's climate targets and contributing to climate change mitigation.[44]

The development of prosumer markets is based on two pillars. The first regards hardware (infrastructure); the other concerns software (the associated legislation and regulation). In this vein, the European Commission made a handful of important steps forward. Firstly, it recognized consumers' right to self-consumption. This will lead to all national jurisdictions gradually embracing self-consumption. Moreover, prosumers are explicitly encouraged to sell their energy surplus to other energy actors, adding in this way to the energy market's resilience and becoming active stakeholders in the energy transition.[45] Secondly, the European Commission explicitly referred to energy communities, granting the right to prosumers to group together and join the market.[46] Finally, the European Commission strongly recommended advancing energy performance-related information as well as information regarding the sources of district heating and cooling systems. This will further empower prosumers and energy communities to improve their energy performance (including production consumption and trading). In addition, the quality of information that consumers obtain will come under the scrutiny of regulatory authorities. This also includes the refinement of the Guarantees of Origin system for energy resources.[47]

The advent of prosumer markets entails the commercialization, rationalization, and economization of consumer behavior. Through demand response, the European Commission expects prosumers to take full control of their energy usage. Prosumers will be able to adjust their patterns and be economical and efficient. The inflow of relevant information will allow them to adjust, conserve, and choose flexible contracts. Switching off unnecessary appliances or turning down the thermostat at peak hours not only provides monetary benefits, but also contributes to balancing the grid. Conversely, consumers are incentivized to use electricity when it is cheap (e.g., doing the laundry at late hours).[48]

Smart applications can substantially enhance energy efficiency. Instructing the washing machine to wash the clothes at the lowest price of electricity during the day can lead to optimal results for both the consumer and the grid. Dynamic price

---

[44]The International Energy Agency (2016).

[45]European Parliament, Electricity "Prosumers" (Nov. 2016), http://www.europarl.europa.eu/RegData/etudes/BRIE/2016/593518/EPRS_BRI%282016%29593518_EN.pdf.

[46]*See Commission Proposal for a Directive of the European Parliament and of the Council on Common Rules for the Internal Market in Electricity,* COM (2016) 864 final (Feb. 23, 2017). The proposal defines the concept of local energy community as "an association, a cooperative, a partnership, a non-profit organisation or other legal entity which is effectively controlled by local shareholders or members, generally value rather than profit-driven, involved in distributed generation and in performing activities of a distribution system operator, supplier or aggregator at local level, including across borders." *Id.* at 52.

[47]*Commission Proposal for a Directive of the European Parliament and of the Council on the Promotion of the Use of Energy from Renewable Sources,* COM (2016) 767 final (Feb. 23, 2017).

[48]*See id.*

8.2 Progress on Decentralization

contracts are also a useful tool for demand management. Based on their consumption patterns, consumers are encouraged to negotiate suitable contracts with electricity suppliers. From the side of utilities, well-targeted, flexible contracts should increasingly become part of their corporate strategy to cater to customers' individualized needs. Competition forces can work well in this sector and lead to a wave of easily adjustable contracts.

Moreover, a number of pricing mechanisms (e.g., real-time pricing, time-of-use pricing, critical-time pricing, and variable peak pricing) can also be put to good use. They not only reflect market fundamentals, they also render consumers more aware of price variations according to market dynamics.[49] Thus, last resort solutions like load-shedding and self-rationing can be altogether abandoned. However, dynamic pricing contracts entail several difficulties. It is hard for utilities to create spot-on abstract models of "representative agents," taking the heterogeneity in the energy use patterns of different consumers into account.[50] Devising effective contracts is also challenging from the supply side, since different utilities face different costs in the energy they buy to respond to their customers' needs. This is especially true when it comes to buying flexibility packages themselves. It is natural then to anticipate that they may remain averse to making even more sophisticated contracts.[51]

An important aspect of the deployment of smart grids lies in revisiting the philosophy behind their functioning rather than borrowing the one underpinning the functioning of the conventional grid. The conventional grid has been premised on the worst-case dispatch philosophy.[52] With the supply side being a priori known, utilities focused their efforts on balancing it every second with demand. The danger lay in an imbalance occurring either due to a supply disruption (e.g., an accident in a generation plant) or an unpredictable surge in electricity demand (e.g., a heat wave). To avert such mishaps, utilities retained large reserve capacity to ensure that electricity dispatch would still be possible when demand exceeded predictions or supply was decreased. Such a policy was neither sustainable nor cheap but at least hedged against the danger of power cuts and load-shedding.[53]

These principles and rationale are unsuitable for smart grids. The dynamic nature of both supply and demand in the new electricity landscape calls for a new philosophy.[54] The increase of intermittent solar and wind energy, the lack of storage capacity as of now, the development of micro-grids, the increased variability regarding consumer preferences, and the way consumers will operate smart appliances result in increased uncertainty in both supply and demand. Smart meters, sensors, and demand response mechanisms can mediate and manage the variability and unpredictability of power markets by providing both mechanisms for controlling

---

[49] Rodríguez-Molina et al. (2014), pp. 6142–6171.

[50] Boscan and Poudineh (2016), p. 10.

[51] *See id.*

[52] Varaiya et al. (2010), pp. 40–57.

[53] *See id.*

[54] Boscan and Poudineh (2016), p. 10.

energy use and precise information on the state of the power system and the supply-demand equilibrium.[55]

It is thus essential to redefine the risks in the operation of the power markets and their management. What is considered acceptable risk now must be adjusted to the new operating conditions of smart grids and power markets. The demand response of all consumers will need to be factored into a probabilistic demand curve, which will be analogous to the generation availability curve of intermittent renewable energy.[56] The focus will continually be on the movements in the net load, the difference between aggregate demand (load) and variable generation. The capacity, ramp rate, duration, and lead time for increasing or decreasing supply will have to be factored into such analyses as well, to optimize the smart grids' responses to the fluctuating supply-demand dynamics.[57]

Finally, it is necessary to integrate cross-border markets and capacity into risk management analysis. The EU has managed to establish a functional cross-border power market through its day-ahead market with many national markets now coupled.[58] This has been instrumental in fomenting price competition, providing further leverage for load balancing, optimizing back-up capacity, and increasing resilience.[59] A handful of physical barriers such as congestion, lack of transmission capacity, and/or underutilization remain, leading to sub-optimal transmission returns and hub market differentials.[60] These block, rather than enhance, cross-border trade. A further step regards the extension of such schemes into Energy Community members that are not EU members as well as to neighboring states outside the Energy Community. A more critical challenge regards the adjustment of the cross-border market to the new reality of "real-time" intra-day trade.[61]

8.2.2 The Situation in Certain EU Member States

Europe has been working towards liberalized energy markets since the 1990s, with a series of Directives to this effect.[62] Many member states have made significant progress in this regard, with increased competition amongst energy suppliers, better

---

[55] *See id.*
[56] Varaiya et al. (2010), p. 40.
[57] Boscan and Poudineh (2016), p. 10.
[58] International Energy Agency (2016).
[59] *See id.*
[60] Boscan and Poudineh (2016), p. 10.
[61] *See id.;* Buchan and Keay (2016).
[62] First Energy Package, electricity (1996) and gas (1998); Second Energy Package 2003, and Third Energy Package (2009). See http://www.europarl.europa.eu/factsheets/en/sheet/45/internal-energy-market.

8.2 Progress on Decentralization

services, and lower energy prices. Decentralization seems to be the next logical step, as it will further help achieve a more democratic and participative energy market.

Belgium started to liberalize electricity at the beginning of the 2000s, in keeping with EU direction. By 2007, the country's three regions (the Brussels-Capital region, the Flemish region and the Walloon region) had legally opened their electricity markets. Thereafter, an amendment to the Law of 29 April 1999 was passed. This amendment, the Law of 8 January 2012, strengthened the role of the federal authority (CREG), and made it a separate entity from the Directorate-General for Energy. Additionally, the Constitutional Court decided on 7 August 2013 that the federal level would have sole competence over the application, determination, and exemption of tariffs.

However, the Special Law of 6 January 2014 changed things so that the central government would set transmission tariffs, but regional authorities could establish distribution tariffs. So, Brugel, the VREG and CWaPE are the deciding authorities for setting distribution tariffs in the Brussels-Capital region, the Flemish region and Walloon region, respectively.[63] Another shift occurred with the Law of 8 January 2012, which added to regional regulator competences and called for opening competition by unbundling electricity markets.[64] In keeping with the new regulation, Belgium's electricity transmission system operator (TSO), Elia, along with the various distribution system operators (DSOs), unbundled completely from utilities.

In general, central government regulators handle electricity transmission for systems that have a voltage higher than 72 kV, while the regional level has competence over the local distribution and transmission of electricity for systems that have a voltage equal to or less than 70 kV. With the exception of offshore wind energy, regional authorities also have competence over renewable energy and over the rational use of energy.

Each region has its own regulatory framework for the management of its electricity market: The Ordinance of 19 July 2001 regarding the organization of the electricity market in the Brussels-Capital region; the Decree of 8 May 2009 on energy policy (also known as the Energy Decree) for the Flemish region; and the Decree of 12 April 2001 regarding the organization of the regional electricity market for the Walloon region.[65]

Reaching a cohesive energy strategy is a key challenge for the country. While currently, Belgium's central government and its three regions share competence for energy and climate change, in reality the regions enjoy jurisdiction over policy related to these issues. The situation sometimes results in lack of clarity when it comes to dividing competences between the federal and regional levels. Further, the system as it stands leads to a lack of coordination amongst the entities managing energy and climate policies. For effective energy governance, the various state

---

[63]International Energy Agency (2016).

[64]Idem.

[65]Elia Group, "Legal Framework," Elia Group, [Online]. Available: http://www.elia.be/en/aboutelia/legal-framework.

players need to work together in a cohesive and integrated way.[66] To this end, it was decided in 2015, to create an energy pact for the country that would encompass a long-term outlook and to identify tangible steps to achieve energy and climate goals both within Belgium and at the EU level.[67] However, political disagreements got in the way of the project fulfilling its ambitions.

In Spain, a process of decentralization has been ongoing since the 1978 Constitution, which officially sealed the country's shift to a democracy. The Constitution facilitated this decentralization by establishing autonomous regions or "*Comunidades Autónomas.*" These regions have gained increasing competence, demonstrated by their growing share in total public spending,[68] a fact not necessarily beneficial to the central government. The Constitution allocates some competences exclusively to the state but the rest are left to the autonomous regions, and most bask in an equal level of political autonomy.[69] The central government has sole competence to authorize electricity networks when their usage impacts another autonomous region or when electricity transmission lines extend further than the relegated territorial allotment.[70] Moreover, the state has the authority to confirm and establish groundwork for mining and energy sectors,[71] which, of course, ties in with the government's role of planning overall economic activity,[72] in which energy plays a key part. Lastly, the government has the authority to lay down regulation for environmental protection, with autonomous regions having competence to establish accompanying regulations.[73]

Autonomous regions have been claiming authority in residual areas related to energy by making sure to not impinge on the central government's prerogative. The regions must ensure that any energy policy only affects their region and not any of the others. Furthermore, autonomous regions have sole authority over industry in their regions while complying with overall government frameworks for economic initiatives. Overall, the competences of the autonomous regions must fit within the framework of the central government's authority on matters related to national security, health or the military, along with sectors confined by regulations on mining, hydrocarbons and nuclear energy.[74]

---

[66]European Environment Agency (2014).

[67]International Energy Agency (2016).

[68]Balaguer-Coll (2010), pp. 571–601.

[69]European Committee of the Regions, "Division of Powers," European Committee of the Regions, 2012. Available at: https://portal.cor.europa.eu/divisionpowers/countries/MembersLP/Spain/Pages/default.aspx.

[70]Article 149, paragraph 1, point 22 Spanish Constitution.

[71]Article 149, paragraph 1, point 25 Spanish Constitution.

[72]Article 149, paragraph 1, point 13 Spanish Constitution.

[73]Article 149, paragraph 1, point 23 Spanish Constitution.

[74]Navarro Rodríguez (2012).

Overall, autonomous regions play an important part in the country's energy policy. They have the authority to decide a host of matters connected to national energy policy. For example, they may confirm power plant operations as long as capacity is below 50 megawatts (MW), which applies to the majority of solar and wind plants. Further, autonomous regions can authorize electricity and gas distribution networks within their geographical scope. And they play a key role in deciding and executing policy related to climate change, energy efficiency and renewable energy on the regional front. It is worth exploring links between Spain's deeply embedded decentralization and the principle of subsidiarity within EU law.[75]

In Italy, there has been a steady move towards liberalization in the country's electricity market since the EU issued its first electricity-related directive (96/92/EC). The EU directive followed the Bersani Decree,[76] which delineated the way forward for liberalizing the market and identified basic tenets for future steps in this direction. Thus far, liberalizing wholesale and retail markets has met with much success. Decentralizing electricity generation and using micro-grids for distribution would be the next major steps.

Before the electricity market was liberalized, it was organized as a vertically integrated monopoly, with Enel S.p.A. the main actor in charge. But legislation was passed to promote competition, in order to benefit end consumers. This was achieved by establishing a vertical distinction among companies in a way that distinguishes free activities from competitive ones. In the liberalization process, demand is addressed incrementally, with customers placed into two categories: suitable and eligible. Eligible customers refer to those who signed contracts with particular producers or distributors. The rest are obliged to sign contracts with those producers or distributers servicing their location. Directive 2003/54/EC was the final step in the market being opened completely, from 1 July 2004 for non-residential customers and from 1 July 2007 for residential.

Overall, liberalizing the country's electricity market has been a legal obligation, in keeping with EU directives and taking place in two phases, broadly speaking: opening markets at a national level and integration of national markets. The grid operator role emerged from the Bersani decree. It is a limited company whose shares are owned by the Ministry of Treasury, Budget and Economic Planning (Enel S.p.A having handed them over for free). The grid operator must provide a public utility service by transmitting and dispatching electricity and managing, in a streamlined way, the high and very high voltage grids. Law No 290/2003 of 27 October 2003 was passed to protect public interests, through unifying the ownership and operation of the network, with the goal of maximizing efficiency and safety. The law delineates the steps for unifying the network's operation and ownership.

Greece started to liberalize its electricity market in 2001, and the process was finalized in July 2013. However, the state-owned Public Power Corporation (PPC) still has a high level of control over electricity production. Entering the market is

---

[75] Article 5.3 of the Treaty on European Union (TEU).
[76] International Energy Agency (IEA) (2016).

challenging due to high levels of capital investment along with licensing processes being plagued by red tape.[77]

Consumers in Greece's interconnected system have been able to choose their energy supplier since 2004, and residential consumers have been able to do so since July 2007. Since 2016, consumers based in Crete and Rhodes have also had the ability to choose their supplier. However, this right does not extend to consumers based in the smaller, non-interconnected islands, since the PPC is the only available supplier in those isolated microsystems. In this sense, 2007 represents a large step forward for the country's energy market in terms of being the starting point for its liberalization. The year 2011 stands out as a large amount of supplier changing took place on the retail market. While PPC still held its ground as the main energy provider, a sizeable number of consumers transitioned to alternate providers, demonstrating that there is room for competition to evolve in the retail market.

According to a 2017 report by the Hellenic Electricity Market Operator (LAGIE)—which is in charge of operating and maintaining the country's energy market—20 companies are registered as energy suppliers and 25 as energy traders, in addition to PPC. However, the majority of suppliers play a greater role in energy trade than in supply. Additionally, seven companies are registered as producers and are recorded in LAGIE's archives,[78] while some of the producers are also active as suppliers and traders. There is potential for energy retailers and local communities and cooperatives to avail of the WiseGRID application WiseCOOP to more efficiently manage their customers and assets, and also help domestic and small businesses, consumers and prosumers to attain better energy deals. WiseCOOP could help them provide their member or customers with better services and prices, especially in conjunction with demand response programs.

Greece's natural gas market is also undergoing a process of liberalization, but more gradually than the electricity market. In 2011, Law 4001 was established, which provided for separating networks from supply and production activities. The law has undergone numerous modifications since its establishment.

According to Law 4336, passed in 2015, the process of liberalizing the gas market derived from the need to trigger strong competition and lower energy costs. New provisions have since followed, largely related to updating natural gas infrastructure in Greece, the need to legally separate distribution and supply activities, and broadening the eligible customer qualification, which includes electricity customers and several industrial customers. Restrictions related to the supply and monopoly of the Public Gas Corporation of Greece (DEPA) were removed so that public service commissions and large industrial consumers and generators received the right to purchase gas directly from international suppliers.

---

[77]Anagnostopoulos and Papantonis (2013).

[78]ΛΑΓΗΕ Operator of the electricity market, "www.lagie.gr / DAS monthly report December 2016."

The opening of the retail market, which started with the industrial sector, continued in 2017 for professional customers, and is to be completed by integrating domestic customers in addition. Many have displayed a keen interest, and by the end of 2016, more than 40 companies had applied for a license to supply gas with the aim of reaching the earliest eligible customers. Opening up the natural gas market can lead to more profitable conditions for the use of CHP. The combination of the WiseGRID STaaS/VPP and WiseCORP applications could be help to integrate CHP systems in the tertiary sector and in industries, to provide a greater incentive to taking part in the energy market, resulting in more services and lowering their overall energy expenses.

Despite the strength of Germany's industrial sector, the country has a strong background in environmental protection.[79] It is said that the concept of "sustainability" itself (*Nachhaltigkeit*) was inspired by the highly influential methods adopted in the forestry sector pioneered in Prussia and in Saxony.[80]

At the moment, the current political coalition has supported the transition from fossil fuels to renewable energy known as *Energiewende*. This "energy revolution" has been the main driver for the deployment of smart grids in Germany. Germany is indeed changing is energy policy, aiming at decarbonising its economy and reaching its international and European goals in terms of emissions reduction. The transformation is still at a starting stage, but already several positive efforts have been made, especially in the field of renewables, which are increasingly replacing other energy sources that Germany has traditionally used, such as nuclear and lignite.

A 2015 Eurostat study shows that Germany is among the five European Union countries that consume the largest amounts of energy. Its domestic oil and natural gas production is limited, making Germany heavily dependent on imports of oil and coal (61.9%).[81]

Germany decoupled its GHG emissions from economic growth in the 2000s and focused on renewables. The energy transition (*Energiewende*) brought the production of renewable sources to consistently grow in the last decade, increasing from a total production of 8% in 2000 to 28.3% in 2011 and it is expected to reach up to 33.2% in 2030. This percentage will include: biofuels (21.6%), wind (5.6%), solar (3.2%), geothermal (1.9%), and, finally, hydropower, which represents the weakest renewable sector in Germany, at only 1%.[82] The recent 2017 Renewable Energy Sources Act aims at fostering the renewables sector, providing a feed-in tariff to households and businesses to encourage the generation of renewable electricity.[83] As a result, half of all electricity supply in Germany will derive from renewable

---

[79]Brunekreeft et al. (2015), pp. 45–78.

[80]Heal, Bridget, "The Nature of German Environmental History," *German History*, 1 January 2009, 113–130.

[81]International Energy Agency, *Energy Policies of IEA Countries*, 2013 Review.

[82]*Ibid.*

[83]Renewable Energy Sources Act 2017, available at www.bmwi.de/Redaktion/EN/Downloads/renewable-energy-sources-act-2017.pdf?__blob=publicationFile&v=3.

energy sources by 2030. This development in the renewables sector must be complemented by large-scale, cost-efficient investments in the electricity transmission and distribution systems.

The high production of wind power in the more rural states to the north of the country has to provide for the states of the south (Bavaria and Baden-Württemberg), where industry is more developed and demand for electricity is higher. The remote island of Heligoland, located off the coast of the state of Schleswig-Holstein, for example, produces wind power on a large scale and has become popular for its offshore wind farms. However, it has to serve the rest of the mainland, especially, the states of the far south. Also, renewable energy production depends on the variability of the weather, but the energy demand of other regions of the country has to be met regardless of weather conditions.

Therefore, one of the biggest challenges in Germany is to improve the transmission infrastructures that, at the moment, are increasingly congested. If grid expansion is not growing at the same pace as the generation of electricity from renewables, there might be the risk of temporary bottlenecks in transmission that have to be managed by the Transmission System Operators (TSOs). Although there are measures for them to avoid or rectify such "grid congestions" (feed-in management and re-dispatch management), these emergency measures cannot be adopted on a regular basis. In East Germany, the recent completion of the Thüringer Strombrücke, that has connected Saxony-Anhalt and the South of the country, has helped its TSO 50 Hz to significantly lowered its intervention, but TSOs in the rest of the country have reported that their stabilising interventions on the grid were needed for 329 days in 2017.[84] The expansion of the grid has become necessary to ensure a continuous transmission flow and to guarantee the security of supply, consequently avoiding the costs for feed-in or re-dispatch measures.[85]

The Energy Concept was adopted in September 2010 by the federal government. This new long-term strategy aims at integrating energy pathways to 2050, focusing on the development of renewable energy sources, power grids and energy efficiency.[86]

After the 2011 nuclear accident in Japan (Fukushima), Germany accelerated the phase-out of nuclear power by 2022, starting with the closure of the eight oldest plants. That was only the first step of a process that will progressively phase out nuclear energy in Germany, until, eventually, all nuclear reactors will be shut down by 2022.[87]

---

[84]Appunn, Kerstine, "Energiewende hinges on unblocking the power grid," *Clean Energy Wire*, 2018.

[85]Netz Entwicklungs Plan Strom, *Energy Networks*, available at www.netzentwicklungsplan.de/en/background/energy-networks.

[86]Federal Government, Energy Concept, available at cleanenergyaction.files.wordpress.com/2012/10/german-federal-governments-energy-concept1.pdf.

[87]International Energy Agency, *Energy Policies of IEA Countries*, 2013 Review.

The total final consumption (TFC) of energy Germany was 221 Mtoe in 2011. The sector with the highest impact on TFC is industry (35.6%, or 78.8 Mtoe). This has increased from 33.3%, following the economic crisis in 2009. Other sectors are: residential (25.4%, or 56.1 Mtoe), transport (24.5%, or 54.2 Mtoe), and commercial and other services (32 Mtoe).

Since 2011, some important changes have been happening. The most evident change is the decrease of nuclear energy and oil in the energy mix. Around 40% of the total production of energy in 2018 comes from renewable sources.[88] The ambitious energy transformation set out in the suite of new policy measures (*Energiewende*) will result in a decline in TFC over the years until 2030, though the government forecasts that shares per sector will be maintained.[89]

When it comes to natural gas, Germany's natural gas market is the biggest in Europe in terms of consumption, although it has a very little production capacity. The main exporter is the Netherlands. The overall annual demand represents about 24.3%, which is potentially a risk in terms of energy security and makes the switch to renewables even more important. In recent years, in fact, gas production has been in constant decline. Also, an animated political discussion related to environmental protection and water quality has made Germany very unlikely to start focusing on the production of unconventional gas, as has happened in the United States and Canada.

Some of the main actors that operate in the energy sector will have a crucial role in achieving the *Energiewende* and dealing with a changing energy mix. In order to promote fair competition among market-players, the EU Third Energy Package, and, consequently, Germany's Energy Industrial Act (EnWG), have introduced the rule that energy supply and generation have to be separated from the operation of transmission networks (unbundling, as per Electricity Directive 2009/72/EC). This has resulted in the creation of a multitude of market-players.

The four large electricity producing companies that traditionally operate in Germany are RWE, E.ON/Uniper, Vattenfall and EnBW. They have all been thoroughly restructured as a consequence of the *Energiewende*. Their earnings from producing and supplying fossil (especially lignite) and nuclear power have been declining, making them experience a drop in their economic value, rather than a swift adjustment to the new focus on renewables.[90] An example of how the new transition to a zero-emission economy can impact these businesses is well illustrated by the case of RWE: 42% of business depends on lignite, which is a very high-emission source of energy.[91]

Germany's four transmission system operators (TSOs) are 50 Hz, formerly Vattenfall Europe Transmission (publicly-owned, ownership unbundling), Amprion,

---

[88]*Ibid.*

[89]International Energy Agency, *Energy Policies of IEA Countries*, 2013 Review.

[90]Clean Energy Wire, "German utilities and the Energiewende," available at www.cleanenergywire.org/factsheets/german-utilities-and-energiewende.

[91]*Ibid.*

a subsidiary of RWE (vertically integrated undertaking), TenneT (publicly-owned, ownership unbundling), and TransnetBW, a subsidiary of EnBW (vertically integrated undertaking).[92]

As the percentage of electricity from renewables will increase in the coming years, the demand for expanding the grid will also increase. The Energiewende will indeed require Germany's energy infrastructure to be dramatically improved, and this is a task the TSOs are responsible for. TenneT covers Germany's electricity demand in a part of the country that stretches from north to south. One of its main challenge will be the transmission of energy from the wind turbines in the north to the industries in the south.

In 2017, 879 distribution system operators (DSOs) operated in Germany,[93] not only with the role of distributing electricity from the national transmission network to homes and businesses, but also for realizing new mechanisms for optimization, such as platforms that connect the different participants in the system.

8.3 Regulatory Framework for the Electricity Market

Spain accumulated a large tariff deficit from 2001 to 2012. This stemmed from the fact that costs of the electricity system were greater than its revenues. In 2012, when the debt skyrocketed, the government enacted emergency measures. These included terminating subsidies for renewable energy installations; lowering payments to TSOs and DSOs; raising access tariffs; and placing a 7% tax on electricity generation.[94] A new regulatory framework followed in 2013, aimed at ensuring the electricity's industry long-term health by tackling the ongoing issue of tariff deficits. The changes helped improve the electricity market's overall structure and reflected the measures taken by the European Commission's Third Energy Package of 2009. The regulation addresses a range of issues including unbundling ownership; regulatory oversight and cooperation; network cooperation; transparency; data tracking and access to storage facilities and LNG terminals by regulating transmission network ownership; more effective regulatory oversight; better consumer protection; regulating third party access to gas storage; and promoting cooperation among EU Member States.[95]

The Electricity Sector Law 24/2013 of 26 December was passed to address these key factors while assessing factors relevant to Spain, such as the need to increase

---

[92]Bundesnetzagentur, "Regulatory challenges of Germany's 'Energiewende,'" available at www.renewable-ei.org/images/pdf/20160309/NadiaHorstmann_REvision2016.pdf.

[93]Ante, Ulrich Scholz and Johann, "Electricity regulation in Germany: overview," available at uk.practicallaw.thomsonreuters.com/5-524-0808?transitionType=Default&contextData=(sc.Default)&firstPage=true&comp=pluk&bhcp=1.

[94]International Energy Agency (2015).

[95]Leal-Arcas et al. (2015a).

8.3 Regulatory Framework for the Electricity Market

renewable energy usage and the country's position as an "energy island" compared to other EU Member States.[96] It heralds a move towards greater flexibility when it comes to remunerating regulated projects such as those related to renewable energy, so as to keep up with changes in the electricity sector. A full review is possible every 3 years and the rationale behind the new framework is budgetary: the revenues of the electricity system must be enough to offset the system's total costs.[97]

Renewable energy, combined heat and power (CHP) and waste-to-energy (WtE) facilities must compete with traditional electricity sources and remuneration for them is based on their market presence. To ensure market fairness, the Electricity Sector Law 24/2013 sets out a plan of support for renewable energy power plants and those producing electricity from cogeneration and waste. The law allows these power plants to regain the investment and operating costs (which they cannot recover on the market), thereby providing them the opportunity for a decent return.[98] In addition, the law allows for further support schemes for these power plants as long as they meet EU-established goals. Autonomous regions may approve renewable energy (RE) power plants with an output below 50 MW.[99] Moreover, RE, cogeneration and waste plants have priority access and use of grids when it comes to both connection and dispatch, as long as their actions do not negatively impact the grid.[100]

The Electricity Sector Law 24/2013 is also significant in that it places a fee on self-consumption. Self-consumption is not to be confused with net metering. Net metering refers to a system in which RE generators such as solar panels are connected to a public-utility power grid and surplus power is transferred onto the grid, allowing customers to offset the cost of power drawn from the utility. Alternatively, this surplus can be recovered when consumption exceeds generation. Self-consumption also occurs when the generating facility is connected with a consumer's network, but unlike net metering, it does not enable consumers to transfer surplus energy to the public grid.[101]

Customers availing of self-consumption must pay for associated grid operation costs the way that direct grid consumers do.[102] The regulation conceptualizes self-consumption as "any electricity consumption originating from generation facilities connected in a consumer's household," assuming full or partial connection of the

---

[96]International Energy Agency (2015).

[97]Rojas and Carreño (2014).

[98]International Energy Agency, "Electricity sector regulation (Electricity Law 24/2013)," International Energy Agency, 2 May 2017. Available at: https://www.iea.org/policiesandmeasures/pams/spain/name-130502-en.php.

[99]Cuatrecasas, "Legal Update I Energy. Act 24/2013, of December 26, on the electricity sector," January 2014. Available at: http://www.cuatrecasas.com/media_repository/gabinete/publicaciones/docs/1388678102en.pdf.

[100]Nachmany (2015).

[101]The Mediterranean Energy Regulators (2016).

[102]Leal-Arcas et al. (2015a).

network to the grid.[103] The payment is likely the most disputed part of the Electricity Sector Law 24/2013, and various RE companies have been protesting self-consumption, arguing that they may stop consumers from setting up their own RE generation systems,[104] such as photovoltaic panels on household rooftops. However, the government justifies self-consumption fees on the grounds that market generation fees comprise only about 40% of the cost of electricity. Thus, all consumers connected to the grid, irrespective of whether or not they are self-consumers, must share the costs of annual tariff deficits and support schemes for renewable energy, cogeneration and waste.[105]

The Royal Decree 900/2015 of 9 October addresses self-consumption in further detail, requiring all self-consumers to register with the Registry for Electrical Energy Self-Consumption (except for a few isolated facilities).[106] The Royal Decree 900/2015 defines two kinds of self-consumption[107]:

- *Type 1: Supply with self-consumption*
  Consumers with an installed capacity below 100 kW fall into this category, which does not count as a production facility as the energy is only produced for self-consumption and the consumer receives no compensation for transmitting surplus electricity to the grid.
- *Type 2: Generation with self-consumption*
  This category comprises consumers in a single supply point or facility linked with one or several production facilities, on record as a production facility, connected within its network, or which share connection infrastructure with it. In contrast with type 1, consumers supplying surplus electricity to the grid receive remuneration.[108]

Legally, type 1 self-consumers are considered to be solely consumers, whereas type 2 self-consumers are considered both consumer and producer. Generally speaking, commerce and industry sector actors can sell power surplus in the same way as other producers, particularly because they have a greater chance to qualify as type 2 self-consumers. In this sense, the legal system in Spain does not help private households to turn into prosumers. The Royal Decree 900/2015 raises financial and other barriers, hindering citizens from selling surplus electricity to the grid. In fact,

---

[103]International Energy Agency, "Electricity sector regulation (Electricity Law 24/2013)," International Energy Agency, 2 May 2017. Available at: https://www.iea.org/policiesandmeasures/pams/spain/name-130502-en.php.

[104]Cuatrecasas, "Legal Update I Energy. Act 24/2013, of December 26, on the electricity sector," January 2014. Available at: http://www.cuatrecasas.com/media_repository/gabinete/publicaciones/docs/1388678102en.pdf.

[105]Rojas and Carreño (2014).

[106]Articles 19–21 Royal Decree 900/2015.

[107]Article 4 Royal Decree 900/2015.

[108]International Energy Agency, "Royal Decree 900/2015 on self-consumption," International Energy Agency, 10 May 2017. Available at: https://www.iea.org/policiesandmeasures/pams/spain/name-152980-en.php.

current regulation generally only enables citizens and households to supply surplus electricity to the grid for free.[109] Also, the Royal Decree 900/2015 bans a single installation from supplying power to a range of different end-users.[110] This makes it hard to spread PV technologies in urban areas as it prevents a single installation from supplying electricity to a neighborhood, for example.

Another area of the regulation that has resulted in controversy is the back-up tolls that make up the "sun tax." These fees result from "charges related to the electricity system costs" and "charges for other services of the system".[111] According to the Spanish Photovoltaic Union, it is unfair that PV self-consumers pay a "sun tax" for the entire power capacity installed (the power that their provider supplies as well as the power generated from their own PV facility). Moreover, PV users who have installations greater than 10 kW must pay for the electricity they generate and self-consume from their own PV technology.[112] Some facilities (e.g. off-grid installations; installations smaller than 10 kW; all installations in the Canary Islands, Ceuta and Melilla) do not have to pay this additional "sun tax," and installations with co-generation are temporarily exempted until 2020, while the islands of Mallorca and Menorca will pay a reduced rate.[113] The Royal Decree 900/2015 aligns with the Electricity Sector Law 24/2013 as it emphasizes the need to ensure electricity industry's long-term financial viability. This means that keeping costs down has become the main goal, perhaps at the cost of more farsighted policy regarding self-consumption.

The goal of keeping costs in check aligns with the overall basis for the reshaping of the electricity industry. But effective policy for self-consumption is an evolving area and the current restrictions on self-consumers are not conducive to decentralizing energy, in that they prevent the consumer to prosumer transition. This is at odds with the government's keenness to increase RE and electricity from cogeneration and waste. The financial and administrative barriers to self-consumption described above can ultimately keep smart grid applications from effective market penetration in Spain. Decentralizing energy via self-consumption would have many benefits for Spain. It would help consumers receive direct gains and participate in energy markets, becoming prosumers. It would help to lower costs of energy facilities, reduce system losses, and help pay for the energy transition.[114] Self-consumption also promotes technological progress and energy security.

---

[109]López Prol and Steininger (2017).

[110]Article 5, paragraph 1, point c) Royal Decree 900/2015.

[111]Articles 17 and 18, respectively, Royal Decree 900/2015.

[112]I. Tsagas, "Spain Approves 'Sun Tax,' Discriminates Against Solar PV," Renewable Energy World, 23 October 2015. Available at: https://www.renewableenergyworld.com/articles/2015/10/spain-approves-sun-tax-discriminates-against-solar-pv.html.

[113]I. Tsagas, "Spain Approves 'Sun Tax,' Discriminates Against Solar PV," Renewable Energy World, 23 October 2015. Available at: https://www.renewableenergyworld.com/articles/2015/10/spain-approves-sun-tax-discriminates-against-solar-pv.html.

[114]Commission Staff Working Document. Best practices on Renewable Energy Self-consumption, p. 3, COM(2015) 33 final (15 July 2015).

Unfortunately, most power companies are threatened by energy decentralization. Prosumers, as new players on the field, will result in greater competition in a reduced market.[115] It is important to overcome inaccurate assumptions about self-consumption and to not allow the self-interest of various actors, as well as cautious government policy, to prevent prosumers from emerging. More robust legal frameworks are needed, to remove fees and bans and other barriers to energy decentralization, to help electricity markets to take the plunge in the inevitable energy revolution. With that goal in mind, regulations to come must examine net-metering to help promote decentralized RE.

The Electricity Sector Law 24/2013 also created a significant change in the form of increased powers allocated to the National Regulatory Authority (NRA), that is, the National Commission of Markets and Competition (CNMC), which now establishes methodologies for calculating transmission and distribution tariffs; grants access to cross-border infrastructure; provides balancing services; levies penalties; and devises enforceable measures for relevant companies.[116] The CNMC's empowerment resulted from the EC calling for greater regulatory oversight in the Third Energy Package. At the EU level, the NRA is to cooperate with other regulatory bodies, and does so via the Council of European Energy Regulators (CEER) as well as the Agency for Cooperation of Energy Regulators (ACER). The CEER and the ACER are platforms where the various regulatory entities can collaborate on the internal energy market and create network codes.[117]

While the NRA's empowerment is a positive step, there remains a lot to be done. As mentioned above, the CNMC devises the methodology for calculating network tariffs for transmission and distribution. But the Ministry of Industry, Energy and Tourism is the entity with the ultimate authority to set the tariffs. Granting complete independence to the NRA is a key step towards market confidence.[118] In fact, optimal levels of transparency are still so out of reach as to have caused frequent market and investment disturbances within Spain's electricity system. The government has full authority to tweak prices for political motivations, e.g., to curb inflation. Full NRA independence would create market certainty and could be achieved by giving it exclusive authority to determine network tariffs, among other things.[119] One thing is for certain: a stable and transparent electricity system is conducive to promoting self-consumption.

In Italy, a collection of decrees, directives and decisions shape the country's electricity market, which, combined, replace relevant EU Directives. They also dictate state policy with regard to the electricity market. The laws include:

---

[115]Donoso (2015).
[116]International Energy Agency (2015).
[117]Ibid.
[118]Roberts and Skillings (2015).
[119]International Energy Agency (2015).

8.3 Regulatory Framework for the Electricity Market

- Law No. 481 of 14 November 1995, establishing the Regulatory Authority for Electricity, Gas and Water (AEEGSI) and elaborating its role;
- Legislative Decree No. 79/99 of 16 March 1999, implementing Directive 96/92/ that addresses common rules for the internal electricity market. The decree grants GME the authority to handle the Italian Electricity Market in keeping with principles of neutrality, transparency, objectivity and competition among producers;
- The Integrated Text of the Electricity Market rules, which govern the operation of the wholesale electricity market;
- Law No. 239/2004, which aims to harmonize the energy sector and oversee its liberalization;
- AEEGSI Decision 111/06, creating a process to record forward electricity purchases/sale contracts on the Italian Power Exchange's OTC Registration Platform;
- Legislative Decree No. 93/2011, launching the Third Energy Package, with a goal of improving energy security and protecting low-income consumers.

Approving licenses to build and run energy generation facilities takes place at the regional level. Regions also regulate such plants. Given the decentralization of energy in the country, AEEGSI only addresses legal, technical and financial matters. Even though Legislative Decree No. 112/1998 (as amended by Legislative Decree No. 443/1999) holds that passing region-specific is allowed, local governments must still abide by AEEGSI's overall direction.

Since the EU issues its first electricity-related directive (96/92/EC), there has been a steady move towards liberalization in Italy's electricity market. The EU directive followed the Bersani Decree,[120] which delineated the way forward for liberalizing the market and identified basic tenets for future steps in this direction. Thus far, liberalizing wholesale and retail markets has met with much success. Decentralizing electricity generation and using micro-grids for distribution would be the next major steps.

A day-ahead market and an intraday market comprise the wholesale electricity market in Italy. In 2013, Italy's wholesale electricity market comprised 42% of total national supply. GME is in charge of both. An ancillary services market (MSD) exists as well, for which TERNA is the key actor. Two segments comprise the MSD: the ex-ante MSD for energy trades and harmonization to prevent backing-up and to build reserves; and the balancing market for trades in real-time balancing services, in order to rebuild reserves and keep a grid balance. GME is also responsible for a forward electricity market (MTE) that is a negotiating platform for forward electricity contracts and withdrawals. While the Italian Power Exchange (IPEX) manages these markets, participants are not limited to IPEX—sellers and buyers can also conduct bilateral trade, circumventing the exchange. Such bilateral contracts must be recorder in the Energy Accounts Platform.

---

[120]International Energy Agency (IEA) (2016).

Three retail markets operate in the country: the safeguarded market, the enhanced protection market, and the open or free market. The safeguarded market is the default one, targeted at end consumers who are ineligible for the enhanced protection market, and would not have any electricity supplier. The enhanced protection market must by law supply electricity to those who have not opted to change suppliers. AU S.p.A, a state-owned company manages this market. It buys electricity on the wholesale market and then sells it to retailers who, in turn, sell it to customers at a regulated price. The retail market is the open, or free market, with more than 336 retailers, the largest of whom is Enel.[121] Customers on the can choose from a broad array of competitive suppliers.

Pricing on the wholesale electricity market is unregulated, and demand and supply are the deciding factors. On the retail market, AEEGSI establishes transmission tariffs for 3-year timeframes, applying a cost cap approach. This approach sets a tariff based on how much capital was invested and the costs of operation. The amount that customers pay depends on both network charges and the rate of their particular supplier.

Italy is working towards increasing renewable energy usage and decreasing fossil fuel dependency. To this end, several schemes are in place. One of the main approaches has been to provide electricity stemming from RE sources dispatching preferences. But since RE sources tend to be irregular, relying on RE for a sizeable proportion of power jeopardized the electricity network's functioning. To address this, AEEGSI established fines for falling short.[122] Another scheme has been to create a feed-in-tariff for solar energy in 2008, after which a feed-in-tariff was placed on all RE sources in 2012. Additionally, the GSE has established more streamlined buying and reselling procedures since 2008, which helps small producers sell electricity generated to GSE, instead of having to sell via bilateral contracts or on the Italian Power Exchange. Eligibility requires producing less than 10 MVA through RE or hybrid plants.[123] Until recently, the GSE was also granting "green certificates," which producers of RE (except for those using photovoltaic) could be eligible for. But this scheme is on the way out and will be replaced by the feed-in-tariff regime.

As of yet, Italy lacks clear policy on rolling out micro-grids, in spite of having a strong emphasis on making the national grid smart as well as integrating distributed energy resources (DERs). While a few micro-grid projects do exist, to truly achieve the goal of electricity decentralization, authorities would need to address several challenges, including establishing grid connection fees, transmission approaches and developing incentives for micro-grids.

---

[121]N. Rossetto, "An oversized electricity system for Italy," Italian Institute for International Political Studies, 22 January 2015. Available at: https://www.ispionline.it/it/energy-watch/over sized-electricity-system-italy-12135.

[122]Bersani Decree, Legislative Decree No. 79, 1999.

[123]International Energy Agency (IEA) (2016).

8.3 Regulatory Framework for the Electricity Market

To generate, buy, or supply electricity derived from RE sources, no paperwork or license is needed in Italy.[124] Moreover, energy derived from RE sources in incorporated into the national grid. Thus, Italy has a pro-prosumer environment, perhaps epitomized by its *ritiro dedicato* policy, which helps small-scale prosumers such as households with market access. One challenge yet to be addressed, though, is how "pooled loads" from different DERs can qualify to have access to wholesale markets. On the surface of it, since the wholesale market is ostensibly open, legal entities should have access. But technical requirements may stand in the way. For example, in the MSD, bid offers are accepted on the basis of factors related to generating units and pumped-storage units—facilities that small-scale producers are not likely to have. In short, existing restrictions may prevent the whole move towards helping prosumers from transpiring in practical terms.

In the case of Germany, it is moving fast towards decarbonising its economy in terms of laws and regulations for renewables, but it is still quite cautious when it comes to rolling out smart meter technologies, especially due to concerns related to data protection, as the right to privacy is perceived as a very significant matter by German citizens.

Despite Germany's energy industry being liberalized in 1998 with the Energy Industry Act (EnWG),[125] the TSOs control for the most part the retail electricity market, except for a few small electricity and gas distribution companies (mostly municipalities owned). The responsibility of the TSOs is to ensure the expansion of the German energy grid and to guarantee the security and stability of the power supply system of the country, depending on their specific geographic area of coverage. As the TSOs have to plan and maintain Germany's ultra-high voltage grid and regulate grid operations, they must guarantee the uninterrupted exchange of electricity across all regions. They reach this goal using power lines and ensuring that generation and consumption levels are balanced at all times.

In recent years, the electricity sector has experienced a decrease of electricity derived from conventional sources and an expansion of electricity generation from renewables.[126] In September 2010, the federal government adopted the Energy Concept, as part of the *Energiewende*, as an elaboration of Germany's energy policy until 2050. The four transmission system operators have a crucial role in the transition towards renewables. A very recent challenge for the TSOs was to approve the Grid Development Plan 2030 (2017),[127] which is the plan for the onshore transmission network.

---

[124]M. Cicchetti and G. Fabbricatore, "Electricity Regulation in Italy: Overview," Practical Law Company, 1 May 2014. Available at: https://uk.practicallaw.thomsonreuters.com/Document/Ieb49d7bb1cb511e38578f7ccc38dcbee/View/FullText.html?originationContext=docHeader&contextData=(sc.Defadult)&transitionType=Document&needToInjectTerms=False&firstPage=true&bhcp=1.

[125]Energy Industry Act, available at: www.gesetze-im-internet.de/enwg_2005/EnWG.pdf.

[126]Bundesnetzagentur, Annual Report 2016.

[127]Grid Development Plan 2030, available at: www.netzentwicklungsplan.de/en/front.

The Plan is a comprehensive depiction of the revised Renewable Energy Act (EEG 2017),[128] which came into force at the beginning of 2017. The Grid Development Plan 2030 describes measures that satisfy both the legal requirements and the underlying framework imposed by the Federal Network Agency. The plan highlights the transmission requirements between a starting and an ending point. The Grid Development Plan 2030 is closely linked to the Offshore-Network Development Plan 2030 (Offshore-Netzentwicklungsplan, O-NEP),[129] as required by the Germany's Energy Industry Act (EnWG). The main goal of the Plan is to create a change towards a consistent and efficient offshore grid expansion. This is a biannual binding plan that will improve the coordination of grid connections and offshore wind farms[130] within the framework set by the Energy Industry Act, which introduced a compensation regulation for the construction and operation of grid connections to offshore wind farms.

The Renewable Energy Sources Act is an important legislation that promotes the generation of electricity from renewable sources and plays a crucial role in the *Energiewende*.[131] Despite one important case under the European General Court and the criticism it has received over the years, the Act has enabled Germany to accomplish an increase of 32% in its electricity consumption deriving from renewables since 2015.

The Renewable Energy Sources Act makes the whole electricity supply more sustainable and helps Germany achieve its climate targets. The Act has been in force since the year 2000. Over the years, it has gone through a number of amendments and faced various criticisms. In 2012, the European General Court ruled that the funding under the Act constituted state aid. In particular, it was decided that (1) the supports for undertakings producing electricity from renewables and (2) the reduction in the EEG surcharge for certain electricity-intensive undertakings were state aid.[132]

In 2017 it was strongly amended, introducing a real shift in the paradigm for renewable energy. First of all, the amended Renewable Energy Sources Act introduced a revised mechanism for the price of electricity from renewables. The Act is now based on competitive auctions to set the price of electricity generated from renewable sources, making obsolete the old system where the fees were decided by the government. This is one reason why competition entities such as the Federal Network Agency for Electricity, Gas, Telecommunications, Post and Railway and the Federal Cartel Office have an integral position within the *Energiewende,* in order to ensure that competition rules are applied and respected.

---

[128]Renewable Energy Sources Act 2017, available at: www.bmwi.de/Redaktion/EN/Downloads/renewable-energy-sources-act-2017.pdf?__blob=publicationFile&v=3.

[129]Grid Development Plan 2030, available at: www.netzentwicklungsplan.de/en/front.

[130]*Ibid.*

[131]Renewable Energy Sources Act 2017, available at: www.bmwi.de/Redaktion/EN/Downloads/renewable-energy-sources-act-2017.pdf?__blob=publicationFile&v=3.

[132]Judgment in Case T-47/15, 2016.

8.3 Regulatory Framework for the Electricity Market

In order to maintain a high level of market-players diversity and to guarantee the possibility for citizens to participate in auctions along with other stakeholders, the Renewable Energy Sources Act introduced an interesting concept in 2017: The idea of a "citizens' energy company". Point 15 of section 3 of the Law (Definition) defines this as:

> every company a) which consists of at least ten natural persons who are members eligible to vote or shareholders eligible to vote, b) in which at least 51 percent of the voting rights are held by natural persons whose main residence has been registered pursuant to Section 21 or Section 22 of the Federal Registration Act for at least one year prior to submission of the bid in the urban or rural district in which the onshore wind energy installation is to be erected, c) in which no member or shareholder of the undertaking holds more than 10 percent of the voting rights of the undertaking, whereby in the case of an association of several legal persons or unincorporated firms to form an undertaking it is sufficient if each of the members of the undertaking fulfils the preconditions pursuant to letters a to c (Point 15 of section 3 of the Law).[133]

There are special auctioning rules for citizens' energy companies, so that they can participate in the auctions on simplified terms that are listed in article 36g of the Law.[134] Auctions can be held for funding onshore or offshore wind energy, PV energy, and biomass projects. This is intended to guarantee competition, in terms of keeping the price of electricity fair. It also aims at an improvement of the democratization of the energy sector, where citizens can have a role in setting the price of electricity and support renewables projects.

The Act was originally introduced to guarantee renewable energy producers high returns on investment, lowering the costs of the installation of renewable power capacity. In addition to this, the reform aimed at bringing the price down for the consumers, although there are different points of view regarding the effective benefits to both the costs and the impact on emissions reduction and, consequently the ability of the country to meet its international targets to tackle climate change.[135]

In terms of energy efficiency, the Renewable Energy Sources Act also links the expansion in renewables capacity with the grid expansion. This guarantees energy efficiency and guarantees that there is no waste in the energy grid: clean power can therefore reach consumers without dispersion. Each type of technology has been assigned a specific expansion volume. For example, the expansion of onshore and offshore wind power, photovoltaics, and biomass have been set in line with the available grid capacity.

The new Renewable Energy Sources Act intervenes also in grid congestions, which represent a critical issue in Germany, especially when referring to the flow of wind power from the north of the country to the industrial areas of the south. A new ordinance (2017) designates areas within which the pace at which wind power

---

[133]Renewable Energy Sources Act 2017, available at: www.bmwi.de/Redaktion/EN/Downloads/renewable-energy-sources-act-2017.pdf?__blob=publicationFile&v=3.

[134]*Ibid.*

[135]Clean Energy Wire, "Germany's energy transition revamp stirs controversy over speed, participation," available at: www.cleanenergywire.org/dossiers/reform-renewable-energy-act#controversial.

capacity is being developed will be limited to a maximum of 58% of the average capacity added over the last 3 years (Section 36c(4)).[136] This restriction applies until the grid has been sufficiently upgraded. Additional installations that cannot be built in certain areas because of grid congestion are being built in other parts of the country instead.[137]

Another legislation recently adopted by the Federal Government is the Electricity Market Act (26 July 2016), which contains provisions on how to efficiently use the surplus of renewable energy, because a large amount of renewable energy power is lost due to curtailment. To avoid this loss, the new Act enables system operators to purchase the surplus of energy. Consumers will not have to pay anything (any levy, tax or the costs for generation and distribution), resulting in a purchasing price that will be lower than usual electricity prices.[138]

In a wider context, regarding climate change and the commitments that Germany made within the framework of the Paris Agreement on Climate Change,[139] the Climate Action Plan 2050 (Klimaschutzplan 2050) provides guidance for different areas, including, among others, energy, buildings, transport, trade and industry.[140] Finally, a relevant action plan that inspires German's policy on energy efficiency is the German National Action Plan on Energy Efficiency (NAPE). This Plan will shape the energy efficiency sector focusing on three cornerstones: "*1. Stepping up energy efficiency in the building sector; 2. Establishing energy efficiency as an investment and business model; 3. Increasing individual responsibility for energy efficiency.*"[141]

The goals of the *Energiewende*—nuclear phase-out, replacement of fossil fuels with renewable energies—imply a real revolution of the entire energy system and a consequent destabilisation of the current established system. This is due to the volatility of renewables. The risk of intermittency requires that the new policies have to maintain a balance between generation and consumption. So far this has not represented a problem for Germany, especially in its long-term policies. However, especially in the short-term, TSOs have to take responsibility for maintaining the balance and this is possible only when adequate infrastructures are in place.

---

[136]Renewable Energy Sources Act 2017, available at: www.bmwi.de/Redaktion/EN/Downloads/renewable-energy-sources-act-2017.pdf?__blob=publicationFile&v=3.

[137]*Ibid.*

[138]German Federal Ministry for Economic Affairs and Energy, Electricity Market 2.0, available at: www.bmwi.de/Redaktion/EN/Artikel/Energy/strommarkt-2-0.html.

[139]UNFCCC, Text of the Paris Agreement 2015, available at: unfccc.int/sites/default/files/english_paris_agreement.pdf.

[140]Bundesministerium für Umwelt, Naturschutz, Bau und Reaktorsicherheit (BMUB), *Klimaschutzplan 2050*, available at: www.bmu.de/fileadmin/Daten_BMU/Download_PDF/Klimaschutz/klimaschutzplan_2050_bf.pdf.

[141]Federal Ministry for Economic Affairs and Energy (BMWi), National Action Plan on Energy Efficiency 2014, available at: www.bmwi.de/Redaktion/EN/Publikationen/nape-national-action-plan-on-energy-efficiency.pdf?__blob=publicationFile&v=1.

8.3 Regulatory Framework for the Electricity Market

Although there is a good legal framework that potentially can guarantee energy security in the country, the TSOs will need to work hard to guarantee that adequate infrastructures will be able to move electricity for different parts of the national territory and also, when necessary, abroad to neighbouring countries. The main law focusing on energy security in Germany is the Energy Industry Act (EnWG).[142] It originally set the legal framework for the liberalisation of the energy market in 1998. As opposed to other sectors (such as, for example, telecommunications) and many European countries, the liberalization of energy markets inspired by the EU Commission has not brought so far too many benefits in Germany in terms of lower prices and providing better services for consumers.[143] Although each retailer has non-discriminatory access to all customers in each regional market, German household customers show a great degree of inertia.[144] In 2005 the Energy Industry Act was amended to include sustainable energy production and enhance competition. It requires electricity labelling that depends on different types of energy sources. This provides greater information on electricity sources, allowing consumers to therefore make informed decisions about suppliers.[145] In December 2012 the EnWG was further amended, especially in order to improve its rules in the sectors of offshore grid connection lines.

A big challenge that Germany will face in the next decade is the decommissioning of about 12 GW of nuclear power by 2022. This means that Germany has to fill this potential gap with other sources of energy in order to avoid risks to energy security in the country. This gap will be filled with the main focus of the *Energiewende*: renewable energy. Although it is likely that renewables will play a crucial role in guaranteeing energy security in Germany, other aspects have to be taken into account, such as changes in electricity demand and the rate of network capacity expansion. Three reports by the Federal Network Agency have analysed the consequences of the shift from nuclear to renewables, the impact on the transmission systems and the security of energy supply. System operation appears to be the most important area where network operators have to intervene. In case of a threat, section 13(1) of the EnWG assigns responsibility to the four TSOs. TSOs are indeed the entities that deal with authorisations and are requested to adopt any necessary remedy to guarantee energy security. Similar responsibilities, but in a different setting, are imposed to electricity distribution system operators (DSOs), that have liability for the security and reliability of the electricity supply within their own networks (Section 14(1) of the EnWG).

The Federal Ministry for Economic Affairs and Energy (BMWi) is the institution responsible for monitoring security of energy supply. Every 2 years it publishes a report on the situation. The last report confirmed that supply is secure for the period of reference (up to 2025). This result has to be considered in an international context.

---

[142]Energy Industry Act, available at: www.gesetze-im-internet.de/enwg_2005/EnWG.pdf.

[143]Joskow et al. (2008), pp. 9–42.

[144]Duso and Szücs (2017), pp. 354–372.

[145]Energy Industry Act 2018, available at: www.gesetze-im-internet.de/enwg_2005/EnWG.pdf.

The report indeed also stressed the importance of the international dimension of the monitoring and the effects of cross-border electricity trading. Germany is part of a network of many countries in Europe that are linked in order to trade electricity, ensuring security of supply and lower costs. These countries are: Norway, Sweden, Denmark, Netherlands, Belgium, Luxembourg, France, Poland, Czech Republic, Austria, Switzerland, and Italy. The growing importance of this network has been reflected by a change in the methodology that the latest reports have adopted[146]; rather than considering Germany alone, an international approach is now preferred and adopted, because it is able to give a better perspective on what is actually happening in terms of European and regional cooperation on electricity market issues.[147]

Alongside the challenge of guaranteeing proper infrastructures to ensure the flow of clean energy (mostly wind power) from the north of the country to the industries in the south (Bavaria and Baden-Württemberg) and to tackle the issue of congestion, another challenge is to guarantee the flow of energy with neighbouring countries. Regulation 347/2013 of 17 April 2013[148] sets the guidelines for trans-European energy infrastructure (TEN-E Reg) and aims at the timely development and interoperability of priority corridors and areas of trans-European energy infrastructure. This Regulation has been implemented in Germany along with other measures such as the Pentalateral Energy Forum. This latter initiative has promoted collaboration between Germany, France, Belgium, the Netherlands, Luxembourg, Switzerland, and Austria.

This emphasis on interconnection has linked up markets across borders, involving regulators, grid operators, power trading exchanges and market players and simplifying cross-border trading.[149] As a result of an increase in cross-border collaborations in recent years, for example, German energy regulator (Federal Network Energy) has tasked Germany's TSOs to implement congestion management on the German-Austrian border, which, creating a single price zone, will contribute to energy security in both countries.[150] Recently, another interconnection has been improved between Germany and the Netherlands. The Netherlands is the biggest importer of Germany's surplus power at the moment and the new agreement between Grid operators TenneT and Amprion in August 2018 has further improved

---

[146]Consentec GmbH, System Adequacy for Germany and its Neighbouring Countries, 2015.

[147]Federal Ministry for Economic Affairs and Energy, Security of Supply, available at: www.bmwi.de/Redaktion/EN/Artikel/Energy/security-of-supply.html.

[148]The European Parliament and Of the Council, Regulation (EU) No 347/2013.

[149]Federal Ministry for Economic Affairs and Energy, Competition and Regulation, available at: www.bmwi.de/Redaktion/EN/Textsammlungen/Energy/wettbewerb-energiebereich.html?cms_artId=255904.

[150]Argus, "Germany-Austria power zone split special report," available at: www.argusmedia.com/-/media/Files/white-papers/germany-austria-zone-split-white-paper.ashx.

8.3 Regulatory Framework for the Electricity Market

the capacity for exports from Germany and the Netherlands.[151] Considering Germany's geographic location, it is expected that more of such agreements and collaborations will be needed among Germany and Austria, Switzerland, France, Luxembourg, Belgium, the Netherlands, Denmark, Poland, and the Czech Republic.

Renewable energy reduces Germany's dependency on energy imports. This is particularly important when imports of a large section of an entire source of energy (e.g. gas) depend on one single foreign country. Russia, for example, provides 40% of German gas consumption. If this dependence can be replaced by nationally produced renewable energy, it would make the country less vulnerable in the face of fluctuations in prices of fossil fuels and their unpredictability, as well as less politically dependent on other countries. This would change the types of contracts between countries. Rather than unidirectional supply contracts, the focus will switch to multilateral agreements to refer to only when energy independence cannot be guaranteed nationally.

In the Netherlands, to support the energy transition, the Dutch Government has involved various stakeholders in the decision-making process.[152] Such an approach appears to be rather productive in the energy transition, where many stakeholders are called to contribute to creating a proper circular economy. In April 2016, the government opened a dialogue with people and organizations from all over the country to discuss the energy transition. As a result, in December 2016 the Energy Agenda was published, mapping out the route to a low-carbon energy supply by 2050.[153] In particular, the aim is to produce the following percentage of energy from RES: 14% by 2020, 16% by 2023, and almost 100% by 2050, with CO_2 emissions being 80–95% lower than in 1990.[154]

Soon after the Energy Agenda, in March 2017, the Dutch energy-intensive industries agreed with the Dutch Government to invest in energy efficiency, aiming at saving a total of 9 petajoules (PJ) by 2020.[155] Despite the commitment of the Government and its historical tradition for windmills, a 2016 report from Eurostat shows that the Netherlands was almost last among the EU countries in the share of energy from renewable sources between 2004 and 2016.[156] Only in the last few years has the percentage of renewable power increased, reaching 14.5% in 2018. In particular, solar panels accounted for about 0.2 PJ, onshore wind electricity

---

[151] Energy Live News, "Power interconnector between Germany and Netherlands goes live," available at: www.energylivenews.com/2018/08/25/power-interconnector-between-germany-and-netherlands-goes-live.

[152] See Verkade and Hoeffken (2018), pp. 799–805.

[153] Ministry of Economic Affairs of The Netherlands, "Energy Agenda: Towards a low-carbon energy supply," 2017.

[154] Ibid. p. 21.

[155] https://www.government.nl/topics/renewable-energy/central-government-encourages-sustainable-energy.

[156] https://ec.europa.eu/eurostat/statistics-explained/index.php/File:Figure_1-Share_of_energy_from_renewable_sources_2004-2016.png.

generation was 3.5 PJ, and offshore wind electricity generation was 1.5 PJ, 40% higher than the year before.[157]

It is expected that the legal and regulatory framework would evolve too. The main Dutch law on electricity is the 1998 Electricity Act (E-Act). The E-Act assigns regulatory powers to the Ministry of Economic Affairs and Climate Policy (MEA), including the power to adopt secondary legislation. Every 4 years, an energy report is issued by MEA to lay down the energy policy in the short and long term.

The E-Act has been amended several times. After a controversial proposal for amendments in 2015, the so-called "STROOM", did not find majority in the Senate, the latest major amendment was adopted in 2018 under the name of Energy Transition Act (wet VET).[158] Part of such an Act came into effect on the 1st January 2019.

The Energy Transition Act focuses mostly on removing existing obstacles and bottlenecks that are detrimental to the energy transition. Some important provisions of the E-Act deal with how the grid operator can connect, use, and develop the energy grid. The connection to the grid for plant operators happens through the grid operator, who must enter into such contracts.[159] This obligation is not applied when there are problems with the capacity of the grid. In the case of insufficient capacity, the access can be denied. At the same time, the grid operator is obliged to expand his grid according when it is needed.[160] Because renewable energy has not been a top priority for the Netherlands in the last decades, but is only developing recently, Articles 23(2) and 24(3) of the E-Act set a non-discriminatory principle for every energy source. Therefore, renewables do not benefit from any preferential treatment in accessing or using the grid.[161]

Many provisions aim at keeping commercial activities of the network company separated from the legal responsibility of the network operators.[162] This was necessary because the Netherlands has opted for the full ownership unbundling,[163] but the DSOs can still be part of a larger network company that performs commercial activities. Article 17 of the E-Act delineates strict rules regarding the activities that the DSOs can carry on: their commercial activities are to perform their traditional tasks, such as the operation and maintenance of the network, safeguarding its safety and reliability and building, repairing, upgrading, and expanding the network.[164]

Network companies can only engage in activities that relate to infrastructure-related projects or are referred to in the legislation (Article 17c of the E-Act) if

---

[157] https://www.en-tran-ce.org/.

[158] https://www.eerstekamer.nl/behandeling/20180130/gewijzigd_voorstel_van_wet/document3/f=/vklim6u3j6zt_opgemaakt.pdf.

[159] Art. 23, para. 1 of the E-Act.

[160] Art. 16 para. 1(c) of the E-Act.

[161] http://www.res-legal.eu/search-by-country/netherlands/.

[162] https://www.sciencedirect.com/science/article/pii/S0301421518308553.

[163] Article 10b of the E-Act.

[164] Edens and Lavrijssen (2019), pp. 57–65.

8.3 Regulatory Framework for the Electricity Market

related to network operations. Both network companies and DSOs can perform temporary activities that need to last no longer than 5 years and that need not to be performed by other market operators.[165] On the one hand, this clear demarcation of activities allows new actors to access the energy market. On the other hand, however, the law gives the chance for the incumbent parties in the market (such as the DSOs and network companies) to perform some of the activities that otherwise could be attractive for some other new market parties that wish to engage in new initiatives.

The new Act, aware of the imminent need for new actors to access the renewable-energy market, allows DSOs and network companies a short-term extension of their responsibilities.[166] Renewable sources are not very developed in the Netherlands at the moment, being natural gas the most used source of energy in the country. In 2016, RES represented only the 6% of the energy mix. This proportion is expected to grow to 12.4% by 2020 and to 16.7% by 2023. Although it is too late for the country to reach its 2020 goal,[167] it is expected that the 2023 goal will eventually be reached.

This optimism is due to the increase in investments, subsidies, and other mechanisms for RES that the Netherlands is currently supporting. For instance, the SDE+ (*Stimulering Duurzame Energieproductie*) is a premium feed-in scheme that promotes renewable-energy sources for electricity, renewable gas, and heating purposes. As for electricity, RES-E is promoted through investment subsidy for PV installations, net-metering, and tax benefits. For example, an exemption from tax can be granted to generators of electricity from RES that use the electricity they produce. The Act on the Environmental Protection Tax (WBM) regulates the Netherlands' consumption of electricity and natural gas, which will be widely applied when the use of RES will be spread consistently among all businesses and households. This is the example of how a consumer that becomes a prosumer can save on electricity bills. If the consumer, rather than only consuming electricity, is also the producer of that same electricity (own consumption clause) and becomes a "prosumer", he/she is exempt from paying the Environmental Protection Tax.[168]

Investment in renewable-energy plants can benefit from tax credits up to 54.5% of the total investments made in renewable energy or energy-efficiency technologies within 1 year.[169] Because technology innovation evolves very quickly in the energy sector, and to reflect the different priorities of the Dutch Government, every year, the Netherlands Enterprise Agency publishes a list (namely, the Energy List) of RES-E technologies that are allowed to receive credits. The eligible investments have been reduced since 2018 and there are some restrictions regarding wind energy, which is

---

[165]Idem.

[166]Idem.

[167]https://www.reuters.com/article/us-netherlands-climatechange/dutch-will-miss-2020-green-energy-climate-targets-report-idUSKBN1CO2EV.

[168]Art. 59 and 64(1), in conjunction with Art. 50(4)–(5) WBM.

[169]Art. 3.42 (3) Wet IB 2001.

not generally eligible, except for wind/watermills (namely the traditional Dutch system that pumps water using wind force).

Loans can be arranged with a special interest rate for projects relating to renewables. According to Regulation Green Projects 2016 (RGP 2016),[170] a project that has a positive impact on the environment can apply for a declaration, valid for 10 or 15 years,[171] on the basis of which a reduction of the interest rate in the order of 1% is recognised.[172] Except for projects on biomass and biogas, all other RES-E-related projects are eligible.[173]

The Offshore Wind Energy Act[174] is another important piece of legislation in the Netherlands. It entered into force in July 2015. Already in 2013, the Dutch Government and different actors of the energy market signed an Energy Agreement to develop five wind farms in the North Sea, to be completed by 2023. The aim is to reach a total installed capacity of 4.5 GW (considering also the production of the other existing wind farms).[175] The Offshore Wind Energy Act focuses on the planning aspects of the required infrastructures to achieve this goal. Rather than the project developer, the Dutch Government is responsible for spatial planning (choosing the location of the plants), the pace of development, and the offshore grid connection to the electricity network of TenneT.

The Netherlands Enterprise Agency deals with the offshore wind energy subsidy and permit tenders on behalf of the Ministry of Economic Affairs and Climate Policy. The award for subsidies and permits for the development of the wind farms is given to the bid that meets the required criteria. Soon after the adoption of the Offshore Wind Energy Act, it became clear that a decreasing need for subsidies was creating the world's first successful subsidy-free tender. This means that a full exposure to market prices could create uncertainties and distortions. Therefore, at the end of 2018, an amendment was presented, containing some changes of the tender system, to make sure that subsidies can still be granted in some cases.[176]

To conclude the analysis of the Dutch regulatory framework, an overview of some Dutch legal mechanisms and policy programs deserves being mentioned. The program that mostly supports such mechanisms is the Energy Research Subsidy Program (Energie Onderzoek Subsidie, EOS). Since 2005, interesting concepts have been developed for smart-grid architectures under such program, especially regarding infrastructures.[177]

---

[170]Regulation Green Projects 2016 (RGP 2016).

[171]Art 16.1(a) and (d), RGP 2016.

[172]http://www.res-legal.eu/search-by-country/netherlands/tools-list/c/netherlands/s/res-e/t/promotion/sum/172/lpid/171/.

[173]Art. 7, RGP 2016.

[174]https://wetten.overheid.nl/BWBR0036752/2015-07-01.

[175]https://english.rvo.nl/subsidies-programmes/offshore-wind-energy.

[176]https://www.lexology.com/library/detail.aspx?g=df69140a-e443-4386-9d6b-3280108f72b5.

[177]https://www.rvo.nl/sites/default/files/Smart%20Grids%20and%20Energy%20Storage.pdf, p. 4.

8.3 Regulatory Framework for the Electricity Market

In 2011, the Ministry of Economic Affairs, Agriculture and Innovation approved the Energy Report (Energierapport 2011). Some pilot projects included in this report were allowed to deviate from the existing laws and regulations (on a local and temporary basis) for the purpose of experimenting with smart grids. This exception is issued under the condition that the Netherlands Competition Authority (or sometimes even the European Commission) would approve it.[178]

Considering that the Netherlands is traditionally inclined to deal with new societal and policy challenges in a collective manner, a good example of such approach is the Smart Grids Knowledge & Innovation Top Consortium (Topconsortium Kennis & Innovatie, TKI). This is a public-private partnership (PPP) that puts together companies, research institutes, and non-government organisations in the energy sector. The PPP is a form of business model that could be widely adopted in the energy sector, which is a sector where there are many different types of stakeholders. Given the increased need of flexibility in the energy supply, these collaborative types of tools could be useful for businesses and end-users.

A special type of Dutch public-private partnership is the Collectief Particulier Ondernemerschap (CPO). This new strategy for smart, integrated, decentralized energy systems works for small realities and at the local level. However, one can solve major legal issues regarding the implementation of collective projects that mainly use microgrids as a way to connect to the energy system. One example is the Aardehuizen, a small housing project in the east of the Netherlands, where the concept of CPO was applied. As the Netherlands is not new to collective decision-making processes, this concept fits in such an approach. It is a form of collective private partnership that allows a more inclusive form of participation of the new actors ("prosumers") in the energy sector and in the decision-making. The residents of the community collectively act as the client for their housing development project. They gain full ownership of the land and responsibility over how to design and construct the buildings. They can choose the company that manages the construction of the housing complex and team up with them for major decisions. By doing so, they manage to cut costs by an estimated 10–20%, and take a large degree of control.[179]

This model of collective partnership in the energy sector could potentially bring many benefits not only to the Netherlands, if it starts to be implemented more widely in the country, but it could be replicated in other countries where, as in the case of the Netherlands, the public is often involved in the development of local realities.

---

[178] https://www.rvo.nl/sites/default/files/Smart%20Grids%20and%20Energy%20Storage.pdf.

[179] https://www.metabolic.nl/wp-content/uploads/2018/09/SIDE-Systems-Report.pdf.

8.4 Smart Grids and Meters

We are witnessing an energy democratization (i.e., more democratic access to energy) in the decentralization of energy security governance and creation of new actors such as prosumers.[180] In other words, the fact that citizens have less dependence on energy companies for their energy security. These are all mega-trends of the twenty-first century. This means we are witnessing a paradigm shift in the governance of international economic law, broadly defined, and how citizens can play a greater role to make this transition more solid.[181] In other words, there is a shift from the core (i.e., centralized approaches to governance) to the crowd (i.e., decentralized, self-organizing approaches to governance).[182] One outcome of this shift is the deployment of smart grids.

The twentieth century electrical grid that we are using in the twenty-first century has three main components: (1) power plants that responsible for electricity production, (2) transmission lines that carry electricity across distances, and (3) distribution networks that deliver electricity end users.[183] However, the twentieth century grid is struggling with a shift to clean, renewable energy.

In the twenty-first century, we need an adaptive grid that can accommodate the fluctuations of solar and wind energy. The emerging smart grid, which is "a digital refashioning of the traditional grid with the needs of a clean energy economy in mind,"[184] engages in two-way[185] communication between the supplier and the consumer of energy to predict and adjust power supply and demand. Thanks to the internet, intelligent software, and responsive technologies, it is possible to manage the electricity flow. One great advantage of smart grids is that they can reduce energy consumption while running away from centralized fossil-fuel power plants that produce GHG emissions. In the view of the International Energy Agency, smart

---

[180]Leal-Arcas et al. (2017), pp. 139–172.

[181]See for instance the development at the sub-national level in the US, where cities and states, via their mayors and governors, are determined to implement the Paris Agreement on Climate Change, despite the decision of the federal government to withdraw from it. *See* David Lumb, *61 US Cities and Three States Vow to Uphold Paris Climate Agreement*, Engadget (June 1, 2017), https://www.engadget.com/2017/06/01/61-us-cities-and-three-states-vow-to-uphold-paris-climate-agreem/. See also an open letter to the international community and parties to the Paris Agreement from US state, local and business leaders by a bottom-up American network called *We Are Still In*, at http://wearestillin.com/. Similarly, see the role of the United States Alliance at *United States Climate Alliance*, https://www.usclimatealliance.org/, or *America's Pledge* at https://www.bloomberg.org/program/environment/americas-pledge/, both platforms committed to fight climate change. Other ways in which citizens can have a greater involvement in the energy-transition phenomenon is in solar energy, where people could install solar panels on the roof of their houses. This option would solve the delicate debate over where to place wind farms as part of the energy-transition phenomenon.

[182]For a similar approach to explain how work happens, see McAfee and Brynjolfsson (2017).

[183]"Coming attractions: Smart grids," in Hawken (2017), p. 209.

[184]Idem.

[185]Unlike the twentieth century electrical grid, which was a one-way system.

8.4 Smart Grids and Meters

grids could help to achieve net annual emissions reductions of 0.7–2.1 gigatons of CO_2 by 2050.[186]

Demand for electricity is variable depending on the season and time of the day, peaking usually in the late afternoon and in the hottest and coldest months of the year. Here is where smart grids could revolutionize the current situation: they could activate the charging of plug-in electric cars at night when demand is lowest. Doing so would reduce GHG emissions and both end users and utilities save money. An enormous investment will be required, but it will be worth it thanks to GHG emissions reduction, financial savings, and grid stability. For instance, in the case of the US, an investment of $338–476 billion in an intelligent grid system would provide a net benefit of $1.3–2 trillion over 20 years.[187]

The evolution of smart grids presents formidable challenges. The load in the electricity networks must be continually balanced to store electricity surplus during low demand spells and release it when demand increases. This can be achieved in two ways. Either through the maintenance of the supply and demand balance via market mechanisms or by means of adequate storage capacity. The low-carbon transition has been based on the proliferation of solar and wind energy. Both are intermittent in nature (which means that one would need storage capacity), thus raising the issue of what happens at the times when they underperform.[188] Moreover, it is necessary to have large empty areas to produce solar energy at a large scale, especially because solar energy is already competitive with fossil fuels in sunny places.[189]

Smart grids are energy networks that can keep track of energy flows and respond to fluctuations in energy supply and demand as needed. When smart metering systems are also installed, consumers and suppliers can receive information regarding real-time consumption. Smart meters thus enable consumers to adapt their energy uses to a range of energy prices during the day, saving money by consuming more energy when prices are lower.[190]

---

[186] IEA (2011).

[187] M. Kanellos, "Smart grid price tag: $476 billion; benefits: $2 trillion," 8 April 2011, available at https://www.greentechmedia.com/articles/read/smart-grid-price-tag-476-billion-benefits-2-trillion#gs._aWoaGw.

[188] Scientific American, "Renewable Energy Intermittency Explained: Challenges, Solutions, and Opportunities," Scientific American, 11 March 2015. [Online]. Available: https://blogs.scientificamerican.com/plugged-in/renewable-energy-intermittency-explained-challenges-solutions-and-opportunities/.

[189] See for instance the case of a floating solar farm in China, which is the largest in the world. Daley, J. "China turns on the world's largest floating solar farm," Smithsonian.com, 7 June 2017, available at http://www.smithsonianmag.com/smart-news/china-launches-largest-floating-solar-farm-180963587/. Other places where there would be potential for solar mega-farms would be the Arabian and Sahara deserts because there is a lot of sunlight and they are not cloudy.

[190] European Commission, "Smart grids and meters," available at: https://ec.europa.eu/energy/en/topics/market-and-consumers/smart-grids-and-meters.

Smart grids are also technically equipped to better incorporate renewable energy, and thus contribute to environmental protection. Overall, therefore, in pursuing its goals of decentralization, energy efficiency, and energy security, it is in the EU's interests to explore smart grids and meters. This is indeed the case, as the European Commission is promoting the modernization of electricity networks via intelligent metering systems.[191] In fact, one of the key areas of energy cooperation in the EU concerns the deployment of smart grids.

Smart grids in the EU may be the way forward to reach sustainable energy. However, the energy security, regulatory, and social and ethical aspects of smart grids in the EU first need to be assessed. We ask the question whether the level of deployment of smart grids, the degree of their current regulation, and their social and ethical dimension are adequate to make the transition to a low-carbon economy happen. There is still a long way to go before we reach a desirable outcome.

Some of the benefits of smart grids are that they create the conditions for the proliferation of renewable energy generation. They allow for the self-consumption of energy. They boost energy efficiency via demand response. They alleviate energy poverty. They lead to decreases in fossil fuel imports. They decrease dependence on unreliable oil and gas suppliers and volatile prices and they promote low-carbon energy security. However, the transition to the new energy architecture may also generate adverse results, such as higher prices, abuse of market power, and an increase in overall energy consumption.[192]

In broad terms, the planned deployment of around 200 million smart meters by 2020 in EU countries' electricity sectors fits within the abovementioned goals related to energy efficiency and sustainability. The EU is aiming to reach 20% energy savings by 2020 and 27% by 2030. The rationale is that energy efficiency has multiple benefits, including lowering energy costs for citizens, reducing GHG emissions and boosting energy security by reducing dependency on external energy suppliers.[193] To this end, where do the various member states stand when it comes to smart meter regulation and deployment? This section explores the status in Belgium, Spain, Italy, and Greece.

---

[191] Article 3, paragraph 11 Directive 2009/72/EC concerning common rules for the internal market in electricity.

[192] Since humans have a geological impact, a way to tackle the issue of increase in energy consumption is the so-called Pigou effect. To have less of something, you need to tax it in order to deal with the unsustainability problem. For an analysis of how humans have damaged the environment and how it can be fixed, see Carson (1962); Georgescu-Roegen (1971); Naess (2010); Baudrillard (1998); Yudina (2017); OECD (2011).

[193] European Commission (2017).

8.4.1 The EU Legal Basis

Historically, the first legally binding instrument mentioning smart grids was the Measuring Instruments Directive.[194] It established the requirements for the deployment and use of instruments for measuring water, gas, electricity, and heat.[195] More recently, the Third Energy Package, adopted in 2009, which seeks to further integrate the EU energy market, set out a more detailed agenda for the development of smart grids.[196] It enjoins Member States, subject to a positive cost-benefit analysis, to ensure the roll-out of smart meters. The implementation of intelligent metering systems aims to facilitate the active participation of consumers in electricity markets. Directive 2009/72/EC states that subject to an economic assessment of all the long-term costs and benefits to be conducted by September 2012, the Member States or any competent authority they designate shall prepare a timetable with a target of up to 10 years for the roll-out of smart meters.[197] Where the assessment is positive, at least 80% of consumers shall be equipped with smart meters by 2020.[198]

While the Directive is not an obligation on Member States to introduce smart grids, Article 3(10)–(11)[199] represents the legal foundation on which Member States can facilitate the development and deployment of smart grids. The Directive also includes rules designed to benefit European energy consumers and protect their rights. One of these rights is the right to choose or change suppliers without extra charges. To make this a reality, a review of the existing technical and operational landscapes, together with their attendant regulatory framework is required.

---

[194]Parliament and Council Directive 2004/22/EC, On Measuring Instruments, 2004 O.J. (L 135) 1.

[195]Papakonstantinou and Kloza (2015), pp. 41 and 46.

[196]Parliament and Council Directive 2009/72/EC, Concerning Common Rules for the Internal Market in Electricity and Repealing Directive 2003/54/EC, 2009 O.J. (L 211) 55; Parliament and Council Directive 2009/73/EC, Concerning Common Rules for the Internal Market in Natural Gas; Parliament and Council Regulation 714/2009/EC, On Conditions for Access to the Network for Cross-Border Exchanges in Electricity; Parliament and Council Regulation 715/2009/EC, On Conditions for Access to the Natural Gas Transmission Networks; Parliament and Council Regulation 713/2009/EC, Establishing an Agency for the Cooperation of Energy Regulators.

[197]Parliament and Council Directive 2009/72/EC, Concerning Common Rules for the Internal Market in Electricity, annex 1(2), 2009 O.J. (L 211) 55.

[198]Id.

[199]Id. art. 10 ("Member States shall implement measures to achieve the objectives of social and economic cohesion and environmental protection, which shall include energy efficiency/demand-side management measures and means to combat climate change, and security of supply, where appropriate. Such measures may include, in particular, the provision of adequate economic incentives, using, where appropriate, all existing national and Community tools, for the maintenance and construction of the necessary network infrastructure, including interconnection capacity."); art. 11 ("In order to promote energy efficiency, Member States or, where a Member State has so provided, the regulatory authority shall strongly recommend that electricity undertakings optimise the use of electricity, for example by providing energy management services, developing innovative pricing formulas, or introducing intelligent metering systems or smart grids, where appropriate.")

The patchwork of binding directives set out in the Third Energy Package is further supplemented by several non-binding policy instruments, opinions, and recommendations issued by various EU institutions, including the Digital Agenda for Europe (2010),[200] the European Commission's policy document "Smart Grids: from innovation to deployment," and the Commission's recommendation on the preparation for the roll-out of smart metering.[201]

8.4.2 Current Status Across Europe

According to a 2014 study conducted in 27 European states by the CEER, 42% of participating countries already had a strategic roadmap to implement smart grids.[202] Expressed in absolute numbers, 10 countries had established such plans, while 17 had not.[203] Table 8.1 provides an overview of smart grid implementation plans across European States. Specifically, Austria, Cyprus, Denmark, Finland, France, Greece, Luxembourg, and Norway published national implementation plans. In 11 of the countries, these plans were established at the national level, while in Belgium, this plan is being developed at local levels.[204] Implementation plans were not created, for example, in the Czech Republic, Slovenia, and Spain. Although Great Britain had not established an implementation plan, it did develop a high-level route-map, which is the responsibility of the national GB Smart Grid Forum.[205] There is no convergence across Europe in terms of timeframe for the implementation of smart grids. In most of them, national governments and DSOs are responsible for

---

[200] See Commission Communication to the European Parliament, the Council, the European Economic and Social Committee and the Committee of the Regions: A Digital Agenda for Europe, COM (2010) 245 final (May 19, 2010).

[201] Commission Recommendation of 9 March 2012 on Preparations for the Roll-out of Smart Metering, pp. 9–22, COM (2012) 1342 final (Mar. 13, 2012).

[202] Council of European Energy Regulators, *CEER Status Review on European Regulatory Approaches Enabling Smart Grids Solutions ("Smart Regulation")*, p. 7 (C13–EQS-57-04, Feb. 18, 2014).

[203] Since the publication of the CEER Report, Greece and Romania have implemented national programs for the roll out of smart grids. *See also* European Technology Platform SmartGrids (2016).

[204] For instance, the Flemish government approved the concept note, "Digital meters: Roll-out in Flanders," on February 3, 2017. The Flemish regulatory body VREG was asked by the Flemish government to update its earlier cost-benefit analysis on the basis of the principles of the new concept note. VREG concluded that the roll-out of the smart meters in Flanders would be a correct policy decision. *See, Kosten-batenanalyse slimme meters,* VREG (July 11, 2017), https://perma.cc/VW6J-U45C.

[205] Council of European Energy Regulators, *CEER Status Review on European Regulatory Approaches Enabling Smart Grids Solutions ("Smart Regulation")*, p. 17 (C13–EQS-57-04, Feb. 18, 2014).

Table 8.1 Development of smart grids implementation plans in European States

Country	National or local level	Details
Austria	National level	National Smart Grids Technology Platform (www.smartgrids.at), published roadmap in 2010
Belgium	Local level	Wallonia: http://www.cwape.be/?dir=4&news=122 Flanders: http://www.vreg.be/nl/nieuws/kosten-batenanalyse-slimme-meters
Croatia	No	
Cyprus	National level	
Czech Republic	No	Under construction
Denmark	National level	http://www.kebmin.dk/sites/kebmin.dk/files/klima-energi-bygningspolitik/dansk-klima-energi-bygningspolitik/energiforsyning-effektivitet/smart/smart%20grid-strategi%20web%20opslag.pdf
Finland	National level	http://energia.fi/sites/default/files/haasteista_mahdollisuuksia___ja__hiilineutraali_visio_vuodelle_2050_20091112.pdf and http://www.emvi.fi/files/Tiekartta%202020%20-%20hankkeen%20loppuraportti_15_11_2011%20(2).pdf
France	National level	Published by the Energy Agency (ADEME), current version is available at: http://www2.ademe.fr/servlet/getDoc?sort=-1&cid=96&m=3&id=84680&ref=&nocache=yes&p1=111
Germany	No	
Great Britain	No	High-level route map has been developed
Greece	National level	
Hungary	No	
Italy	National level	Incentives were deliberated by the energy authority (AEEG-SI) in 2010: http://www.autorita.energia.it/it/docs/10/039-10arg.htm The latest update concerns the second generation of smart meters, published in August 2016: http://www.autorita.energia.it/it/docs/dc/15/416-15.jsp
Lithuania	No	
Luxembourg	National level	For smart meters: http://www.eco.public.lu/documentation/etudes/2012/Etude_ComptageIntelligent.pdf
Norway	National level	www.nve.no/ams
Poland	No	
Portugal	No	
Romania	National level	http://www.anre.ro/ro/legislatie/smart-metering
Slovenia	No	Under construction
Spain	No	
Sweden	National	A roadmap with recommendations on how to stimulate the deployment of smart grids for the years 2015 to 2030 is currently under construction by the Swedish Coordination Council for Smart Grid (http://www.swedishsmartgrid.se). Due date December 2014

(continued)

Table 8.1 (continued)

Country	National or local level	Details
Switzerland	No	
The Netherlands	No	There is a vision document from the Taskforce Smart Grids established by the Ministry of Economic Affairs: http://www.rijksoverheid.nl/documenten-en-publicaties/rapporten/

Source: Adaption and update of CEER Status Review on European Regulatory Approaches Enabling Smart Grids Solutions ("Smart Regulation"). C13-EQS-57-04, 18-Feb-2014, pp. 42–43

implementation, while National Regulatory Authorities (NRAs) have monitoring functions.[206]

As far as actual implementation is concerned, Italy is a forerunner. Italy has completed smart metering implementation covering 99% of electronic metering points.[207] The DSO is the owner and responsible party for implementing the smart grid and for guaranteeing power quality.[208] Remarkably, the Italian implementation is not merely aimed at achieving a roll-out of AMIs, but envisages their progressive improvement. For instance, given that the low voltage remote control meters that were first rolled out in 2001 have a lifespan of 15 years, the first replacement campaign was launched in 2016.[209] These first generation (1G) meters have since reached their end-of-life. True to expectation, some companies have started installing 2G meters. The Italian experience is also a regulatory paragon because the law laid down functional specifications for 2G meters and identifies some crucial criteria. The requirements include: 2G meters, once installed, shall remain in operation, presumably, for another 15 years; and, over this period, they must be able to support every electric system transformation, such as the new distributed production paradigm and the changes of the electricity market.[210]

Other countries, such as Spain, have not developed an implementation plan for smart grids. Yet, the roll-out of smart meters is ongoing and is planned to be completed by 2018.[211]

---

[206] *Id.* at 7, 17.

[207] *See* Carmen Gimeno, GEODE Workshop Presentation: From Theory to Reality: Overview of Smart Meters Roll-Out Across Europe (Mar. 20, 2014); *Cost-Benefit Analyses & State of Play of Smart Metering Deployment in the EU-27*, p. 33, C (2014) 189 final (June 17, 2014).

[208] The metering activity in Italy is regulated by the Regulation ARG/elt 199/11 (TIT).

[209] Press Release, Enel S.p.A., Enel Presents Enel Open Meter, The New Electronic Meter (Sept. 5, 2016), https://www.enel.com/en/media/press/d201606–enel-presents-enel-open-meter-the-new-electronic-meter.html.

[210] Italian legislative decree 102/2014; Autorita per l'energia elettrica il gas e il sistema idrico (AGEESI), *Smart Metering Second-Generation Systems for the Measurement of Electricity in Low Voltage* (Aug. 6, 2015) http://www.autorita.energia.it/it/docs/dc/15/416–15.jsp#.

[211] *Cost-benefit analyses & state of play of smart metering deployment in the EU-27*, p. 35, C (2014) 189 final (June 17, 2014).

8.4 Smart Grids and Meters

With a view to promoting smart grids, many Member States have adopted regulatory incentives. In the CEER study, 79% of the countries were found to use tools for price regulation and 63% use performance indicators. In contrast, tools to regulate the provision of information, charges, and licensing are used significantly less. In most of the countries (76%), regulatory instruments will need to be adapted to facilitate the deployment of smart grids.[212] For example, in Belgium, as of 2018, Atrias will provide a new clearing house with new MIG6 market protocol implementation. This means that from 2018 onwards, new market models for prosumers with PV < 10 kW peak will be established, making dynamic tariffs and sale of injection possible.[213] In Great Britain, the value of demand side flexibility for the electricity system will have to be reflected in the incentives to invest in smart grids.[214] In Lithuania, reaping the benefits of smart grids and managing related data privacy issues will require amendments to the current regulatory framework. In Italy, an "input-based" type of incentive regulation has been used for the transmission network as well as to support smart grid pilot projects in distribution networks. In Poland, in order to assess the benefits of smart metering for consumers, two new performance indicators were introduced. In Spain, the deployment of smart meters is ongoing, and it is viewed as a necessary step towards the development of smart grids. As part of Spain's efforts, the low voltage code has been proposed to be changed and a new discriminatory tariff that, thanks to smart meters, promotes charging of EVs at times of lower demand and prices has been established.[215] Despite what appear to be wide-spread attempts at regulatory reform within the continent, some actors in some of these market believe that regulatory reform may not be necessary as the current regime already provides an enabling ground for smart grids.[216] While this may be true in some cases, the reality is that the existing regimes for electricity regulation are skewed towards the traditional grid and do not take into account the dynamic nature of smart grids.[217] Consequently, if smart grids are to be afforded an opportunity to enter what is currently often an oligopolistic market, regulatory reform will be essential.

Given that smart grids are largely experimental, demonstration projects have played a pivotal role in the development and deployment of the new technologies developed. Different countries in Europe have adopted various approaches towards

---

[212]Council of European Energy Regulators, *CEER Status Review on European Regulatory Approaches Enabling Smart Grids Solutions ("Smart Regulation")*, p. 14 (C13–EQS-57-04, Feb. 18, 2014).

[213]*Atrias and MIG6.0: Towards a New Energy Market Model in Belgium,* Energy Outlook by Sia Partners (July 12, 2017) https://perma.cc/D5CA-CJ9J.

[214]Council of European Energy Regulators, *CEER Status Review on European Regulatory Approaches Enabling Smart Grids Solutions ("Smart Regulation")*, p. 14 (C13–EQS-57-04, Feb. 18, 2014).

[215]*Id.*

[216]*Id.*

[217]Veldman et al. (2010), pp. 287–289; *see also* Adrian de Hauteclocque & Yannick Perez, *Law & Economics Perspectives on Electricity Regulation* (EUI Working Paper RSCAS 21, 2011).

promoting these demonstration projects. 61% of countries which participated in the CEER study use a combination of sources for funding.[218] 56% of the countries have been funding demonstration projects through industry funding, public funding institutions, the European Commission, and integrated municipal energy suppliers.[219] In 61% of the countries, governments are responsible for making decisions about granting funds.[220] For example, Finland passes costs onto consumers to a certain extent, but also adopts efficiency targets for companies.[221] Italy uses a cost-benefit indicator to select projects.[222] Austria finances demonstration projects through a combination of funding from industry, public institutions, and the national budget.[223] The federal government established the Climate and Energy Fund (Klima- und Energiefonds—KLIEN) to support the implementation of the climate strategy. KLIEN is responsible for providing most of the funds for demonstration projects.[224] Remaining costs are audited and covered through network charges during the regulatory period, with the application of efficiency targets. Great Britain does not apply efficiency targets to demonstration projects.[225] However, a key criterion for awarding funding is the project's value for consumers and its long-term efficiency. The NRA, rather than the government, is responsible for most decisions.[226]

Regarding more general incentives to encourage DSOs to adopt and fund smart grid innovation projects and how they are funded, most European countries use a combination of regulatory mechanisms, national government initiatives, and European initiatives. 63% of the countries assessed by CEER use general incentives not specific to smart grids to promote the development of smart grids.[227] For example, Austria incentivizes cost reductions through efficiency targets that do not distinguish between traditional and smart grids. As a result, regulated companies favor smart solutions when they are more cost efficient than other alternatives.

---

[218]Council of European Energy Regulators, *CEER Status Review on European Regulatory Approaches Enabling Smart Grids Solutions ("Smart Regulation")*, p. 19 (C13–EQS-57-04, Feb. 18, 2014).

[219]*Id.*

[220]*Id.* at 20.

[221]*Id.*

[222]*Id.*; World Energy Council, "World Energy Perspective: Smart Grids: Best Practice Fundamentals for a Modern Energy System," 14–15 (2012).

[223]Council of European Energy Regulators, *CEER Status Review on European Regulatory Approaches Enabling Smart Grids Solutions ("Smart Regulation")*, p. 20 (C13–EQS-57-04, Feb. 18, 2014); Energy Research Knowledge Centre, *SETIS Energy Research – Austria*, European Comm'n (Sept. 5, 2017), https://perma.cc/NEW8–UYGQ.

[224]*Id.*

[225]Council of European Energy Regulators, *CEER Status Review on European Regulatory Approaches Enabling Smart Grids Solutions ("Smart Regulation")*, p. 20 (C13–EQS-57-04, Feb. 18, 2014).

[226]*Id.*

[227]*Id.* at 21.

8.4 Smart Grids and Meters

Belgium has not yet specifically defined incentives, while Cyprus currently has no incentives in place. In the majority of countries, incentives for DSOs to innovate are funded through distribution network charges. National and European funding is also used to a significant extent. Many European countries adopt a combination of sources of funding. For instance, Austria, Finland, Italy, and France use network charges, national funding, and European funding. The Netherlands, Poland, and Norway use network charges as well as national funding. Lithuania and Slovenia use network charges and European funding. Greece and Spain use European as well as national funding.[228]

Finally, with regard to issues of data privacy and security, it is a commonly held view that the technology associated with smart grids poses significant risks to data privacy and cybersecurity; both require concerted regulatory reform if these risks are to be adequately managed.[229] However, according to the CEER status review on European regulatory approaches enabling smart grids solutions, there is no clear consensus about whether NRAs for the energy sector will and should be responsible for data security regulation in relation to smart meter data.[230] Be that as it may, different countries are considering different proposals and approaches for dealing with the problem of data protection and security for smart grids. For example, in the UK, data aggregation plans will be proposed by the DSO and then approved by the NRA, and data privacy requirements will be regulated in the context of license conditions.[231] In Slovenia, a cost-benefit analysis carried out by the NRA will also look at security issues.[232] In Spain, energy suppliers are precluded from having access to any information other than that of their own customers.[233] In contrast, in the Czech Republic, the Office for Personal Data Protection is responsible for data security.[234] Similarly, in France there is a separate and dedicated agency with competence over data security. In Germany, this is the responsibility of the Federal Office for Information Security.[235] Finally, in countries such as Belgium and the Netherlands, the NRA for the energy sector and the Data Protection Authority will work jointly on data security issues.[236]

---

[228] *Id.*

[229] Int'l Energy Agency, Technology Roadmap: Smart Grids 16 (2011) https://www.iea.org/publications/freepublications/publication/smartgrids_roadmap.pdf; Hörter et al. (2015), pp. 291 and 297.

[230] Int'l Energy Agency, Technology Roadmap: Smart Grids 15–16 (2011) https://www.iea.org/publications/freepublications/publication/smartgrids_roadmap.pdf.

[231] *Id.* at 16.

[232] *Id.*

[233] *Contadores Inteligentes y Protección de Datos*, EnerConsultoría (Dec. 8, 2015) http://www.enerconsultoria.es/BlogLeyesEnergia.aspx?id=36002236&post=Contadoresinteligentesyprotecciondedatos.

[234] Council of European Energy Regulators, *CEER Status Review on European Regulatory Approaches Enabling Smart Grids Solutions ("Smart Regulation")*, p. 16 (C13–EQS-57–04, Feb. 18, 2014).

[235] *Id.*

[236] *Id.*

Table 8.2 Data privacy and security regulation in European countries

	In control and informed	In control and not informed	No control over data	Not available
Free	AT, BE, DK, FI, FR, DE, GB, IE, IT, LU, NO, PL, NL		CY, CZ, EE, IS, RO, SI, ES, SE	LT, PT
Not free				

Source: CEER Status Review of Regulatory Aspects of Smart Metering. C13-RMF-54-05, 12 September 2013, p. 16

The ERGEG Guidelines of Good Practice on regulatory aspects of smart metering recommend that:

> it is always the customer that chooses in what way metering data should be used and by whom, with the exception of metering data required to fulfil regulated duties and within the national market model. The principle should be that the party requesting information shall state what information is needed, with what frequency and will then obtain the customer's approval for this. Full transparency on existing customer data should be the general principle.[237]

From the CEER status review of regulatory aspects of smart metering, Table 8.2 shows that many European countries indeed provide customers with information about, and ensure control over, their metering data, free of charge. However, the same table also shows that, in a number of countries, customers are not given control over their own data.[238]

8.4.3 The Situation in Certain EU Member States

Belgium is still at the planning stage when it comes to smart meter systems, as can be seen by the fact that little relevant regulation exists at this point. Indeed, Belgium lacks a strategic plan for their widescale deployment. Currently, smart grid projects are few and far between. Thus far, the country's Flemish region appears to have made the most progress on the smart meter front. When a full roll-out does occur, the application of smart grids may prove attractive in an electricity market seeking cost effective alternatives in the face of the anticipated transition away from nuclear energy.

In the Flemish region, the Decree of 14 March 2014 transposing Directive 2012/27/EU and amending the Decree of 8 May 2009 on energy policy regulates smart

---

[237]European Regulators Group for Electricity & Gas, *Final Guidelines of Good Practice on Regulatory Aspects of Smart Metering for Electricity and Gas*, p. 12 (E10–RMF-29–05, Feb. 8, 2011).

[238]Council of European Energy Regulators, *CEER Status Review on European Regulatory Approaches Enabling Smart Grids Solutions ("Smart Regulation")*, p. 16 (C13–EQS-57–04, Feb. 18, 2014).

8.4 Smart Grids and Meters

meters in open-ended terms. Article 4.1.22/2 of the Decree of 8 May 2009 sets out the basic principles. Firstly, the Flemish government will identify when DSOs can deploy smart meters. Secondly, in case of a smart meter being installed, DSOs are responsible for providing consumers with detailed information regarding their rights, obligations, and the technology's full scope. Thirdly, the Flemish government will determine the mandatory criteria for smart meters. Fourthly, the Flemish government will decide how to share data from smart meters. Lastly, the decree states that parties receiving data from smart meters are responsible for conforming with relevant data protection regulation.

In spite of the lack of progress on smart meter rollout across the country, an interesting and relevant project in the works is a federal clearing house for energy. The goal behind the clearing house is to create a new energy market model that will factor in emerging technologies including, of course, smart meters and decentralized approaches to energy provision.[239] The clearing house will be instrumental in streamlining regional smart meter initiatives and will thus contribute to more coherent energy governance overall in the country.

In a further push towards energy efficiency by the EU, two European Commission directives—Directive 2009/72/EC and Directive 2009/73/EC—require cost benefit analyses (CBA) by each Member State to determine whether a countrywide rollout of smart metering systems would be economically viable, in which case Member States must propose timeframes for proceeding and, in fact, provide 80% of consumers with intelligent metering systems by 2020.[240] Belgium's CBA—for which each region carried out individual CBAs—showed generally negative findings.[241] Below follows a detailed look at the CBA for the Flemish region.

The VREG was in charge of the Flemish region's CBA, which considered a joint smart meter deployment for electricity and gas. The CBA took place in 2012 and, in the case of electricity, led to a positive outcome. However, it was updated in 2014 at which time it projected a net cost of €157 million.[242] The viability of smart meters is still under consideration, however, and local authorities are operating pilot projects across the country, which are providing valuable insights with regard to the technology. This data will help with the success of an eventual widescale rollout of smart meters.[243]

Currently, the Flemish region has implemented around 50,000 pilot smart meter projects, installed by Eandis and Infrax, the DSOs for the region. These pilot projects have taken on board input from various stakeholders and are seeking ways to

---

[239] USmart Consumer Project (2016).

[240] Annex I, point 2 Directive 2009/72/EC concerning common rules for the internal market in electricity.

[241] Commission staff working document country fiches for electricity smart metering. Report form the Commission benchmarking smart metering deployment in the EU-27 with a focus on electricity, p. 108, SWD(2014) 188 final (17 July 2014).

[242] ICCS-NTUA and AF-MERCADOS-EMI, "Study on Cost Benefit Analysis of Smart Metering Systems in EU Member States," ICCS-NTUA and AF-MERCADOS-EMI, 2015.

[243] VREG, "The future of smart metering in Flanders/Belgium," VREG, 2013.

decentralize electricity generators and find the most appropriate grid areas in which to integrate RE.[244] On 14 July 2017, the Flemish government released a draft decree calling for the segmented deployment of smart grids. Starting 2019, a progressive rollout of intelligent metering systems will start in the region, replacing all conventional meters (up to 56 kV). As of September 2017, advisory councils were assessing the draft decree, with a final version to be submitted to the Flemish Parliament.[245] At the time of writing, therefore, the Flemish region's legal framework for intelligent metering systems is still work in progress.

Spain has a broad range of legislation addressing smart metering systems. The Royal Decree 809/2006 of 1 July, for one, made it mandatory, starting 1 July 2007, for new electricity meters to allow differentiation of consumption in different timeframes, as well as remote management for consumers with a contracted power below 15 kW. In addition, the Royal Decree 1110/2007 of 24 August ruled that all household electricity meters must be smart meters. It additionally delineated the functions of smart metering systems.

The Royal Decree 1634/2006 of 29 December and the Ministerial Order ITC/3860/2007 of 28 December map out the smart metering deployment plan in Spain (2011–2018). The legislation establishes how smart meters will be deployed for around 28 million supply points. DSOs will undertake the roll-out (with Endesa, Iberdrola, Gas Natural Fenosa, EDP and Viesgo the main operators). The steps will impact electricity consumers with a contracted power below 5 kW. Smart meters were anticipated to replace conventional meters by the end of 2018. The NRA estimates that by June 2016, around 17.54 million smart meters had already been installed,[246] which means that smart meters already comprise more than 62% of existing meters in the country.

The Royal Decree 216/2014 of 28 March is also significant in this context, as it establishes final prices for electricity consumers in terms of real metering on an hourly basis. The decree requires distributors to submit hourly energy consumption data. The legislation addresses two situations: (1) where the household already utilizes a smart meter and (2) where the household still utilizes a conventional meter. With a smart meter, consumers can opt for billing based on their real hourly energy consumption data. However, some experts believe that pricing based on hourly real consumption data is too advanced for today's smart meters in Spain. Such billing will involve huge quantities of data while also handling much more data submission and processing than before.[247] With a conventional meter, bills reflect the average national hourly cost, with variable prices. The Royal Decree 216/2014

---

[244]Brugel (2013).

[245]Vlaanderen, "De digitale energiemeter," Vlaanderen. Available at: https://www.vlaanderen.be/nl/bouwen-wonen-en-energie/elektriciteit-aardgas-en-verwarming/dedigitale-energiemeter.

[246]Comisión Nacional de los Mercados y la Competencia, "El 62% de los contadores analógicos ya han sido sustituidos por contadores inteligentes," Nota de Prensa, 2017.

[247]Leiva (2016).

poses an alternative to both situations: a fixed price for 1 year of energy consumption[248] (as long as the household's contracted power is less than 10 kW).

Spain's legislation is in keeping with the EC's approach to smart metering. The Third Energy Package encourages Member States to benefit consumers by installing smart metering systems. The Energy Efficiency Directive[249] calls for modernizing energy services by data from smart meters, demand response and dynamic prices. In fact, Commission Recommendation 2012/148/EU[250] guides Member States on the transition to smart meters, identifying steps to achieve technical and commercial interoperability or to incorporate them in the future.[251]

Spain has kept pace with all the functionalities that the EU has identified toward achieving technical and commercial interoperability, with the exception of recommended functionality.[252] In this case, the functionality requires a sufficient frequency at which consumption data can be updated and made available to the consumer (and a third party the consumer identifies) to in order to save energy. The recommendation calls for an update rate of every 15 min. The European Commission admits that this recommended functionality is the most challenging, but also labels it the most effective, in that it enables consumers to make informed decisions on their consumption patterns.[253] It offers a bottom-up system that gives consumers control while they do their part towards energy efficiency and security.

Smart metering and billing are essential for maximizing demand response, which rises in significance when one factors in the projected rise in the share of RE on the national grid. As opposed to maintaining or building new power plants and network lines to balance demand and supply, energy efficiency and demand response offer more effective approaches.[254] Thus, real-time energy consumption data is essential for maximizing energy efficiency and consumer savings. Through increased awareness of their own consumption patterns, consumers can alter their habits to conserve

---

[248]USmart Consumer Project (2016).

[249]Directive 2012/27/EU of the European Parliament and of the Council of 25 October 2012 on energy efficiency, amending Directives 2009/125/EC and 2010/30/EU and repealing Directives 2004/8/EC and 2006/32/EC.

[250]Commission Recommendation 2012/148/EU of 9 March 2012 on preparations for the roll-out of smart metering systems.

[251]Report from the Commission "Benchmarking smart metering deployment in the EU-27 with a focus on electricity", p. 6, COM (2014) 356 final (17 June 2014).

[252]Commission staff working document country fiches for electricity smart metering accompanying the document Report form the Commission benchmarking smart metering deployment in the EU-27 with a focus on electricity", SWD(2014) 188 final (17 July 2015).

[253]Report from the Commission "Benchmarking smart metering deployment in the EU-27 with a focus on electricity", p. 6, COM (2014) 356 final (17 June 2014).

[254]Communication from the Commission to the European Parliament, the Council, the European D1.1 Legislation, business models and social aspects 152 Economic and Social Committee and the Committee of the Region, p. 5, COM (2015) 339 final (15 July 2015).

energy. Estimates show that household energy consumption could decrease by up to 9% through smart meter deployment.[255] Smart meters are revolutionizing the electricity market by placing consumers at the forefront and enabling them to lower their bills via demand response.

In Italy, the country's DSO handles metering activity, and also owns and manages smart meter deployment. Additionally, it is responsible for releasing meter data to third parties. Italy has already successfully implemented smart meters, with 99% of electronic metering points being covered. Regulation in December 2006 called for the installation beginning in 2008 for electronic meters with minimum functional requirements for all DSOs and LV consumers, to be complete at 95% by 2011.

Regulation dating back to 2006 identifies the minimum functionality requirements for electric meters. It called for all DSOs and LV consumers to install new meters starting 2008. Thus, by 2016, Italy had a smart meter coverage of 99%. While the NRA has overall responsibility for smart deployment, DSOs have direct responsibility as they own and operate metering facilities. DSOs are also in charge of maintenance, meter reading and data management.

In Italy, deploying smart meters was a step towards achieving overall goals related to boosting competition in the electricity market, increasing access to remote meters, and better data collection for cost-benefit analyses.[256] To this end, the AEEGSI stated that, at minimum, meters should:

- Record active energy consumption;
- Include alarms that notify customers when they go beyond contractual limits before circuit breakers are activated;
- Have the capacity to prevent data loss;
- Include technology (ideally GPS-based) to coordinate with clocks and calendars;
- Undertake remote transactions such as periodic reading of consumption data, software upgrades and remote management of customers' contractual details;
- Include display screens supplying information such as consumption, and pricing details.

Specifying the above minimum requirements via regulation meant that AEEGSI could guarantee a high quality of functionality and customer service. For example, smart meters let customers continue using minimal electricity amounts (0.5 kW for households) even when not paying, before service cancelation. Upon payment, the smart meter enabled suppliers to reconnect the customer remotely, generally within 24 h of payment. If reconnection does not occur within 24 h, customers receive automatic compensation.[257] Moreover, Italian smart meters also use time-of-use (ToU) pricing for LV customers, which places household consumption within

---

[255]European Commission, "Smart grids and meters," European Commission. Available at: https://ec.europa.eu/energy/en/topics/market-and-consumers/smart-grids-and-meters.
[256]Zhou and Brown (2017), pp. 22–32.
[257]F. Villa, "Regulation of Smart Meters and AMM Systems in Italy," in 19th International Conference on Electricity Distribution (CIRED), Vienna, 2007.

8.4 Smart Grids and Meters

three bands: peak, mid-level and off-peak, each with a different price. Thus, consumer bills are based on how much electricity they consume at different times of the day. Such an approach to pricing impacts only the energy component of the bill, and not network charges. It provides customers with more power to lower bills; this is even more relevant when one considers that the energy component of bills makes up three-quarters of total billing for many.[258]

In short, Italy has made impressive progress on smart meter deployment, mainly thanks to the DSOs complying with the NRA's plans after numerous consultations. In fact, Enel D, one of the DSOs, had initiated its own smart meter deployment strategy prior to the NRA's. The NRA also established a fine for DSOs that did not comply with the regulation, which was a key factor toward their cooperation. Establishing minimum functional requirements has also been extremely effective, not only guaranteeing that consumers receive standardized service, but also helping towards interoperability and uniformity. clearly, the NRA noted the role of smart meters towards progress on smart grids and made sure that technical disadvantages did not block the smart grid rollout.

Furthermore, Italy has been exploring the potential for smart homes systems. Given the move towards decentralized energy systems based on variable RE generation, the rising cost of energy and the rapid proliferation of smart grid innovation, there is a need to promote demand response options. One crucial player in this context could be the advancement of smart home systems.

When it comes to demand response, HEMS (home energy management systems) have a few chief objectives:

- Adjusting consumption or generation of energy to keep an equilibrium in generation, e.g., via photovoltaic (PV) or micro-CHP (combined heat and power) units and in usage (e.g., via smart lighting and home appliances). Maintaining such an equilibrium could impact home generation and consumption (for instance, to boost self-consumption from RE) or that at the distribution grid level. Energy storage, although costly, could also help with load balancing. Unfortunately, though, it is impossible to prevent storage losses. Thus, availing of existing storage (such as electric vehicles, heat capacity of buildings) is a better option. Using flexible loads offers an alternative to energy storage but, once more, electric loads with the most potential for balancing generally include a type of storage (e.g., heat pumps).
- Reducing energy consumption through device adjustments, for instance by altering temperatures in unused rooms to lower energy consumption, and by switching off elements of heating/cooling systems when possible to decrease stand-by losses.

User acceptance is a key factor for HEMS, in contrast to energy management in the commercial and industrial sectors. HEMS users are generally those living in the home, who are not likely to have technical know-how. Thus, HEMS functionality

---

[258]Idem.

must be cost effective, but also reliable and easy to use for all. In addition, user habits and needs should be factored when planning appropriate devices.

HEMS have a range of market drivers, including:

- Incentives for promoting energy efficiency and sustainable development, incorporated into national and EU laws and policy.
- Strong demand side management that incorporates household loads and generators, decreasing the need to expand the network and building the network's ability to integrate RE.
- Promoting environmental sustainability and decreasing GHG emissions.
- Boosting household energy self-reliance while also increasing openness and fairness.
- Lowering energy prices by lessening energy consumption.
- Promoting two upcoming and highly linked areas, i.e., next-generation media and smart grids.

HEMS also face crucial challenges at this time, which include:

- High cost of investment and operation.
- The need for smart metering facilities to facilitate connection with the grid operator.
- No widescale deployment thus far for "Home area automation controllers."
- Few available interoperable smart appliances with open control functionality.
- No capacity to communicate between smart home and smart grid actors such as the DSO, virtual power plant and market aggregators, that would help the end consumer in automatic energy procurement.
- A need for different manufacturers to coordinate on solutions to be controlled by the HEMS.
- A need to tackle data privacy through effective measures towards confidentiality, integrity and data transfer between end user and all market actors.
- An overall low interest in smart home technicalities on the part of potential users.
- The fact that end users lose control and decision capability when allowing grid operators to control their loads.
- The risk of HEMS being hacked and the resulting loss of privacy.

In Greece, the Hellenic Electricity Distribution Network Operator (HEDNO) is managing the smart meter deployment project, in keeping with a 5-year national strategy to "smarten" the country's grid. Smart meters have already been placed at important low-voltage (LV) customer locations and also at medium-voltage (MV) customer sites. HEDNO has also installed two telemetering centers, one to collect remote meter readings from all MV customers and RES producers, and the other to collect remote meter readings from all major LV customers (>55 kVA) including photovoltaics (PV). Following the success of initial rollout projects, HEDNO is assessing bids for the pilot installation of around 200,000 smart meters for residential and small commercial customers across selected areas of the country. HEDNO is legally obliged to ensure 80% of consumers are part of a telemetering system by the end of 2020. However, according to current projections, this does not

8.4 Smart Grids and Meters

appear to be possible. Thus, a new timeline is being planned, to adopt a new legislation.[259] Notably, telemetering meters that have already been set up in Greece cover 45% of the installed power, at all voltage levels.

Commission Recommendation 2012/148/EU and Ministerial Decision, GG B 297/13.2.2013 contain legal requirements for minimum technical functions of smart meters, to be complied with during pilot rollouts and smart meter procurement. The aim of the rollouts is that the installed meters have the capacity to communicate with a central system using general packet radio services (GPRS). The pilot rollout for the 200,000 residential and small commercial customers calls for communication between the central system and the meters to employ a combination of GPRS/3G and powerline (PLC). HEDNO's rollouts, on the other hand, are based on International IEC standards, EU Directive 2004/22/EC (MID and available meter technologies at tender time and substantially different customer types. Thus, they employ slightly different technical specifications. For key LV customers power consumption must be measured using current transformers (CT), for small LV customers power consumption must be measured using direct connection (DC), and for MV customers power consumption must be measured using current transformers (CT) and voltage transformers (VT).

In keeping with Commission Recommendation 2012/148/EU,[260] the smart meters that have been deployed in Greece are equipped with functionalities aimed at making interoperability easier and maximizing benefits to consumers. These include pulse outputs for real-time consumption monitoring; remote reading via AMR/AMI; two-way communication; interval metering at 15-min intervals; and remote metering amongst others. Future plans, within the pilot rollout, include alarm capability, as well In-Home display, to alert consumers of exceptional energy use.

While HEDNO is the main body in charge of the smart meter roll-out in Greece, the program does, in certain situations, allow for private participation. For example, customers with PV installations may, in order to participate in the net metering scheme, purchase and install PV meters from a selection of HEDNO-approved meters and modem models. The suppliers of such equipment are approved beforehand by HEDNO based on technical requirements. Similarly, PPC Technical Guideline 5/1974 permits property managers to purchase and install sub-meters for individual tenants. While the PPC rule entered into force before the smart meter rollout, it is worth extending it to cover smart meter rollout. This approach would not only involve the public in the rollout, but would help to liberalize the smart meter market.

HEDNO's stance on standardization, although commendable, does not appear to address future growth in the sector. For instance, the minimum technical standards described above appear to relate only to the pilot stages of the rollout. Thus, they

---

[259]Regulatory Authority for Energy, "Hellenic Electricity Distribution Network Management Code," Ministerial Decision 395/2016, Official Gazette 78/ B /2017.

[260]Ministry of Environment, Energy & Climate Change, Ministerial Decision, GG B 297/13.2.2013 "Intelligent Systems Development of Measurement to HEDNO."

only require that the equipment meet a certain standard. However, as recommended by the European Regulators Group for Electricity and Gas (ERGEG),[261] the standardization of metering should cover the system level and not only deal with the equipment level. Granted, ERGEG's recommendation is based on the market's liberalization, and aspires to a situation where there is freedom to employ a range of different technologies while ensuring interoperability. Still, arguably, HEDNO is working towards determining ideal and workable systems, so at this stage system level standardization is not feasible. Also, given that HEDNO is the only meter operator, equipment level requirements are necessary to make sure the current and future AMR/AMI systems and smart meters are compatible. In any event, the fact that there is no system level standard means that developers of the smart grid architecture will have plenty of flexibility when it comes to planning the design of relevant telecommunication systems, hardware and software. However, it is important that the design choices related to the project promote the overall aims of HEDNO as well as Greece's national smart grid implementation policy.

It is worth pointing out that consumer-centric minimum functionality requirements are absent from HEDNO's technical requirements. Regardless, several aspects of HEDNO's technical requirements happen to guarantee data protection and security and help towards building competition in the retail markets. It is likely that the pilots will result in important lessons that can help towards designing an appropriate level of consumer-centric standards, such as requirements that would help the wide-ranging deployment of demand response tools in the future smart grid of the country. The fact that Greece's market is still evolving at various stages presents a unique opportunity to shape the country's market by providing evidence-based examples to help identify the relevant functionality requirements.

Smart metering is a technology that could help promote additional market liberalization in the country and facilitate participation of the end-user in demand response schemes. As of now, smart meters have been set up for all MV customers and are being installed for large LV customers (between 55 kVA and 250 kVA). Simultaneously, authorities have set up the necessary AMI infrastructure to help support their functioning. In keeping with EU goals, HEDNO has started pilot projects to test the new technologies and assess their benefits, so as to further broaden the installation of smart meters on customer premises. In addition, a 200,000 LV customer pilot rollout is in the planning stages. However, this has been delayed due to legal matters arising out of tendering procedures.

Greece is also exploring the potential of smart homes systems. In homes across the country, there is a slow build-up of automated systems and smart home devices, with the goal of boosting the structure's energy efficiency and improve the quality of life for those residing there. The main appliances in this context include heating and HVAC systems, security systems (and alarms), appliances to track indoor environmental quality, shutters, lighting control systems, smart entertainment devices and

---

[261]"Commission Recommendation of 9 March 2012 on preparations for the roll-out of smart metering systems (2012/148/EU)."

smart household electrical appliances. According to the Applied Code of Energy Efficiency in Buildings, houses must have some type of automation system for heating and HVAC systems functioning in order to score a high rating. Still, setting up and utilizing smart appliances—whether smart thermostats, wifi-controlled air-conditioners, smart coffee-makers, etc.—depends on how willing end users are to avail of such technologies, as well as on their preferences and lifestyles. More advanced smart homes systems even use advanced protocols that can allow combinations of scenarios in a home, to monitor its performance in different rooms remotely. In Greece, certified engineers are working on the implementation and integration of such systems. Still, although detailed data is not available, it appears that, depending on the application, smart device usage in Greek households is still largely at a preliminary stage.

The WiseHOME application, in conjunction with smart devices and systems, would permit household customers (consumers and prosumers) to play an active role in monitoring and controlling their energy consumption, participate in the market, and thus better manage their energy costs. WiseHOME could thus act as a key incentive to encourage residents to actively manage their consumption, and so reduce their energy bills, supporting self-consumption by means of real time data, demand response and load optimization schemes. This scenario, of course, depends on activating demand control contracts at low-voltage levels. On this basis, prosumers would receive RES forecasts that would help them plan their energy usage for the following day in a more efficient manner.

Progress on smart grids is crucial from another standpoint, too: the widespread promotion of electric vehicles (EVs). While the potential proliferation of EVs is environmentally interesting, the main concern relates to their charging, in that a large number of consumers charging their EVs may place a heavy demand on the grid. However, a smart grid has the capability to respond to and mitigate load impacts and allows consumers to understand better the costs and impacts of charging. Integrating EVs with smart grids, thus, will be an important step in ensuring the widespread diffusion of EVs across the EU.

In the case of France, the introduction of intelligent metering systems is one of the measures identified in Article 3, paragraph 11 of Directive 2009/72/EC concerning common rules for the internal market in electricity as contributing to energy efficiency at the EU level. Both Directive 2009/72/EC and Directive 2009/73/EC establish a framework providing for: (1) the performance of a cost-benefit analysis (CBA) to ascertain whether the national deployment of smart metering systems would be economically sound in the different Member States; (2) where suitable, the drafting of a timeframe for their roll-out; and (3) where the CBA leads to positive results, the establishment of a goal of attaining the supply to 80% of users with smart meters by 2020. France is one of the few EU Member States (along with Spain) where the main distribution system operator (DSO) has chosen to engage in a large scale roll-out of smart meters between December 2018 and 2021.[262]

As opposed to the situation in other EU Member States, such as the United Kingdom, France is running on schedule as regards European small-metering deployment obligations[263]: France has rapidly engaged in the deployment of smart-meters, and between 2015 and 2021, ERDF (*Electricité Réseau Distribution France*, the French grid manager) will deploy, through its subsidiary ENEDIS, over 28 million units of "Linky," France's smart meter.[264] Projections indicate that France could reach 95% of its digital meter deployment by 2020.[265] As regards the underlying infrastructure model, in France:

> the Linky smart meters will communicate with data concentrators through powerline carrier technology. Data will then transfer to a central information system using telecommunications network such as GPRS. ERDF, the DSO responsible for the electricity distribution activities of 95% of French municipalities, is responsible for the implementation and ownership of the rollout as well as for third-party access to metering data.[266]

Thus, the overall distribution-network landscape in France works under the following coordinates: "*1.285 million km of MV and LV power lines, 35 million meters, 345 TWh of electricity distributed in 2011 of which 18.9 TWh are injected directly into the distribution network from distributed sites including wind, solar PV, biomass and small hydroelectric stations.*"[267]

The response to this panorama, from the industry and (private) research sectors has been led by "SmartGrids French Clusters," a consortium of nine French business and research clusters, specialized in the area of energy and ICT, which intends to "foster collaboration and exchange with the view to building up and highlighting the value of the French SmartGrids sector."[268] They self-identify as the first consortium of its type in Europe, and assert that they intend to "*convey the vision and the strategic orientations of the business and research clusters, which have a regional*

---

[262]European Parliament, Briefing *Smart Electricity Grids and Meters in the EU Member States* (Sep. 2015), http://www.europarl.europa.eu/RegData/etudes/BRIE/2015/568318/EPRS_BRI%282015%29568318_EN.pdf.

[263]Metering & Smart Energy International, available at: www.metering.com/features-analysis/smart-meters-101-frances-linky-electricity-meters.

[264]Metering & Smart Energy International, available at: www.metering.com/features-analysis/smart-meters-101-frances-linky-electricity-meters and *The Guardian,* "Is your smart meter spying on you?" available at: www.theguardian.com/money/2017/jun/24/smart-meters-spying-collecting-private-data-french-british.

[265]Metering & Smart Energy International, "Smart meters 101: France's Linky electricity meters," available at: www.metering.com/features-analysis/smart-meters-101-frances-linky-electricity-meters.

[266]*Ibid.*

[267]*Ibid.*

[268]Global Smart Grid Federation, June 2013 newsletter, available at: www.globalsmartgridfederation.org/wp-content/uploads/2014/03/13-05-juni.pdf.

8.4 Smart Grids and Meters

base, regarding the future of energy grids and energy management, and the notions of smart grid and smart City."[269]

In Germany, because of their dependence on collecting large amount of data, smart meters are a delicate issue. According to the EU 2009 Third Energy Package, wherever it is cost-effective, all of the EU member states have to implement smart meters and replace 80% of electricity meters by 2020.[270] Although Germany had initially decided against deploying smart meters, in 2011 the government stated that *"the smart grid is part of an evolutionary, not a revolutionary, development."*[271] In 2016, the Act on the Digitisation of the Energy Transition set for January 2017 the beginning of the rollout of smart grids, smart meters and smart homes in Germany,[272] setting the legal and technical basis for an intelligent energy network. This Act does not imply that Germany has opted for a full rollout, like most of the countries in the EU. Germany has in fact been defined as an "ambiguous mover," meaning that although there is a legal and regulatory framework set in place, the adoption of this technology is still developing and only some of the DSOs have started with the installations of smart meters.[273] Since the EU package allows different approaches according to a cost-benefit analysis, Germany has decided for a progressive adoption of smart meters, depending on the consumption of the consumer and on the sector.

The full rollout of smart meters is still in progress and will be structured in two phases:

- In the first stage, only large consumers, with a consumption of over 10,000 kilowatt-hours (kWh) and renewable energy generators distributed generation above 7 kWh of installed capacity must have smart meters installed.
- In the second stage, the obligation will be extended to households and companies with a consumption higher than 6000 kWh, which applies to approximately 15% of electricity consumers.[274]

---

[269]*Ibid.*

[270]European Union, Third Energy Package 2009, Directive on electricity 2009/72/EC, repealing Directive 2003/54/EC and Directive on gas 2009/73/EC, repealing Directive 2003/55/EC.

[271]Energy Industry Act, available at: www.gesetze-im-internet.de/enwg_2005/EnWG.pdf.

[272]Act on the Digitisation of the Energy Transition, available at: www.bundesrat.de/SharedDocs/drucksachen/2016/0301-0400/349-16.pdf?__blob=publicationFile&v=1.

[273]USmartConsumer Project, European Smart Metering Landscape Report 2016, USmartConsumer, Madrid.

[274]Act on the Digitisation of the Energy Transition, available at: www.bundesrat.de/SharedDocs/drucksachen/2016/0301-0400/349-16.pdf?__blob=publicationFile&v=1.

Until 2020	After 2020	Final result
Large consumers, with a consumption >10,000 kWh a year	Households with a consumption >6000 kWh	– Installation of 7.5 million smart meters – 50 million metering end-points – The majority of German households will remain unaffected
Renewable energy generators with installed capacity between 7 kW and 100 kW	Companies with a consumption >6000 kWh	

The decision for the German conditional rollout came after a cost-benefit analysis commissioned in 2013 by the Federal Ministry for Economic Affairs and Energy to Ernst & Young (updated in 2014).[275] The conclusion of the analysis was that a nation-wide rollout of smart meters was not recommended until 2020. Even after 2020, the majority of German households will remain unaffected, because an average four-person household consumes around 3500 kWh and therefore it is far from reaching the threshold of 6000 kWh.

Alongside the Act on the Digitisation of the Energy Transition, the Metering Point Operation Act sets more technical rules on the installation and operation of smart metering systems. It also deals with data protection and IT security. The Metering Point Operation Act sets protection profiles and technical guidelines with the aim of achieving security and enables development in areas such as, for example, e-Mobility.[276]

The German approach to smart grids has been quite cautious. In contrast to the way the country has embraced the adoption of renewable energy, the same enthusiasm has not been applied to the need for a new, smart, and interconnected energy grid. After 2020, when the rollout of smart meters will finally start, many people will still not have to be connected to the smart grid with a smart meter. Considering that 2020 is already approaching and considering that smart grids are at the moment the only technology that will efficiently support the switch to renewable energy, it will be interesting to see whether this caution will prevent the government from adopting more ambitious rules regarding the need to install smart meters.

As for the Netherlands, since it has such an important production and consumption of natural gas, most houses have a smart meter that measures gas and electricity at the same time.[277] A cost-benefit analysis has been completed to determine whether the national deployment of smart metering systems will be economically sound. The result of this analysis was positive in the Netherlands and even recorded expected benefits of €770 million as well as energy savings between 3% and 6%.[278] Despite these results being similar to those in other European countries, and despite the fact that such countries have mostly adopted prevaricated approaches to smart metering

---

[275]Ernst & Young, Report 2013.

[276]Metering Point Operating Act 2016, Article 1, available at: www.gesetze-im-internet.de/messbg/MsbG.pdf.

[277]https://www.cs.ru.nl/E.Poll/papers/smartmetering_short.pdf.

[278]https://www.engerati.com/article/netherlands-smart-meter-rollout-goes-large-scale.

8.4 Smart Grids and Meters

systems, the Netherlands, quite on the contrary, decided to adopt a large-scale rollout.

In 2006, the introduction of smart metering systems began, in compliance with Directive 2006/32/EC. The Dutch Government opted for the installation of smart meters in every home. Before the amendments introduced by the 2010 *novelles* (i.e., new bills) as a reaction to some privacy concerns, consumers could risk a fine (€17,000) or even imprisonment (6 months) if they refused to install a smart meter.

Considering that already in 2016 almost 3 million households were equipped with a smart meter, the roll-out may reach the target of having smart meters in 80% of households by 2020 (as per EU Directive 2009/72/EC).[279] However, in terms of savings for the consumers, the Netherlands is far from reaching the projected targets. In contrast with expected savings of around €4.1 billion, consumers are actually only saving 1% on their energy bill (2017).[280] This might be due to the fact that the energy transition has not yet reached its full development and potential. This fact raises the question of whether the use of smart meters means greater energy efficiency and a change in consumption patterns. When a proper integration of RES and electric vehicles into the smart grid will be achieved and when the consumer will actually become an active part of the energy transition (thereby becoming a prosumer), only then will the real benefits in terms of savings be maximized.

DSOs are the entities responsible for smart meters and they need to comply with the Dutch Smart Meter Requirements.[281] One of the issues that have been reported by the operators refers to the efforts made in installing the latest type of smart meter devices available.[282] As technology progresses, the applications that an appliance is able to perform will evolve. While DSOs are installing a certain type of device, new types of more innovative "next-generation" devices have been constantly made available on the market, with the risk that the benefits may outweigh the costs. For example, some smart meters started to be installed in 2007, using a 2G mobile network, which is a technology that will soon be phased out.[283]

Although the Netherlands has the reputation for being a progressive and innovative country with a good social policy, there seem to be some contradictions in the energy sector for such a wealthy nation. Its energy-transition projects are small and not on a scale to make a significant difference towards decarbonization. However, it is a country where citizens have been very critical of their Government's passive role in climate change mitigation. In fact, in 2015, an environmental organization won a lawsuit against the Dutch Government,[284] who was obliged to adhere to its CO_2

---

[279] https://www.cs.ru.nl/E.Poll/papers/smartmetering_short.pdf.

[280] https://utilityweek.co.uk/where-smart-meter-rollouts-are-going-wrong-and-how-to-put-it-right/.

[281] https://www.netbeheernederland.nl/_upload/Files/Slimme_meter_15_32ffe3cc38.pdf.

[282] https://www.engerati.com/article/dutch-smart-meter-experience-lessons-mass-rollout.

[283] https://www.power-technology.com/features/building-smarter-grid-netherlands/.

[284] N. Geiling, "In landmark case, Dutch citizens sue their government over failure to act on climate change," 14 April 2015, available at https://thinkprogress.org/in-landmark-case-dutch-citizens-sue-their-government-over-failure-to-act-on-climate-change-e01ebb9c3af7/. See also D. van Berkel,

emissions reduction target: by 2020, its CO_2 emissions must decrease by 25% (compared to the 1990 levels), as dictated by the famous Urgenda ruling.[285] Nonetheless, research from the Netherlands' Environmental Assessment Agency (PBL) shows that the Netherlands will not reach its GHG targets by 2020.[286]

As a reaction to the potential risk of missing its target, one would have expected from the Netherlands a stronger focus on the development of renewables and greater investment in smart grids. However, the need for this radical change does not seem to be happening in the Netherlands and the developments in terms of incentives and investments could and should have been much greater. A good step forward in the right direction is the amendment of the Energy Transition Act of 2018. As the energy transition will involve the consumer at a greater scale than has been the case in the past, thereby making the consumer a prosumer, a potentiality for the Netherlands could derive from the use of public participation, which is already quite advanced, in comparison with other EU countries.

8.4.4 Smart Grids Within a Circular Economy

A threat to EU economic security and growth is access to raw materials. Increasing energy efficiency is part of a broader goal to increase resource efficiency in the EU.[287] One strategy is to develop a circular economy. This section outlines the concept and the reason why it is needed, especially in relation to the smart grids. The discussion covers themes of design obsolescence, extended product responsibility and e-waste management. The section considers how responsibilities should be allocated and to whom during the life cycle and value chain of products in a decentralized digital energy system. The life cycle assessment is a process used to evaluate the environmental burdens that come with a product, production process or activity throughout its entire life cycle from the phase of raw material extraction to final disposal.[288]

"Urgenda: Climate case explained," available at https://www.urgenda.nl/en/themas/climate-case/climate-case-explained/.

[285] J. Pieters, "Netherlands' 2020 climate goals "out of reach"," 25 January 2019, *NL Times*, available at https://nltimes.nl/2019/01/25/netherlands-2020-climate-goals-reach-planning-office-says.

[286] Ibid.

[287] Commission Staff Working Paper, Analysis associated with the Roadmap to a Resource Efficient Europe, Part I, Accompanying the document Communication from the Commission to the European Parliament, the Council, the European Economic and Social Committee and the Committee of Regions, Roadmap to a Resource Efficient Europe, SEC(2011) 1067 final, Brussels, 20.9.2011.

[288] See for instance life cycle assessment of energy and environmental impacts of LED lighting products, at https://www1.eere.energy.gov/buildings/publications/pdfs/ssl/lca_factsheet_apr2013.pdf.

8.4 Smart Grids and Meters

The transition to a low-carbon economy will not be without waste. It is imperative that forethought goes into business modelling and resource management for the entire lifecycle of the product to limit impacts on the environment and contribute to increasing energy efficiency. Today, much is wasted in three key resources: materials, food and energy. Around 60% of energy in the US economy is wasted[289]; about 40% of food produced in the US in never eaten,[290] and up to 18% of water treated in the US is wasted.[291] The situation is not much better in the rest of the world: around 33% of energy is lost[292]; between 30% and 50% of all food produced is wasted,[293] and up to 60% of water is lost through leaky pipes worldwide.[294] Researchers in Austria are currently studying the notion of socio-metabolism, which will help us describe and understand the transition to a new kind of society, namely the concept of a circular economy. In their words, 'socio-economic systems depend on a continuous throughput of materials and energy for their reproduction and maintenance. This dependency can be seen as a functional equivalent of biological metabolism, the organism's dependency on material and energy flows'.[295] For instance, the metabolism of a city implies the transformation from raw materials, water and fuel into goods, human biomass, and waste. It has been defined as "the technical and socioeconomic processes that occur in cities, resulting in growth, production of energy, and elimination of waste."[296] The goal, therefore, is to move towards an industrial ecosystem, where 'the consumption of energy and materials is optimized, waste generation is minimized, and the effluents from one process serve as the raw material for another.'[297]

The global growth in renewable energy capacity will soon bring end-of-life cycle waste management issues to the fore. First, planning ahead is necessary to manage the existing waste stream from established renewables. Second, it is necessary to promote a circular closed-loop approach to the whole life-cycle of products and contribute to a green economy.[298] Countries need to undertake reforms of existing laws and develop innovative policy and regulation to meet these challenges. The risks are high, primarily because renewable energy is far from being "clean."

The EU energy targets promote energy efficiency, renewable energy, as well as decentralisation, but these goals also need to fit within the broader 2030 EU Agenda

---

[289]Lawrence Livermore National Laboratory 2014.

[290]*Wasted: How America Is Losing Up to 40 Percent of Its Food from Farm to Fork to Landfill*, Natural Resources Defense Council.

[291]*The Case for Fixing the Leaks,* American Water Works Association.

[292]IIEA 2014.

[293]Global Food Report, Institution of Mechanical Engineers, January 2013.

[294]*Is the world thirsty for water management?* IBM.

[295]Alpen-Adria University, "Institute of social ecology," at https://www.aau.at/en/social-ecology/research/social-metabolism/.

[296]Kennedy et al. (2007), pp. 43–59.

[297]Frosch and Gallopoulos (1989), p. 144.

[298]Morgera and Savaresi (2013), pp. 14–28.

for Sustainable Development[299] and the Circular Economy Action Plan to increase resource efficiency and decrease waste.[300] The rising costs, driven by the growing demand for primary resources, including those needed for smart grid systems, requires new approaches to resource management along the entire life cycle value chain. The EU is increasingly recognising that the current economic model dependent on the linear use of materials is no longer viable. This is the reason why closing the material loop is prioritised.[301]

8.4.4.1 The Circular Economy Concept and the EU

The circular economy, also known as a "closed loop" economy, aims to reach holistic sustainability goals and is based on the concept of "no waste"[302] and is related to the concept of dematerializing. Circular economy is part of the relatively new science of industrial ecology,[303] which is critical to sustainable development. The concept of circular economy has the great advantage that, if you are re-using something, you do not need to go back to the extraction of natural resources and the production process when making a product.[304] Instead, in a circular economy, the end-of-life stage of products and materials must be replaced by restoration.[305] In other words, it is about the notion 'cradle to cradle.'[306] Even Mother Nature uses a circular-economy approach. Reducing waste is therefore at the core of the circular economy model.[307] It is a concept that recognises the continuous potential value of materials to reduce resource inefficiency in both production and consumption, showing thereby that efficiency is an important resource. These must be the objective of a profound transformation. Consequently, the standard approach to creation, fabrication, and commerce of products must change as well.

---

[299]Communication from the Commission to the European Parliament, The Council, The European Economic and Social Committee and the Committee of the Regions Next Steps for a Sustainable European Future European Action for Sustainability, COM(2016) 739 final (22 November 2016).

[300]Communication from the Commission to the European Parliament. Closing the Loop—An EU action plan for the Circular Economy, COM(2015) 614 final (2 December 2015).

[301]Boulos et al. (2015).

[302]De los Ríos and Charnley (2017), pp. 109–122. See also the Zero Waste Europe initiative at http://www.zerowasteeurope.eu/category/products/epr-extended-producer-responsibility/.

[303]Industrial ecology examines "the influences of economic, political, regulatory, and social factors on the flow, use, and transformation of resources." See White (1994). "The aim of industrial ecology is to restructure the industrial system, inspired by our understanding of biological ecosystems (cyclic use of resources, food webs, etc.)..." See Erkman and Ramaswamy (2003).

[304]See the work of the Ellen MacArthur Foundation on circular economy, https://www.ellenmacarthurfoundation.org/.

[305]De los Ríos and Charnley (2017), pp. 109–122.

[306]Braungart (2009).

[307]Communication from the Commission to the European Parliament. Closing the Loop—An EU action plan for the Circular Economy, COM(2015) 614 final (2 December 2015).

The EU is heavily dependent on imported raw materials, especially metal ores and non-metallic minerals that are found in electrical and electronic equipment (EEE).[308] Since the design of a product directly influences the way a value chain is managed, building circular, globally sustainable value chains inevitably implies a fundamental change in the practice of design.[309] Recently, the EU waste law has become part of a wider policy discourse on sustainable production and consumption, moving towards the adoption of a circular economy. For example, as part of the Circular Economy Package, the European Commission has proposed the addition of an obligation to ensure that, by 2030, the amount of municipal waste put into landfill is reduced to 10% of the total amount of such waste.[310]

The EU Commission has committed to analyse the current situation of critical raw materials in the context of the circular economy with a focus on material-efficient recycling of electronic waste, waste batteries and other relevant complex end-of-life products.[311] With the transition to renewable energy systems set by the 2020 EU Climate and Energy Package and the 2030 EU Climate and Energy Framework, greater efforts are required to incorporate the Circular Economy principles into systems infrastructure design. The implications of this new approach are yet to be fully be appreciated. It is clear, however, that existing waste regulation needs to be revised and all actors throughout the supply chain of products need to assume new responsibilities to change the EU current production system and accomplish to close the loop, as required by the Circular Economy.

8.4.4.2 EU Waste Regulation: Key Principles for Renewable Energy and Smart Energy Grids

The EU has an extensive legal framework on waste management.[312] The 1975 Framework Directive on Waste (FDW) lays the foundation for the EU waste law. It defined key concepts, established major principles such as the waste hierarchy, and allocated responsibilities between different actors: authorities, producers and households.[313] Another important Directive is the 1999 Landfill of Waste Directive which introduced the end-of-life cycle principle. Such a Directive requires EU Member

---

[308] Directive 2012/19/EU of the European Parliament and of the Council, 4 July 2012 on Waste Electrical and Electronic Equipment (WEEE), L 197/38, 24.7.2012.

[309] De los Ríos and Charnley (2017), pp. 109–122.

[310] Proposal for a Directive of the European Parliament and of the Council amending Directive 1999/31/EC on the landfill of waste, COM(2015) 594 final (2 December 2015).

[311] See Annex 1 Communication from the Commission to the European Parliament. Closing the Loop—An EU action plan for the Circular Economy, COM(2015) 614 final (2 December 2015).

[312] Langlet and Mahmoudi (2016).

[313] Framework Directive on Waste 75/442/EEC [1975].

States to draft a national strategy for the implementation of measures aiming at developing a whole life-cycle approach to waste management and landfills.[314] It "sets targets to progressively reduce the level of biodegradable waste going to landfill and bans the landfilling of certain hazardous wastes, such as liquid waste, clinical waste and used tyres."[315] The overall goal within the EU is to reduce the percentage volume of waste being discarded in landfills. Additional Directives included the Packaging and Packaging Waste Directive,[316] the End-of-Life-Vehicles Directive,[317] and the Waste Electrical and Electronic Equipment (WEEE) Directive.[318] Each of such Directives took forward the FDW waste hierarchy and extended responsibility principles.

In 2008 a new FDW developed the waste hierarchy and extended responsibilities, especially for producers.[319] The Directive was based on Article 192(1) of the Treaty on the Functioning of the European Union, which aims "to protect the environment and human health by preventing or reducing the adverse impacts of the generation and management of waste and by reducing overall impacts of resource use and improving the efficiency of such use."[320] The 2008 Directive explains the concept of life-cycle of products and materials, encourages the recovery of waste and the use of recovered materials and develops end-of-waste criteria for specified waste streams.[321] Under the 2008 FDW, top priority is given to prevention, followed by preparing for re-use, recycling, and other recovery, including energy recovery. Disposal is the least desirable option and it is at the bottom of the hierarchy.

Furthermore, the principle of responsibility is expanded. The FDW places responsibility for waste treatment upon the original waste producer. Under Article 15 of the FDW, EU Member States can specify the conditions of responsibility and decide in which cases the original producer is to retain responsibility for the whole treatment chain or in which cases the responsibility of the producer and the holder can be

---

[314]See Article 1 Council Directive 1999/31/EC of 26 April 1999 on the landfill of waste.

[315]Cherrington et al. (2012), p. 14.

[316]European Parliament and Council Directive 94/62/EC of 20 December 1994 on packaging and packaging waste.

[317]Directive 2000/53/EC of the European Parliament and of the Council of 18 September 2000 on of-of-life vehicles.

[318]Directive 2002/96/EC of the European Parliament and of the Council of 27 January 2003 on waste electrical and electronic equipment (WEEE)—recast as Directive 2012/19/EU of the European Parliament and of the Council of 4 July 2012.

[319]Directive 2008/98/EC of the European Parliament and of the Council of 19 November 2008 on waste and repealing certain Directives, L 312/3, 22.11.2008, where the principle of "extended producer responsibility" is introduced for the first time.

[320]See Article 1 Directive 2002/96/EC of the European Parliament and of the Council of 27 January 2003 on waste electrical and electronic equipment (WEEE).

[321]Directive 2008/98/EC of the European Parliament and of the Council of 19 November 2008 on waste and repealing certain Directives.

shared or delegated among the actors of the chain.[322] This includes the scenario in which the original waste producer bears the cost of waste management.

The trend in the EU is towards recognising an extended producer responsibility to new products, product groups and waste streams such as electrical appliances and electronics.[323] However, the effectiveness of extended producer responsibility within the EU Member States is variable. Having different national extended producer responsibility interpretations for waste electrical and electronic equipment hampers the effectiveness of recycling policies. For this reason, in 2012, the Commission proposed that essential criteria needed to be decided by the European Union and minimum standards for the treatment of waste electrical and electronic equipment should be developed.[324]

The EU is taking steps to address the impacts of renewable energy and smart grids—including the upscaling of solar PV,[325] wind turbines and batteries for EVs. One substantive initiative in this regard is the amendment of the WEEE Directive for the collection and recycle of solar PV panels.[326] Most of the EU Member States have revised EEE waste regulation to include solar PV in national law (e.g. Spain[327] and Italy[328]).

The principle of producer responsibility could be extended for manufacturers to recycle wind turbine blades in the same way it has so effectively been done with the WEEE Directive amendment.[329] If legislation is introduced within the wind energy industry it is likely to be similar to the end-of-life vehicles legislation that introduces set recycling and recovery targets for manufacturers. This would require the producer to have more responsibilities. Some EU Member States have adopted measures to deal with the problem of wind turbine blades landfill dumping. Since 2005, Germany has banned all types of untreated municipal solid waste from its landfills.[330] Consequently, materials with a high organic content (e.g. wind turbine

---

[322] See Article 15 Directive 2008/98/EC of the European Parliament and of the Council of 19 November 2008 on waste and repealing certain Directives.

[323] Organisation for Economic Co-operation and Development, "Extended producer responsibility," OECD, [Online]. Available: http://www.oecd.org/env/tools-evaluation/extendedproducerresponsibility.htm.

[324] See Recital 6 Directive 2012/19/EU of the European Parliament and of the Council of 4 July 2012 on waste electrical and electronic equipment (WEEE).

[325] Beyond the EU boundaries, research shows that "solar PV systems are now at or approaching retail electricity prices in many markets, across both residential and commercial user segments." See Report: Solar at grid parity in 80% of world by 2017, available at http://www.utilitydive.com/news/report-solar-at-grid-parity-in-80-of-world-by-2017/370346/.

[326] Directive 2012/19/EU of the European Parliament and of the Council of 4 July 2012 on waste electrical and electronic equipment (WEEE).

[327] Royal Decree 110/2015 of 20 February.

[328] Legislative Decree No. 49, 14 March 2014, Implementing Directive 2012/19/EU on Waste Electrical and Electronic Equipment (WEEE).

[329] Cherrington et al. (2012), pp. 13–21.

[330] 2006 Municipal Solid Waste Management Report, Municipal solid waste management in Germany TASi one year on—no wastes landfilled without pretreatment in Germany since 1 June

blades) need to find different end-of-life routes. R. Cherrington *et al.* state that "landfill bans effectively divert waste from landfill and drive towards energy recovery."[331] EU legislation increasingly discourages the disposal of waste to landfill, setting steeper reduction targets, for example the reduction of 10% by 2030 included in the Circular Economy Plan.[332] Wind turbine manufacturers could take the initiative. Investing in solutions now will provide time to develop efficient systems and reduce technology costs.[333]

The amendments to the WEEE Directive to increase recycling of solar PV panels and proposals to limit the discarding of wind turbine blades in landfills are important steps to manage the end-of-life waste from these renewable energy sectors.

8.4.4.3 New Concepts and Principles to Close the Smart-Grid Loop

Extended producer responsibility (EPR) was intended to incentivise manufacturers to increase waste management efficiencies through better product design. EPR's rationale is that financial and/or physical responsibility on producers brings them to internalize waste management considerations into their product strategies.[334] Reports illustrate, however, that extended product responsibility remains a distant goal within the EU.[335] A new model is needed. The EU Circular Economy Action Plan[336] goes in such a direction as it tackles one of the main obstacle to a fair management of the life-cycle of EU products: the planned obsolescence.

This term dates to the Great Depression, when Bernard London recommended the strategy as a means to foster economic recovery.[337] London perceived the economic value of stimulating repetitive consumption. Lightbulbs were the first items to be designed with planned obsolescence in mind.[338] On the contrary, the circular economy is based on the principle of planned durability, of which manufacturers

---

2005—A new era has dawned in municipal solid waste management, 1 September 2006, [On-line]. Available http://www.bmub.bund.de/fileadmin/bmu-import/files/pdfs/allgemein/application/pdf/bericht_siedlungsabfallentsorgung_2006_engl.pdf.

[331]Cherrington et al. (2012), pp. 13–21.

[332]Proposal for a Directive of the European Parliament and of the Council amending Directive 1999/31/EC on the landfill of waste, COM(2015) 594 final (2 December 2015).

[333]Ortegon et al. (2013), pp. 191–199. Cherrington et al. (2012), pp. 13–21.

[334]Kalimo et al. (2014), pp. 40–57.

[335]Report from the Commission to the European Parliament, the Council, the European Economic and Social Committee and the Committee of the Regions on the Thematic Strategy on the Prevention and Recycling of Waste, COM(2011) 13 final (19 January 2011).

[336]Communication from the Commission to the European Parliament, the Council, the European Economic and Social Committee and the Committee of the Regions, Closing the loop—An EU action plan for the Circular Economy, Brussels, 2.12.2015, COM(2015) 614 final.

[337]London (1932).

[338]M. Krajewski, "The Great Lightbulb Conspiracy," IEEE Spectrum, 24 September 2014. [Online]. Available: http://spectrum.ieee.org/geek-life/history/the-great-lightbulb-conspiracy.

have full responsibility. Improving product durability and reparability is important to reducing pressure on natural resources, reducing import costs for manufacturers and saving money for consumers.[339]

There is no legal definition of durability, but, so far, the European Commission has proposed the following:

> Durability is the ability of a product to perform its function at the anticipated performance level over a given period (number of cycles/uses/hours in use), under the expected conditions of use and under foreseeable actions. Performing the recommended regular servicing, maintenance, and replacement activities as specified by the manufacturer will help to ensure that a product achieves its intended lifetime.[340]

The practicalities of delivering planned durability are numerous and challenging.[341] Manufacturers generally want to restrict access to spare parts, and limit repair and reuse of old products.[342] Key issues include not only the cost of spare parts but also access to information and skills development. The EU has produced reports exploring the potential for using regulations to stimulate durability, reparability and reusability of products.[343] It has also developed rules to increase design durability for some products, such as lighting and vacuum cleaners. Several EU Member States have introduced national legal measures to reduce planned obsolescence and increase reparability.[344] France, for instance, introduced a law to address planned obsolescence. Article L. 213-4-1 of the Consumer Code now reads: "Planned obsolescence is forbidden and is defined by all the techniques by which a person that places goods on the market seeks to deliberately reduce the lifespan of a product to increase the substitution rate."[345] Although limited in scope due to pressure from manufacturers lobbying in the negotiation of the law, interpretation by the courts could provide positive developments to reduce design obsolescence. Another example is Norway, which requires companies to extend consumer guarantees on certain products, increasing the responsibility of the manufacturer.[346]

---

[339] *Ibid.*

[340] Boulos et al. (2015), p. 4.

[341] European Environmental Bureau (2015). Ardente and Mathieux (2014), pp. 126–141.

[342] Dalhammar (2016), p. 155.

[343] A. M. Bundgaard et al., "Ecodesign Directive 2.0. From Energy Efficiency to Resource Efficiency," Environmental project No. 1635, 2015, Aalborg, 2017; Ardente et al. (2014), pp. 158–171; RREUSE, "Improving Product Reparability: Policy Option at the EU Level," RREUSE, 2015.

[344] RREUSE, "Improving Product Reparability: Policy Option at the EU Level", RREUSE, September 2015, [On-line]. Available http://www.rreuse.org/wp-content/uploads/Routes-to-Repair-RREUSE-final-report.pdf.

[345] Law No. 2015-992 on Energy Transition for Green Growth (Energy Transition Law); "I.- L'obsolescence programmée se définit par l'ensemble des techniques par lesquelles un metteur sur le marché vise à réduire délibérément la durée de vie d'un produit pour en augmenter le taux de remplacement." https://www.legifrance.gouv.fr/affichCodeArticle.do?cidTexte=LEGITEXT0000 06069565&idArticle=LEGIARTI000031053376 [translation from French by Rafael Leal-Arcas].

[346] Maitre-Ekern and Dalhammar (2016), pp. 378–394.

To overcome excessive and unnecessary consumption, product designers need to factor durability and reparability to product design. Spare parts should be made easily available at an affordable price that incentivises repair. Design models should be able to incorporate old components into newer versions of a product. Regarding software, making new software compatible with older models can deter consumers from upgrading to new versions. The electronic equipment industry notoriously exploits incompatibility across new models and fosters design obsolescence. This has driven a global e-waste disposal crisis, especially in several developing countries such as Nigeria.[347] Apple even has the battery built into its computers and phones. Batteries are a component that is easily replaced, but are a high-level toxic waste that requires safe disposal using the best available technology.

The Eco-design Directive is a key instrument in order to promote durability.[348] Already used to set binding minimum energy efficiency requirements, the Directive is being used to develop new eco-design requirements to manufacturers. The Directive obligates manufacturers to provide mandatory information on proper disposal, disassembly and recycling at the end-of-life stage, especially for product groups with toxic content (e.g. mercury).[349] Extension of lifetime is specifically listed in the Directive and for certain products is "expressed through: minimum guaranteed lifetime, minimum time for availability of spare parts, modularity, upgradeability, reparability."[350] A different way of tackling the issue is through an indirect approach, where voluntary based agreement with manufacturers are signed. Even if those agreements are not compulsory, they imply the interest of the manufacturers to commit to these issues. This sometimes makes such a voluntary approach even more effective than legal or regulatory rules, leading to better and longer term results in terms of contribution to the circular economy.[351]

The definition of durability does not refer to reparability. Design for reparability is difficult to measure and can lead to legal complexities if not addressed.[352] Durability and reparability are two sides of the same coin.[353] The circular economy opens opportunities for small and medium scale enterprises to provide reparability and recycling services. Remanufacturing and repair industries need rules that clarify that the repairer, or anyone putting the product into re-use, should not be considered the manufacturer/producer of the repaired/re-used product. Re-manufacturers will seek to avoid becoming a "producer" in the meaning of some EU Directives because

---

[347]Pickern (2014), pp. 403–423.

[348]Communication from the Commission to the European Parliament. Closing the Loop—An EU action plan for the Circular Economy, COM(2015) 614 final (2 December 2015).

[349]European Environmental Bureau (2015).

[350]See Part 1.3, point (i) of Annex I Directive 2009/125/EC of the European Parliament and of the Council of 21 October 2009 establishing a framework for the setting of ecodesign requirements for energy-related products.

[351]Directorate General for Internal Policies, Policy Department, A Longer Lifetime for Products: Benefits for Consumers and Companies, IP/A/IMCO/2015-11 June 2016.

[352]Boulos et al. (2015).

[353]Maitre-Ekern and Dalhammar (2016), pp. 378–394.

they would be otherwise economically responsible for the collection and the recycle of the product. They would also need to comply with the requirements of "new" products, such as respecting the rules on energy efficiency.[354] Similarly, being a re-manufacturer implies being carbon-negative, which is a desired outcome.

The complexity of smart grid systems will undoubtedly lead to demand manufacturers and service providers to offer support services to consumers. It will benefit consumers, the collaborative economy and the environment as well as future generations if the legal and regulatory framework are in place to ensure this occurred in a circular economy where all the loops are closed.

8.4.5 Regulatory Policy Recommendations

The most relevant issues at the moment revolve around network planning, priorities about grid reinforcement, and the ways DSOs are incentivised by national regulation to invest in smart grids. In simplified terms, a crucial issue concerns how to convince DSOs to test and innovate more. The 'obvious' answer seems to lie in the regulatory incentives set by the NRAs. Yet, these agencies also have to protect consumers from potentially excessive charges that natural monopolists such as DSOs could charge. This problem might be made even more acute when DSOs are state-owned and a major sources of public revenue. Therefore, a balance has to be struck between incentivising DSOs to invest in smart grids and avoiding the imposition of high tariffs on consumers.

Another important concern is the possibility of conflicts of interest between DSOs and, for instance, self-producers. The desire of DSOs to optimize the economic benefit of grid utilization, inherently conflicts with the idea of self-production. Consequently, without regulatory interventions, DSOs would be opposed to the development of technology which potentially affects their bottom line.[355] To achieve this, the support of the DSOs, who have historically benefitted from the status quo is required.[356] Indeed, as has been demonstrated in Italy, DSOs are capable if the enabling environment is created to spearhead the desired change.

The European Commission as well as the Council of the European Energy Regulators (CEER) and the European Regulators Group for Electricity and Gas (ERGEG) hold that DSOs should be "market facilitators".[357] The notion of market facilitator in this context means that DSOs should play a crucial role in setting up and

---

[354]*Ibid.*

[355]Union of the Electric Industry (Eurelectric) (2011), p. 14.

[356]de Hauteclocque and Perez (2011), p. 5.

[357]Council of European Energy Regulators (CEER), "CEER Status Review on European Regulatory Approaches Enabling Smart Grids Solutions ("Smart Regulation")," CEER, C13-EQS-57-04, Brussels, 2014; European Regulators Group for Electricity & Gas (ERGEG), "Position Paper on Smart Grids: An ERGEG Public Consultation Paper," ERGEG, E09-EQS-30-04, 2009.

managing the infrastructure necessary to perform new services, for example, demand side and load aggregation functions. But they should not be directly involved in the provision of such functions, which instead should be left to actors competing against each other (*e.g.* suppliers, aggregators, and Energy Service Companies (ESCOs)).

An additional set of regulatory challenges relates to the use of, and access to, smart meter data for smart grids. In most EU Member States, smart grids will make use of, and indeed rely on smart meter data and infrastructures. In general, how consumers' data will be managed and by whom will have to be clearly explained. Otherwise, an anxiety about privacy issues will be inevitable. Indeed, access to, as well as ownership of data appear to be the key issues. These are not specific to energy sector alone, but represent challenges that have been discussed thoroughly in other domains from which lessons may be drawn, such as "big data". While the regulatory nature of data protection for smart grids still remains unclear, it seems likely that national bodies (*e.g.* independent regulatory agencies for energy), will play a central role. Regulators and policymakers more generally can learn from other sectors which had to face already similar issues (*e.g.* internet search engines).

It is also important to consider the standardization of smart grid technology with the view to improving security and integrity of the infrastructure. Although the various components of smart grids are at various levels of development, the concept envisages the interconnection of various components. Consequently, the absence of minimum technological requirements might result in, or facilitate the development of vulnerabilities such as cyber-attacks. Similarly, it is not inconceivable to envisage situations where sub-standard assets which interface with a smart grid network inhibit the smooth operation of the network or damage it. Granted that standardisation may occur at different levels, differing national standards increases costs, which are often passed on to consumers. It may therefore be prudent for the Agency for the Cooperation of Energy Regulators (ACER) to take lead on standardisation efforts to provide an international framework to guide national, local or enterprise based standardisation and perhaps delineate the relevant levels of standardisation. This will go a long way to facilitate international interoperability and the market integration efforts of the EU.[358]

Furthermore, a significant barrier to smart grid deployment would be insufficient/ lack of consumer demand for such technology. Given fears associated with cyber security, government espionage and data protection, as well as public scepticism on the utility of such technology, concerted action has to be taken to create sufficient awareness to tackle this barrier. It is therefore critical that more information is provided to citizens about the benefits of smart grids, and more specifically about why smart meters are being deployed. This would increase consumers' awareness and engagement in energy markets, and, in turn, facilitate the development of smart grids.

---

[358]Swora (2011), p. 15; Eisen (2013), p. 123.

8.5 Electric Vehicles (EVs)

At the consumer level, energy storage can contribute to the integration of decentralized production.[359] This benefit is further augmented when EVs are integrated into a smart grid design. EVs have traditionally been lauded as climate-friendly alternatives to internal combustion engines, which emit greenhouse gases. However, more recently, the lithium-ion batteries used in EVs have been recognized as a potential storage device that can be used to provide reserve capacity to a grid, under what has come to be known as the Vehicle to Grid (V2G) system.[360] Further, the integration of EV charging infrastructure with the appropriate management systems will allow the charging of EVs to become a controllable load. This would go a long way towards improving the reliability of the distributed power system,[361] while ensuring that the EV is charged at the most convenient time. Despite the inability to store large volumes of electricity to meet traditional modes of supply in traditional electricity markets, current storage technology could play an important role in VPPs.

VPPs aggregate energy produced by diverse distributed generation sources, including small scale generators. Consequently, unused electricity stored in batteries from small scale RES could be fed-into a VPP. Similarly, energy stored in the lithium-ion[362] batteries used in EVs could also be fed-into VPP or grids under the V2G system.

There are predictions that electric vehicles will make up 14% of total car sales by 2025, up from 1% in 2017.[363] The Organization for Petroleum Exporting Countries expects 266 million EVs to be on the street by 2040, up from 46 million.[364] Regulations are getting tighter to the extent that the UK and France, among other European countries, have announced that all new cars must be zero-emission by 2050.[365] If implemented in other jurisdictions beyond Europe, this sort of policy will have serious implications. For instance, in the US, around 85% of workers commute by car[366] and around 65% of oil consumption comes driving on roads.[367] China,

---

[359] Stoppani (2017), pp. 10 and 17.

[360] Masson et al. (2016), p. 63; Changala and Foley (2011), pp. 108–109.

[361] Veldman et al. (2010), p. 287 at 300.

[362] It will be interesting to see whether Chile, a very rich country in lithium, will end up a new Saudi Arabia as a result of large amounts of lithium.

[363] Peter Campbell, "Electric car costs forecast to hit parity with petrol vehicles," Financial Times, 19 May 2017.

[364] "OPEC Drastically Increases 2040 Electric Vehicle Forecast," *Manufacturing*, 18 July 2017, available at https://mfgtalkradio.com/opec-drastically-increases-2040-electric-vehicle-forecast/.

[365] The Economist, "Roadkill," 12 August 2017, pp. 7–8.

[366] Idem.

[367] Idem.

which accounted for about 50% of the electric vehicles sold in 2016, aims at 2 million electric and plug-in hybrid cars on China's roads by 2020 and 7 million by 2030.[368] Most of the nearly 1 billion cars on the road today are powered by fossil fuels.[369] Moreover, existing electric cars reduce CO_2 emissions by 54% compared with petrol-powered cars.[370]

However, two major concerns seem to arise for electric car buyers: where can one charge an electric car and how long will it take? Currently, over 90% of charging is done at home.[371] However, in the US, public electric-vehicle charging stations have been growing steadily since 2011.[372] Carmakers such as Mercedes, BMW, Volkswagen, and Ford have said that they will together install a total of 400 public charging points in Europe, which will deliver 350 kW.[373] In Europe, countries such as Germany, France, the Netherlands, and Norway are committed to improving access to public charging.[374] In 2017, China installed 800,000 public charging points, including semi-public charging points for taxis and commercial vehicles and at workplaces.[375] As for how long it will take, with a standard residential electricity supply and a 3.5 kW charger, the owner of a small electric car can have its battery charged in 8 h.[376] An acceptable solution to these two concerns is crucial for the EV revolution to take off.

From an environmental point of view, it goes without saying that EVs have many advantages over traditional gasoline-powered cars. EVs offer significant energy efficiency and reduced emissions. EU member states are at different stages when it comes to the widespread proliferation of EVs. Yet, predictions are that "around 85% of vehicles [in the world] are still expected to use internal-combustion engines in 2030."[377]

8.5.1 The EU Legal Basis

The EU has set for itself an ambitious target of reducing the use of internal combustion engine vehicles by 50% by 2030 and phasing them out entirely by

---

[368]The Economist, "Electrifying everything," 12 August 2017, pp. 13–15 at 13.

[369]The Economist, "Roadkill," 12 August 2017, pp. 7–8.

[370]Idem (citing the National Resources Defense Council in the USA).

[371]The Economist, "Charge of the battery brigade," 9 September 2017, pp. 63–64 at p. 64.

[372]Idem, citing the US Department of Energy.

[373]The Economist, "Charge of the battery brigade," 9 September 2017, pp. 63–64 at p. 64.

[374]Idem.

[375]Idem.

[376]Idem.

[377]The Economist, "Crude awakening," 9 February 2019, p. 9.

8.5 Electric Vehicles (EVs)

2050 as part of efforts to reduce GHG Emissions.[378] The alternative fuels directive[379] encourages Member States to develop systems that enable EVs to feed power back into the grid. In addition, the Commission has recently published a strategy for low-emission mobility, which seeks to promote the removal of obstacles to the scaling up of the use of EVs.[380]

8.5.2 Current Status Across Europe

Regarding EVs, the European Environment Agency (EEA) reports that in 2015, 150,000 new EVs were sold in the EU. However, 90% of these sales were in the Netherlands, the UK, Germany, France, Sweden, and Denmark.[381] Despite a steady growth in the number of EVs sold in the EU over the years, the 2015 numbers represent only 1.2% of total vehicle sales. Figure 8.1 shows the trend of EV sales since 2010.

In countries such as Norway and the Netherlands, where EV sales are very high, regulatory incentives have played a large role in promoting consumer interest.[382] These incentives include tax exemptions on EV purchases, one-off grants, and the imposition of taxes on fossil fuels. Figure 8.2 summarizes the use of incentives for EVs across Europe.

In Belgium, Greece, Hungary, Latvia, and the Netherlands, there is a full registration tax exemption on EV Purchases, while Denmark and Finland provide a partial exemption.[383] Other financial schemes employed by governments are fixed grants, as employed in France and Portugal for replacing an end-of-life vehicle with a new electric vehicle.

Beyond promoting consumer interest, many countries also support research and development with a view to promoting innovation in the EV sector. Finland, for instance, instituted the Electric Vehicles Systems Programme in 2011 with a budget of €100 million to support the growth of the EV sector.[384]

---

[378] *Commission White Paper: Roadmap to a Single European Transport Area – Towards a Competitive and Resource-Efficient Transport System*, p. 9, COM (2011) 144 Final (Mar. 28, 2011).

[379] Parliament and Council Directive 2014/94/EU, On the Deployment of Alternative Fuels Infrastructure, 2014 O.J. (L 307) 1.

[380] *Commission Communication to the European Parliament, The Council, The European Economic and Social Committee and the Committee of the Regions: A European Strategy for Low-Emission Mobility*, COM (2016) 501 final (July 20, 2016).

[381] European Environment Agency, Report on Electric Vehicles in Europe 47 (2016).

[382] Paul Hockenos, *With Norway in Lead, Europe Set for Surge in Electric Vehicles*, Yale Env't 360 (Feb. 6, 2017) https://e360.yale.edu/features/with-norway-in-the-lead-europe-set-for-breakout-on-electric-vehicles; *European Vehicle Market Statistics, 2015/2016*, Int'l Council on Clean Transp. (Nov. 25, 2015) http://www.theicct.org/european-vehicle-market-statistics-2015–2016.

[383] European Environment Agency, Report on Electric Vehicles in Europe 60 (2016).

[384] *Id.* at 62.

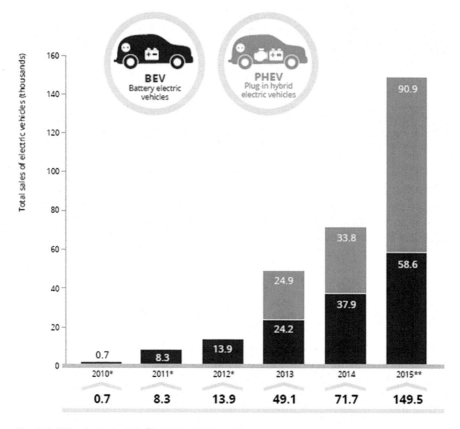

Fig. 8.1 EV sales in the EU. *In 2010, 2011, and 2012, only statistics for battery electric vehicles are available. **The data for 2015 are provisional [Source: European Environment Agency, Electric Vehicles in Europe (EEA Report, 2016), p. 49]

Governments have also taken various actions to support the development of infrastructure, particularly charging points. France, for instance, set up a special fund, for the construction of charging infrastructure, which led to the construction of 5000 charging points in 2015.[385] In Sweden, individuals who installed charging points in their homes obtained a tax reduction for the associated labor cost.[386] However, an emerging barrier to the large-scale deployment of charging infrastructure is that new, fast charging technology is not only expensive to install, but also requires high voltage input. The associated consumption fee is therefore high.[387]

---

[385] *Hybrid & Electric Vehicle Technology Collaboration Programme*, Int'l Energy Agency, http://www.ieahev.org/; Ken Seaton, *The Push for Electric Cars,* The Connexion (Sept. 19, 2013) https://www.connexionfrance.com/Archive/The-push-for-electric-cars.
[386] European Environment Agency, Report on Electric Vehicles in Europe 23 (2016).
[387] *Id.* at 26.

8.5 Electric Vehicles (EVs)

	PURCHASE SUBSIDIES (purchase-related tax exemptions or reductions, registration tax, import tax, co-funding or other finantial purchase support)	OWNERSHIP BENEFITS (annual tax exemption, reduction of electricity or energy costs)	BUSINESS AND INFRASTRUCTURE SUPPORT (business development or infrastructure support)	LOCAL INCENTIVES (free parking, access to bus lanes, no toll fees, free charging, access to restricted areas in city centres)
AUSTRIA	✓	✓	✓	✓
BELGIUM	✓	✓	✓	
BULGARIA	✓	✓		✓
CROATIA	✓		✓	
CYPRUS		✓		✓
CZECH REPUBLIC	✓	✓	✓	
DENMARK	✓	✓	✓	✓
ESTONIA			✓	✓
FINLAND	✓	✓	✓	
FRANCE	✓	✓	✓	✓
GREECE	✓	✓		✓
GERMANY	✓	✓	✓	✓
HUNGARY	✓	✓		✓
ICELAND	✓	✓	✓	✓
IRELAND	✓	✓	✓	✓
ITALY	✓	✓	✓	✓
LATVIA	✓			✓
LIECHTENSTEIN				
LITHUANIA	✓			✓
LUXEMBOURG	✓		✓	
MALTA	✓	✓	✓	✓
NETHERLANDS	✓	✓	✓	✓
NORWAY	✓	✓	✓	✓
POLAND		✓		
PORTUGAL	✓	✓	✓	✓
ROMANIA	✓	✓		
SLOVAKIA		✓		
SLOVENIA	✓			✓
SPAIN	✓	✓	✓	✓
SWEDEN	✓	✓	✓	✓
SWITZERLAND	✓	✓	✓	✓
TURKEY	✓	✓	✓	
UNITED KINGDOM	✓	✓	✓	✓

Fig. 8.2 Use of incentives for EVs across Europe [Source: European Environment Agency, Electric Vehicles in Europe (EEA Report, 2016), p. 65]

Non-financial measures, particularly at the local government level, have also been instrumental towards the promotion of EVs in Europe. In the UK, for instance, some local councils have adopted a procurement policy that requires at least one EV amongst their fleet of vehicles.[388] In Bulgaria, the National Action Plan for the promotion of EVs gave EVs free parking in all its cities.[389] In other countries like Spain and Norway, road toll exemptions and discounts apply to EVs.[390]

As national responses to climate change and air pollution continue to increase in response to EU Directives, it is expected that many more countries will adopt policies that would enhance EVs and storage technology.

8.5.3 The Situation in Certain EU Member States

In 2008, revenue in Belgium's transport sector amounted to €15.9 billion, making it one of the country's key sectors.[391] In fact, greenhouse gas (GHG) emissions in transportation rose by 31% from 1990 to 2010. Belgium responded by various actions at both the federal and regional levels, and managed to lower GHG emissions in the transportation sector by 7% from 2010 to 2014.[392] These steps included placing high taxes on fuel, creating low-emission zones barring more polluting vehicles from city centers, and enacting policies to promote alternative modes of transport.[393] Moreover, Belgium has kept pace with the EV "revolution."

In 2017, EVs had obtained a market share of nearly 2% in the country. In fact, Belgium has one of the EU's largest fleets of electric buses.[394] Plans are afoot to ensure the use of EVs continues to rise in the country. For example, a joint-stakeholder platform—the Belgian Platform on EVs—has been established to create a national strategy for electric transport, which has produced a policy paper titled "Roadmap 2030 for the Stimulation of Electric Mobility in Belgium." Numerous institutes are researching EVs and hybrid vehicles in Belgium, and their work is driving the EV trend. These include Flanders' DRIVE, "*Katholieke Universiteir Leuven*" (K.U. Leuven), the Limburg Catholic University College (LCUC), and University of Ghent, "*Vrije Universiteit Brussel*" (VUB). Some of this research explores the idea of integrating EVs with smart grids as an option for charging EVs,

---

[388]Pete McAllister, *Huge New Study Compares Every UK Council's Electric Vehicle Usage*, Intelligent Car Leasing (Jan. 23, 2015) http://www.intelligentcarleasing.com/blog/new-study-compares-every-uk-council-electric-vehicles.

[389]Lewis Macdonald, *Bulgarian City Introduces Free Parking for Electric Cars*, Eltis: The Urban Mobility Observatory (Nov. 12, 2014) http://www.eltis.org/discover/news/bulgarian-city-introduces-free-parking-electric-cars.

[390]European Environment Agency, Report on Electric Vehicles in Europe 62 (2016).

[391]International Energy Agency: Hybrid & Electric Vehicle Technology Collaboration Programme, "Belgium," International Energy Agency. Available at: http://www.ieahev.org/bycountry/Belgium.

[392]National Climate Commission, "Greenhouse gas inventory for Belgium," Brussels, 2016.

[393]Boussauw and Vanoutrive (2017), pp. 11–19.

[394]European Alternative Fuels Observatory (EAFO), "Belgium," EAFO, available at: http://www.eafo.eu/content/belgium; European Environment Agency (EEA) (2016).

8.5 Electric Vehicles (EVs)

ideally using RE. Other research is looking into ways to use EVs as a solution towards energy storage.

When it comes to regional efforts to promote EVs, the Flemish government has, since 2010, invested more than €16 million in support of EV testing sites. Authorities have also instituted specific tax policies to encourage EV usage, such as tax breaks for businesses using electric or hybrid vehicles and subsidies for those buying EVs domestically, among others.[395]

Large-scale EV deployment relies on charging infrastructure. Identifying the appropriate regulation and policy regarding the installation and running of public charging stations entails a number of challenges. For example, regulation would need to consider the type of charging technology, charging station locations and ownership, safety, standardization, and pricing. For Belgium, the fact that competence over energy-related regulation is divided across the federal and regional levels exacerbates these challenges. Still, the Belgian Platform on EVs is a move in the right direction, as it promotes trans-regional initiatives towards creating appropriate regulation. In any case, largely thanks to policies supporting the private sector in building charging infrastructure, Belgium has around 1500 public charging stations, even without a tangible regulatory framework in place.

In Spain, road transport consumes more energy by far than other modes of transport (e.g., airlines, railways, and marine transport), comprising 80% of the country's total energy consumption. The transport sector in the country has several key features, including significant vehicle usage, an ageing fleet, and low proportion of freight transport via railways. The transport sector depends on oil for more than 90% of its energy use. It therefore has negative environmental impacts and has much room for improvement on the energy efficiency front.[396] Given all these issues, EVs seem like the best way forward. The government has been working toward this goal since 2003, releasing various legislations and policies aimed at a "smarter" and sustainable transport sector.[397]

Several strategies exist in Spain aimed at increasing purchase and usage of EVs. The Ministry of Industry, Energy and Tourism, along with IDEA, has launched two schemes to this effect, which have had significant impact over the last years. These are the Efficient Vehicle Incentive Program (PIVE) and MOVELE, which promotes electric mobility. PIVE was framed within the 2008–2012 Action Plan of the 2004–2012 Energy Savings and Efficiency Strategy (E4), and aims to replace out-of-date vehicles with new, cleaner, and more efficient ones. Those already receiving government subsidies for purchasing vehicles are not eligible for PIVE grants.[398]

---

[395] European Environment Agency (EEA) (2016).

[396] International Energy Agency (2015).

[397] The International Energy Agency. Hybrid and Electric Vehicle Technology Collaboration Programme, "Spain Policies and Legislation," The International Energy Agency. Hybrid and Electric Vehicle Technology Collaboration Programme. Available at: http://www.ieahev.org/by-country/spain-policy-and-legislation/.

[398] Martínez Lao (2017).

The MOVELE program was launched in order to bring home the feasibility of EVs in urban environments. To this end, MOVELE pushed for introducing up to 2000 EVs in the Spanish fleet, and for manufacturing more than 500 EV charging points in different cities across Spain.[399]

Along with the above plans, in April 2010 the government established the "Integral Plan for the Promotion of Electric Vehicles," which included an "Integrated Strategy for EVs 2010–2014," with a goal of reaching one million Hybrid and Electric Vehicles (H&EVs) in Spain by the end of 2014.[400] More recently, the "Impulse to the Alternative Energy Vehicle Plan" (VEA) came into effect in June 2015. As of 2017, the 2017 MOVEA plan is the latest development under the VEA strategy with a total budget of €16.6 million managed by the Ministry of Economy, Industry and Competition.[401] The VEA strategy aims to harmonize the various domestic policies while working towards the goal of increasing the number efficient vehicles.[402] The ultimate aim is to maximize the various national initiatives so that Spain may meet its EU-level 2020 climate and energy goals.[403]

The Sustainable Economy Law 2/2011 of 4 March supports research, development and innovation in renewable energy, energy conservation and energy efficiency for transport and sustainable mobility.[404] It calls for the central government, autonomous communities and local municipalities to take steps to advance the usage of plug-in H&EVs.[405] Thus, all public regulators are legally required to facilitate the implementation of H&EVs by, among other things, providing them with the RE applications and infrastructure linked with such vehicles. The Royal Decree 647/2011 of 9 May identifies the load system manager's role in terms of EV charging facilities. Additionally, Royal Decree 647/2011 addresses the rights and obligations associated with EV energy charging services. The Electricity Sector Law 24/2013 goes further by regulating the role of the load system manager.

The Royal Decree 647/2011 of 9 May defines the role of the load system manager, which is then further detailed and regulated in the Electricity Sector Law 24/2013. A load system manager refers to a commercial entity that, while itself a consumer, has the right to resell electricity for charging services.[406] The rights and

---

[399] Ávila Rodríguez (2017).

[400] The International Energy Agency. Hybrid and Electric Vehicle Technology Collaboration Programme, "Spain Policies and Legislation," The International Energy Agency. Hybrid and Electric Vehicle Technology Collaboration Programme. Available at: http://www.ieahev.org/by-country/spain-policy-and-legislation.

[401] Corriente Eléctrica, "Se activa el Plan MOVEA 2017 para coches y vehículos eléctricos," Corriente Eléctrica, 17 June 2017. Available at: http://corrienteelectrica.renault.es/asi-sera-el-plan-movea-2017-para-coches-y-vehiculos-electricos/.

[402] Ávila Rodríguez (2017).

[403] Ministry of Industry, Energy and Tourism, "Estrategia de Impulso del vehículo con energías alternativas (VEA) en España (2014–2020)," Ministry of Industry, Energy and Tourism, 2015.

[404] Martínez Lao (2017).

[405] Article 82, paragraph 2 Sustainable Economy Law 2/2011 of 4 March.

[406] Article 6, point h) Electricity Sector Law 24/2013 of 26 December.

obligations attached to energy charging services for EVs are delineated in the Royal Decree 647/2011.

The Royal Decree 1053/2014 approved the Complementary Technical Instruction (ITC) BT-52 "Facilities for special purposes. Infrastructure for recharging EVs." Royal Decree 1053/2014 requires public areas to be provided with requisite facilities for installing charging points corresponding to the number of anticipated parking spaces in both municipal and supra-municipal sustainable mobility plans. The decree also requires newly-constructed buildings and car parks to be equipped with specific electric facilities towards EV charging. These charging facilities have different requirements, contingent upon the type of parking lot. In collective parking lots for private use, a minimum level of pre-installations is obligatory, to ensure parking space owners can charge their EVs without having to pay. In the case of public car parks or private fleet car parks, necessary facilities must be installed so that a charging point exists for every 40 parking spaces.[407] Electricity Sector Law 24/2013 also refers to electricity charging services, identifying their primary purpose as providing power through vehicle charging services and storage batteries while ensuring that charging takes place efficiently and at minimum cost to both consumer and the electricity system.[408]

Royal Decree 639/2016 of 9 December is the latest regulatory development regarding charging infrastructure. It adopts a range of steps to facilitate alternative fuel facilities. The legislation defines alternative fuels as options to replace, at least somewhat, traditional fossil fuels as power sources for transport, which could potentially make the transport sector more environmentally-friendly. These alternative sources include electricity, hydrogen, biofuels, synthetic and paraffinic fuels as well as natural gas.[409] The legislation also addresses two matters regarding charging stations. Firstly, all public charging points must allow users to charge their EVs on an *ad hoc* basis, without the need to enter into a contract with the electricity supplier or the load system manager. Secondly, energy suppliers other than the one providing the electricity to the building or premises of the charging point should have the right to contract the power supply for the charging point.[410]

Regulatory measures to increase EVs in Spain, it thus appears, focus largely on advancing charging infrastructure. This is the right move, as research shows that the number of available charging stations is a strong predictor of EV adoption.[411] Of course, appropriate financial schemes, such as direct grants, play a strong role, too. But steps that are specific to EVs are more reliable predictors of EV adoption rates than more general socio-demographic variables such as income, education level or

---

[407] Ávila Rodríguez (2017).

[408] Article 48, paragraph 1 Electricity Sector Law 24/2013 of 26 December.

[409] Article 2, paragraph 1 Royal Decree 639/2016 of 9 December.

[410] Article 4, paragraphs 4 and 5 Royal Decree 639/2016 of 9 December.

[411] Bjerkan (2016).

environmentalism.[412] Thus, building the requisite charging infrastructure, with appropriate legal frameworks, is necessary for promoting EVs.

Spain has several schemes in place, such as PIVE and MOVELE, to provide financial incentives towards purchasing H&EVs. However, overlaps exist amongst these schemes that can create confusion. For example, PIVE aims to replace out-of-date vehicles with more efficient ones, while MOVELE focuses on increasing EVs in Spain. Moreover, these schemes are being implemented by different actors, and would perhaps be more effective through stronger coordination and a more cohesive approach.[413]

It is worth looking into alternative policies to help increase EVs, for example by looking into the approaches of other countries. Countries such as Korea, the Netherlands, Portugal, and the United States have implemented EV-specific entry rules to urban access areas. In France, Sweden and the United Kingdom, EVs have access to preferential parking areas. Other states such as Denmark or Germany incentivize the use of batteries.[414] On the other hand, Spanish regulation mainly focuses on direct economic incentives and on strengthening the charging infrastructure. Still, the country's existing approaches are reasonable, given that the low number of charging points is believed by many to be a priority issue to tackle.[415] As Spain progresses on this front, it may well look into adopting additional regulatory incentives for EVs, such as preferential parking areas.

EVs are projected to have a key part to play in future electricity systems, especially when it comes to distributed energy storage systems, as EVs are able to integrate storage capacity in smart grids. In fact, vehicle-to-grid (V2G) systems enable EVs to both power and be powered by the grid. Moreover, V2G allows electricity to be stored, for usage during times of low production.[416] Increasing EVs promises a range of positive impacts. Smart charging during valley hours helps flatten the demand curve; EVs can contribute to optimizing electricity grid surpluses and will lead to more RE sources being integrated in the domestic grid. Also, wide adoption of EVs will help reduce CO_2 emissions and will lower dependency on external energy suppliers. It will also reduce both air and noise pollution in urban areas.[417] Spanish TSO, Red Eléctrica de España, estimates that in the near future, if smart charging is effectively conducted during valley hours, the national electricity system will be able to power an EV fleet equal to one fourth the total number of

---

[412]Martínez Lao (2017).

[413]Ávila Rodríguez (2017).

[414]Martínez Lao (2017).

[415]Energías Renovables, "La escasez de puntos de recarga frena la compra de coches eléctricos más que el precio," Energías Renovables. El periodismo de las energías limpias, 22 June 2017. Available at: https://www.energias-renovables.com/movilidad/la-escasez-de-puntos-de-recarga-frena-20170622.

[416]Amsterdam Vehicle2Grid, "The Solution to Sustainable Urban Mobility and Energy," Amsterdam Vehicle2Grid, Available at: http://www.amsterdamvehicle2grid.nl.

[417]Red Eléctrica de España, "Electric vehicle," Red Eléctrica de España. Available at: http://www.ree.es/en/red21/electric-vehicle.

8.5 Electric Vehicles (EVs)

vehicles in Spain without additional costs to the transmission grid.[418] Thus, EVs and their smart charging have a tremendous potential to strengthen the national grid. In this sense, certain technological products catered by the WiseGRID project could play their part by rendering EV-specific challenges more user-friendly. For example, Wise EVP (Electrical Vehicle Platform) is an application which will be used by vehicle-sharing companies and e-vehicles fleet managers (*e.g.* taxi companies) to optimize activities related to smart charging and discharging of the EVs and reduce energy billing. More importantly in this context, the application WG Fast V2G makes it possible to use EVs as dynamic distributed storage devices, feeding electricity stored in batteries back into the energy system when needed.

The need to promote sustainable transport in cities is crucial in the face of the high level of hydrocarbon-fueled transport in Spain. The transition to EVs will play a key role in decarbonizing the transport sector in Spain. Meeting the EU's 2020 and 2030 targets for CO_2 emissions will require ambitious efforts in the country. In fact, estimates reveal that Spain will need 300,000 EVs by 2020 and up to 6 million in 2030 to lower its CO_2 emissions in keeping with EU goals.[419] Achieving this will entail a long-drawn and progressive evolution, with significant investment. To place matters in context, in 2015 Spain had 6500 EVs, comprising a market share of 0.2%, far beneath the EV numbers in Norway (23%) and the Netherlands (10%) which are leading in this area.[420]

Italy's EV market is large, and is on the rise thanks to the emergence of a complete production chain, stretching all the way from research until finished vehicles.[421] While currently the market is not regulated, two proposals have been reviewed by the government. The proposals aim at facilitating the development and deployment of electric vehicles. The first, "Law no. 2844 of 2009: Measures to favor the development of mobility by using vehicles without CO_2 emissions," proposes steps such as subsidizing purchases, installing public battery charging systems, exemptions from property tax, the right to restricted public areas, and free parking in reserved parking areas. Additionally, the proposal suggests placing a tax on plastic bottles. The second proposal, "Law no. 3553 of 2010: Measures for the realization of infrastructure aimed at assisting the broad introduction of electric vehicles," aims at incorporating EV charging infrastructure into strategies at both the national and regional levels that address energy and GHG emissions. It also emphasizes the need to create a national strategy that will promote EV development. Unfortunately, neither of these laws has been passed. As a result, in 2010 AEEGSI took steps to

---

[418] Idem.

[419] Deloitte, "¿Cuántos coches eléctricos necesita España?" Deloitte. Available at: https://www2.deloitte.com/es/es/pages/strategy/articles/Cuantos-coches-electricos-necesita-Espana.html.

[420] Idem.

[421] A. A. &. C. Scamoni, "Lexology," Globe Business Media Group, 1 September 2016. Available at: http://www.lexology.com/library/detail.aspx?g=4bf3dda1-44ba-47c3-a860-af76ca74cbe4.

help promote EVs, including liberalizing the supply of electric meters for EV charging systems.[422]

The government developed a national strategy for creating better EV infrastructure in 2012. The plan aligns with 2014/94/EU Directive, and since 2013, 50 gigaeuros (G€) of incentives have been put in place to catalyze the creation of proper infrastructure. As of the end of 2016, 19 projects in 19 Italian regions had been approved, involving a total amount of 5 G€. The plan also includes rules and incentives for buyers of EVs.

Any plan for developing EV charging infrastructure, especially public charging stations, must factor in the potential impacts on competition. An appropriate regulatory framework should address issues related to the ownership of public charging infrastructure. For example, can electricity run their individual charging stations? This would naturally affect customers' ability to access charging stations, because the need to always locate a charging station run by one's supplier would become a problem. It also poses challenges for urban planning and how to allocate charging stations, particularly if all 365 of the country's suppliers are to have equal access to the EV charging market. Alternatively, public stations could permit open access to all energy suppliers. Such a scenario would require creating technology and guidelines to keep track of customer consumption for various suppliers. It would also call for creating measures to calculate volumes and payments in V2G scenarios. Thus far, the approach has been to leave the matter to AEEGSI, which has approved various projects that will test different charging infrastructure ownership models, with the goal of ultimately choosing the most appropriate.[423]

Since EVs are still limited to a relatively small circle, with uneven usage globally, there has not been much emphasis on standardizing the technical aspects related to them, such as batteries and charging technology. With international standards still in progress, Italy, along with several other countries, has established intermediate standards. For example, in 2010, the Italian EV Association, a committee of Italy's Electrotechnical standardization body CEI adopted "Safety requirements for charging stations for electric road vehicles (CEI 312-1)."

Along with other European states, Italy is working towards developing V2G capability, for which numerous projects are in development phases, including the WiseGRID project. The goal is to use bidirectional charge management that allows EVs to store unused power and supply it to the grid. A few challenges that arise out of this is the need for effective compensation schemes as well as re-defining the traditional understanding of "storage" in the electricity network context, to recognize distributed generation through battery technology. When it comes to EVs, any compensation mechanism must also factor in the wear and tear on the EV owner's

---

[422]E. Comelli, "ItalyEurope24," Il Sole 24 ORE, 23 February 2017.

[423]F. Villa, "Regulation of Smart Meters and AMM Systems in Italy," in 19th International Conference on Electricity Distribution (CIRED), Vienna, 2007.

8.5 Electric Vehicles (EVs)

battery stemming from grid supply.[424] Creating effective compensation schemes might entail identifying an EV-specific compensation mechanism that is separate from a compensation scheme created for distributed energy resources (DERs). The numerous pilot projects currently underway may shed light on appropriate steps forward.

In Greece, the market for EVs is in a very early phase, although there has been some progress in this regard. Given the lack of detailed data, providing an accurate picture of the market overview regarding the number of EVs and charging stations in the country is not possible. The Hellenic Association of Motor Vehicle Importers Representatives issues monthly reports of vehicle registrations in Greece, according to which report five new electric cars were registered in May 2016 out of a total of 10,660 new cars (including 209 hybrid cars). Moreover, in 2016, 32 out of 78,873 cars that were sold were EVs, while 1556 were hybrid cars. EV sales remain somewhat low and end-users attribute this to their high price, limited autonomy and the lack of widespread charging infrastructure. To this end, companies are attempting to promote EVs, dating as far back as 2011 when an oil and fuel trading company and an electrical energy production and trading company collaborated to install EV charging stations in three of the gas stations operated by the former.

Additionally, a cooperation between an EV producer and an electrical energy supplier involved promoting an EV model by selling it at a significant discount along with a lowered electricity rate for residential charging purposes. The partnership has employed this business model twice by now. Moreover, various public and private projects have been implemented to broaden charging infrastructure in the country, as well as a few associated pilot projects. Business plans for electric vehicle supply equipment (EVSE) operators are being created, too. While the European Research Project MERGE8 was underway, three varying schemes for EV penetration by 2030 in Greece were planned. The more realistic of these schemes anticipated around 300,000 EVs in Greece by that time, the optimistic anticipated 600,000 EVs, and the extremely optimistic around 1,200,000 EVs. This study was based on projections and market data from 2010 and need updating to allow for Greece's financial status today. Lastly, e-mobility in public transport is becoming more relevant. Aside from the existing trolley and tram services, the authorities in Athens are planning a pilot project for an e-bus servicing a heavily populated part of the city center, in partnership with the Athens Urban Transport Organisation.

There are numerous ways to boost EVs in the country, such as WiseEVP and WG FastV2G, which could strengthen EV management, boost existing projects and help them succeed and increase their market share. The WiseEVP application could additionally help operators and EV fleet managers to optimize activities related to smart charging and discharging of EVs and reduce their energy consumption and lower their energy bills, while factoring in the renewable generation profile, tariffs,

---

[424]Hybrid & Electric Vehicle Technology Collaboration Programme, "Hybrid & Electric Vehicle Technology Collaboration Programme," International Energy Agency. Available at: http://www.ieahev.org/by-country/italy-policy-and-legislation.

and the EV users' needs. The WG FastV2G application could facilitate the use of EVs as dynamic distributed storage devices and feeding electricity stored in their batteries back into the system, as long as a stronger regulatory framework related to energy storage emerges. Thus, WG FastV2G could lower electricity system costs by providing a cost-effective way to operate reserve and the ability to reduce power consumption during periods of maximum demand (i.e., peak-shaving).

A roadmap developed by the Greek Regulatory Authority of Energy (RAE) addresses the need to achieve electrification of transport. However, electric vehicle penetration in Greece remains relatively low, with a market share of 0.03% in the passenger car market.[425] To this end, the government has initiated a number of steps to improve electric vehicle usage. These include an exemption from registration, annual circulation and luxury taxes for both electric and hybrid vehicles[426] and access to certain restricted areas of city centers. Despite these incentives, only a few charging positions are installed in Greece, 33 to be exact.[427] Thus, there remains much room for improvement. Additionally, there is no specific provision in the regulation around providing e-mobility V2G services. Existing legislation merely addresses the installation of charging infrastructure in gas stations and parking lots, as well as the energy pricing performed by the operators of such stations.

Regarding France, through a so-called "bonus-malus" system, the French Ministry for Ecologic Transition and Solidarity aims at promoting the acquisition of "low-emission vehicles" ("*véhicules peu polluants*"), defined as new cars and vans whose emissions range between 0 and 20 g of CO_2 per kilometre ("*voitures ou camionnettes neuves émettant de 0 à 20 grammes de CO_2 par kilomètre*").[428] In practice, as the system stands nowadays, low-emission vehicles are, in fact, all-electric vehicles (as of 1 January 2018 hybrids are excluded), whose batteries are not lead-based.[429]

As its name indicates, the system is twofold: on the one hand, it directly incentivizes the purchase of low-emission vehicles through the "bonus." On the other hand, it discourages the purchase of the more polluting ones via the imposition of a "*malus*" (i.e., a supplementary tax on top of the standard/common one) upon the official registration of any new vehicle emitting more than 119 g of CO_2 per

---

[425]Endergiewende Team, "Greece's first battery storage system under way in the Aegean Sea," *Energy Transition*, 2 May 2017.

[426]Idem.

[427]Idem.

[428]Ministère de la Transition écologique et solidaire, "Bonus-malus écologique, prime à la conversion et bonus vélo", 10 January 2018, available at www.ecologique-solidaire.gouv.fr/bonus-malus-ecologique-prime-conversion-et-bonus-velo.

[429]Ministère de la Transition écologique et Solidaire, *Bonus écologique*, available at www.ecologique-solidaire.gouv.fr/sites/default/files/bonus%202-3%20roues-misenpage.pdf.

8.5 Electric Vehicles (EVs)

kilometre.[430] The "Ecologic malus" ranges from €50 for vehicles emitting 120 g of CO_2 per kilometre to €10,500 for vehicles emitting 185 g and more.[431]

The bonus side of the scheme works as follows, making a distinction depending on the net maximal power of the engine[432]: (1) for vehicles displaying an engine net maximal power of 3 kW or more, the bonus is fixed at €250 per kilowatt/hour of battery power, but it is capped at the lowest of the two following amounts: (a) 27% of the purchase price (comprising all taxes), potentially increased by the cost of the battery, if the latter is rented; or (b) €1000; (2) for vehicles displaying an engine net maximal power of less than 3 kW, the bonus is fixed at 20% of the purchase price (comprising all taxes), but it is capped at €200.

Irrespective of political and policy considerations regarding the means chosen, overall, the existence of this bonus-malus scheme clearly points to the conclusion that the acquisition of electric vehicles is actively supported by the French government. This has led France to become the second country in the European continent, after Norway, in terms of development of the plug-in electrified-vehicles market.[433]

Additionally, in France, electric vehicles are expected to play a significant role in demand response schemes, via their use as stockade units, and their ulterior contribution, whenever needed, to smart grid stability. It will therefore be possible, during periods when the vehicle will be connected to the electricity grid, to use the stored electricity to inject it on the network during periods of high demand or, conversely, to charge the vehicle battery in off-peak hours. This is the concept of "vehicle-to-grid", or V2G, which uses the batteries of electric vehicles as a mobile storage capacity.[434]

As for Germany, although the regulatory framework still remains underdeveloped, electric vehicles technology in Germany is quite developed in comparison to other countries. This is due to the strong position that Germany's industry has in the car sector. As part of the *Energiewende*, Germany recognizes the importance of an emission-free transportation system.[435] This determination has been reflected in the allocation of about €2.2 billion for research and development on electric mobility.[436]

---

[430]The malus is "*une taxe additionnelle à la taxe sur les certificats d'immatriculation*", specifically, "*Le malus écologique est une taxe à la première immatriculation concernant les véhicules les plus émetteurs de dioxyde de carbone.*" See www.ecologique-solidaire.gouv.fr/bonus-malus-ecologique-prime-conversion-et-bonus-velo.

[431]*Ibid.*

[432]*Ibid.*

[433]*France Becomes Fifth Nation To Buy 100,000 Plug-in Vehicles*, Hybrid Cars (Oct. 10, 2016), https://www.hybridcars.com/france-becomes-fifth-nation-to-buy-100000-plug-in-vehicles/.

[434]*L'utilisation du Véhicule électrique Comme Moyen de Stockage*, Smart Grids-CRE, http://www.smartgrids-cre.fr/index.php?p=stockage-vehicule-electrique.

[435]Taefi et al. (2016), pp. 61–79.

[436]Federal Minister for Economic Affairs and Energy, "Regulatory environment and incentives for using electric vehicles and developing a charging infrastructure," available at www.bmwi.de/Redaktion/EN/Artikel/Industry/regulatory-environment-and-incentives-for-using-electric-vehicles.html.

The Federal Government of Germany has also launched some advanced programmes in this sector, such as the "Electric Mobility Showcases" (promoting new solutions for electric mobility)[437] and a large-scale project to rollout fast chargers, called "SLAM" (German acronym for "network of fast chargers for transport corridors and major cities").[438]

The Charging Station Ordinance entered into force on 17 March 2016 and contains provisions on socket standards and minimum requirements for the establishment and operation of publicly accessible charging stations for electric vehicles. The adoption of the Ordinance places Germany ahead of all other EU countries in implementing the EU Directive on the deployment of alternative fuel infrastructures (Directive 2014/94/EU).[439]

Recent developments—that can inspire also the regulatory framework on electric vehicles, because of the use of a similar type of vehicle-to-vehicle (V2V) technology—have been developed within the communication sector. A draft bill of 2015[440] for a revised German Road Traffic Act[441] has been at the centre of an ongoing debate regarding autonomous (i.e. driverless) driving.[442] It has not yet been passed, but this bill alone could help reduce energy consumption and improve traffic flow.

This development in V2V communication can be beneficial for fostering V2V technology in the energy sector too, especially considering two factors: the strong focus that the German government has on the *Energiewende,* and the fact that a plan on similar technology, V2G (vehicle-to-grid), has also been recently developed. The "ICT for electric mobility II" programme, issued by the Federal Ministry for Economic Affairs and Energy, provides funding for those technologies that bring together e-vehicle technologies and connect them to: (a) smart traffic (improving traffic flow allows cars to drive more efficiently in terms of energy consumption) and (b) the grid in general (e-vehicles can be charged in a way that contributes to the stability of the grid, because of the connection to smart homes and renewable generation facilities).[443]

---

[437]Schaufenster Elektromobilitaet, available at schaufenster-elektromobilitaet.org/en/content/index.html.

[438]Slam Projekt, available at www.slam-projekt.de.

[439]European Parliament and the Council, Directive 2014/94/EU of the European Parliament and of the Council of 22 October 2014 on the deployment of alternative fuels infrastructure Text with EEA relevance, available at eur-lex.europa.eu/legal-content/EN/TXT/?uri=CELEX%3A32014L0094.

[440]Federal Ministry of Transport and Digital Infrastructure, Strategy for Autonomous and Connected Driving 2015, available at www.bmvi.de/SharedDocs/EN/publications/strategy-for-automated-and-connected-driving.pdf?__blob=publicationFile.

[441]German Road Traffic Regulations, available at www.bmvi.de/SharedDocs/EN/publications/german-road-traffic-regulations.pdf?__blob=publicationFile.

[442]Fulbright, Norton Rose, Autonomous vehicles, 2017.

[443]Federal Ministry for Economic Affairs and Energy, R&D funding provided by the Federal Ministry for Economic Affairs and Energy, available at www.bmwi.de/Redaktion/EN/Artikel/Industry/electric-mobility-r-d-funding.html.

8.5 Electric Vehicles (EVs)

In the Netherlands, the transport sector is still currently dominated by fossil fuels. However, according to the Energy Agreement, the Netherlands may experience a reduction in GHG emissions from transport of at least 60% by 2050 compared to the 1990 levels. The plan is to support the use of electric vehicles for smaller vehicles, such as cars; and to support the use of liquid biofuels and biogases for heavier and longer distance transport by road, water, and air.[444]

To reach these goals, the Renewable Fuels Long-term Plan (duurzame brandstofvisie) was adopted. This plan represents another example of Dutch collective decision-making, as many organisations in the transport and sustainability sector were involved in drafting the plan. In comparison with other European countries, the Dutch commitment to CO_2 emissions requirements for road transport is stricter, and this can only be justified by a strong commitment in investing in the transport sector, especially in terms of new technology and infrastructures.

There are reasons to believe that the Netherlands will make its transport sector one of the most innovative in Europe (together with Norway and Denmark[445]) and a crucial feature of its energy transition. There are already good examples of progress in this direction. In February 2019, the Dutch Government opened a roundtable discussion with the aim to adopt a law to ban the sale of fuel-based vehicles by 2025.[446]

The EV market surged in the country in 2015 in view of some new tax cuts introduced in early 2016. The Dutch legislator would need to do more if the country wants to keep the record in the EU of having one of the lowest levels of CO_2 emissions from new cars.[447] Norway could represent a good source of inspiration, as the Scandinavian country has adopted some financial incentives (exemption from VAT and purchase tax almost half the price of a new EV car) that have allowed Norway to beat its own selling targets.[448]

In the Netherlands, however, the tax incentive for all-electric cars remains at 4% tax. A new development in terms of incentives for new purchases is a tax cut for plug-in hybrids, which has moved from 7% to 15%. Since the 2025 target of a fossil fuel-free transport sector is not far, this focus on hybrids, rather than on electric cars, is not up to par.

If the Netherlands wants to be one of the leaders in the E-mobility sector, several challenges need to be addressed. For instance, the development of charging infrastructures, an issue that would need to grow rapidly and in accordance with the increase in the number of EVs expected in the near future. To this end, the Dutch

---

[444] See generally Ministry of Economic Affairs of the Netherlands, "Energy Report: Transition to sustainable energy," 2016, available at https://www.government.nl/documents/reports/2016/04/28/energy-report-transition-tot-sustainable-energy.

[445] Smart Energy International, "Netherlands aims to ban conventionally-fueled vehicles by 2050," 4 February 2019, available at https://www.smart-energy.com/industry-sectors/electric-vehicles/netherlands-aims-to-ban-conventionally-fueled-vehicles-by-2050/.

[446] Ibid.

[447] Ibid.

[448] Ibid.

Government has made €7.2 million available to subnational authorities for the installation of charging infrastructure via the 'Green Deal Publicly Accessible Electric Charging Infrastructure.'[449] This government investment together with the reduction (although only temporary) of the energy tax for electricity used in public charging stations represent a step forward to guarantee that public charging infrastructure will be able to develop without further governmental support.[450]

Another challenge is the interoperability of the charging infrastructure. This regards not only the infrastructures within the national boarders, but interoperability is also important when relations with neighbouring countries are considered, as EV-drivers need to be able to use charging stations in their country and abroad.[451]

A further challenge is support for public-private partnerships (PPPs). PPPs have proven to be an efficient tool in the Netherlands because they facilitate the collaboration of businesses, knowledge institutions, and the government. Examples of PPP are the various Green Deals.[452] They can be used in different sectors and involve different actors. In the EV sector, there is the Electric Transport 2016–2020 Green Deal[453] and the Publicly Accessible Electric Charging Infrastructure Green Deal.[454]

A final challenge is promoting the development of EV infrastructures at the local level. Focusing on the local level and adopting a bottom-up approach allow to make full use of the well-developed participatory and collective approach to policy challenges in the Netherlands.[455] Some cities and regions are particularly pioneers (e.g., Amsterdam and The Hague).[456]

Finally, one of the Dutch national icons, namely the bicycle, or, rather, the evolution of the bicycle, i.e., the electric bicycle, has a lot of potential when it comes to the role of EVs in decarbonization. The EU regulation on the e-bike was particularly important in the Netherlands. E-bikes are very common in the country (in fact, one in five bikes is electric) and the technology is quite advanced (some of the models can easily reach a speed of almost 50 km/h). Therefore, there was a strong need to regulate the sector to address major safety concerns. The EU regulation of

---

[449]Netherlands Enterprise Agency, "We are the Netherlands. Your partner in E-mobility," pp. 14–15, 2017, available at https://www.rvo.nl/sites/default/files/2017/10/We%20are%20the%20Netherlands%20your%20partner%20in%20E-mobility%20-%201535_CU_E-mobility%20ENG%20IV%20def%20v1.0.pdf.

[450]See generally Netherlands Enterprise Agency, "We are the Netherlands. Your partner in E-mobility," 2017.

[451]Ibid, pp. 18–19.

[452]Netherlands Enterprise Agency, "Electric transport in the Netherlands, Highlights 2017," pp. 32–34, 2018, available at https://www.rvo.nl/sites/default/files/2018/04/Highlights%20EV%202017%20English.pdf.

[453]Ibid. at pp. 33–34.

[454]Ibid at p. 34.

[455]See generally Netherlands Enterprise Agency, "We are the Netherlands. Your partner in E-mobility," 2017.

[456]Ibid. p. 49.

2017 treats e-bikes as mopeds, with all the consequences in terms of age limit, helmet, and insurance.[457]

To conclude, there is a high level of innovation in the business of EVs in the Netherlands, both in terms of investment in new technology (mostly vehicle-to-grid technology) and in charging infrastructures.

8.5.4 Regulatory Policy Recommendations

Policy-makers should create incentives for consumers and companies to use EVs, in addition to the construction and operation of electric vehicle charging facilities. Such incentives might include lower taxes for EVs, higher taxes for vehicles using gasoline, the possibility for EVs to use exclusive taxi or bus lanes, and support for research and development activities.

There are potential concerns. One is how realistic it is to expect states under financial and budgetary distress to pursue measures such as those enumerated above. Another is whether pursuing such measures could go against the State aid regime at the EU level. A further issue is under what conditions these support measures could be accepted and/or whether it would be desirable to amend the current State aid regime (e.g., through State aid guidelines that the Commission regularly produces over time across domains).

It is also worth noting that the increase in the use of EVs will contribute to the increase in demand for electricity. The IEA research scenarios estimate that the transport sector will make up 10% of total electricity consumption by 2050, owing largely to the increase in EV and plug-in electric vehicle use.[458] Therefore, it is critical that EV deployment is done as part of a larger smart grids strategy to ensure strategic low-cost vehicle charging.

8.6 Demand Response

Demand response refers to the process by which consumers can have control over the electric grid by lowering or altering their electricity usage at peak times, based on financial incentives. Demand response programs have numerous practical implications, for instance by reducing the risks of overload and power failures thanks to sensors that can perceive load problems and respond as necessary. From the point of view of the EU's climate and energy goals, demand response can play a significant role as it encourages energy efficiency.

---

[457] http://www.loc.gov/law/foreign-news/article/netherlands-new-rules-pending-for-e-bikers/.

[458] Int'l Energy Agency, Technology Roadmap: Smart Grids 12 (2011) https://www.iea.org/publications/freepublications/publication/smartgrids_roadmap.pdf.

Demand response is defined by ACER as "[c]hanges in electric usage by end-use consumers from their normal load patterns in response to changes in electricity prices and/or incentive payments designed to adjust electricity usage, or in response to the acceptance of the consumer's bid, including through aggregation."[459] It has increasingly gained prominence as a tool to improve energy efficiency and the reliability of grids through the lowering of demand, especially during peak periods.

Demand response programs can be divided into two types: implicit and explicit demand response.[460] In price-based (implicit) demand response, consumers choose to become exposed to time-varying prices that reflect the value and cost of electricity at different time periods. Thus, consumers do not pay fixed prices but rather respond to wholesale market price variations and/or dynamic grid fees. Such flexible prices for consumers do not necessarily require "aggregators."[461]

In contrast, in incentive-based (explicit) schemes, consumers receive direct payments to change their consumption patterns upon request. This can be triggered by activation of balancing energy, differences in prices of electricity, or grid constraints. Consumers may earn from their consumption flexibility by acting individually or by contracting with an aggregator, which in turn might be either a third party or the customer's supplier. Aggregated demand side resources are then traded in the wholesale, balancing, and/or capacity markets.

Aggregators are new actors within the European electricity markets, occasioned by the new market design heralded by the Third Energy Package. They are service providers that employ demand facilities to sell pooled loads of electricity. As their name suggests, they perform the function of "aggregating" flexibility. They agree with industrial, commercial, and/or residential consumers to aggregate their capacity to reduce energy and/or shift loads on short notice. They then create a "pool" of aggregated controllable load, made up of smaller consumer loads. Finally, they sell the pooled load as a single resource to system operators, which use it for their technical needs. Aggregation allows smaller consumers who are excluded from the markets due to the size of their loads to participate in the markets.[462] It should be noted that while load aggregators are new actors emerging in several power markets in Europe, load aggregation is a service which might be performed by a variety of actors. This goes well beyond load aggregators to include "traditional" suppliers or other new companies (e.g., ESCOs). It is important to note that the two distinct forms of demand response are not necessarily substitutes. Indeed, customers might well participate in incentive-based demand response through either an aggregator or a "traditional" supplier and, at the same time, participate in a price-based demand

---

[459] Agency for the Cooperation of Energy Regulators, *Framework Guidelines on Electricity Balancing*, p. 8 (FG-2012–E-009, Sept. 18, 2012); *see also* Murthy Balijepalli et al. (2011).

[460] *See generally* Smart Energy Demand Coalition, Mapping Demand Response in Europe Today (2015).

[461] *Id.* at 21.

[462] Philip Baker & Mike Hogan, *The Market Design Initiative: Enabling Demand-Side Market*, Regulatory Assistance Project 3 (Mar. 2016).

response program based on time-varying prices.[463] Beyond "aggregating" consumers (demand), aggregators also have a role to play in "aggregating" prosumers (consumption, production, and storage).

Given that demand response gives rise to complex relationships between energy suppliers, customers, aggregators, and BRPs, a critical examination of the implications of these relationships is necessary to develop a suitable regulatory framework that enables and facilitates market participation for these actors and ensures that the full benefit of demand response mechanisms are reaped.

8.6.1 The EU Legal Basis

The Third Legislative Package provides a supranational legal foundation for the development of demand response in Europe. Article 3(10) of the directive on the common rules for the internal market enjoined Member States to adopt, amongst others, "demand-side management" measures as part of efforts to combat climate change and improve energy security. Further progress was made with the Energy Efficiency Directive (2012/27/EU),[464] Article 15(4) of which requires Member States to:

> ensure the removal of those incentives in transmission and distribution tariffs that are detrimental to the overall efficiency (including energy efficiency) of the generation, transmission, distribution and supply of electricity or those that might hamper participation of demand response, in balancing markets and ancillary services procurement.[465]

It also requires member states to:

> ensure that network operators are incentivised to improve efficiency in infrastructure design and operation, and, within the framework of Directive 2009/72/EC, that tariffs allow suppliers to improve consumer participation in system efficiency, including demand response, depending on national circumstances.[466]

Furthermore, Article 15(8) of the Directive, establishes that:

> Member States shall ensure that national regulatory authorities encourage demand side resources, such as demand response, to participate alongside supply in wholesale and retail markets. Subject to technical constraints inherent in managing networks, Member States shall ensure that TSOs and DSOs, in meeting requirements for balancing and ancillary services, treat demand response providers, including aggregators, in a non-discriminatory manner, on the basis of their technical capabilities. Subject to technical constraints inherent in managing networks, Member States shall promote access to and participation of demand response in balancing, reserves and other system services markets, inter alia by requiring national regulatory authorities [...] in close cooperation with demand service providers and

---

[463]*Id.* at 7.
[464]Parliament and Council Directive 2012/27/EU, On Energy Efficiency, 2012 O.J. (L 315) 1.
[465]*Id.*, art. 15(4).
[466]*Id.*

consumers, to define technical modalities for participation in these markets on the basis of the technical requirements of these markets and the capabilities of demand response. Such specifications shall include the participation of aggregators.[467]

The set of rules ("network codes") drafted by the European Network of Transmission System Operators for Electricity (ENTSO-E) also emphasizes the importance of promoting demand response.[468] These rules are based on Framework Guidelines from ACER, which are based on priorities set by the European Commission. Specifically, the ACER Framework Guidelines on Electricity Balancing provide that "[t]hese terms and conditions . . . including the underlying requirements, shall, in particular, be set in order to facilitate the participation of demand response, renewable and intermittent energy sources in the balancing markets."[469]

Finally, the Commission Guidelines on State aid for environmental protection and energy 2014–2020, in clarifying the conditions under which Member States are allowed to introduce "capacity remuneration mechanisms," requests Member States to consider alternatives such as demand response.[470] Specifically, the Guidelines state that:

> Member States should therefore primarily consider alternative ways of achieving generation adequacy which do not have a negative impact on the objective of phasing out environmentally or economically harmful subsidies, such as facilitating demand side management and increasing interconnection capacity.[471]

Furthermore, "the measure should be open and provide adequate incentives to both existing and future generators and to operators using substitutable technologies, such as demand-side response or storage solutions."[472] In addition:

> the measure should be designed in a way so as to make it possible for any capacity which can effectively contribute to addressing the generation adequacy problem to participate in the measure, in particular, taking into account the following factors: the participation of generators using different technologies and of operators offering measures with equivalent technical performance, for example demand side management, interconnectors and storage.[473]

---

[467] *Id.*, art. 15(8).

[468] Article 8(6) of Regulation 714/2009/EC, On Conditions for Access to the Network for Cross-Border Exchanges in Electricity of the Third Energy Package set out the areas in which network codes are to be developed. They include balancing rules including network-related reserve power, data exchange and settlement rules, interoperability rules, network connection rules, network security and reliability rules, operational procedures in an emergency, amongst others. Commission Regulation 714/2009/EC, On Conditions for Access to the Network for Cross-Border Exchanges in Electricity, art. 8(6), 2009 O.J. (L 211) 15.

[469] Agency for the Cooperation of Energy Regulators, *Framework Guidelines on Electricity Balancing*, pp. 12–13 (FG-2012-E-009, Sept. 18, 2012).

[470] *Commission Guidelines on State Aid for Environmental Protection and Energy 2014–2020*, 2014 O.J. (C 200) 1.

[471] *Id.*

[472] *Id.*

[473] *Id.*

These supranational frameworks are designed to ensure that fundamental modalities required for the successful deployment of demand mechanisms are possible. These modalities fall into three categories: the legal recognition of demand response, thereby allowing consumer loads to compete with other generation assets in all markets; the legalization and enablement of aggregation services in the markets; and the adjustment of technical specifications in recognition of consumer capabilities and requirements.[474] The transposition period for the Energy Efficiency Directive expired in June 2014.[475] The expectation was that, by this date, the modalities necessary for implementation across Member States would have been in place.

8.6.2 Current Status Across Europe

The CEER's study on regulatory approaches for smart grids revealed that, in order to promote demand response, 71% of the European countries sampled use static time of use tariffs and 58% of them use load control to incentivize demand side response.[476] In countries such as Italy, load control is limited to large industrial customers through remote means.[477] In countries such as Belgium, different types of load control are used by the TSO in the tertiary reserve ancillary services of TSO Elia. In countries such as Greece, there are differential tariffs for peak and off-peak consumption for households.[478] However, not all European States apply "price signals" to induce customers to change their consumption patterns.

Figure 8.3 maps the status of incentive-based (explicit) demand response in Europe as of 2015. The assessment carried out by the Smart Energy Demand Coalition (SEDC)[479] was based on the following four criteria: enabling consumer participation and aggregation, appropriate program requirements, fair and standardized measurement and verification requirements, and equitable payment and risk structures.[480] Overall, the SEDC suggests that, in Europe, incentive-based (explicit) demand response is still in its early development.[481] In a few cases, the SEDC

---

[474] Paolo Bertoldi, Paolo Zancanella, & Benigna Boza-Kiss, JRC Science for Policy Report: Demand Response status in EU Member States 6 (2016).

[475] Parliament and Council Directive 2012/27/EU, On Energy Efficiency, art. 28, 2012 O.J. (L 315) 1.

[476] Council of European Energy Regulators, *CEER Status Review on European Regulatory Approaches Enabling Smart Grids Solutions ("Smart Regulation")*, p. 12 (C13–EQS-57-04, Feb. 18, 2014).

[477] *Id.*

[478] *Residential Night Tariff*, Hellenic Public Power Company SA (July 4, 2017), https://perma.cc/H64V-DMC3.

[479] Smart Energy Demand Coalition, Mapping Demand Response in Europe Today 8–12 (2015).

[480] Fernandez et al. (2016), p. 458.

[481] *See generally* Smart Energy Demand Coalition, Mapping Demand Response in Europe Today 8–12 (2015).

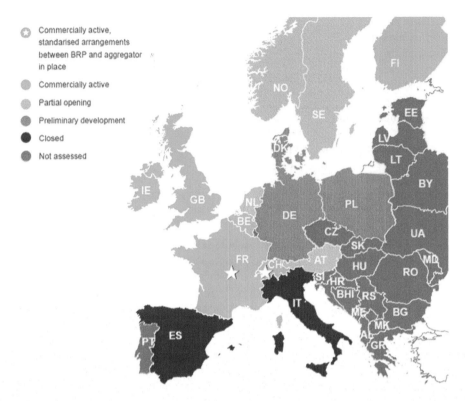

Fig. 8.3 Map of incentive-based (explicit) demand response development in Europe today [Source: Smart Energy Demand Coalition (SEDC) Mapping Demand Response in Europe Today—2015, p. 9]

suggests that markets do not permit consumer participation and are therefore "closed" to explicit demand response.[482] European countries have widely varying regulatory frameworks, each with its own participation requirements and rules. There generally are no standardized contractual arrangements governing the roles and responsibilities of the distinct actors involved. Furthermore, it is often impossible, or even illegal, to aggregate consumers' flexibility in practice.[483]

In some countries, demand response is a commercially viable product. For example, in Belgium, demand response can participate in a number of balancing markets, namely the primary and tertiary reserves.[484] However, a key obstacle is the

---

[482]*Id.*

[483]*Id.* at 11.

[484]*Id.* at 47–54; *Belgian TSO Elia in Demand Response First*, Restore (Sept. 6, 2017), https://www.restore.eu/export/pdfNews/113.

8.6 Demand Response

requirement for aggregators to get the prior agreement of the customer's supplier or BRP[485] in order to be able to contract with the customer.[486] There are at least two private aggregators active on the market ("Restore.eu" and "Actility") as well as a tertiary off-take reserve scheme specifically for aggregators ("Dynamic Profile").[487]

Great Britain is deemed to have competitive energy markets and open balancing markets, though the emerging capacity market has raised uncertainties for demand response. Great Britain was the first EU Member State to open many of its electricity markets to the demand side.[488] Currently, all balancing markets allow the participation of demand response in general and aggregated load in particular.[489] However, according to the SEDC, the UK's measurement, baseline, bidding, and other procedural and operational requirements are not appropriate. Thus, even though the markets are formally open, in practice, results in terms of demand-side participation have been worsening over time.[490] Furthermore, the capacity remuneration mechanism introduced in 2014 is said not to place demand-side resources on a "level playing field" with generation resources. Indeed, only one demand-side aggregator out of around 15 operating in the market managed to secure a contract in the first capacity market auction.[491]

France and Switzerland have redrafted their program requirements and defined clear roles and responsibilities precisely to allow independent aggregation.[492]

Other European countries still present important regulatory barriers, notably program participation requirements not yet tailored for both generation and demand-side resources. For example, Austria requires consumers to install a secure and dedicated telephone line in order to participate in the balancing market.[493] Norway requires TSO signals to be delivered over the phone, thus making the minimum bid-size high.[494] As a result, the participation of consumers other than large industrial consumers is hindered.[495] Similarly, technical and organizational rules do not consider some of the requirements for the provision of balancing services in sufficient detail.[496] This includes the negative impact of complex and

---

[485] Given that market players have an implicit responsibility to balance the electricity system, the balance responsible parties are financially responsible for keeping their own position (sum of their injections, withdrawals and trades) balanced over a given timeframe.

[486] Smart Energy Demand Coalition, Mapping Demand Response in Europe Today 47 (2015).

[487] Id. at 51.

[488] Id. at 85.

[489] Id.; see also PA Consulting, OFGEM: Aggregators—Barriers and External Impacts (2016).

[490] Smart Energy Demand Coalition, Mapping Demand Response in Europe Today 85 (2015).

[491] Id.

[492] Id. at 10.

[493] Id.; Paolo Bertoldi, Paolo Zancanella, & Benigna Boza-Kiss, JRC Science for Policy Report: Demand Response status in EU Member States 31 (2016).

[494] Smart Energy Demand Coalition, Mapping Demand Response in Europe Today 10 (2015).

[495] Id.

[496] Id. at 45.

lengthy approval procedures and their associated costs on market entry and participation.

In still other European countries, aggregated demand response is either illegal or its development is seriously hindered due to regulatory barriers. For example, in Italy, the notion of load aggregator is not formally recognized and no regulatory framework currently exists.[497] Poland and Spain do not seem to be taking the steps required to foster the development of incentive-based (explicit) demand response.[498] Indeed, load aggregators do not exist in every EU Member State. The analogous consideration applies to regulatory frameworks governing their operation.

Italy relies mostly on hydro and gas generation to satisfy its flexibility requirements, while the framework governing consumer participation in balancing markets has not been set up yet. Interruptible contracts are a partial exception and constitute a dedicated demand response program.[499] Load aggregation is not allowed, nor is there currently any regulatory framework in place to govern such activity.[500] Yet, the strategic guidelines for the period of 2015–2018 published by the NRA included an evaluation of demand-side mechanisms and hence might reflect the possible opening of balancing markets to demand response.[501]

Like Italy, Spain also uses mainly hydro and gas generation for its flexibility needs.[502] Even though some smart grid pilot projects are currently being developed, incentive-base (explicit) demand response is currently modest. Even though there is one interruptible load program that allows incentive-based (explicit) demand response, the scheme is only open to large consumers and has not been used for years. Importantly, load aggregation is illegal. Yet, proposals to open balancing markets to demand response could prompt changes in 2016–2018, especially in light of the smart meter roll-out expected by 2018.[503]

---

[497]Paolo Bertoldi, Paolo Zancanella, & Benigna Boza-Kiss, JRC Science for Policy Report: Demand Response status in EU Member States 69 (2016).

[498]Smart Energy Demand Coalition, Mapping Demand Response in Europe Today 10–11 (2015).

[499]*Id.* at 98.

[500]*Id.* at 151.

[501]AEEG, DCO 528/2014/A, Consultation Document: Schema Di Linee Strategiche Per Il Quadriennio 2015–2018 (Oct. 30, 2014) http://www.autorita.energia.it/allegati/docs/14/528–14.pdf.

[502]Smart Energy Demand Coalition, Mapping Demand Response in Europe Today 131 (2015); Paolo Bertoldi, Paolo Zancanella, & Benigna Boza-Kiss, JRC Science for Policy Report: Demand Response status in EU Member States 81 (2016).

[503]Smart Energy Demand Coalition, Mapping Demand Response in Europe Today 131 (2015).

8.6.3 The Situation in Certain EU Member States

Belgium is one of a few EU countries with a commercially sound demand response system.[504] Demand response is eligible for the primary and tertiary reserves, as well as the interruptible contracts program. In 2014, the country increased its demand response capacity to guarantee a secure energy supply in cold weather. As a result, demand response comprises 10% of strategic reserve,[505] and a pilot project is currently exploring the use of demand response in the secondary reserve. If the pilot yields positive results, Elia plans to open the secondary reserve to demand response in 2019.[506]

Although Belgium has made significant progress regarding demand response, some obstacles remain. Several of them related to broadening the scope of existing demand response to household consumers, either individually or via independent aggregators. However, aggregators require prior arrangements with the customer's supplier because when it comes to flexible loads, the seller must be the customers' Balance Responsible Party (BRP). Since the supplier is the customer's default BRP, the right to pool the customer's excess load for onward sale on the power exchange has to be transferred to the aggregator. Moreover, the threshold for being a BRP is providing a performance guarantee of €4000 per MWh,[507] which prevents customers from selecting aggregators and gives suppliers an unfair advantage. Another hindrance to demand response in Belgium is that prequalification requirements, in practice, eliminate all but big industrial consumers. For example, BRP customers must be connected to high, medium, and low voltage grids and must go through the DSO's approval procedure. In other words, household customers, are, in effect, prevented from taking part in balancing markets.

Until now, no regulation in exists in the country that defines aggregators or specifies their role in the electricity market and this could explain the challenges with providing ancillary services and serving customers independently. However, a law passed in July 2017, which is yet to be ratified, addresses this gap by defining the functions of independent aggregators. In addition, the law acknowledges that all customers in the electricity value chain have the right to flexibility without restraints imposed by retailers.[508]

In Spanish regulation, an aggregator for demand response is missing. At this time, a single scheme allows explicit demand response: the interruptible service. This means that a TSO may block energy consumption by placing a power reduction order on large industrial consumers providing this service. The interruptible service is thus a demand-side scheme managed by Red Eléctrica de España. The plan is an

---

[504]Smart Energy Demand Coalition (2017).

[505]Smart Energy Demand Coalition, "Mapping Demand Response in Europe Today," Brussels, 2015.

[506]Smart Energy Demand Coalition (2017).

[507]Idem.

[508]Idem.

emergency approach for situations of imbalance between generation and demand. The mechanism aspires to flexibility and a rapid response to TSOs needs in such a situation.[509] Although Spanish regulation does not officially recognize aggregators, the role of "representatives" has been acknowledged. These "representatives" sell energy on behalf of their "representees" and build balancing perimeters, thus reducing deviations from program and the ensuing penalties.[510] Still, Red Eléctrica de España and other industry stakeholders are entering discussions around the future provision of these services to flexible demand.[511]

Aggregated demand response may not access balancing markets. Only consumers with contracted power greater than 5 MW may access the interruptible demand service that Red Eléctrica de España manages. It is limited to large industrial consumers that are connected to the high voltage grid. Industrial energy consumers participating in this scheme tend to come from the construction sector (such as steel, concrete and glass), other material factories (e.g., paper and chemicals) and desalinization plants (in the Canary Islands). Participants must have an ICT system that connects them to the TSO and not to the DSO where they may be connected, as the DSO does not participate in such schemes.[512]

In Italy, now that smart meters have been successfully deployed and Italy has complied with the EU's goals of liberalizing electricity markets, the next major step will be demand response. As of now, demand response in the country is limited to an interruptible contracts program, which is available to customers with a capacity of at least 1 MW and who qualify as Balance Response Parties (BRPs). The mainland network as well as those in Sicily and Sardinia can access this program, however the TSO rarely avails of it for balancing. Moreover, TSO contracted out all existing capacity until 2018, so no newcomers can join unless an existing participant withdraws.[513] While aggregation is not allowed to take part in interruptible contracts, consortiums or cooperatives may do so. In fact, currently, two consortiums are participating.[514]

There is currently no mechanism for demand response and taking part on the wholesale market. While wholesale market operators may act as demand aggregators, there are no independent "aggregators" within the Italian market. Demand response must be promoted in the country, especially when it comes to allowing aggregators to participate in the market. Aggregators count as third-party service providers but their role in the market is on the rise and they are active player in areas with more developed demand response markets. The EU's Energy

---

[509]Red Eléctrica de España, "Interruptibility Service," Red Eléctrica de España. Available at: http://www.ree.es/en/activities/operation-of-the-electricity-system/interruptibility-service.

[510]Smart Energy Demand Coalition (2017).

[511]IndustRE (2016).

[512]Smart Energy Demand Coalition (2017).

[513]Smart Energy Demand Coalition, "Mapping Demand Response in Europe Today," Brussels, 2015.

[514]Idem.

8.6 Demand Response

Efficiency Directive calls upon Member States to encourage the participation of aggregators in demand response and other ancillary markets.[515] In fact, in countries where aggregators participate on the market, one can see significant infrastructural improvements. This is because aggregators have to undertake whatever improvements are necessary in order to fulfill their service provision responsibilities.[516] Such improvements, in turn, help encourage smart technology diffusion and also help promote competition.

Greece's Fundamental Energy Markets law recognizes that an important goal for the country's internal market is the adoption of demand response mechanisms. However, since the transposition of the EU Directive, very little by way of legislation has been initiated with regard to developing demand response in the country. HEDNO played a key role in attempting to design a demand response framework and has, to this effect, taken part in numerous relevant initiatives. Some steps have also been taken to encourage consumers to participate in the energy market, with the goal of helping the market evolve into a new format, in keeping with the so called "target model."

Some progress towards demand response is underway. For example, in January 2016, the Interruptible Load Service was instituted under Law 4203/2013, which allows the Greek TSO (Independent Power Transmission Operator—IPTO) to sign specific types of contracts with electricity consumers, based on which consumers then must provide interruptibility services upon receiving a relevant direction from the TSO. The service can be offered by consumers connected to the electricity transmission and MV network of the interconnected system via their participation in auctions. The TSO has launched a bidding process for interruptible load contracts for customers connected on high-voltage (HV) and MV networks, and currently 29 companies are registered in the interruptible load archive (with total offered interruptible load of 2191 MW). The final list of participants and price per MW are determined at an auction taking place every 3 months.[517]

The TSO can proceed to temporarily decrease the active power of interruptible counterparties up to an agreed value in return for financial compensation. The Ministerial Decision (ΑΠΕΗΛ/Γ/Φ1/οικ. 184898, Official Gazette Β' 2861/ 28.12.2015) contains information about which consumers are eligible to sign an interruptibility contract, the requirements and preconditions to do so, the reasons behind the establishment of the service, as well as the manner, timing and preconditions for providing compensation to those who participate. Additionally, demand control contracts are in place for customers connected to the MV and LV

---

[515] R. Panetta and A. D'Ottavio, "Data protection in Italy: overview," Thomson Reuters, 1 12 2015. Available at: https://uk.practicallaw.thomsonreuters.com/9-502-4794?transitionType=Default&contextData=(sc.Default).

[516] Smart Energy Demand Coalition, "Mapping Demand Response in Europe Today," Brussels, 2015.

[517] ΛΑΓΗΕ Operator of the electricity market, "www.lagie.gr / DAS monthly reports May 2017."

network of the interconnected system and in the non-interconnected islands, as long as they have the necessary telemetering equipment.

Moreover, there are contracts for residential customers offering lower tariffs during the night and interruptible load contracts for "agricultural customers." In 2017, 2.7% of all LV customers and 4.4% of all MV customers were participating in these special contracts.

Law 4342/2015 also states that the market codes, currently being drafted by the RAE, must contain provisions that oblige the TSO and distribution network operator to treat persons who provide demand response services in an equal and objective way, based also on their technical infrastructure and potential. The law also contains the first definition of "Aggregator."

Law 4425/2015 addresses integrating demand response into the balancing market, in keeping with the goal of helping to incorporate the Greek wholesale market into the European electricity market. The idea is that market codes, which, as mentioned above are currently being drafted by the RAE, are set to contain exact details regarding the demand response mechanism.

Demand response schemes in Greece are unfolding at a rather gradual pace largely because the infrastructure is not ready. For example, smart meters, which are mandatory to accurately record consumption and to allow consumers to control and adjust consumption, are still in the preliminary rollout phase. It is hoped that the numerous research projects currently underway will help demand response to launch quickly, once all the required infrastructure is in place. In Greece, when it comes to market services and demand response schemes, while a preliminary framework exists, and special contracts are in use, the current market is not ready for a widespread deployment of demand response mechanisms.

Not long ago, the legislative framework added provisions allowing the DSO to issue demand control contracts (interruptible load) with any customer on the LV network (upon approval by national regulatory authorities), as long as the customer's facilities are equipped with telemetered load technology and satisfy the necessary technical requirements set by the DSO. However, no such contract has been entered into as yet. Once such contracts start being issued, relevant WiseGRID tools will facilitate the participation of residential and business end-users in demand response campaigns and thus in the overall energy market.

France is said to be amongst the *"European countries that currently provide the most conducive framework for the development of Demand Response."*[518] Along with Switzerland, France is one of the countries that *"have detailed frameworks in place for independent aggregation, including standardised roles and responsibilities*

---

[518]Smart Energy Demand Coalition (2017) available at http://www.smarten.eu/wp-content/uploads/2017/04/SEDC-Explicit-Demand-Response-in-Europe-Mapping-the-Markets-2017.pdf, p. 10.

8.6 Demand Response   383

of market participants."[519] Moreover, France is to be counted amongst those countries that have also enabled aggregated load to participate.[520]

Interestingly, "*a new draft decree being reviewed by the Conseil d'Etat in early 2017 could provide for a new financial settlement framework whereby a significant of the payment to retailers with curtailed customers would be charged to retailers rather than to demand response providers. However, issues persist around a standardised baseline methodology.*"[521]

It is contended that users/drivers of electric vehicles can make a significant contribution to the management of energy, due to the potential of electric vehicles as mobile storage units. Accordingly, the "Commission de regulation de l'energie" (self-defined as the "*Independent administrative body in charge of regulating the French electricity and gas markets*")[522] considers the arrival of these vehicles as a "key element" in the management of electric grids, and encourages, through one of its web sites[523] the use of electric vehicles in such a way. Two facts constitute the basis of the approach taken by the Commission de regulation de l'energie: on the one hand, the Commission asserts that a vehicle is out of use for 95% of its useful life and, on the other hand, it assets that the average use in everyday itineraries of an electric vehicle will require less than 80% of the battery's capacity.

Interestingly, the approach is being intensely supported from the industry side: Enel, an energy multinational corporation, significantly present in the French market, graphically asserts that the "*car of the future is a battery with wheels*", and takes pride in having invested in five pilot projects in this sense, one of them being located on French soil.[524] In a very customer-friendly fashion, the company explains that the:

> V2G technology turns e-cars into large mobile batteries that interact smartly with the power grid, enabling, among other things, the stabilisation of power flows to promote renewable generation. Cars can accumulate energy at lower fuel consumption times and return any excess quantities. In fact, the new technology is based on bidirectional charge management:

---

[519] Smart Energy Demand Coalition (2017) available at http://www.smarten.eu/wp-content/uploads/2017/04/SEDC-Explicit-Demand-Response-in-Europe-Mapping-the-Markets-2017.pdf, p. 10. Moreover, p. 12: "*France and Switzerland are still currently the only countries to have a clear framework on the status of independent aggregators and their role and responsibilities in the market [. . .]*".

[520] This is per opposition to other European countries such as Slovenia and Poland, which, despite opening their markets to load participation, have chosen not to open them to aggregated load "*therefore disqualifying all except the largest industrial consumers from accessing these markets.*" See http://www.smarten.eu/wp-content/uploads/2017/04/SEDC-Explicit-Demand-Response-in-Europe-Mapping-the-Markets-2017.pdf, p. 31.

[521] http://www.smarten.eu/wp-content/uploads/2017/04/SEDC-Explicit-Demand-Response-in-Europe-Mapping-the-Markets-2017.pdf, p. 10.

[522] Commission de regulation de l'energie. See: www.cre.fr/en.

[523] Commission de regulation de l'energie, "L'utilisation du véhicule électrique comme moyen de stockage'" available at: www.smartgrids-cre.fr/index.php?p=stockage-vehicule-electrique.

[524] Enel, "V2G, the car of the future is a battery," available at: corporate.enel.it/en/stories/a/2017/05/V2G-the-car-of-the-future-is-a-battery.

it draws energy for example from home renewable systems, and feeds it into e-cars. If necessary, it transfers it from the car to another plug-in structure or simply to the network, guaranteeing a financial return for the "giver."[525]

Fully embracing this philosophy, the company states on its web site that, in France, its "Gridmotion" project boosted its sale of rechargeable vehicles by 42% between 2015 and 2016.[526]

Notwithstanding these considerations, even if the French "approach" described in previous paragraphs is to be welcomed, it needs to be embraced with caution. The key element in making the whole scheme fit together is "educating" users adequately; the potential success of the scheme depends on the users' behavioural patterns and developing in them an appropriate sense about finding the right moment for recharging their vehicles. Indeed, the Commission de regulation de l'energie warns that it is extremely relevant that the status of the grid be taken into account by users when charging and/or discharging the batteries: recharging an electric vehicle during the winter peak-consumption hour (7:00 PM) would entail a significant additional burden for the balance of the grid, which would need to be avoided.[527]

Thus, while electric vehicles may represent a significant addition to energy storage capacities, the Commission de regulation de l'energie considers that it needs to be pursued cautiously. This is especially true as analyses are required to determine whether this user-involvement in energy management is technologically and economically appropriate. As opposed to what happens with energy mass-storage, this use of the batteries would require the potential for very numerous and very fast charge and discharge cycles, as well as a very high energy density.[528]

Demand response tools are useful as capacity mechanisms to make sure that electricity supply is guaranteed and electricity is not wasted. When the tools are not correctly implemented, this can cause distortions of competition mechanisms, of the electricity flows across borders and of the final electricity price for consumers.[529] TSOs play a crucial role in guaranteeing demand response mechanisms. In some cases, however, ACER (the Agency for the Cooperation of Energy Regulators) had to intervene in response to calls for an improved coordinated capacity allocation procedure on the German-Austrian border.

In a country like Germany, where the *Energiewende* is increasingly introducing intermittent renewable sources in the energy mix, flexibility is crucial to make sure network operators are able to react promptly to volatility in the grid. Also, both the safety of supplies and efficiency have to be guaranteed. For these reasons, Germany

---

[525]*Ibid.*

[526]*Ibid.*

[527]Commission de regulation de l'energie, "L'utilisation du véhicule électrique comme moyen de stockage"' available at: www.smartgrids-cre.fr/index.php?p=stockage-vehicule-electrique.

[528]*Ibid.*

[529]European Commission, State aid, Brussels, 24 October 2016.

8.6 Demand Response

has introduced the *Verordnung zu abschaltbaren Lasten* (AbLaV),[530] a measure that helps network operators to control consumption according to customer needs. This can be done at short notice and remotely, through flexible, weekly competitive auctions, with customers paying a fee. These contracts must involve a total of 1500 megawatt (MW) of capacity and electricity customers consuming more than 10 MW.[531]

Contracts are indeed changing in Germany: the contract between users and a large utility company is no longer the norm in a smart grids system that instead creates a marketplace for supply and demand of electrical power at the local level. This enables smart contracts to balance demand-response, setting the price per kWh and peer-to-peer trade.

In the Netherlands, demand response mechanisms are generally a responsibility of DSOs. At the moment, there is no specific legislation/regulation in place in the Netherlands, as the country is still focusing on safeguarding availability and affordability of the distribution network. This means that the Netherlands is focusing on infrastructure development, which is part of the path towards the energy transition. The Netherlands still has work to do before it reaches the next stage of managing its energy resources for the benefit of society. Evidence of that is the fact that consumers are currently charged a network tariff, without any saving being calculated on the basis of their usage pattern, such as time, location, or electricity price. This means that prosumers pay similar tariffs to all other consumers, although the former category has a higher impact on network capacity.

Nonetheless, new projects are starting to be developed in the field of managing congestion, which is an introductory step to further adopt proper demand response mechanisms. One example of such projects involves Dutch regional DSOs (Stedin, Liander, Enexis Groep and Westland Infra) and the national TSO, namely TenneT. The project is called GOPACS (Grid Operators Platform for Congestion Solution) and aims at creating a flexible solution to reduce congestion in the electricity grid.[532] When market parties with a connection in the relevant area place orders, GOPACS makes sure that the order will not cause disruptions in the electricity grid, offering large and small market parties an opportunity to generate revenues.[533]

In conclusion, these types of projects support the move to the next step in the energy transition, as they can solve practical issues (such as the congestion of the grid in the case of GOPACS); they can represent interesting platforms for collaboration between regional DSOs that will strengthen their relationships and the capacity to work together; and they can also promote aggregators, which can allow even small-end parties to be involved in the energy market.

---

[530]Verordnung über Vereinbarungen zu abschaltbaren Lasten (Verordnung zu abschaltbaren Lasten, AbLaV, available at: www.gesetze-im-internet.de/ablav_2016/BJNR198400016.html.

[531]*Ibid.*

[532]https://gopacs.eu/.

[533]https://www.tennet.eu/news/detail/dutch-grid-operators-launch-gopacs-a-smart-solution-to-reduce-congestion-in-the-electricity-grid/.

8.6.4 Regulatory Policy Recommendations

Overall main regulatory barriers found repeatedly across European countries include:

1. **Demand response might not be accepted as a flexibility resource:** in some European countries wholesale, balancing and/or capacity markets do not accept aggregated demand as a flexibility resource.[534]
2. **Inadequate and/or non-standardized baselines:** in some European countries, standardized measurement and baseline methodologies are absent. Current methodologies are designed for generators and, consequently, do not accurately measure changes in consumption. This could hinder demand response, because consumers might not receive adequate payment for their flexibility.[535]
3. **Technology-biased program requirements:** program participation requirements, historically designed for national generation, might not include demand side resources.[536] Power markets more in line with demand response timeframes have to be established (e.g., based on 15- rather than 60-min timeframes).[537]
4. **Aggregation services are not fully enabled:** prequalification, registration, and measurement may still be conducted at the level of individual consumers, rather than at the level of pooled loads brought together by the aggregator, which hinders entry by placing heavy administrative and legal burdens on individual consumers.[538] Moreover, there is often no real definition of load aggregators. To promote the possibility for consumers to contract with aggregators, load aggregators must be legally acknowledged as facilitators of demand side flexibility.
5. **Aggregators, where existing, are currently active at the high and medium voltage levels, rather than the low voltage level:** load aggregators exist in some countries, such as France and Belgium. Yet, their activities are currently focused on the high and medium voltage levels, namely at transmission and dealing with TSOs. We therefore must learn how these activities might be translated, if at all, at the low voltage level, namely at distribution and when dealing with DSOs.
6. **Lack of necessary infrastructure:** while there is much discussion about the emergence of load aggregators, it must not be forgotten that aggregators rely on

---

[534]*Id.* at 11.

[535]*Id.*

[536]*Id.*

[537]*Id.* at 82; Paolo Bertoldi, Paolo Zancanella, & Benigna Boza-Kiss, JRC Science for Policy Report: Demand Response status in EU Member States 54 (2016).

[538]Smart Energy Demand Coalition, Mapping Demand Response in Europe Today 11 (2015).

certain infrastructures to provide load aggregation services. The key step here is to install smart meters, which in some of the European member states are not yet deployed.[539]

7. **Lack of standardized processes between consumers, BRPs, and aggregators:** it is important that standardized processes protect the relationship between customers and aggregators, and govern bidirectional payment of sourcing costs as well as compensation between the BRPs (often the traditional suppliers) and the aggregators.[540] In other words, it is crucial to put contracts in place between DSOs, load aggregators, and customers. It is vital that the right of consumers to offer their flexibility on the market be acknowledged, while guarantees are put in place so that consumers maintain their rights when they sign up for demand response. There should be a provision for the network side to ensure some minimum balancing support through demand response schemes. Thus, demand response schemes could really contribute in reducing other capacity mechanisms.

8. **Provision of information to consumers:** this relates not only to energy prices and how much customers could save by changing their consumption patterns, but also to other kinds of information. Consumers could feel more motivated to engage in demand response programs and choose among suppliers and aggregators depending on the mix of energy sources from which the electricity they consume is produced. Consumers could prefer a program and service provider that produces energy from cleaner sources, even if the monetary gains they could make were limited.

9. **Differences across consumers that could hinder their participation**: in addition to different monetary incentives and regulatory frameworks primarily set at the national level, consumers within same countries could, de facto, find themselves facing different possibilities for joining demand response schemes. Just as in the case of the installation of micro-generation renewable plants (e.g., solar panels on the rooftop), it might be that consumers living, say, in a flat, rather than in a house with a garden, do not have the same possibility to engage in demand-side flexibility solutions. Hence, it might be appropriate for the relevant authorities at the national level and, if appropriate, also at the EU level, to consider how to create a more level playing field on the consumer side.

10. **Lack of financial incentives for consumers, especially through automatic adjustments within comfort levels**: it is now well-known, especially thanks to studies from the discipline of economics, that the efforts of policymakers to empower consumers are often frustrated by the fact that consumers do not react to efforts to alter their consumption patterns.[541] Ironically, perhaps this is because they do not see the financial gain as sufficient reward for altering

---

[539] *Commission Report Benchmarking Smart Metering Deployment in the EU-27 with a Focus on Electricity*, COM (2014) 356 Final (June 17, 2014).

[540] *Id.*

[541] *See generally* Nolan and O'Malley (2015), p. 1.

their consumption. Considering this difficulty, in addition to increasing financial incentives and promoting more cost-reflective tariffs that provide price signals for customers to adjust their consumption patterns, regulation could also consider providing fiscal incentives. Governments might consider putting in place policies that, through taxation, support demand-side adjustments. Another aspect that could be considered is a stronger use of "negative" financial incentives. These could manifest as increases in the penalties, rather than rewards for changing consumption patterns, which might be more effective than "positive" incentives.

11. **Automatization of demand response mechanisms**: consumer participation in demand response programs should be made as easy as possible. In addition to concentrating on the rewards side of the equation, attention should be devoted also to the cost side. Consumers should have to invest as little time and effort as possible, so that they might engage in demand response even if the financial rewards are not very high in absolute terms. Automatization of responses appears to be crucial in this context. Consumers will not have to do anything, because adjustments in their consumption patterns will be automatic. The North American market is more experienced in the automatization of changes in consumption patterns within customers' "comfort zone."[542] For example, changes in the intensity of lighting within a flat that will not be noticed by its residents and will be activated automatically when appropriate. The US power markets are also more experienced with load aggregators. Hence it appears desirable to look at these experiences and learn from them.

8.7 Storage

One outcome of promoting EVs is the creation of electricity storage systems. While solutions to the problem of large capacity energy storage are still in experimental stages of development,[543] the importance of energy storage in future energy management systems cannot be underestimated. Current storage systems meet the temporary storage needs of small to medium-scale generation, usually from RES. Despite the lack of technological advancement, energy storage is beneficial to all

---

[542]*See generally* Smart Energy Demand Coalition, Mapping Demand Response in Europe Today (2015); Paolo Bertoldi, Paolo Zancanella, & Benigna Boza-Kiss, JRC Science for Policy Report: Demand Response status in EU Member States 42 (2016).

[543] 1. Fluid storage, particularly pumped hydroelectric plants are the most common. They use off-peak electricity to pump water from a low reservoir uphill into an elevated reservoir. The water is then released through turbines to generate power at very short notice. 2. Compressed Air Energy Storage (CAES) operates similarly although still experimental and not developed enough as a commercial storage application—electricity is used to compress air in underground caverns, then tapped later to drive gas turbines. 3. Hydrogen storage, which involves the hydrolysis of water to produce hydrogen gas, is compressed and stored, then converted to power when needed. However, the high explosion risk associated with the technology has impeded its viability.

levels of the electricity market. First, they provide an option to redress the problem of the intermittence of RES generation.[544] Further, the ability to store energy when prices are low and possibly sell when prices increase presents an opportunity for arbitrage.[545]

8.7.1 The EU Legal Basis

The legal framework governing electricity storage in Europe is provided at the EU level by the Third Energy Package.[546] At the same time, laws are under development at the national level that will regulate electricity storage applications.[547] It is important to note that Directive 2009/72/EC[548] does not expressly mention energy storage. However, the proposal for a new directive on common rules for the internal electricity market of February 2017 does regulate energy storage. The text of the proposal clarifies that DSOs should not be allowed, directly or indirectly, to own storage facilities.[549]

8.7.2 Current Status Across Europe

After conducting an overview of the distinct electricity storage technologies used in Europe at the end of 2012 and their expected increase in the ensuing 5 years, the CEER memo on development and regulation of electricity storage applications

---

[544]Luo et al. (2015), p. 511; Masson et al. (2016), p. 61.

[545]Committee on Industry, Research and Energy, *Energy Storage: Which Market Designs and Regulatory Incentives Are Needed?* at 17 (PE 563.469, Oct. 2015).

[546]Parliament and Council Directive 2009/72/EC, Concerning Common Rules for the Internal Market in Electricity, 2009 O.J. (L 211) 55; Parliament and Council Directive 2009/73/EC, Concerning Common Rules for the Internal Market in Natural Gas, 2009 O.J. (L 211) 94; Parliament and Council Regulation 714/2009/EC, On Conditions for Access to the Network for Cross-Border Exchanges in Electricity, 2009 O.J. (L 211) 15; Parliament and Council Regulation 2009/715/EC, On Conditions for Access to the Natural Gas Transmission Networks, 2009 O.J. (L 211) 36; Parliament and Council Regulation 2009/713/EC, Establishing an Agency for the Cooperation of Energy Regulators, 2009 O.J. (L 211) 1.

[547]Council of European Energy Regulators, *CEER Memo on Development and Regulation of Electricity Storage Applications*, p. 3 (C14–EQS-54-04, July, 21, 2014).

[548]Parliament and Council Directive 2009/72/EC, Concerning Common Rules for the Internal Market in Electricity, 2009 O.J. (L 211) 55.

[549]*Commission Proposal for a Directive of the European Parliament and of the Council on Common Rules for the Internal Market in Electricity*, p. 82, COM (2016) 864 final/2 (Feb. 23, 2017).

concluded that hydro-pumping storage is currently the most commonly used electricity storage technology.[550] This picture is not expected to change considerably in the next several years. Although other technologies will be employed (e.g., flywheels, compressed air electricity storage, and electrochemical storage), they will still represent less than 3% of installed power.[551] Even if they increase in number of applications, the associated growth in energy capacity will be minor. It is expected that electrochemical storage will increase by up to 100 MW thanks to new demonstration projects.[552] However, this stands in contrast with hydro-pumped storage, which represents about 37 GW in storage capacity in the CEER member states.[553] Of course, the situation might change, even dramatically, thanks to breakthrough technologies.[554]

The regulation of storage assets faces many conceptual and practical challenges. Conceptually, there is no consensus on the definition of storage assets. The question of whether they should be treated as generation assets or consumption units is particularly unresolved. This lack of clarity stems from the fact that, while storage assets can generate electricity in the literal sense of "generation," the amount of electricity generated is typically not enough to provide a net positive flow to the electricity system.[555] On the other hand, they cannot be properly classified as consumption units because they do not actually consume the energy that they take up. Could they also be classified as part of a transmission or distribution network, given that they can be a bridge asset between generators and final consumers? The answers to these questions is fundamental to the development of an appropriate regulatory regime because they impact, inter alia, ownership, pricing, and the imposition of taxes and levies.

Regarding issues of ownership, the CEER memo shows that in most European countries, storage applications are owned by generators even though, in some countries, network operators may, to a certain degree, own storage applications.[556] In most European countries, storage can provide services to both network operators and generators and its primary users are owners.[557] The ownership of storage assets is one of the challenges that impinges the development of appropriate regulation. While there is no doubt that market operators such as TSO would benefit from owning storage assets, their unique position in the market presents an information

---

[550]Crossley (2013), p. 268.

[551]Council of European Energy Regulators, *CEER Memo on Development and Regulation of Electricity Storage Applications*, pp. 2–3 (C14–EQS-54-04, July, 21, 2014).

[552]*Id.*

[553]*Id.*

[554]*Id.*; *Commission Staff Working Document on Energy Storage – the Role of Electricity*, SWD (2017) 61 final (Jan. 2, 2017).

[555]Gissey et al. (2016).

[556]Council of European Energy Regulators, *CEER Memo on Development and Regulation of Electricity Storage Applications*, p. 3 (C14–EQS-54-04, July, 21, 2014).

[557]Gissey et al. (2016), p. 3.

asymmetry which would operate unfairly to their advantage against other market players. This is particularly true if stored energy is participating in the balancing and ancillary markets. It is in response to this problem that current proposals for the Electricity Directive seek to proscribe the ownership of storage assets by the owners or operators of network infrastructure.[558] The proposed proscription is in keeping with the EU's unbundling policy as a bid to prevent counter-competitive activity in electricity markets.

In Spain, although there is no general regulatory framework for electricity storage, there are hydro-pumped storage power plants that perform the function of providing power during hours of peak consumption.[559] The only exception relates to regulation of storage for small self-consumption systems. Under the Electricity Sector Law 24/2013, battery owners do not only have to pay an additional tax, but are also not allowed to reduce the maximum power they have under contract with their supplier.[560] While it may be argued that this is intended to maintain grid integrity, when coupled with the high self-consumption tax, the regulatory regime for self-consumption and storage appears to be ill-considered.

In some cases, the regulatory framework not only does not promote, but actually hinders the development of storage. For example, in some countries, taxation is not favorable to storage, as typified by the "Grid Fee System."[561] Ordinarily, grid fees are paid by the final consumers of power as a fee for the transportation of electricity through the grid network.[562] In the case of storage, operators of storage assets are first charged for charging the storage asset. The operators are then also charged for discharging it because of the notional double flow of electricity. In real terms, the storage asset is neither a producer nor consumer. Therefore, the strict application of the traditional grid fee model should not extend to storage assets. Often, this double taxation is higher than power prices and results in a very strong disincentivization of electricity storage.[563]

8.7.3 The Situation in Certain EU Member States

In Belgium, this is an important prospect as it prepares to transition away from nuclear energy and increase RE usage. At this time, storage facilities in the country

---

[558]Parliament and Council Directive 2009/72/EC, Concerning Common Rules for the Internal Market in Electricity, 2009 O.J. (L 211) 55.

[559]Masson et al. (2016), p. 26.

[560]Jason Deign, *Spain's New Self-Consumption Law Makes Batteries Impractical for Homeowners*, GreenTech Media, October 16, 2015. *available at* https://www.greentechmedia.com/articles/read/spanish-self-consumption-law-allows-batteries-at-a-cost; *Ibid*.

[561]Crossley (2013), p. 268.

[562]European Parliament Committee on Industry, Research and Energy (ITRE), Study on Energy Storage: Which Market Designs and Regulatory Incentives Are Needed? (PE 563.469, 2015).

[563]*Id*.

are limited, with only two hydropower plants that have a total capacity of around 1.3 GW. Although the initial aim was to use the plants to regulate generation from the Tihange nuclear plant, they are in fact being employed to balance load in the grid.[564] Storage capacity needs in the country are likely to rise from 7 GW to 12 GW by 2020. To meet this need, a manmade offshore pumped-storage facility is being planned, to support offshore wind power generation.[565] One of the hydropower plants may also be upgraded to increase storage capacity. However, while these planned infrastructure upgrades may help, regulatory challenges must be addressed in order to successfully develop storage technology in the country.

To help build and run offshore pumped hydro storage projects, Belgium enacted an amendment in 2014 to the Federal Act of 29 April 1999.[566] Unfortunately, the law's scope was rather narrow, and it did not address the country's regulatory gap regarding energy storage solutions. For example, in Belgium, as in several other European states, storage facilities result in a double payment of grid charges, because storage technology is not categorized correctly in the electricity value chain. A strong regulatory framework should also consider specific tariffs and subsidies for storage infrastructure, which would incentivize investment.

In Spain, electricity storage is negligible compared to current installed power capacity. Reversible hydraulic reservoirs is the primary source of electricity storage in the country, accounting for more than 90% of the country's electricity storage.[567] Total hydro pump storage capacity in Spain is 4749 MW, which comprised 4% of installed capacity in 2016.[568] The government is aiming to reach a storage capacity of 8100 MW by 2020.[569]

Red Eléctrica de España has introduced the "*Almacena*" project to explore other options for electricity storage. The project entails an electrochemical energy storage solution connected to the grid, and the installation of a prototype flywheel in the Canary Islands.[570] Additionally, Endesa has created a pioneer project launching three kinds of energy storage infrastructures (superconductor, flywheel and

---

[564]SIA Partners, "Energy Outlook: Energy Storage in Belgium," SIA Partners, 10 May 2013. Available at: http://energy.sia-partners.com/energy-storage-belgium.

[565]I.-S. Brouhns, "CMS Guide to Energy Storage: Belgium," Global Business Media Group, 1 September 2016. Available at: https://www.lexology.com/library/detail.aspx?g=4ebed335-f539-46dc-95f2-bdf5d7bc705f.

[566]Idem.

[567]Smart Energy Demand Coalition, "Mapping Demand Response in Europe Today," Brussels, 2015.

[568]Energías Renovables, "La escasez de puntos de recarga frena la compra de coches eléctricos más que el precio," Energías Renovables. El periodismo de las energías limpias, 22 June 2017. Available at: https://www.energias-renovables.com/movilidad/la-escasez-de-puntos-de-recarga-frena-20170622.

[569]Stoppani (2017), p. 17.

[570]Red Eléctrica de España, "Energy Storage," Red Eléctrica de España. Available at: http://www.ree.es/en/red21/energy-storage.

electrochemical) in three different Canary Islands. Such storage types may present a range of advantages for the grid.[571]

In Italy, storage capacity is mostly made up of pump hydro storage units, which mainly used to be owned by Enel and other state-owned electricity companies. But now that services have been unbundled and the market liberalized, TSOs own the storage units. Moreover, the law allows DSOs to own storage units.[572]

Regulation around electricity storage is somewhat disjointed and fails to address all relevant aspects. The main rules around it are contained in Legislative Decree no 28/2011. However, the decree simply hands off the right to develop and manage storage units the TSO and DSOs. The AEEGSI's Decision on Provisions related to the Integration of Energy Storage Systems for Electricity in the National electricity System (Decision 574/2014/eel of 10 November 2014) contains provisions regarding the connection of storage systems to the grid by non-regulated participants such as prosumers. It defines energy storage as power generation and places connection, transmitting and metering responsibilities on energy storage, making it necessary for storage facilities to pay connection fees.[573]

The TSO, Terna S.p.A., runs a range of programs with the objective of boosting Italy's storage capacity, especially with regard to creating better battery systems in Sicily and Sardinia. Additionally, in the region of Puglia, a hydrogen storage initiative is anticipated to link supply and demand and help towards the network's more reliable supply overall.

Although there seems to be an overall desire to boost Italy's storage capacity, little support exists for DSOs and other bodies exploring research and development (R&D) for storage technology. Indeed, developing Italy's storage capacity is hindered by the high price of batteries and storage related technologies. The lack of certainty around suitable revenue streams for distributed generation does not help, either. So far, the only means to tackle these issues takes the form of R&D funds under the management of AEEGSI.[574] There is a lot of room to involve the private sector, especially DSOs, to put money into R&D for storage technology. More important, Italy needs appropriate legislation, as right now there is little legal acknowledgement of the important role of storage in the electricity value chain. Effective incentives, too, could help towards progress in the sector.

In Greece, a gap exists in both the market and in legislation regarding batteries, which poses a key challenge at this time. Energy storage is a useful tool that can be applied in various situations to reduce usage at peak times or to stabilize the grid frequency by charging or discharging power. Due to its flexibility, though, there is a

---

[571]Stoppani (2017), p. 17.

[572]European Commission, Directive 2012/27/EU, on energy efficiency, amending Directives 2009/125/EC and 2010/30/EU and repealing Directives 2004/8/EC, 25 October 2012.

[573]Committee on Industry, Research and Energy (ITRE) European Parliament, "Energy Storage: Which Market Designs and Regulatory Incentives Are Needed?" European Commission, 2015.

[574]European Commission, Directive 2012/27/EU, on energy efficiency, amending Directives 2009/125/EC and 2010/30/EU and repealing Directives 2004/8/EC, 25 October 2012.

lack of clarity around its role in the electricity market. Is it a generator or is it a load, can the DSO own it, what are the revenue streams and who qualifies for them? Moreover, at times the battery works without benefitting the grid but provides financial incentives to the end user. At other times, the battery might even work against the grid and still earn financial benefits.

Moreover, Greece generally struggles with unclear regulation when it comes to energy storage. Right now, battery deployment could be integrated with that of large hybrid stations catering to the MV level. However, no provisions exist for batteries in residential or public buildings, for electric vehicles, V2G technology on how to communicate with the power grid, or for how to sell demand response services. Moreover, in the non-interconnected island electrical systems (NIIES), the current operating procedures are dominated by conventional generation work under the assumption that only conventional generation units can provide the ancillary services necessary for grid stability. This precludes the ability of aggregating a battery or an EV fleet to supply the same service.

A National Renewable Energy Plan (NREAP) was launched in the country in 2010, with the goal of lowering Greece's reliance on energy imports as well as increasing renewable energy usage, thereby lowering carbon dioxide emissions. Following the NREAP, the National Energy Strategy Committee, in 2012, developed an energy strategy for 2050. Acknowledging the intermittent status of electricity from RES, Greece is aware of the need to develop appropriate storage technology to support RES penetration. The 2050 roadmap points out that advancing pump hydro energy storage (PHES) units play a key role in attaining large RES electricity production, since it represents the most advanced and reliable technology for large-scale electricity storage, and is suitable to Greece's topography.[575]

Several projects are in the works with the aim of creating better improve storage systems in both the interconnected system and in the NIIES. One such project is the installation of a battery storage system on the island of Tilos in the Aegean Sea. This battery project is expected to power the entire island.[576]

Although the mainland interconnected system lacks regulation that addresses energy storage, Laws 3468/2006, 3851/2010 and 4414/2016 have detailed provisions dealing with the operation of hybrid stations in the NIIEs and within the interconnected system. The most recent activity pertaining to the operation and the building of storage facilities is a public consultation that RAE conducted around 2013. RAE suggested a ruleset to address the involvement of storage in the country's power system. The RAE proposal looks at all types of storage technology, and, with the objective of better integrating RES in the electricity grid, at removing current barriers to RES and boosting RES integration capacity. Moreover, the expectation is that storage would result in ancillary benefits when RES penetration is high. In the RAE proposal, the TSO would be in charge of scheduling storage units equally.

---

[575]Anagnostopoulos and Papantonis (2013).

[576]Garanis-Papadatos and Boukis (2016).

8.7 Storage

Storage units would take part in the market through bidding in the day-ahead and intraday energy market. Lastly, RAE's document permits bilateral contracts between the storage owner and RES stations.

RAE additionally puts forth a pricing system to encourage storage owners to store energy from RES not from thermal units. Additional costs stemming from the participation of storage in the market will be charged to the RES station, given that that the RES station will enjoy increased production owing to the removal of barriers.

Another barrier to developing storage capacity in Greece relates to the grid fees regime in the country. Greece is one of three EU Member States that charge grid fees for charging and discharging storage units by way of treating them as generation assets. As a result, owners of storage units have to pay grid fees as generators when charging units and as consumers when discharging them.[577] The regulatory treatment of storage units in this manner follows the true nature of these assets, and it is clear that regulatory reform will be needed to address this challenge as it poses a disincentive towards investing in storage technology.

Putting aside storage capacity arising from the use of vehicles as storage units, with respect to storage in general in France, EDF, one of the key players in the French energy landscape, is heavily promoting its *"Game-changer solar energy storage solution"* in Guiana, one of France's overseas departments, as a part of its *"solutions for the climate"* strategy.[578] Their Toucan project is defined as a *"photovoltaic power plant equipped with thin-film panels,"* which comprises *"a battery storage system that contributes to the grid's stability by absorbing any surplus solar energy and feeding it into the grid when required."*[579] EDF advertises the Toucan project as being the first of its kind in the world, and asserts that *"by storing electricity and smoothing output, [...] is providing an answer to the question of how to integrate renewable energy into grids. As a consequence, it is paving the way to increasing the share of renewables in the overall energy mix."*[580]

EDF has recently announced that it intends to invest US$9.93 billion in electricity storage by 2035, not only in France, but across the European market, and also in Africa, specifically *"including battery storage and storage plus solar projects in Ghana and the Ivory Coast."*[581] Reportedly, EDF already has five gigawatts of installed grid-scale storage, and calculates that its new 10-billion-dollar energy

---

[577] Gissey et al. (2016). Committee on Industry, Research and Energy (ITRE) European Parliament, "Energy Storage: Which Market Designs and Regulatory Incentives Are Needed?" European Commission, 2015.

[578] EDF, "Smoothing and storing photovoltaic generation," available at: www.edf.fr/sites/default/files/contrib/groupe-edf/premier-electricien-mondial/cop21/solutions/pdf/cop21-solutions_toucan_va.pdf.

[579] Ibid.

[580] Ibid.

[581] William Pentland, *French Nuclear Giant Gambles Big On Energy Storage*, Forbes (Mar. 27, 2018), https://www.forbes.com/sites/williampentland/2018/03/27/french-nuclear-giant-gambles-big-on-energy-storage/#404b846dd8ff.

storage plan will add 10 extra gigawatts of storage, amounting to a total of 15 GW by 2035. Out of those 10 GW, six will be devoted to industrial-scale projects (including pumped hydro storage and batteries), and the remaining four gigawatts will comprise of individual batteries, devoted to retail customers, municipalities, and companies.[582]

In the case of Germany, the storage sector has room for improvement, as Germany has been quite reluctant to build new storage facilities. Establishing an effective market framework for storage would be very useful in the near future. TSOs play an important role in the storage sector. They are responsible for the qualification procedures for any new storage system. Energy storage systems benefit from connection privileges for RES plants to the public grid (section 58 of the EEG 2017).[583]

Although energy storage systems are integral to the *Energiewende,* German energy law does not provide regulatory incentives for storage applications. The Renewable Energy Act focuses more on the production of renewable energy rather than on storage. Section 19 rules that for electricity in temporary storage the TSO's entitlement to payment allowed by the Renewable Energy Act shall not apply. This happens only when energy is removed from the electricity storage installation solely for the purpose of re-feeding electricity into the grid system.[584]

The EEG 2017 guarantees that surcharges are not imposed twice on electricity stored in a storage system. The charge is imposed only when the electricity is fed from storage into the public grid, but not when a consumer uses that stored electricity.[585]

Along with typical contracts related to energy storage (supply contracts for the battery storage system and construction contracts for the planning and installation of the storage system), the *Energiewende* will support the diffusion of new types of contracts, such as innovative leasing and rental models, to keep up with the diffusion of small battery storage and household or commercial small-scale PV plants.[586]

In the Netherlands, one of the main actors in the storage sector is the Dutch Ministry of Economic Affairs. It has the legislative power to adopt new legislation and has recently developed the Netherlands Energy Storage Roadmap. TenneT is also expected to have a major role in the storage sector. In terms of commercial actors that are developing energy-storage projects and exploring possibilities for energy storage, companies such as AES, NUON, SUEZ, and Cofely are very active.

Although the development of energy storage in the Netherlands is still at an early stage, there are a few important projects that put the Netherlands in a position of

---

[582]CleanTechnica, French Utility Company EDF Plans Energy Storage Push, available at: cleantechnica.com/2018/04/04/french-utility-company-edf-plans-energy-storage-push.

[583]Renewable Energy Sources Act 2017, available at www.bmwi.de/Redaktion/EN/Downloads/renewable-energy-sources-act-2017.pdf?__blob=publicationFile&v=3.

[584]*Ibid.*

[585]*Ibid.*

[586]CMS Germany, CMS Guide to Energy Storage: Germany. 2016.

8.7 Storage

leadership in the storage sector. Considering the potential further development of the sector (most projects are still on an explorative phase), one can envisage that more entities such as those described above will start focusing on storage, especially in the business sector. Two examples of very successful projects are the AES battery project of 2016 and the Amsterdam Arena project of 2018.

The AES battery project is designed to support the integration of renewable energy into the grid and improving the grid reliability by balancing energy supply and demand. The array, which provides 10 MW of interconnected energy storage,[587] uses fast response balancing services to the grid.[588] It is the largest and most advanced energy storage platform operating both in Europe and globally.[589]

As for the Amsterdam Arena project, old electric car batteries that have passed their peak performance are used to power the Arena in case of major football matches or other big events. Almost 150 Nissan LEAF batteries are used to store the energy produced by 4200 solar panels on the roof of the stadium and the grid.[590] As lithium-ion batteries can store electricity for many years after they have become useless for vehicles, the new energy system at Amsterdam Arena represents a good example of circular economy.[591]

In the near future, the number of Dutch prosumers able to give back energy to the grid will only increase, as will the need for an efficient integration of renewables. Such a scenario requires that the development of technology and investment in the storage sector will have to be supported by an adequate legal/regulatory framework, which is absent at the time of writing. The amendments on the Electricity Act that entered into force between 2018 and 2019 allow some room for R&D projects. Tools such the "Green Deals" have been used occasionally at the local level to provide incentives to storage projects, but wider reforms at the central government level would be more impactful, such as the adoption of a discount on the entry and exit connection tariffs, as was the case in the gas storage sector.[592]

---

[587] C. van der Weijden, "CMS Guide to Energy Storage: The Netherlands," available at https://www.lexology.com/library/detail.aspx?g=5a326284-f6aa-4439-b733-05d4eecc70f5.

[588] AES, "Fleet accelerates energy storage market in Europe," *Dutchreview*, 17 February 2016, available at https://www.aes.com/default.aspx?SectionId=5cc5ecae-6c48-4521-a1ad-480e593e4835&LanguageId=1&PressReleaseId=cd734f50-3398-4a5b-8ca1-8b2926a9590b.

[589] Idem.

[590] V. Licheva, "The Dutch just launched the largest energy storage system in Europe," 17 July 2018, available at https://dutchreview.com/news/innovation/the-dutch-just-launched-the-largest-energy-storage-system-in-europe/.

[591] Idem.

[592] C. van der Weijden, "CMS Guide to Energy Storage: The Netherlands," available at https://www.lexology.com/library/detail.aspx?g=5a326284-f6aa-4439-b733-05d4eecc70f5.

8.7.4 Regulatory Policy Recommendations

Given the importance of unbundling of energy suppliers under the Third Energy Package, a definition of storage is necessary. Particularly, a clear delineation of which operators in the market can own, operate, or control these assets.

Regulatory intervention would also be required to incentivize investment in the development of storage technologies. In the case of prosumers, given that they arguably contend with a double economic hurdle typified by the high cost of storage technology as well as uncertain and sometimes unfavorable market structures for self-generated electricity, the need for investment incentives must be coupled with favorable policies related to demand response mechanisms and self-generation/consumption of renewables. Ultimately, the impact of storage on electricity markets hinges largely on the economics of storage solutions. Therefore, the institution of appropriate regulatory incentives is critical to ensuring the desired level of storage solutions.

A review of grid fees structure is also necessary to avoid situations where storage assets pay double grid fees. Better consideration should be given to the kind of service provided by storage assets in determining the applicability or otherwise of grid fees or other similar taxes. In the grander scheme of facilitating the development of smart grids and electricity markets, the regulatory framework should not discriminate between DERs, thereby ensuring that storage resources are granted equal access to flexibility markets to enable them to compete equally with fossil-fuel based generation units.

8.8 Interconnection

Better interconnections among EU member states will help boost energy security and contribute to better integration of renewables in energy markets. To this end, numerous projects are underway across the EU.[593]

Belgium's electricity network has a transmission capability of 3500 megawatts (MW). With its position at the heart of Europe, the country's electricity network connects with France, the Netherlands and Luxembourg at an interconnection rate of 17%.[594] This means that Belgium is complying with the EU's electricity interconnection plan for the near future, which aims for interconnection of 10% by 2020[595]

---

[593]European Commission, "Electricity interconnection targets." Available at: https://ec.europa.eu/energy/en/topics/infrastructure/projects-common-interest/electricity-interconnection-targets.

[594]Communication from the Commission to the European Parliament, the Council. Achieving 10% electricity interconnection target. Making Europe's electricity grid fit for 2020, p. 4, COM (2015) 82 final (25 February 2015).

[595]European Commission, "Electricity interconnection targets," Available at: https://ec.europa.eu/energy/en/topics/infrastructure/projects-common-interest/electricityinterconnection-targets.

and recommends reaching 15% by 2030.[596] Belgium is taking further steps to boost its interconnection with neighboring EU Member States, in a number of cross-border collaborations called Projects of Common Interest (PCI).[597] For example:

- *Nemo*: an electrical interconnector between Belgium and the UK, to be operational in 2019. With a capacity of around 1000 MW, and a span of 140 km, it will connect the two countries via subsea and subterranean cables. Nemo is projected to fulfill energy needs for half a million homes. It will also boost efforts towards increasing renewable energy usage.[598]
- *Brabo*: this north border interconnector with the Netherlands has a transfer capability of 1000 MW. A three-phase project envisioned to reach completion by the end of 2023, Brabo is expected to boost the economy around Port of Antwerp, help create new generation plants, and build the transfer capacity between Belgium and the Netherlands.[599]
- *ALEGrO*: projected to launch in 2020, ALEGrO is the first interconnector between Belgium and Germany. With a capacity of 1000 MW, it will span around 100 km between the two countries and will help secure electricity supply for both. It will also help secure supply for the overall EU electricity market and will help towards adopting renewable energy.[600]
- *Creos*: this project—an interconnector between Belgium and Luxembourg—is split into two phases and is expected to reach completion in 2020. The new interconnection will contribute to the overall integration of the European electricity market.
- *Avelin-Avelgem*: This overhead link covers 21 km between Belgium and France. The project involves significant upgrading to existing infrastructure and is expected to facilitate the creation of a European energy market and promote the transition to renewable energy.[601]

The fact that the abovementioned projects promote EU cooperation on matters of mutual interest, means they are eligible for a number of EU funding schemes, e.g., Connecting Europe Facility (CEF).[602] The projected infrastructures will significantly boost the capacities of external electricity exchange and will enable domestic

---

[596]Communication from the Commission to the European Parliament, the Council. European Energy Security Strategy, p. 10, COM (2014) 330 final (28 May 2014).

[597]European Commission, "Projects of Common Interest," European Commission. Available at: https://ec.europa.eu/energy/en/topics/infrastructure/projects-common-interest.

[598]Nemo Link, "Overview. Why an interconnector?" Nemo Link. Available at: http://www.nemolink.com/the-project/overview/.

[599]Elia Group (2016).

[600]TSCNET Services, "Interconnecting Belgium and Germany," TSCNET Services, 30 September 2016. Available at: http://www.tscnet.eu/interconnecting-belgium-and-germany/.

[601]Elia Group (2017).

[602]European Commission, "Financing trans-European energy infrastructure – the Connecting Europe Facility," 5 March 2015. Available at: http://europa.eu/rapid/press-release_MEMO-15-4554_en.htm.

grids to hold an increasing ratio of renewable energy (RE). In recent times, electricity derived from RE sources has seen a sizeable increase. As an example, from 2009 to 2014, the share of renewables in national electricity production went up by more than 10% (from 7.8% to 19%). This sudden rise stemmed from the implementation of a range of policies, such as green certification schemes and offshore wind projects.[603]

This upswing in RE is likely to rise for several reasons. For one thing, with Belgium set to transition away from nuclear energy by 2025, renewables and natural gas are the most plausible alternatives. Also, the abovementioned projects will facilitate the country's interconnection and help integrate RE into the grid. Lastly, with the launching of EU-wide goals and roadmaps such as the EU 2020 Climate and Energy Package or the 2030 Climate and Energy Framework, Belgium is obliged to keep promoting RE usage. Continuing to strive for a cleaner energy mix will help towards energy security in the country while contributing to decarbonization at the same time.[604]

In Spain, the electricity system interconnects with that of Portugal to create the Iberian electricity system. Spain's electricity system also interconnects with the North African system, through Morocco and with the Central European system through France. The Central European system, for its part, connects with the Nordic countries, countries in Eastern Europe, as well as the British Isles. This makes it the world's biggest electricity network.[605] At the EU level, the long-standing goal is to accomplish the European internal energy market, but doing so will require tackling the remaining fragmentation within Member State energy markets. In fact, the European Commission has stated a goal to achieve an electricity interconnection rate of 10% by 2020.[606] It has since been recommended to raise the interconnection target to 15% by 2030.[607] A strongly interconnected energy grid across Europe would reap significant benefits for citizens, saving them as much as €12–40 billion annually by 2030.[608]

Spain walks a rather fine line when it comes to cross-border interconnections. According to the European Network of Transmission System Operators for Electricity (ENTSO-E), Spain has one of the lowest interconnection capacities in the

---

[603]International Energy Agency (2016).

[604]Communication from the Commission to the European Parliament, the Council. Achieving 10% electricity interconnection target. Making Europe's electricity grid fit for 2020, p. 4, COM (2015) 82 final (25 February 2015).

[605]Red Eléctrica de España, "Strengthening interconnections," Red Eléctrica de España. Available at: http://www.ree.es/en/red21/strengthening-interconnections.

[606]European Commission, "Electricity interconnection targets." Available at: https://ec.europa.eu/energy/en/topics/infrastructure/projects-common-interest/electricity-interconnection-targets.

[607]Communication from the Commission to the European Parliament, the Council. European Energy Security Strategy, p. 10, COM (2014) 330 final (28 May 2014).

[608]Booz & Company et al. (2013).

8.8 Interconnection

EU—at less than 5%,[609] it is far below the aspired level of 10%—on a par with the Baltic States, Cyprus and Poland.[610] The government will need to invest in new cross-border connections in order to comply with the EU's goals, and the benefits of doing so are greater than the projected costs. In fact, a strongly interconnected grid is fundamental for sustainable development, as, by integrating a greater proportion of energy from renewable sources, we further decarbonize the energy mix. Moreover, greater use of renewable energy helps towards achieving energy security. And, of course, it gives a healthy impetus to the renewable energy sector, increasing employment and innovation. Overall, effective interconnection lowers electricity prices because of greater market efficiency, leads to a more stable electricity supply, and helps protect the environment.[611]

Spain has good interconnections with Portugal and represents on of the few portals between Europe and Africa via a subsea interconnector with Morocco. On the other hand, the French border is a bottleneck zone for electricity exchanges, and hinders the Iberian Peninsula from an effective connection with the European internal energy market. Despite its strong energy infrastructure, Spain is therefore unable to take advantage of being connected to the rest of the European grid. Still, some changes have transpired recently that may alleviate this situation.

The EU has several new energy projects in the works, of which the Santa Llogaia—Baixàs power line is a prototype. February 2015 saw the launching of the Santa Llogaia—Baixàs power line, as part of the Energy Union project. The power line doubled the interconnection between Spain and France,[612] and raised the transfer capacity between the two countries from 1400 MW to 2800 MW. Its key role has earned it the status of Project of Common Interest (PCI),[613] which means it is eligible for various funding schemes, such as the Connecting Europe Facility (CEF).[614] The subterranean interconnection will power the high-speed train on the Spanish side, and will enable the grid to integrate a greater proportion of RE, in particular wind energy from the peninsula. This PCI is the first interconnector to be

---

[609] Red Eléctrica de España, "Strengthening interconnections," Red Eléctrica de España. Available at: http://www.ree.es/en/red21/strengthening-interconnections.

[610] European Network of Transmission System Operators for Electricity, "Scenario Outlook and Adequacy Forecast 2014," 2014.

[611] Communication from the Commission to the European Parliament, the Council. Achieving 10% electricity interconnection target. Making Europe's electricity grid fit for 2020, p. 4, COM (2015) 82 final (25 February 2015).

[612] European Commission, "Building the Energy Union: Key electricity interconnection between France and Spain completed," European Commission—Press release, 2015.

[613] European Commission, "Projects of common interest," European Commission. Available at: https://ec.europa.eu/energy/en/topics/infrastructure/projects-common-interest.

[614] European Commission, "Financing trans-European energy infrastructure – the Connecting Europe Facility," European Commission—Fact Sheet, 5 March 2015. Available at: http://europa.eu/rapid/press-release_MEMO-15-4554_en.htm.

established between France in Spain in nearly 30 years.[615] Strategic cross-border interconnectors are being prioritized when it comes to political and economic support. Increasing Spain's interconnection levels will depend on support from the Energy Union and various EU financial instruments.

Another relevant project is Biscay Bay, which uses an undersea cable to connect the Biscay Bay in Spain to Aquitaine in France. The project under technical review and, as a PCI, will receive significant CEF funding (around €3.25 million). Given the projected benefits of the project it, it is likely to receive further EU subsidies. The Biscay Bay project is expected to rebalance electricity flows and may raise the transfer capacity between France and Spain. In fact, the electricity exchange capacity is projected to rise up to 5000 MW.[616]

Two further initiatives are being planned through the Pyrenees. One would operate from Basque country or Navarra in Spain to Cantegrit in France. The other will connect Aragón in Spain to Marsillon in France. Along with the Biscay Bay project, these projects are projected to increase the transfer capacity between France and Spain to 8000 MW in 2020[617] Overall, planned projects following the Santa Llogaia—Baixàs power line may almost triple the current transfer capacity.

The supranational considerations depicted above seem to constitute the ideal juncture for the market penetration of the set of solutions and technologies brought by the WiseGRID project in the Spanish electricity industry. The EU's goal for a 10% interconnection rate should transpire between Spain and the rest of the European grid by 2020. This would lead to a stronger and more stable electricity market in Spain, and one that would better incorporate renewable energy to the national grid. In turn, consumers may start looking into smart grid applications for cost effective solutions to energy. Self-consumption and net metering are likely to follow, and given the right regulatory environment, prosumers will be able to sell their energy surplus with a domestic grid that is able to integrate higher proportions of renewable energy, thanks to more effective transfer ratios with neighboring states.

In Italy, energy is comparatively more expensive than in other EU countries.[618] Various steps are being taken to address this situation. For example, the TSO was unbundled, which helped increase competition. But supply during peak times remains low. TERNA (the national TSO) has been working on upgrades to the system, particularly to address congestion, and this improved connectivity between Sicily and Sardinia. In terms of connections to Europe, Italy's grid is linked through France Switzerland, Austria, Slovenia, Corsica, Greece and Malta. Prices have tended to vary significantly between the north and south, but TERNA's recent

---

[615]Red Eléctrica de España, "Spain-France underground interconnection," Red Eléctrica de España. Available at: http://www.ree.es/en/activities/unique-projects/new-interconnection-with-france.

[616]*Madrid Declaration,* Madrid: Energy Interconnections Links Summit, 2015.

[617]Idem.

[618]European Commission, "European Commission Country Reports," 2014. Available at: https://ec.europa.eu/energy/sites/ener/files/documents/2014_countryreports_italy.pdf.

8.8 Interconnection

upgrades have contributed toward alleviating this, by resulting in supply surpluses. The market has also been helped by the surge in renewable energy usage.

In Greece, hydroelectric power plants accounted for 19% of electricity generated from the interconnected system in 2016, with 28% deriving from natural gas. Following a push for more renewable energy (RE) usage, Greece achieved a 29% share of RE in its energy mix in the interconnected system (with the islands, not interconnected, achieving a 21.8% share of RE).[619]

In terms of consumption, in 2016 the country's total electricity consumption reached 50.1 TWh, which was less by about 10% from consumption in 2015. The fall in consumption is likely a result of the economic challenges in the country, which have led to all sectors and individual consumers finding ways to reduce expenditure.

Greece belongs to the Union for the Coordination of the Transmission of Electricity (UCTE). Dating back to 2004, its transmission system is synchronized its transmission system with the overall European one. Its transmission system has interconnections with all bordering states, including an underwater cable to Italy. Greece has also created a sole authority to oversee licensing Projects of Common Interest (PCIs), which is in keeping with the Trans-European Networks—Energy (TEN-E) Regulation.[620] The country has to projects that qualify as PCIs: the AC 400 kV interconnection between Maritsa East 1 in Bulgaria and Nea Santa in Greece, and the DC 600 kV underwater connection between Israel, Cyprus and Greece.[621] Another project currently underway is a submarine interconnection of the Cyclades islands, which aims to promote overall energy security in the Aegean islands, boost RE usage and lower costs. The TSO authorized this project, which is anticipated for completion in 2017. Another project in the works is an interconnection in Crete, which is due for completion by 2022.

Greece has yet to capitalize on its closeness to Russian gas supply, to enhance its gas interconnection system. The country is working on a few regional initiatives, such as the Trans-Antolian Pipeline, Trans Adriatic Pipeline and the Southern Caucasus Pipeline.[622] The goal for these pipelines to become part of Europe's Southern Gas Corridor for the Caspian supply of gas. Other projects include upgrading the LNG terminal Revithoussa, building interconnections with Bulgaria through the reverse flow in Kula-Sidirokastron and the Interconnector Greece-Bulgaria pipeline.[623] Greece is also working towards interconnections with

---

[619] Energypedia, "Greece Energy Situation," 26 April 2017.

[620] Official Journal of the European Union, "On guidelines for trans-European energy infrastructure and repealing Decision No 1364/2006/EC and amending Regulations (EC) No 713/2009, (EC) No 714/2009 and (EC) No 715/2009." Available at: https://eur-ex.europa.eu/LexUriServ/LexUriServ.do?uri=OJ:L:2013:115:0039:0075:en:PDF.

[621] European Commission, "European Commission Country Reports," 2014. Available at: https://ec.europa.eu/energy/sites/ener/files/documents/2014_countryreports_italy.pdf.

[622] For an analysis of these pipelines, see Leal-Arcas (2016), pp. 53–74; Leal-Arcas et al. (2015a), pp. 291–336; Leal-Arcas et al. (2015b), pp. 122–162; Leal-Arcas et al. (2015c), pp. 38–87.

[623] European Commission, "European Commission Country Reports," 2014.

Italy through the Trans-Adriatic Pipeline, possibly via IGI Poseidon, and with Turkey by improving the compressor station in Kipi.[624] Considering Greece faces a deficit in energy supply, improving gas infrastructure, and thereby supply, would certainly facilitate the energy challenge.

Several non-interconnected island electrical systems (NIIES) exist in Greece—independent power networks that are separate from that of the mainland. They are made up of 60 islands, with 32 independent electrical systems, mostly in the Aegean Sea. They serve 15% of the country's population and are responsible for nearly 14% of total national annual electricity consumption (around 42,300 GWh per year). The NIIES face higher electricity costs in contrast to the mainland, due to their heavy reliance on diesel or fuel oil as well as challenges to integrating RE into the local energy mix.

The Regulatory Authority for Energy publishes a set of provisions titled Non-Interconnected Islands Management Code (NIIMC), which governs the electricity market in the non-interconnected islands. The the Hellenic Electricity Distribution Network Operator (HEDNO S.A.) is in charge of the operation and management of NIIES. In addition to its automatic position as distribution network operator, it also acts as TSO for the islands as well as market operator.

The smallest of the NIIES is that of Kythnos—it has an installed thermal capacity of nearly 6 MW, supplemented by 665 kW wind and 238 kW PV stations. The demand for electricity on Kythnos reaches up to 5 MW during hot months, and goes down to an average of 600 kW in the winter.

EU regulation has a large impact on the electricity market of NIIES, in particular Directive 2003/54/EC and Directive 2009/72/EC.[625] These directives address market liberalizing and the establishment of new conventional production stations and anticipate exemption from the provisions for micro-isolated systems that consumed less than 500 GWh in 1996, which applies to Kythnos and all the other NIIES other than Crete. Hellenic Republic responded to the directives by applying for derogation for all NIIES other than for Crete and Rhodes. According to the EC Decision 2014/536/EU[626] the derogation is allowed only for refurbishment, expansion and upgrading of existing conventional units (derogation for new conventional capacity, RES and CHP is not approved) for micro isolated systems until 2021, and derogation for market opening until 2019, or earlier if the infrastructures described in NII Code are ready. The Greek Law 4414/2016,[627] has adjusted Greek legislation to the provisions of Decision 2014/536/EE.

The case of the Netherlands differs from the other European member states from an energy mix perspective because it is an important producer and exporter of natural gas. This business started in 1959, when the Groningen gas field was discovered in the North-East of the country. The 2014, links were found between earthquakes in

---

[624]Idem.
[625]Verbong et al. (2013), pp. 117–125.
[626]Eid et al. (2016).
[627]Bressand (2013), pp. 15–29.

8.8 Interconnection

that region and gas extraction activities. This finding made the Dutch Government decide to drastically reduce the production of natural gas.[628] It is expected that this reduction will be compensated by an increase in the use of RES. To facilitate a smooth transition to RES and manage the increase in energy demand, infrastructures and interconnections with neighbouring countries are of crucial importance.

TenneT, the only Dutch TSO, plays a crucial role in such a challenge. The TSO is working closely with other foreign TSOs to expand and improve the grid so as to guarantee energy security throughout the whole year, despite weather conditions and unexpected energy demand rises. The level of interconnection in the country is already well developed due to the size of the Netherlands and its geographic location. In 2014, the European Council asked all EU member states to reach a level of interconnection of at least 10% of their installed electricity production capacity by 2020[629] and of 15% by 2030.[630] By 2014, the Netherlands was already at 17%.[631]

Therefore, TenneT is currently adopting a preventive and long-term approach, which takes into consideration further developments for the near future due to the increase of renewables that will be integrated to the grid. For example, the interconnection between the Netherlands and West Denmark may connect new offshore wind farms; moreover, the use of new interconnections between the Netherlands and Germany at Doetinchem-West may be a precautionary measure in case of insufficient interconnection in that region. This is expected if/when the energy price in Germany will drop because of high renewables production.[632] Other important interconnections are between the Netherlands and Belgium as well as between England and Norway (NorNed, the world's longest high-voltage subsea direct current link).

In general terms, Dutch interconnections play a central role in the entire North European electricity grid. This fact implies that the Netherlands plays a crucial role when it comes to EU projects regulated by the Trans-European Network-Energy (TEN-E) Regulation (347/2013).[633] Such Regulation refers to trans-European energy infrastructure projects. In December 2018, the European Parliament proposed to revise the Regulation to better match the challenges of the European Union's current climate and energy targets. The proposal has been very positively welcomed by EDSO, i.e., the European Distribution System Operators' Association for Smart Grids. European DSOs will be the major actors of this revised Regulation.

---

[628]http://www.energyboardroom.com/article/european-energy-security-through-gas-the-netherlands-and-the-viability-of-the-gas-roundabout/.

[629]https://ec.europa.eu/energy/en/topics/infrastructure/projects-common-interest/electricity-interconnection-targets.

[630]http://www.europarl.europa.eu/legislative-train/theme-resilient-energy-union-with-a-climate-change-policy/file-15-electricity-interconnection-target.

[631]https://renews.biz/50642/dutch-issue-security-of-supply-report/.

[632]Idem.

[633]https://eur-lex.europa.eu/LexUriServ/LexUriServ.do?uri=OJ:L:2013:115:0039:0075:en:PDF.

It is expected that an innovative concept of cross-border projects will be adopted so as to include local, decentralized, and participative Dutch smart grid projects (alongside large transmission projects that are already regulated).[634] Considering the convenient location of the Netherlands, the tendency in its political approach to collective participation, and the fact that the DSOs are responsible for the operation of the electrical grid at a regional level, it would be expected for the Dutch DSOs to be in a very good position to participate in innovative projects under the revised Regulation.

8.9 Concerns About Data Protection

8.9.1 Background

Information and communications technologies (ICT), especially new digital applications for smart grids, play a central role in enabling new energy providers to monitor and process data and create opportunities to meet the various EU energy policy goals, including efficiency, security and sustainability. New digital technologies have made it possible to re-design the traditional analogue electricity power system infrastructure that has dominated the energy landscape in Europe since World War II. A transformation in the energy system will provide new opportunities not only to energy suppliers, but also to consumers.[635] Through advanced sensing technologies, it is now feasible to provide predictive information and bespoke recommendations based on almost real-time data to all stakeholders (*e.g.* utilities, suppliers, and consumers). Smart grids refer exactly to this new digital networked energy infrastructure.[636]

Smart grids services regard intelligent appliance control for energy efficiency and better integration of distributed energy resources to reduce carbon emissions. They offer the potential of higher level capabilities to meet current and future energy demands.[637] Smart grids could deliver improved performance related to concepts such as reliability, resiliency, environmentally friendly generation, transmission, and distribution, which will help the EU to achieve strategic economic, environmental and social goals.[638] These changes, which ultimately make energy systems more

---

[634]https://www.edsoforsmartgrids.eu/joint-statement-from-the-dso-associations-on-the-proposal-to-revise-the-ten-e-guidelines/.

[635]The Climate Group, GeSI, "SMART 2020: Enabling the low carbon economy in the information age", 2008. [On-line]. Available: http://gesi.org/files/Reports/Smart%202020%20report%20in%20English.pdf.

[636]European Commission Directorate-General Information Society and Media, ICT for Sustainable Growth Unit, "ICT for a Low Carbon Economy", Smart Electricity Distribution Networks, July 2009.

[637]Liu et al. (2012), pp. 981–987.

[638]Polinpapilinho et al. (2016), pp. 1–20.

8.9 Concerns About Data Protection

complex, have led to concerns regarding potential cyber-attacks of critical infrastructure, energy and data theft, fraud, denial of service, hacktivism and design obsolescence adding to energy poverty.[639] Also, the regulation of smart grids and smart meter[640] technologies directly impacts the way data privacy is implemented in technical systems (like smart meters and energy saving services).[641]

Intelligent control and economic management of energy consumption require more interoperability between consumers and service providers. Unprotected energy-related data will cause invasions of privacy in the smart grid.[642] Law and policy-makers need to consider the trade-offs to enable smart grids to deliver low-cost and green energy within locally, regionally and nationally secure networked systems. Given the dependency of smart grids on digital technology, their uptake is intricately interlinked with law and policy on ICT more generally.

This section outlines developments in ICT law that are relevant to smart grids, both internationally and in the EU. It first provides a survey of key law and policy issues related to security and privacy when dealing with smart grids and ICT, including an analysis of concepts such as cybersecurity, cybercrime, and data management. The section is followed by an outline of existing and emerging EU and international legislation addressing the above-mentioned issues and is divided into privacy and data protection and digital systems security. It then provides an analysis of the current situation in certain EU member states.

8.9.2 Smart Grids: Cybersecurity and Privacy Issues

Energy systems are critical infrastructure. The loss of delivery capacity can have multiple impacts on domestic households, public utility services such as hospitals, as well as transport and communication systems. The cyber-hacking of Ukraine's existing large-scale energy systems illustrated new weaknesses in digitalised infrastructures.[643] Smart-grid energy systems will compound existing security threats.

---

[639] C. Wueest, "Attacks Against the Energy Sector," Symantec Official Blog, 13 January 2014. [Online]. Available: https://www.symantec.com/connect/blogs/attacks-against-energy-sector.

[640] Smart meters are advanced metering systems which provide real-time information on consumers' energy use or generation. They use digital technologies, regularly update information and provide two-way electronic communication between consumers and the grid. See Joachain and Klopfert (2014), pp. 89–96.

[641] REScoop, "6.1 European Legislative Environment", REScoop, 2017. European Parliamentary Research Service, "Smart electricity grids and meters in the EU Member States", EPRS, September 2015. [On-line]. Available: http://www.europarl.europa.eu/RegData/etudes/BRIE/2015/568318/EPRS_BRI%282015%29568318_EN.pdf.

[642] Liu et al. (2012), pp. 981–987.

[643] "Ukraine power cut 'was cyber-attack'," *BBC News*, 11 January 2017, available at http://www.bbc.co.uk/news/technology-38573074.

Indeed, smart grids bring risks. Some risks are known, old, and foreseeable issues. Other risks are new and less predictable. Cybersecurity is likely to become more important in the next few years.[644] Cyber-technologies are becoming less expensive and easier to acquire, which allows states and even non-state actors to potentially inflict considerable damage.[645] Cyber-operations may not only be used for industrial espionage or intelligence collection, but also to delete, alter, or corrupt software and data resident in computers. This could entail negative repercussions on the functionality of computer-operated physical infrastructures, including disabling power generators.[646] Smart grids increasingly couple information in the energy sector with digital communication systems. This has created new vulnerabilities and resulted in smart grids becoming a security issue beyond traditional energy security framing and including cybersecurity.[647]

Smart grids are integrated systems that include technologies, information, social and organizational components, policy and political requirements, and legislative and regulatory compliance.[648] Consequently, this increases the risk of compromising the ultimate objective of smart grids: reliable and secure power system operation. In 2008, the European Commission acknowledged that the electricity sector constitutes "an essential component of EU energy security."[649] Some even argue that the current interdependence between the electricity and communication infrastructures is so profound that it could be conceived within an "energy-and-information" paradigm.[650] This inter-dependence becomes even more intricate when considering the energy systems' critical infrastructure status and the potentially catastrophic impact of cyber-attacks.

An effective regulatory framework manages both known and unknown risks, with the latter involving a precautionary approach. Smart grids need to ensure the security of sensitive customer information transmitted over an increasing number of "internet of things" (IoT) devices. Smart grids must also ensure that communication between stakeholders is reliable enough to deliver stable operation. There is a need to develop resilient formulations of risk related to holistic considerations.[651] An integrated multilevel governance approach is required to integrate smart grids

---

[644]Cesar Cerrudo, *Why Cybersecurity Should Be The Biggest Concern Of 2017*, Forbes (Jan. 17, 2017), https://www.forbes.com/sites/forbestechcouncil/2017/01/17/why-cybersecurity-should-be-the-biggest-concern-of-2017/#4a61899c5218.

[645]Roscini (2014).

[646]*Id.*; Ciere et al. (2016).

[647]*See generally* Masera et al. (2016),p. 85.

[648]Katina et al. (2016), pp. 1–20.

[649]*Commission Staff Working Document Accompanying the Communication from the Commission to the European Parliament, the Council, the European Economic and Social Committee and the Committee of the Regions Second Strategic Energy Review An EU Energy Security and Solidarity Action Plan: Europe's Current and Future Energy Position – Demand-Resource-Investments*, SEC (2008) 2871 (Nov. 13, 2008).

[650]Pearson (2011), p. 5211.

[651]Katina et al. (2016), pp. 1–20.

8.9 Concerns About Data Protection

securely within society, although this approach presents new legal challenges for lawyers and policy-makers.

Unlike traditional energy systems, smart grids fully integrate high-speed and two-way communication technologies to create dynamic and interactive infrastructure with new energy management capabilities.[652] Smart grids energy systems are "a literal IoT": networks with billions of interconnected smart objects, such as smart meters, smart appliances, and other sensors.[653] As a cyber-physical system, an IoT-based smart grid presents risks across different domains (i.e., generation, transmission, distribution, customer, service-provider, and operations markets).[654] The EU acknowledges that smart metering systems and smart grids foreshadow this impending IoT. The EU also acknowledges that with this development come potentially increasing risks associated with the collection of detailed consumption data.[655] Sander Kruese, privacy and security adviser at Alliander, a DSO in the Netherlands, noted that "[e]very component in the grid that has become digitized is becoming an attack-point."[656] Providing securitization across the entire system, a system that continuously incorporates new software systems and hardware from a range of providers, is a demanding task. The EU has adopted a strategy on cybersecurity.[657] Operationalizing the goals contained in the strategy will be integral to addressing new potential threats posed by embedding ICT into the European Union's energy system.[658]

Threats that were not possible in the traditional electric grid[659] are now the main concern.[660] When combined with data from other multiple independent data sources,[661] smart meter data becomes part of a broader and more open meta-data

---

[652] Wang and Lu (2013), p. 1344.

[653] K.T. Weaver, *Smart Meter Deployments Result in a Cyber Attack Surface of "Unprecedented Scale"*, Smart Grid Awareness (Jan. 7, 2017), https://smartgridawareness.org/2017/01/07/cyber-attack-surface-of-unprecedented-scale/.

[654] Katina et al. (2016), pp. 1–20.

[655] Article 29 Data Protection Working Party, *Opinion 07/2013 on the Data Protection Impact Assessment Template for Smart Grid and Smart Metering Systems ("DPIA Template")*, 2064/13/EN WP209 (2013).

[656] Anca Gurzu, *Hackers Threaten Smart Power Grids*, Politico (Apr. 1, 2017), http://www.politico.eu/article/smart-grids-and-meters-raise-hacking-risks/.

[657] *See generally Joint Communication to the European Parliament, the Council, the European Economic and Social Committee and the Committee of the Regions on Cybersecurity Strategy of the European Union: An Open, Safe and Secure Cyberspace*, JOIN (2013) 1 final (Feb. 7, 2013).

[658] *Id.*

[659] Such as energy theft and fraud, sensitive information theft, service disruption for the purpose of extortion, cyber-espionage, vandalism, hacktivism and terrorism.

[660] K.T. Weaver, *Smart Meter Deployments Result in a Cyber Attack Surface of "Unprecedented Scale"*, Smart Grid Awareness (Jan. 7, 2017), https://smartgridawareness.org/2017/01/07/cyber-attack-surface-of-unprecedented-scale/.

[661] Such as geo-location data, tracking and profiling on the internet, video surveillance systems, radio frequency identification systems, *et cetera*. See Data Protection Working Party, Opinion 04/2013 on the Data Protection Impact Assessment Template for Smart Grid and Smart Metering

system.[662] In different ways, all users are potential victims of attacks in such a context. In addition, their vulnerabilities could be drawn from previous experience gained in different sectors, such as IT and telecommunications.[663] As an example, automated smart meters rely on tracking, in real time, actual power usage, and allow for two-way communication between utilities and end-users. Hackers targeting this technology may induce disruptions in power flows, create erroneous signals, block information (including meter reads), cut off communication, and/or cause physical damage.[664]

Digital ICT has accelerated the expansion of personal data systems—making them more extensive and consequential in the lives of ordinary citizens.[665] The costs of using personal data in today's computerized record-systems are all but negligible. The result is that all sorts of personal data that would otherwise be "lost" are now "harvested" by different actors that do everything from allocating consumer credit to preventing cyber terrorist attacks.[666] New technologies allow for an unprecedented level of information-integration, "providing the possibility to combine new and existing data and technologies (interoperability) and cope with growing resources and number of users (scalability), through the adoption of distributed systems (cloud computing)."[667]

Information gathered from energy users is integral to empowering individuals, households, and organizations to change their consumption patterns, increasing efficiency, and reducing energy costs and carbon emissions.[668] In 2010, the European Commission noted that "the ICT sector should lead the way by reporting its own environmental performance by adopting a common measurement

---

Systems ('DPIA Template') para 29 prepared by Expert Group 2 of the Commission's Smart Grid Task Force, 00678/13/EN, WP205, 22 April 2013; Recommendation CM/Rec(2010)13 of 23 November 2010 of the Council of Europe Committee of Ministers to Member States on the protection of individuals with regard to automatic processing of personal data in the context of profiling.

[662]Committee of Ministers, Council of Europe, *Recommendation CM/Rec (2010) 13 to Member States on the Protection of Individuals with Regard to Automatic Processing of Personal Data in the Context of Profiling* (adopted Nov. 23, 2010).

[663]Liu et al. (2012), p. 981.

[664]US Dep't of Energy, Transforming the Nation's Electricity System: The Second Instalment of the QER (2017).

[665]Rule and Greenleaf (2010).

[666]Public opinion seems to be meeting these developments with concern, and suggestions are being put forward in order to empower users to gain better control of the situation. See in this sense Brian X. Chen, *How to Protect Your Privacy as More Apps Harvest Your Data*, N.Y. Times (May 2, 2017), https://www.nytimes.com/2017/05/03/technology/personaltech/how-to-protect-your-privacy-as-more-apps-harvest-your-data.html?mcubz=1.

[667]*See generally* Zulkafli et al. (2017), p. 58.

[668]As a practical example, see, for instance, the analysis provided by Cai and Jiang (2008), p. 1667.

8.9 Concerns About Data Protection

framework."⁶⁶⁹ This EU policy is grounded in the positive relationship between data access, processing, and dissemination, and beneficial changes in consumer behavior.

Unlike oil, "a product that does not generate more oil (unfortunately) ... the product of data (self-driving cars, drones, wearables, etc.) will generate more data."⁶⁷⁰ According to a 2012 estimate, "90% of the world's data was created in the last two years alone. In fact, 2.5 quintillion bytes of data are created each day, which is more data than was seen by everyone since the beginning of time."⁶⁷¹ However, a consequence of increased data availability, especially in the form of meta-data, is to narrow the realm of anonymity—so that fewer interactions, relationships, and transactions are possible without identifying one's self.⁶⁷² This leads to questions about privacy and security.⁶⁷³

In 1890, Louis Brandeis and Samuel Warren defined the individual's need for privacy and solitude as a fundamental right, due to the increasing intensity and complexity of life.⁶⁷⁴ Privacy is going to be even further challenged in the digital era.⁶⁷⁵ Computer technologies increasingly make it possible to capture and use personal data in all sorts of settings and for all sorts of purposes that would once have been inconceivable. Leading figures amongst online corporations have argued that privacy is no longer a social norm or even possible: "Facebook and its CEO Mark Zuckerberg have taken the position that sharing of information and connectedness is the new social norm, and that privacy, on the contrary, is now outmoded."⁶⁷⁶ Key questions at stake include what personal information institutions and other non-state actors collect, how it is collected, where it is stored, who can access it, and what actions can be taken on its basis.⁶⁷⁷ It can be argued that this pressure on information privacy is not the result of a new social norm, but the consequence of a desire for profit at the expense of eroding privacy protection.⁶⁷⁸ Governments and citizens in the EU are pushing back on these incursions on

---

⁶⁶⁹*Commission Communication to the European Parliament, the Council, the European Economic and Social Committee and the Committee of the Regions on A Digital Agenda for Europe*, COM (2010) 245 final (May 19, 2010).

⁶⁷⁰Shah (2015), pp. 207 and 208.

⁶⁷¹International Telecommunication Union, ICT Facts & Figures: The World in 2015 (2015).

⁶⁷²Rule and Greenleaf (2010).

⁶⁷³McKenna et al. (2012), p. 807.

⁶⁷⁴Brandeis and Warren (1890), p. 193.

⁶⁷⁵For a brief discussion of this topic, see Somini Sengupta, *U.N. Urges Protection of Privacy in Digital Era*, N.Y. Times (Nov. 25, 2014), https://www.nytimes.com/2014/11/26/world/un-urges-protection-of-privacy-in-digital-era.html?mcubz=1.

⁶⁷⁶Meg Roggensack, *Face It Facebook, You Just Don't Get It*, Huffington Post (May 25, 2010), http://www.huffingtonpost.com/human-rights-first/face-it-facebook-you-just_b_589045.html.

⁶⁷⁷Rule and Greenleaf (2010).

⁶⁷⁸Jennifer Stoddart, Privacy Commissioner of Canada, Remarks at the 2010 Access and Privacy Conference: Why Privacy Still Matters in the Age of Google and Facebook and How Cooperation Can Get Us There (June 10, 2010) (available at https://www.priv.gc.ca/en/opc-news/speeches/2010/sp-d_20100610/).

privacy. The EU has always paid much attention to personal and domestic privacy, unlike Asian countries.[679]

The perceived threat to the security of personal and family life has led to citizen resistance to smart meters: with their personal sphere is at stake, some react with distrust, suspicion, and hostility towards such new systems.[680] "Surveillance" via smart meters and IoT, therefore, results in extortion and fraud of the domestic sphere.[681] Ensuring privacy appears to be crucial in order to address social barriers and support the new energy system technologies.[682] The European Commission has recognized that, in order to achieve its broader energy and climate policy goals, building consumer trust in smart grids and data management must play a central role in its smart grids policy. In 2011, the European Commission advised that:

> [d]eveloping legal and regulatory regimes that respect consumer privacy in cooperation with the data protection authorities . . . and facilitating consumer access to and control over their energy data processed by third parties is essential for the broad acceptance of Smart Grids by consumers.[683]

Increasing consumer and business confidence in smart grids requires good governance and effective regulatory frameworks and laws.[684] In the following section, the legal approaches adopted to deliver increased privacy and reduce cybersecurity risks from ICT and smart grids technologies are critically examined.

8.9.3 International and EU Law

Dependence on energy systems based on smart grids and ICT poses two major risks: one to privacy and data protection; the other to digital systems security. This section considers the evolving legal frameworks, especially within the EU, for providing a reasonable regulatory architecture to ensure the risks are managed effectively.

---

[679]European Comm'n, Flash Eurobarometer 443 (2016).

[680]Balta-Ozkan et al. (2013), p. 363.

[681]Lazar et al. (2015).

[682]*Id.*

[683]*Commission Communication to the European Parliament, the Council, the European Economic and Social Committee and the Committee of the Regions* on *Smart Grids*, COM (2011) 202 final (Apr. 12, 2011).

[684]*See Dow Launches 2025 Sustainability Goals to Help Redefine the Role of Business in Society*, Dow (Apr. 15, 2015) (arguing that "[b]y 2025, Dow will work with other industry leaders, non-profit organizations and governments to deliver six major projects that facilitate the world's transition to a circular economy,"), http://www.dow.com/news/press-releases/dow%20launches%202025%20sustainability%20goals%20to%20help%20redefine%20the%20role%20of%20business%20in%20society.

8.9.3.1 Privacy and Data Protection

Internationally, privacy is embedded in fundamental legal documents. Privacy is included as a normative principle in the post-war Universal Declaration of Human Rights (1948) and the legally binding International Covenant on Civil and Political Rights (1966).[685] Developments in technology in the late 1960s and 1970s led both the US and Europe to recognize the need to guarantee data protection alongside the right to privacy. Each jurisdiction, including the various EU Member States, adopted differing approaches.[686]

The EU has separate legislation and guidelines on data protection. They promulgate data protection and guidelines that are technology neutral. Explicit recognition of the legal basis for data protection is contained in Article 16 TFEU.[687] Article 7 of the EU Charter of Fundamental Rights (the Charter) protects the fundamental right to the respect for private and family life, home, and communications and Article 8 provides specifically for the protection of personal data. EU data protection law has been decentralized in each Member State. The decentralization of this governance structure has led to jurisdictional tensions amongst the relevant public authorities with respect to the identification of both the domestic law applicable to data processing operations and the relevant enforcing national authority.[688] Additionally, the Member States of the Council of Europe have a positive obligation to act in a proactive manner in order to secure the effective enjoyment of those rights protected under the European Convention on Human Rights (ECHR); if it could be established that a State failed to take appropriate measures to protect individuals under its jurisdiction from privacy violations, the State would be liable under the ECHR.[689]

---

[685] European Convention on Human Rights, Sept. 3, 1953, CETS no. 194.

[686] *See generally* Lynskey (2015).

[687] Article 16 TFEU reads as follows:

1. Everyone has the right to the protection of personal data concerning them.

2. The European Parliament and the Council, acting in accordance with the ordinary legislative procedure, shall lay down the rules relating to the protection of individuals with regard to the processing of personal data by Union institutions, bodies, offices and agencies, and by the Member States when carrying out activities which fall within the scope of Union law, and the rules relating to the free movement of such data. Compliance with these rules shall be subject to the control of independent authorities. The rules adopted on the basis of this Article shall be without prejudice to the specific rules laid down in Article 39 of the Treaty on European Union.

TFEU, art. 16.

[688] Lynskey (2016), p. 1.

[689] Izyumenko (2011).

The European Commission has addressed data privacy matters and it has also specifically referred to smart grid technologies. The 1995 Data Protection Directive provides the foundational legal architecture for subsequent regulation.[690] Subsequent regulation and directives added to these initial foundations in an ad hoc manner. The 2002 e-Privacy Directive,[691] which was subsequently amended by Directives 2006/24/EC and 2009/136/EC, has failed to live up to the challenges of technological developments. The 2016 EU General Data Protection Regulation (GDPR), however, repealed the e-Privacy Directive. Other key legal developments included the 2008 Data Protection Framework Decision[692] and the Regulation 45/2001.[693]

The EU data protection framework establishes a number of general principles applicable to the process of any personal data.[694] The 2002 e-Privacy Directive constitutes a layered system consisting of three levels. The first level is the general level that applies to every processing of personal data. The second level, which extends from the first level, applies when sensitive data are being processed. The third level is applicable when personal data are transferred to third countries. Hence, if this happens, all three levels apply.[695] All subsequent data protection legislation at the EU level must to comply with these principles. Courts must also follow them when interpreting related legislation. These principles are incorporated into the GDPR, which has been in force in all EU Member States since May 2018.[696]

It is important to consider how these new principles will be incorporated into the law and policy related to smart grids. Each principle needs to be applied according to certain conditions. Understanding the principles is essential to interpreting data protection laws in any given context, for example, with smart grids.[697] The following pages consider the key principles in greater detail:

---

[690]Parliament and Council Directive 1995/46/EC, On the Protection of Individuals with Regard to the Processing of Personal Data and on the Free Movement of Such Data, 1995 O.J. (L 281) 31 [hereinafter 1995 Data Protection Directive].

[691]Parliament and Council Directive 2002/58/EC, Concerning the Processing of Personal Data and the Protection of Privacy in the Electronic Communications Sector, 2002 O.J. (L 201) 37.

[692]Council Framework Decision 2008/977/JHA, On the Protection of Personal Data Processed in the Framework of Police and Judicial Cooperation in Criminal Matters, 2008 O.J. (L 350) 60.

[693]Parliament and Council Regulation 2001/45/EC, On the Protection of Individuals with Regard to the Processing of Personal Data by the Community Institutions and Bodies and on the Free Movement of Such Data, 2001 O.J. (L 8) 1.

[694]Papakonstantinou and Kloza (2015), pp. 41–129.

[695]Cuijpers and Koops (2012), p. 269.

[696]Lynskey (2014), p. 569.

[697]For a full overview, see European Union Agency for Fundamental Rights (2014).

8.9 Concerns About Data Protection

Lawful processing To understand this principle, it is necessary to refer to Article 52 (1) of the Charter[698] and Article 8(2) ECHR.[699] The processing of personal data is only lawful when it is done in accordance with the law, pursues a legitimate purpose, and is necessary in a democratic society to achieve that legitimate purpose. However, there is no definition of what constitutes "lawful processing" in Article 5 of the Convention 108[700] or in Article 6 of the 1995 Data Protection Directive.[701] The GDPR does not include a definition, either. The obligations to meet the principle fall on the data gatherer and user. As such, it is imperative in developing regulations

---

[698] Article 52 of the Charter of Fundamental Rights of the European Union:

Article 52 Scope of guaranteed rights 1. Any limitation on the exercise of the rights and freedoms recognised by this Charter must be provided for by law and respect the essence of those rights and freedoms. Subject to the principle of proportionality, limitations may be made only if they are necessary and genuinely meet objectives of general interest recognised by the Union or the need to protect the rights and freedoms of others. 2. Rights recognised by this Charter which are based on the Community Treaties or the Treaty on European Union shall be exercised under the conditions and within the limits defined by those Treaties. 3. In so far as this Charter contains rights which correspond to rights guaranteed by the Convention for the Protection of Human Rights and Fundamental Freedoms, the meaning and scope of those rights shall be the same as those laid down by the said Convention. This provision shall not prevent Union law providing more extensive protection.

Charter of Fundamental Rights of the European Union, art. 52, Dec. 12, 2007, 2007 O.J. (C 303) 1.

[699] Article 8(2) of the European Convention on Human Rights reads as follows:

2. There shall be no interference by a public authority with the exercise of this right except such as is in accordance with the law and is necessary in a democratic society in the interests of national security, public safety or the economic well-being of the country, for the prevention of disorder or crime, for the protection of health or morals, or for the protection of the rights and freedoms of others.

European Convention on Human Rights, art. 8(2).

[700] Convention for the Protection of Individuals with Regard to Automatic Processing of Personal Data, Jan. 28, 1981, ETS 108. For further details, see *Complete List of the Council of Europe's Treaties*, Council of Europe, https://www.coe.int/en/web/conventions/full-list/-/conventions/treaty/108.

[701] Parliament and Council Directive 95/46/EC reads as follows:

1. Member States shall provide that personal data must be: (a) processed fairly and lawfully; (b) collected for specified, explicit and legitimate purposes and not further processed in a way incompatible with those purposes. Further processing of data for historical, statistical or scientific purposes shall not be considered as incompatible provided that Member States provide appropriate safeguards; (c) adequate, relevant and not excessive in relation to the purposes for which they are collected and/or further processed; (d) accurate and, where necessary, kept up to date; every reasonable step must be taken to ensure that data which are inaccurate or incomplete, having regard to the purposes for which they were collected or for which they are further processed, are erased or rectified; (e) kept in a form which permits identification of data subjects for no longer than is necessary for the purposes for which the data were collected or for which they are further processed. Member States shall lay down appropriate safeguards for personal data stored for longer periods for historical, statistical or scientific use. 2. It shall be for the controller to ensure that paragraph 1 is complied with.

related to smart grids that lawmakers clearly identify what legitimate purposes might be for data gatherers and users.

Data minimization Data minimization requires that the purpose of processing data be visibly defined before processing is started. This requirement, although part of EU law, is left for Member States to interpret in domestic law. However, there will be less scope for such flexibility under the GDPR. Data specification requirements regulations are designed to limit the accumulation of data gathered and prevent the processing of data for undefined purposes.[702] This is a procedural requirement based upon the principle of transparency. The use of collected data for another purpose needs an additional legal basis if the new processing purpose is incompatible with the original one.[703] An additional legal basis is also necessary if data is transferred to third parties. The onus is placed on the data controller to comply with the obligations. The data controller must specify and make it clear to data providers the purpose for which data is being processed.[704] There is space for flexibility only if data is used for a compatible purpose. Both the Convention 108 and the Data Protection Directive resort to the concept of compatibility: the use of data for compatible purposes is allowed on the ground of the initial legal basis.[705] Neither law defines "compatibility," leaving this open to interpretation when determining if the initial legal basis for collecting the data is valid for a purpose different than the original one for which it was collected. The Data Protection Directive explicitly declares that the "further processing of data for historical, statistical or scientific purposes shall not be considered as incompatible provided that Member States provide appropriate safeguards."[706] There is no requirement on the data controller to obtain the consent of the data subject where collected data is used for a purpose compatible with the original one. This flexibility gives data controllers freedom to use collected data further. This could result in uses that data subjects would, if they were made aware, object to. It is also a way to keep data beyond the time period the original data was gathered for. Despite a lack of reference to consumer rights in even the more recent GDPR, the European Data Protection Supervisor has stated that

---

Parliament and Council Directive 1995/46/EC, On the Protection of Individuals with Regard to the Processing of Personal Data and on the Free Movement of Such Data, art. 6, 1995 O.J. (L 281) 31.

[702] Parliament and Council Regulation 2016/679/EU, On the Protection of Natural Persons with Regard to the Processing of Personal Data and on the Free Movement of Such Data, art. 5.1(b), 2016 O.J. (L 119) 1 [hereinafter General Data Protection Regulation].

[703] *Id.*

[704] Article 29 Data Protection Working Party, *Opinion 03/2013 on Purpose Limitation*, 00569/13/EN WP 203, (2013).

[705] Parliament and Council Directive 1995/46/EC, On the Protection of Individuals with Regard to the Processing of Personal Data and on the Free Movement of Such Data, art 6(1)(b), 1995 O.J. (L 281) 31.

[706] *Id.* rec. 29.

8.9 Concerns About Data Protection 417

consumer protection law has a part to play in data protection, especially on the subject of transparency of data usage.[707]

Data quality, retention, and accuracy All processed data must be "adequate, relevant and not excessive in relation to the purpose for which they are collected and/or further processed."[708] The data controller must ensure that the purpose for gathering the data is clear, that gathering is kept to a minimum, and that the data collected are relevant for processing operations purposes. The data quality principle is aligned with the principle of limited data retention. Data should be deleted as soon as it is no longer needed for the purposes for which it was collected by the data controller. The obligation lies with the data controller to ensure that the principle of retention is met. As with the data minimization principle, exemptions to the principle of data retention must be established in law. Consumers need safeguards to ensure that their data are not used in contravention to the retention principle. Data controllers are obliged to ensure that the data held is as accurate as can reasonably be expected. This is essential for billing purposes, for example.[709]

Fair processing This principle upholds procedural transparency between data subjects and data controllers. Controllers must inform data subjects on whose behalf they are processing their data and whether the controller has any intentions to process the data for other purposes. Fair processing prevents secret or covert processing that may be against the wishes or interests of the data subject. This principle is perhaps the most significant for developing trust between the data subject and the data controller.[710] For this principle to be effective, the terminology used to communicate with data subjects by data controllers must be understandable. Where data subjects have specific needs, these should be taken into account by the data controller in order to meet their transparency principle obligations. Indeed, fair processing also means that controllers are prepared to go beyond the mandatory legal minimum requirements, if the legitimate interests of the data subject so require.[711] Going beyond what it is expected can be demonstrated by adopting data management standards. Data subjects should have free, easy access to their data. Data controllers should be able to demonstrate how their procedures meet data

---

[707] European Data Protection Supervisor, "Privacy and Competitiveness in the Age of Big Data: The Interplay Between Data Protection, Competition Law and Consumer Protection in the Digital Economy," 23–24 (2014).

[708] Parliament and Council Directive 1995/46/EC, On the Protection of Individuals with Regard to the Processing of Personal Data and on the Free Movement of Such Data, art 6(1)(c), 1995 O.J. (L 281) 31.

[709] Juan Shishido, Smart Meter Data Quality Insights 1 (2012).

[710] Fairness of processing is referred to, notably, in recital 45, and paragraphs 2 and 3 of Article 6 ("Lawfulness of processing") of the General Data Protection Regulation. Parliament and Council Regulation 2001/45/EC, On the Protection of Individuals with Regard to the Processing of Personal Data by the Community Institutions and Bodies and on the Free Movement of Such Data, rec. 45, art. 6, 2001 O.J. (L 8) 1.

[711] European Union Agency for Fundamental Rights (2014), p. 75.

protection requirements under EU law. This emphasis on accountability and legitimacy is integral to building secure and trustworthy relations between data generators and data controllers. According to the 2013 OECD privacy guidelines, "a data controller should be accountable for complying with [data management] principles."[712] Also, according to the Article 29 Working Party's opinion,[713] the essence of accountability is the controller's obligation to put in place measures that would—under normal circumstances—guarantee that data protection rules are adhered to in the context of processing operations, and to have documentation ready that proves to data subjects and to supervisory authorities what measures have been taken to comply with data protection rules.[714]

Data Anonymization/Pseudonymization Pseudonymization is central to significantly reducing the risks associated with data processing, while also maintaining the data's utility. The concept of pseudonymization is central to the GDPR. The GDPR defines pseudonymization as "the processing of personal data in such a manner that the personal data can no longer be attributed to a specific data subject without the use of additional information."[715] To pseudonymize a data set, the "additional information" must be "kept separately and subject to technical and organizational measures to ensure non-attribution to an identified or identifiable person." Any "personal data," which is defined as "information relating to an identified or identifiable natural person 'data subject,'" falls within the scope of the Regulation. There are limits to pseudonymization: it is "not intended to preclude any other measures of data protection."[716]

Ongoing interpretation of principles in data protection law is important in considering their relevance for smart grids. All actors involved in the supply and demand of energy via smart grids need to understand and consider how to meet the legal obligations they face. As noted above, the failure to address regulators' and customers' privacy concerns will pose a major obstacle to successfully moving forward with establishing the new systems.[717]

Aware of this significant problem, in 2010, the European Commission established an institution body to examine the multiple regulatory matters relating to smart grids: the Smart Grid Task Force (SGTF). The SGTF brings together eight different

---

[712]Organisation for Economic Co-operation and Development, "OECD Guidelines on the Protection of Privacy and Transborder Flows of Personal Data," 15 (2013).

[713]Article 29 Data Protection Working Party, *Opinion 3/2010 on the Principle of Accountability*, 00062/10/EN, WP 173 (2010).

[714]European Union Agency for Fundamental Rights (2014), p. 76.

[715]Parliament and Council Regulation 2001/45/EC, On the Protection of Individuals with Regard to the Processing of Personal Data by the Community Institutions and Bodies and on the Free Movement of Such Data, art. 4(5), 2001 O.J. (L 8) 1.

[716]Parliament and Council Regulation 2001/45/EC, On the Protection of Individuals with Regard to the Processing of Personal Data by the Community Institutions and Bodies and on the Free Movement of Such Data, rec. 28, 2001 O.J. (L 8) 1.

[717]Liu et al. (2012), pp. 981–987.

8.9 Concerns About Data Protection

Commission Directors General including energy, climate, environment, and justice along with thirty European organizations representing all relevant stakeholders in the smart grids arena, from both the ICT and the energy sector.[718] Given its cross-sectoral representation, the SGTF is key to regulatory development on ICT and energy interconnections.

The SGTF's main purpose is to advise the Commission on policy and regulatory frameworks at the European level and to assist in coordinating initial steps towards the implementation of smart grids under the provision of the Third Energy Package.[719] Four expert working groups were established in April 2011 to explore the key challenges to smart grid deployment.[720] Expert group 2 (EG2) specifically focuses on privacy and security issues, including developing a data protection template and an energy-specific cybersecurity strategy, and identifying minimum security requirements. The mandate of EG2 was to create a Smart Grid Data Protection Impact Assessment (DPIA) template. In 2014, the EG2 published a template for data protection impact assessment for smart grids and smart grid metering systems.[721] The purpose of the DPIA is to provide guidance on how to perform an assessment for smart grid and smart metering systems. The template will help organizations to take the "necessary measures to reduce risks, and as such, reduce the potential impact of the risks on the data subject, the risk of non-compliance, legal actions and operational risk, or to take a competitive advantage by providing trust."[722] The DPIA is intended to help achieve holistic implementation of data protection principles and rules. The SGTF believes this holistic approach will safeguard confidentiality, integrity, and information assets for the smart grid system. Under the GDPR, it is mandatory to conduct a DPIA.

The new regulatory landscape within the EU, dominated by the reform of data protection under the GDPR, is largely considered to provide more effective data protection and privacy arrangements for data subjects than previously. However, concerns remain, especially with the rapid development of technology, including the upscaling of IoT and Big Data, and it seems that legislators are perpetually fighting a losing battle on privacy.[723] The GDPR arguably restrains this slightly but only to a relatively limited degree, and is arguably easily circumvented by procedural formatting over "consent" protocols.[724] Purtova argues that "personal data will be appropriated in proportion to the de facto power of the data market participants to exclude

---

[718] Smart Grid Task Force Group 3, Workshop on Experiences and Conditions for Successful Implementation of Storage 1 (2016).

[719] Id.

[720] *Commission Communication to the European Parliament, the Council, the European Economic and Social Committee and the Committee of the Regions on Smart Grids*, COM (2011) 202 final (Apr. 12, 2011).

[721] Smart Grid Task Force 2012 Expert Group 2, "Data Protection Impact Assessment Template for Smart Grid and Smart Metering Systems" (2014).

[722] Id.

[723] Mayer-Schönberger and Padova (2016), p. 315.

[724] Kosta (2013).

others."[725] It may be that the boundaries within which the legal concept of privacy is interpreted are changing. Schwartz, who considers that the normative function of privacy lies in the formation of community and personal identity, argues that the individual-specific privacy focus is now challenged. Schwartz further argues that privacy should be a condition of social systems instead of a feature of "inborn" autonomy or a means to control personal data.[726] The shifting nature of this debate will no doubt be evident in Court cases brought to interpret the EU GDPR in the coming years. What is certain, however, is that the principles for data protection will provide the foundations upon which the substantive law will continue to evolve.

8.9.3.2 Digital Systems Security

For all actors engaged in delivering a digital ICT-based energy system across Europe, security is a priority. The previous section considered security in data handling by data controllers of data subjects' information with respect to the fundamental right to privacy. This section surveys efforts within the EU to address risks posed by the increasing dependence of all sectors in society on ICT. It frames this within the context of upscaling smart grids energy systems that see a rise in the number of service providers.

In 2013, the EU launched the Cybersecurity Strategy.[727] It was understood that the goal of achieving a Single Digital Market would flounder if cybersecurity issues were not addressed: The strategy acknowledged that "for new connected technologies to take off, including e-payments, cloud computing or machine-to-machine communication, citizens will need trust and confidence" and that this would be undermined by "threats [from] different origins—including criminal, politically motivated, terrorist or state-sponsored attacks as well as natural disasters and unintentional mistakes."[728] The key initiative by the EU to secure critical digital ICT systems, such as banking, energy, health, and transport, is the 2016 Directive on Security of Network and Information Systems (NIS Directive).[729] The NIS Directive strengthens and modernizes the mandate of the European Network and Information Security Agency that was established in 2004.[730] The NIS Directive will apply to operators of "essential services" and to "digital service providers." EU countries have until 9 May 2018 to transpose the Directive into national law. There will be

---

[725]Purtova (2015), pp. 83 and 83.

[726]Schwartz (2000), p. 743.

[727]*Joint Communication to the European Parliament, the Council, the European Economic and Social Committee and the Committee of the Regions on Cybersecurity Strategy of the European Union*, JOIN (2013) 1 final (Feb. 7, 2013).

[728]*Id.* at 3.

[729]Parliament and Council Directive 2016/1148/EU, Concerning Measures for a High Common Level of Security of Network and Information Systems Across the Union, 2016 O.J. (L 194) 1.

[730]Parliament and Council Regulation 460/2004/EC, Establishing the European Network and Information Security, 2016 O.J. (L 77) 1.

8.9 Concerns About Data Protection

some overlaps with the obligations under the GDPR, but organizations, both large and small, will face new requirements. A significant distinction can be made regarding the type of data protected under the NIS Directive and the GDPR. The NIS Directive covers any type of data breaches whereas the data protected under the GDPR is limited to "personal data."[731]

Unlike the GDPR, which revised existing data protection law within the EU, according to the European Commission Vice-President for the Digital Single Market, Andrus Ansip, the NIS Directive is the first comprehensive piece of EU legislation on cybersecurity and a fundamental building block in that area.[732] As a Directive, the NIS Directive requires Member States to adopt legislation to transpose it. This is different from the GDPR, which is a Regulation and, per its very nature, directly applies to all EU Member States.[733] Consequently, there is space for differences in the approaches adopted by Member States in how to meet the NIS Directive's requirements. This could impact its effectiveness in terms of securing transboundary critical energy digital ICT infrastructure, however, the NIS Directive does actively promote network collaboration and cooperation.[734]

The NIS Directive provides guidelines for "essential service operators," for example within the energy, transport, banking, financial market infrastructure, health, drinking water, and digital infrastructure sectors, as well as "digital service providers," including entities such as online marketplaces, online search engines, and cloud computing service providers. National governments are to play a key coordinating role amongst other actors nationally and within the EU as the NIS Directive requires each Member State to set up a Computer Security Incident Response Team Network (CSIRT) to promote swift and effective operational cooperation on specific cybersecurity incidents and to share information about risks.[735] Critical service providers who will need to cooperate with national CSIRTs are defined under the NIS Directive as entities who "provide a service which is essential for the maintenance of critical societal and/or economic activities; that the provision of the service depends on network and information systems and that an incident would have a significant disruptive effect on the provision of that service."[736]

Operators of essential services have obligations to "take appropriate and proportionate technical and organizational measures to manage the risks posed to the security of network and information systems."[737] To achieve this, service providers

---

[731] Day (2016), p. 4.

[732] Id. at 1.

[733] In the European Union's legal system, European Regulations are self-executing; they require no "transposition" into the legal orders of the different Member States, they are directly binding, and can be directly resorted to by individuals. See TFEU, art. 288.

[734] Parliament and Council Directive 2016/1148/EU, Concerning Measures for a High Common Level of Security of Network and Information Systems Across the Union, art. 11, 2016 O.J. (L 194) 1.

[735] Id. art. 9, art. 11, annex I.

[736] Id. art. 5(2).

[737] Id. art 14.

are encouraged to adopt internationally accepted standards and specifications in order to secure networks and information systems.[738] Annex 11 of the NIS Directive lays out the entities considered to be "essential service operators." Electricity is a subsector of the energy sector. The NIS Directive provision applies to several entities as outlined in Article 2 of the 2009 Electricity Directive.[739] These include DSOs[740] and TSOs, who are engaged in an "electricity undertaking," which includes at least one of the following functions: generation, transmission, distribution, supply, or purchase of electricity.[741] The NIS Directive clearly applies to the electricity sector. Providers of the service, whatever the size of the operation, need to comply with the NIS Directive's requirements.

It is important that small-scale energy providers, such as prosumers and energy cooperatives, are given the necessary support to adopt appropriate measures to reduce the risks to their technical and information networks. The focus of the observers is often on large-scale cyberattacks across national systems, however targeted criminal activities on relatively small-scale energy providers could inflict harm on customers (as well as on service providers) in many ways, from loss of power to fraud. The national government as well as service providers have an obligation to ensure this situation does not occur. One area that will require further security risk measures is the financial transactions between service providers and customers. This could become more challenging with the emergence of virtual currencies and smart contracts.[742]

8.9.4 The Situation in Certain EU Member States

Given that smart grids monitor energy consumption and fluctuations in supply and demand, this involves collecting user data including personal information, habits, and time spent at home. Naturally, this raises concerns regarding data protection and privacy, and the EU member states need to have relevant and effective regulation in place before smart grids become widespread.

In Belgium, the Law of 8 December 1992 provides a regulatory framework for data protection and privacy in Belgium and, thanks to a number of subsequent amendments, it has kept pace with technological advancements. In general, Belgium has a progressive stance in this area, which has even caused some to dub it a data protection hub.[743] The national data protection authority—the Privacy

---

[738]*Id.* art. 19(1).

[739]Parliament and Council Directive 2009/72/EC, Concerning Common Rules for the Internal Market in Electricity, 2009 O.J. (L 211) 55.

[740]*Id.* annex II. DSO is defined in art. 2(4) and 2(6).

[741]*Id.* art. 2(35).

[742]*See generally* Romain et al. (2016), p. 133; Mihaylov et al. (2014), p. 58.

[743]Smedt (2015), pp. 213–218.

8.9 Concerns About Data Protection

Commission—ensures the country's compliance with relevant laws, and Belgian regulation complies with the basic requirements of Directive 95/46/EC on the protection of individuals with regard to the processing of personal data and the free movement of such data.

When it comes to electricity, the DSO holds the position of data controller and thus, in keeping with the broad guidelines of data protection, it is responsible for primary data protection. This means that it is required to let the Privacy Commission know before using any kind of automated system for processing personal data. According to the law's rather broad definition, processing includes collection, recording, organization, storage and deletion of personal information, among other things. Arguably, by placing stress on the idea of "automation," the law seems to exclude manual data processing from its scope.[744] Thus, the controller only needs to notify the Privacy Commission when it uses automated systems for data processing.

Moreover, controllers are not required to notify the Privacy Commission when data processing relates to administrative duties, such as billing procedures. This leads to complications, though, because the assumption is that data around consumption is solely used for billing processes. But while consumption data received via smart meters might not be personal *per se*, when combined with other key data it could suffice to identify a customer. Unfortunately, the law does not seem to factor in this situation. Still, since the law came before the onset of smart metering, this gap is perhaps understandable. Moreover, with meter reading being an annual undertaking in the country,[745] it is plausible that the data is processed too infrequently to pose a major risk to consumer privacy. As advanced metering infrastructure (AMI) is currently at the pilot phase and has not been deployed across the country yet, there is still time to enact legal reform.

In addition to notifying the Privacy Commission about automated data processing, the controller must also take suitable steps toward ensuring security, in order to avoid any loss or risks to personal data. The does not set out any exact requirements to this end, though, which leaves data controllers with some freedom in terms of integrating relevant measures as they see fit.

Across the country, the electricity market runs a standard communication platform—the Belgian Utility Market Information Exchange (UMIX). The DSOs retrieve and use information from UMIX, e.g., working orders and forecasts that the TSO provides for the network's smooth functioning. The platform additionally lets suppliers view meter readings recorded by DSOs for billing processes.[746] In this sense, the rules on third-party processing become relevant, where Article 16(1) of the law on data protection calls for a formal agreement between the controller and the

---

[744]T. D'hulst, "Data Protection in Belgium: overview," Practical Law—Thomson Reuters, 1 July 2016. Available at: https://uk.practicallaw.thomsonreuters.com/2-502-2977?__lrTS=20170610192313070&transitionType=Default&contextData=(sc.Default).

[745]Council of European Energy Regulators (CEER), "CEER Benchmarking Report on Meter Data Management Case Studies," CEER, Brussels, 2012.

[746]Idem.

third party processing the data. Such a contract must factor in the required steps to ensure security and must make clear the third party's responsibility vis-à-vis the controller. It is the controller's responsibility to apprise the customer of any data sharing and to get the customer to sign-off on this. The European Commission's Smart Grid Task Force has observed a lack of clarity, though, regarding the role of third-party processors and is pushing for a more rigorous approach.[747]

Currently, the UMIX is being re-worked to accommodate the eventual country-wide deployment of smart meters and the proliferation of DERs in the country's electricity network.[748] The new portal would allow for a much higher number of personal data flows, requiring strong legal mechanisms for protection. Further, a Data Protection Impact Assessment (DPIA) is likely to greatly lower risks to data and privacy.[749]

In Spain, the Data Protection Law 15/1999 of 13 December safeguards individuals in terms of data processing and the free movement of data. The Royal Decree 1720/2007 of 21 December takes the Data Protection Law 15/1999 further. Spain's data protection legislation is among the EU's most stringent. Penalties are on the higher spectrum (up to €600,000 per violation).[750] Moreover, the Spanish Data Protection Agency, which is the country's national data protection authority, is known for its tough enforcement of data protection rules.[751]

The Data Protection Law 15/1999 calls upon data controllers to draft internal security policies identifying the technical and organizational measures that staff will enact. Measures will be devised in accordance with security levels (low, medium, and high), which are determined based on data sensitivity, or the nature of the entity involved. Data controllers, according to the Data Protection Law 15/1999, are individuals or legal persons who control and are in charge of personal data use and storage on a computer or in manual files. The data processor is the entity that processes the data for the data controller. Should personal data processing be outsourced, i.e., should processing be exclusively executed by the data processor, the data controller may be entitled to require an internal security policy of the data processor. The Data Protection Law 15/1999 maps out measures to be enforced in the face of each different security level.[752]

Anyone who desires to make or change a personal data file needs to sign up for free with the Spanish Data Protection Agency. The procedure entails completing a form via the organization's website. The notification of data files is also free. The

---

[747] Smart Grids Task Force, "My Energy Data (Report by Expert Group on Smart Grid Deployment (EG1))," 2016.

[748] SIA Partners, "Atrias and MIG6.0: Towards a new energy market model in Belgium," SIA PArtners, 1 July 2016. Available at: http://energy.sia-partners.com/20160701/atrias-and-mig60-towardsnew-energy-market-model-belgium.

[749] Smart Grids Task Force, "My Energy Data (Report by Expert Group on Smart Grid Deployment (EG1))," 2016.

[750] Article 45, paragraph 3 Data Protection Law 15/1999.

[751] Hogan Lovells (2015).

[752] Azim-Khan (2008).

8.9 Concerns About Data Protection

Data Protection General Registry will approve notifications that they meet necessary requirements. Moreover, if security levels are medium or high, a data protection officer must be appointed.[753] The Spanish Data Protection Agency carefully complies with the requirements set out by Data Protection Law 15/1999 in terms of registration. In fact, the Data Protection Agency has noted a rise in registrations, particularly from medium-sized companies as well as independent professionals.[754]

When it comes to the process of DSOs submitting consumption information to consumers, the Royal Decree 216/2014 of 28 March sets out the relevant obligations. For those with smart meters in place, the legislation calls for DSOs to provide hourly consumption data. In addition, DSOs operate websites that allow customers to access and save their hourly consumption data (after billing). DSOs also enable consumers to download their consumption profiles made available to energy suppliers for billing purposes in comma separated values (CSV) and Excel flat-files. Data received from smart meters are stored in the DSOs' metering managing systems. DSOs submit data to energy suppliers via secure File Transfer Protocol (FTP). Energy suppliers can only access data relating to their customers—to access data for other customers they require prior consent.[755] DSOs own smart metering data but must provide the data to end users for consultation purposes and to energy suppliers for billing.[756]

The Royal Decree 1074/2015 of 27 November is also relevant in this context as it amends some aspects of Article 7 of the Royal Decree 1435/2002, which regulates data to be stored in the Supply Points Information System (SIPS) by establishing the basic conditions for acquiring energy contracts and accessing low voltage networks. The DSOs manage the SIPS database, which only the NRA, the CNMC, and energy suppliers are allowed to access. The database, updated frequently, contains information related to the supply points connected to the networks and transport networks of the areas that each DSO manages. The SIPS was developed in order to boost competition in the reduced electricity supply market. It achieves this by helping create the required consumption data that will encourage new energy suppliers while ensuring consumer privacy.[757] Royal Decree 1074/2015 contains an amendment that prevents the SIPS from collecting data regarding consumer hourly load curves, because consumption data, which DSOs compile via smart meters, are considered personal data and thus requires corresponding protection. On that note, energy

---

[753]Linklaters, "Data Protected. Spain. General. Data Protection Laws," Linklaters. Available at: https://clientsites.linklaters.com/Clients/dataprotected/Pages/Spain.aspx.

[754]Spanish Data Protection Agency, "Spanish Data Protection Agency," Spanish Data Protection Agency, Madrid.

[755]European Smart Grid Task Force (2016).

[756]Leiva (2016).

[757]EnerConsultoría. Derecho de la energía, "contadores inteligentes y protección de datos," EnerConsultoría. Derecho de la energía, 8 December 2015. Available at: http://www.enerconsultoria.es/BlogLeyesEnergia.aspx?id=36002236&post=Contadoresinteligentesyprotecciondedatos.

suppliers cannot access any information—except for that relating to their own customers—which enables the identification of the supply point incumbent (such as supply point location, household address, or first and last names).[758]

Safety of consumer privacy is of high importance, especially when it comes to the European internal energy market. The successful roll-out of smart meters relies on robust management of consumption data. Data protection and consumer privacy form basic requirements for the success of the broad deployment of smart metering systems.

Fortunately, technical advancements allow smart grids today to monitor energy consumption almost in real time. Smart meters avail of the Advanced Metering Infrastructure (AMI) that allow this insight. Considering the volume and sensitivity of the data they process, smart meters must integrate secure storage systems as well as backup and contingency mechanisms. Analyzing end-user smart metering data allows for surprisingly accurate insights into their private lives: time spent at home, working schedules, vacations, use of specific gadgets, hobbies, etc. Such information is naturally valuable to many third parties, thus consumer profiling endangers the privacy of consumers.[759]

The Italian Personal Data Protection Code (Legislative Decree No. 196/2003) regulates data protection in the country, and, as it is general in scope, it can be applied to smart grids. Its definition of personal data does not just refer to data related to a natural person that enables that person to be identified, but also to data that enables a person to be identified indirectly by referring to other types of information.

The code contains all necessary elements of a strong data protection mechanism, including the need for data controllers to be registered, the need to gain consent before storing and handling personal data, and the need for data controllers to follow tight safety protocols to make sure data is protected. Additionally, the law obliges data controllers to notify affected parties of any risks or situations that could jeopardize their personal data, and make sure data subjects are aware of their rights to access or correct their data. Overall, the code aims to prevent unlawful information processing by which data subjects can be identified, either directly or via reference to other information. Italy has a national data protection body, known as the Italian Data Protection Authority (IDPA), which makes sure relevant parties adhere to the principles of the Data Protection Code.

As in other EU countries, Italy's Data Protection Code makes the data controller legally responsible for data processing. When any data processing work deals with genetic or biometric data, geo-localization, or behavioral advertising (barring a few exceptions), data controllers must let the IDPA know before embarking on the activity.[760] Data controllers are also obliged to use systems and software that involve using the least amount of personal data, and when using such data is necessary it

---

[758]EnerConsultoría. Derecho de la energía, "contadores inteligentes y protección de datos," EnerConsultoría. Derecho de la energía, 8 December 2015.
[759]Rubio (2017).
[760]Schiavo et al. (2011).

8.9 Concerns About Data Protection

must done in a way that that the data subject is aware of and has already approved. Controllers should not keep data longer than they need to for processing.

The Italian Code is more detailed than many in that it specifies minimum security measures that data controllers must use for their processing systems. If data is being processed electronically, such measures refer to computerized authentication systems for those who have access to the system, regular updates to the system, and processing for storing backup information and restoring the system and data if necessary. If data is being processed manually, the code calls for making sure certain records are kept in restricted-access locations and that certain people be appointed responsible for processing.

Only electronic communication service providers and those handling biometric data must let the IDPA know in case of security threats. However, in case of any threat, if a data subject's privacy is at risk, the controller must also let the subject know. This does not apply, though, if the compromised data was anonymous or encrypted. The code is enforced by imposing a range of fines, with a maximum of €2,448,000. If individuals are found to be responsible for breaches, they may face criminal charges and up to 3 years imprisonment.

In Italy, DSOs are primarily responsible for data protection, as they handle metering and thus collect customer data. The contract between retailers and customers includes an agreement regarding data collection and handling. The data management system for electricity in Italy, though, is centralized. This means that smart meter data is transmitted to a central database that Acquirente Unico Spa manages,[761] called the Integrated Information System (IIS). DSOs access customer data from the IIS for billing. TSOs also have access to this information, for balancing processes.[762] For the DSO to share information with the TSO and even Acquirente Unico Spa., could perhaps be seen as a breach of the customer's privacy. But according to the definition of personal data in the Italian Data Protection Code, aggregated meter information shared with the TSO does not count as personal data, which means this data sharing should be permissible.

In Greece, no legislation specifically addresses data access and security for smart grids, however, this falls within the country's general data protection laws. This law adheres to EU Directive 95/46/EC of the European Parliament and Council and protects people against the unlawful processing of personal data. Additionally, the general data protection law has undergone several amendments, to address situations related to collecting and processing data via information and communication technology systems.

Greek data protection laws cover the foundational aspects of data protection. In other words, the laws address the registration of data controllers with the Hellenic Data Protection Authority (HDPA); the obligation to obtain prior informed consent when collecting and processing personal data; security obligations on data

---

[761] Smart Grids Task Force, "My Energy Data (Report by Expert Group on Smart Grid Deployment (EG1))," 2016.
[762] Idem.

controllers to ensure data security; the obligation to inform the individuals of breaches that may compromise their personal data; and the data subject's right to access, request rectification and object to the processing of personal data. By extension, these overall laws apply to the electricity sector and, by that very fact, to all the participants in a smart grid situation. It remains to be discovered as to whether these laws adequately cover all potential challenges related to smart grids. Still, it is perhaps worth exploring the possibility of additional legislation to address potential gaps in the existing scenario and likely challenges that could stem from these gaps.

Another important factor to examine is the legal definition of "personal data" when discussing smart grids. Greek Data Protection law defines personal data as information relating to the data subject, barring data related to statistics that does not enable identification of the data subject any more.[763] The Data Protection Authority (DPA) has not released a specific framework to inform the definition of personal data. However, DPA decisions seem to suggest that information is considered to be personal data if it can be combined with other information on a data subject to result in the data subject's identification. Along these lines, there is a strong likelihood that information recorded through smart meters could be categorized as personal data and might thus pose an obstacle towards the rapid rollout of smart grids.

Moreover, interestingly Greek data protection law makes a distinction between data that is "personal" and data that is "sensitive," which is explained as data regarding racial or ethnic origin, political opinions, religious or philosophical beliefs, trade union membership, health, social welfare and sexual life, criminal charges or convictions. While data classified as personal may result in certain obligations when it comes to smart grids, sensitive data will probably not result in any data protection obligation in the context of smart grids. This is because smart grid systems are not likely to collect and process data that would fall within the sensitive category.

In the smart grid context, another area that has yet to be clarified is the role of data controller, in charge of ensuring data protection. According to Law 2472/1997 the controller is a *"person who determines the scope and means of the processing of personal data."* This definition seems to suggest that DSOs should be tasked with the responsibility of controller, since customer consumption data that smart meters record is mainly used for helping with balancing work. This begs the questions, however, as to whether the DSO is not simply acting as a data processor for the Hellenic Electricity Market Operator (LAGIE) when one considers the complicated wholesale market system and additional services for which extra data on customer consumption is needed. Directive 95/46/EC[764] may provide a way to resolve this

---

[763]European Regulators' Group for Electricity and Gas (ERGEG) (2007).

[764]Directive 95/46/EC of the European Parliament and of the Council of 24 October 1995 on the protection of individuals with regard to the processing of personal data and on the free movement of such data.

8.9 Concerns About Data Protection

dilemma, as it calls for a clear delineation of 'controllers' and 'processors' as well as their respective roles and responsibilities in the smart grid context.[765]

Data retention is an ongoing challenge in Greece. Although the European Court of Justice in Digital Rights Ireland[766] invalidated the Data Retention Directive, Greece has not amended Law 3918/2011, which replaced the Data Retention Directive. Thus, the authority to determine data retention durations lies with the Hellenic Data Protection Authority (HDPA). Since the HDPA has not released any framework related to the retention of personal data collected through smart meters, one may infer that data collectors within a smart grid network face no exact obligations regarding the retention of such data. Still, arguably, complying with the general principles of data retention means that retaining personal data beyond periods for which its processing is necessary would count as a violation of the data subject's rights of the data subject.

When it comes to data anonymization, Greek law does not specify which classifications of personal data should be anonymized. In fact, a legal definition of anonymization does not exist. Still, arguably, the law does address the situation to the extent that it states that "personal data in order to be lawfully processed must be...(d) kept in a form which permits identification of data subjects for no longer than the period required, according to the Authority, for the purposes for which data was collected or processed." Questions remain, however, as to what counts as adequate levels of anonymization. The general approach is to employ coded formats for data, but Greek law does not address levels of codification which are considered adequate.[767]

In this context, the EU General Data Protection Regulations (GDPR), which entered into force in May 2018, state that persons whose operations pose a high risk of breaches to data security are obliged to undertake data protection impact assessments. While this has not yet transpired, various data protection authorities across Europe have started taking steps towards complying with the regulation.

Data security is a key aspect of data protection. Laws in Greece related to data protection call upon data controllers to take institutional and technical measures towards guaranteeing security and confidentiality in the data processing process. Considering that smart grids give rise to a high risk of data intrusion, risks to electricity infrastructure and perhaps to national security, it is crucial that data protection laws be combined with establishing standards for the various technological components related to smart grids, including smart meters. As mentioned earlier, the smart meter rollout that has taken place thus far in Greece did not seem to follow specific technical requirements towards guaranteeing data protection and security.

In the case of France, data protection was regulated, until May 2018, as a part of a legal instrument with a broader scope known as "*Loi informatique et libertés*" (Computering and Freedoms Act; Loi n°78-17 du 6 janvier 1978, Loi relative à l'informatique, aux fichiers et aux libertés). With the entry into force of the European

---

[765]Karageorgiou & Associates Law Firm, "Linklaters LLP."

[766]Papakonstantinou and Kloza (2015), pp. 41–129.

[767]Joined Cases C-293/12 and 594/12, Digital Rights Ireland and Seitlinger and Others.

General Data Protection Regulation (GDPR, Regulation (EU) 679/2016), the French parliament voted on 14 May 2018 for a new *"Loi informatique et libertés,"* amending accordingly the basic legal framework in France for this particular area. Among other aspects, the new piece of legislation is meant to deal with three different areas:

i) the national specificities allowed by the GDPR in respect of sensible information (biometrical data, health data, political affiliation data, religious data, and sexual-life data), public-interest-related data, and other specific scenarios (social security number, employment relations, etc.);
ii) the organization and functioning of the *"Commission Nationale de l'Informatique et des Libertés"* (CNIL), the French data protection authority; and
iii) the transposition of Directive (EU) 2016/680 of the European Parliament and of the Council of 27 April 2016 on the protection of natural persons with regard to the processing of personal data by competent authorities for the purposes of the prevention, investigation, detection or prosecution of criminal offences or the execution of criminal penalties, and on the free movement of such data.[768]

However, an appeal against the text was lodged before the Constitutional Council by at least sixty senators on May 16, 2018.[769] The French Constitutional Council rendered its decision on the new law on June 12, 2018 and the Law on the Protection of Personal Data (No. 2018-493) (*Loi n° 2018-493 du 20 juin 2018 relative à la protection des données personnelles*) was promulgated on June 20, 2018.[770] The new law was then complemented with an implementation decree (No. 2018-687 of 1 August 2018).[771] The former Act No. 78-17 of 1978 also remained in force, allowed by the national points of maneuver provided by the GDPR as a complementary regulation to the main one.[772]

It is important to highlight that the French understanding of privacy and individual rights is having an impact on the deployment of "linkys": media have reported number of public demonstrations, campaigns and, in general, civil-society reluctance to the rollout of smart-metering, due to the view that smart metering deployment entails being "spied" upon, and an intrusion into one's private life.[773] This resonates with the recent formal notice issued by the CNIL to *"Direct Energie, societe anonyme"* (France's primary electric supplier) on 27 March 2018, giving it 3 months to obtain specific consent for the collection of customer usage data through smart

---

[768]Commission Nationale de l'Informatique et des Libertés, available at: www.cnil.fr/fr/cnil-direct/question/1254.

[769]*Décision n ° 2018-765 DC du 12 Juin 2018*, Conseil Constitutionnel, https://www.conseil-constitutionnel.fr/decision/2018/2018765DC.htm.

[770]Bryan Cave, *France Adopts Regulations Implementing the GDPR*, Lexology (Aug. 24, 2018), https://www.lexology.com/library/detail.aspx?g=e4de5746-c831-419e-b48a-712cef451985.

[771]*Id.*

[772]*Id.*

[773]Patrick Collison, "Is your smart meter spying on you?" *The Guardian,* 24 June 2017, available at www.theguardian.com/money/2017/jun/24/smart-meters-spying-collecting-private-data-french-british.

8.9 Concerns About Data Protection

meters, or else face a fine of up to 3 million euros: *"CNIL observed that at the time of the installation of smart meters, customers were asked to provide a single consent for the installation of the meter and for the collection of hourly electricity consumption data as a corollary to the activation of the new meter and in order to benefit from certain tariffs. However, as the installation was mandatory, customers were in fact only consenting to the data collection."*[774]

As a consequence, the CNIL considered that the consent thus obtained by the electric supplier was invalid, for not being *"free, informed and specific."*[775] The CNIL's findings further entail that *"the automatic collection of this data, which is particularly intrusive and detrimental to* [the customers'] *privacy, disregards their interests and rights, especially since there are no tariff offers based on their hourly consumption."*[776] Although entrenched in French legislation, and in the French approach to privacy and individual rights, this decision is said to be *"nearly identical to the arguments made in the United States in the court case of Naperville Smart Meter Awareness (NSMA) v. City of Naperville and where failure to obtain a valid customer consent for granular smart meter data collection represents an illegal, unwarranted, and unreasonable search in violation of the Fourth Amendment."*[777]

As for Germany, data protection is a very delicate issue. As smart grids and smart meters approach deployment, certain concerns are still under discussion in terms of a wide rollout of smart meters, as they collect large amounts of data from citizens and businesses that then has to be properly managed and stored. In Germany, the EU Data Protection Directive (95/46/EC)[778] *has been implemented by* the Federal Data Protection Act (*Bundesdatenschutzgesetz*) (*BDSG*),[779] which represents the main law for data protection. In May 2018, when the GDPR (Regulation (EU) 2016/679 on the protection of natural persons with regard to the processing of personal data and on the free movement of such data)[780] came into force, the BDSG was replaced by the new Federal Data Protection Act (*BDSG-neu*).[781]

---

[774] Smart Grid Awareness, *France: No Legal Basis for Smart Meter Data Collection without Valid Consent*, available at: smartgridawareness.org/2018/04/01/no-legal-basis-for-smart-meter-data-collection.

[775] *Ibid.*

[776] *Ibid.*

[777] *Ibid.*

[778] European Parliament and of the Council, Directive 95/46/EC, available at eur-lex.europa.eu/legal-content/EN/TXT/?uri=celex:31995L0046.

[779] Federal Data Protection Act (Bundesdatenschutzgesetz) (BDSG), available at www.gesetze-im-internet.de/englisch_bdsg/englisch_bdsg.pdf.

[780] European Parliament and of the Council, Regulation (EU) 2016/679 on the protection of natural persons with regard to the processing of personal data and on the free movement of such data, and repealing Directive 95/46/EC, available at https://eur-lex.europa.eu/legal-content/EN/TXT/PDF/?uri=CELEX:32016R0679&from=EN.

[781] Norbert Nolte and Christoph Werkmeister, Data protection in Germany: overview, Practical Law, available at uk.practicallaw.thomsonreuters.com/3-502-4080?transitionType=Default&contextData=(sc.Default)&firstPage=true.

The provisions contained in the new act are the framework within which another new legislation, the new Metering Point Operation Act, is contained. The Metering Point Operation Act specifically focuses on smart meters and addresses major concerns for data protection in the smart grid.[782] One of the main goals of the Metering Point Operation Act is indeed data protection and security. In particular, part 3 of the Metering Point Operation Act regulates which party can receive what data and for what purpose. Data transmission is permitted only for the applications required for the energy industry operations. It also regulates when the data received have to be deleted. Any additional data traffic always requires the consumer's consent.[783]

When smart meters are installed in households, the transmission of data is required only on an annual basis. This means that those consumers with an annual consumption higher than 10,000 kW-hours will retain their data to record their consumption. However, the consumer can choose a tariff which requires more detailed metering and data transmission.[784]

As part of the Act, the German Federal Ministry for Economic Affairs and Energy assigned some institutions specialised in data protection the important role of drafting a set of technical protectionary guidelines (*The smart meter gateway*)[785] to ensure data protection. The Federal Office for Information Security (BSI) has been appointed as the leading institution. It collaborates closely with industry representatives, the Federal Commissioner for Data Protection and Freedom of Information, the Federal Network Agency and the PTB (National Metrology Institute). The smart meter gateway makes meters connected to the smart grid in a way that meets data protection and data security standards and only the systems that respect the level of data protection and data security set by the guidelines can receive the BSI's approval.[786]

Data security is certainly one of the biggest challenges of the *Energiewende*. Germany has to face the digitalisation and integration of decentralized, renewable energies. Intelligent network areas needed to link energy generation, storage and consumption. There are many new challenges, such as the increased risks to data security, infrastructures becoming more complex and, of course, exponential increases in the amounts of data being collected, stored, and processed. In fact, one of the reasons smart meters will be not rolled out widely in Germany even after

---

[782]Lisa Alejandro et al., Global Market for Smart Electricity Meters: Government Policies Driving Strong Growth, U.S. International Trade Commission, No. ID-037, 2014.

[783]Federal Ministry for Economic Affairs and Energy, R&D funding provided by the Federal Ministry for Economic Affairs and Energy, available at www.bmwi.de/Redaktion/EN/Artikel/Industry/electric-mobility-r-d-funding.html.

[784]*Ibid.*

[785]Bundesamt für Sicherheit in der Informationstechnik (BSI), Das Smart-Meter-Gateway, Cyber-Sicherheit für die Digitalisierung der Energiewende, 2018.

[786]Federal Ministry for Economic Affairs and Energy, R&D funding provided by the Federal Ministry for Economic Affairs and Energy, available at www.bmwi.de/Redaktion/EN/Artikel/Industry/electric-mobility-r-d-funding.html.

8.9 Concerns About Data Protection

2020 is the attention that the country is giving to the risk of potential violation of citizens' privacy and the need to build new, stronger, and more secure communication infrastructures in order to guarantee data protection in a widely implemented smart grid.

Data protection is also one of the main issues with smart meters in the Netherlands. A debate started on this topic when the introduction of smart meters was envisioned in 2006, and it is still far from being solved. Issues with smart meters and privacy concerns started after the 31374 Bill had been submitted to Parliament. The Dutch Data Protection Authority (DDPA) was asked to verify the conformity of the Bill with Dutch privacy law. It was found that the proposal for the Dutch smart metering Act was in violation of the Dutch Data Protection Act (Wet Bescherming Persoonsgegevens).

The Minister of Economic Affairs amended the proposal, addressing the concern raised about lack of consent and data access, as well as stressing that the conditions of chapter 2 of the Dutch DPA would apply ('purpose specification and use limitation, right of access, data removal after use and relevant security measures'). Only after the amendment, did the DDPA validate the compliance of the Act with the Dutch data protection rules, and the Act was finally passed. Nonetheless, the Dutch Consumer Union commissioned a study to verify the existence of further concerns related not only to data protection, but also the right to inviolability of the home and the right to respect family life. An analysis on the conformity of the Act with Article 8 of the European Convention on Human Rights found that some aspects of the bill were not necessary in a democratic society, e.g., the generation and passing on of quarter-hourly/hourly readings to grid managers.[787]

As less invasive alternatives had not been sufficiently taken into consideration, some new bills (called *novelles*) entered into force in 2011. They contained the following amendments: (1) end-users receive only one bill, rather than separate bills from the grid manager and the energy supplier, enhancing the coherence in the management of end-user data; (2) enhancing transparency and awareness; (3) clarification and codification of the terms and conditions under which personal data can be processed by the parties involved in the process of energy supply; (4) cancelling the obligatory roll-out of smart meters. With all these changes, Dutch law on smart grids is more in line with privacy requirements.

We find that, in the context of smart grids in the EU, there is a need for stronger legislation on data protection and cybersecurity, as well as international standards on data. Setting the rules, however, is not enough. Execution is necessary, for instance, by providing incentives to get things done.

---

[787]Colette Cuijpers and Bert-Jaap Koops, Het wetsvoorstel 'slimme meters': een privacytoets op basis van art. 8 EVRM [The 'smart meters' bill: a privacy test based on article 8 ECHR], Study commissioned by the Dutch Consumers' Association, October 2008.

8.10 Conclusion

Technological advancement is key for a successful decarbonization process, with the aim to provide the highest quality of life with the lowest footprint. The winners of this process are consumers, the environment, and our future. However, technology alone is not enough; we also need the right public policies to reach our decarbonization goals. Smart grids are clearly part of the EU's future economic, social, and environmental policy landscape. Key strategies on the economy, the environment, and technology provide opportunities for the radical transformation in Europe's energy infrastructure to take place through smart grids.[788] It is also evident that the EU needs to work towards the energy transition in a manner that ensures balanced, equitable, fair, and just outcomes for all citizens. The collaborative economy, for example, should not undermine employees' rights or environmental standards. Moreover, the concept of circular economy[789] needs to be embedded in public policy, it needs social acceptance, and private-sector product design and resource management will play a crucial role in the future. All of this will be possible with the right public policies in place and changes in behavior: change is difficult, even when the status quo is bad, but it is necessary. As a result, one may be a short-term pessimist, but a long-term optimist.

This chapter has also analyzed the legal framework related to smart grids in the EU. We find that the EU legal framework on smart grids is fragmented and needs to be streamlined. Although sufficient direction for the roll-out of an "intelligent grid" exists at the regional level, there is still much legislation and policy that needs to be put in place, particularly at the national level. We also find that regulation may exist, but is not in force or is incoherent. The various components envisaged by smart grids are at different levels of development. Consequently, legislative responses towards more ecological regulation have been insufficient or lacking.[790] Although specific legislation, and perhaps standardization, is desirable, the absence thereof should not operate as a hindrance to the successful deployment of smart grids, given that

---

[788] For an analysis of how transformation can happen locally, see Hopkins (2013).

[789] Some commentators question that the industrial economy is circular and argue, instead, that it is entropic and that the concept of 'circular economy' is only an aspiration of the twenty-first century. The industrial economy uses exhaustible resources such as fossil fuels. It burns them for energy. The energy dissipates and produces CO_2 in excessive quantities. The industrial economy deposits waste anywhere it can: the atmosphere, oceans, and rivers. See Haas et al. (2015); Caticha and Golan (2014), pp. 149–163. Arguably, there is no circular economy because of the use of fossil fuels. However, technology will eventually rectify this situation. See Harremoes et al. (2001). The protection of the environment has led to the creation of the concept of eco-compensation, which aims at compensating land users or suppliers of ecosystem services for lost income due to environmental protection policies. See E. Gray and C. Jones, "Eco-compensation in China: Opportunities for Payments for Watershed Services," *World Resources Institute*, 15 May 2012, available at http://www.wri.org/blog/2012/05/eco-compensation-china-opportunities-payments-watershed-services. Other commentators believe in the concept of degrowth, such as Kallis (2018), or wealth without growth, such as Jackson (2016).

[790] See the views of Gorz and Turner (1994) and Gorz (1985).

8.10 Conclusion

sufficient legal bases exist at the regional level, along with apparent political support at the national level. We also find that, in the context of smart grids in the EU, there is a need for stronger legislation on data protection and cybersecurity. Setting the rules, however, is not enough. Execution is necessary, for instance, by providing incentives to get things done.

An additional set of regulatory challenges relates to the use of, and access to, smart meter data for smart grids. In most EU Member States, smart grids will make use of, and indeed rely on, smart meter data and infrastructures. In general, how consumers' data will be managed and by whom will have to be clearly explained. Otherwise, an anxiety about privacy issues will be inevitable. Indeed, access to, as well as ownership of data appear to be the key issues. These are not specific to the energy sector alone, but represent challenges that have been discussed thoroughly in other domains from which lessons may be drawn, such as "big data,"[791] which may be very useful for environmental performance improvement and therefore is a big opportunity. While the regulatory nature of data protection for smart grids remains unclear, it seems likely that national bodies (e.g. independent regulatory agencies for energy), will play a central role. Regulators and policymakers more generally can learn from other sectors which had to face already similar issues (e.g. internet, search engines).

Moving forward, society needs to find a way to make sure that corporations see incentives for green growth, so that they can make a profit and protect the environment (for instance, by selling solar panels or electric vehicles).[792] Short-termism is a great challenge for sustainable development and needs to be avoided at all costs. Short-termism does not allow sustainable policies to be in place because politicians (at least in Western liberal democracies, not dictatorships) are keen to win elections, which are short-term phenomena. Since energy is the driver for much of what we do, clean energy is a sure way to reach sustainability.[793] But the question remains: in the transition to clean energy, can clean energy sources be implemented on a scale that will replace fossil fuels? Ultimately, following the invisible-hand concept introduced by Adam Smith in the eighteenth century, an invisible "green" hand will bring sustainability to the economy.

We must act now to conserve our living environment for future generations. The deployment of smart grids, their improved regulation, and careful consideration of their social and ethical dimensions are all necessary to make the transition to a low-carbon economy happen. Arguably, oil-producing countries may lose out in the

---

[791] Drewer and Miladinova (2017), pp. 298–308; Rubinstein (2013), pp. 74–87; Leonard (2014), pp. 53–68.

[792] One can think, for instance, of the National Industrial Symbiosis Program, Nat'l Indus. Symbiosis Program http://www.nispnetwork.com/about-nisp.

[793] An interesting remark is that there is even a political commitment to clean energy for EU islands. The rationale is that islands could make use of tidal, solar, and wind energy. See "Political declaration on clean energy for EU islands," 18 May 2017, available at https://ec.europa.eu/energy/sites/ener/files/documents/170505_political_declaration_on_clean_energy_for_eu_islands-_final_version_16_05_20171.pdf.

transition to a low-carbon economy because most of their GDP comes from fossil fuels. But similarly, most of these countries are blessed with unique solar irradiance and, therefore, the potential to generate wealth out of renewable natural resources. Carbon capture of fossil fuels will also move forward the agenda of a transition to a low-carbon economy. When it comes to nuclear energy, safety is currently a major issue, as is intermittency of solar and wind energy.[794]

To this end, smart grids offer tremendous potential, with their ability to accommodate the varying nature of renewable resources and integrate them into the grid in a way that is not currently being capitalized on. Moreover, smart grids will change the way we both consume and supply electricity, by facilitating wireless communication regarding consumption and pricing to both suppliers and end users. With their ability to predict, adjust, and coordinate power supply and demand, their potential impact is immense. In fact, generating renewable energy is not the biggest challenge; creating a grid that can integrate their unique nature is the key. Or, to quote a recent book on solutions to climate change, "More green requires a wiser grid."[795]

The rise of civil society's role on the electricity market will have an overall positive impact. A higher number of stakeholders triggers higher competition, bringing energy prices down. Further, new actors exporting electricity to the grid will build domestic electricity security. Thus citizens, cooperatives, and small- and medium-sized enterprises can play a key part in bringing about energy democracy—a phenomenon that reflects the EC's desire for a consumer-centric, bottom-up approach. Such an approach encourages citizen participation, while empowering and protecting citizens. Lastly, demand response means that consumers will be able to help protect the environment by contributing clean energy such as wind or solar to the grid, avoiding the use of fossil fuel plants that activate when electricity prices rise.[796]

Overall, thus, these are exciting times for the EU, as decentralization comes with opportunities for innovation, employment, and consumers becoming prosumers. Greener technologies are on the horizon and the face of energy usage across the EU will likely transform significantly in the years to come. As can be seen from this chapter, progress on decentralized energy is happening rapidly across the EU, even if most member states are currently at different stages of the process.

---

[794] According to Jason Bordoff, the nuclear renaissance stalled because of: a flat electricity demand, cheap shale gas, falling renewable costs and support policies, lack of carbon pricing, declining wholesale electricity prices, and rising nuclear costs. That said, nuclear energy still accounts for most zero-carbon power in many countries. See lecture by Jason Bordoff at Harvard University, 5 October 2018.

[795] Hawken (2017), p. 209.

[796] European Commission, "Clean energy for all. New Electricity Market Design: A Fair Deal for Consumers," 2016.

References

Aissaoui A et al (1999) Gas to Europe: the strategies of four major suppliers. Oxford University Press, Oxford

Anagnostopoulos J, Papantonis D (2013) Facilitating energy storage to allow high penetration of intermittent renewable energy. StoRe

Ardente F, Mathieux F (2014) Identification and assessment of product's measures to improve resource efficiency: the case-study of an Energy using Product. J Clean Prod 83:126–141

Ardente F et al (2014) Recycling of electronic displays: analysis of pre-processing and potential ecodesign improvements. Resour Conserv Recycl 92:158–171

Ávila Rodríguez C (2017) Marco jurídico para la implantación de infraestructuras para las energías alternativas en el transporte en España. Comunicación Proyecto (I+D) de Investigación

Azim-Khan R (2008) New Spanish regulation tightens up data protection requirement. Privacy Data Secur Law J

Balaguer-Coll M (2010) Decentralisation and efficiency in Spanish local government. Ann Reg Sci 45(3):571–601

Balta-Ozkan N et al (2013) Social barriers to the adoption of smart homes. Energy Policy 63:363

Baudrillard J (1998) The consumer society: myths and structures. Sage Publications

Bjerkan S (2016) Incentives for promoting Battery Electric Vehicle (BEV) adoption in Norway. Transp Res Part D: Transp Environ:43

Booz & Company et al (2013) Benefits of an Integrated European Energy Market. Booz & Company

Boscan L, Poudineh R (2016) Flexibility-enabling contracts in electricity markets. University of Oxford, Oxford

Boulos S et al (2015) The durability of products: standard assessment for the circular economy under the eco-innovation action plan. Publications Office

Boussauw K, Vanoutrive T (2017) Transport policy in Belgium: translating sustainability discourses into unsustainable outcomes. Transp Policy 53:11–19

Brandeis L, Warren S (1890) The right to privacy. Harv Law Rev 4:193

Braungart M (2009) Cradle to Cradle: remaking the way we make things

Bressand A (2013) The role of markets and investment in global energy. In: Goldthau A (ed) The handbook of global energy policy. Wiley, West Sussex, pp 15–29

Brugel (2013) Avis relatif à l'introduction des systèmes intelligents de mesure: vision de Brugel por la region de Bruxelles-Capitale. Brugel, Brussels

Brunekreeft G et al (2015) Germany's way from conventional power grids towards smart grids. In: Regulatory pathways for smart grid development in Chin. Springer Vieweg, Wiesbaden, pp 45–78

Buchan D, Keay M (2016) EU energy policy – 4th time lucky? Oxford Institute for Energy Studies

Cai J, Jiang Z (2008) Changing of energy consumption patterns from rural households to urban households in China: an example from Shaanxi Province, China. Renew Sustain Energy Rev 12:1667

Carson R (1962) Silent spring. Houghton, Mifflin

Casier T (2016) Great game or great confusion: the geopolitical understanding of EU-Russia energy relations. Geopolitics 21(4):763–778

Caticha A, Golan A (2014) An entropic framework for modelling economies. Physica A 408:149–163

Changala D, Foley P (2011) The legal regime of widespread plug-in hybrid electric vehicle adoption: a Vermont Case Study. Energy Law J 32(99):108–109

Cherrington R et al (2012) Producer responsibility: defining the incentive for recycling composite wind turbine blades in Europe. Energy Policy 47:13–21

Ciere M et al (2016) Ecrime, "D6.2 Executive Summary and Brief: The Economic Impact of Cyber-Crime on Non-ICT Sectors"

Clastres C (2011) Smart grids: another step towards competition, energy security and climate change objectives. Energy Policy 39:5399
Crossley P (2013) Defining the greatest legal and policy obstacle to "Energy Storage". Renew Energy Law Policy Rev 4:268
Cuijpers C, Koops BJ (2012) Smart metering and privacy in Europe: lessons from the Dutch Case. In: Gutwirth S et al (eds) European Data Protection: coming of age, p 269
Dalhammar C (2016) Industry attitudes towards ecodesign standards for improved resource efficiency. J Clean Prod 123:155
Day J (2016) The new EU Cybersecurity Directive: what impact on digital service providers? p 4
de Hauteclocque A, Perez Y (2011) Law & economics perspectives on electricity regulation. European University Press (EUP), p 5
De los Ríos IC, Charnley F (2017) Skills and capabilities for a sustainable and circular economy: the changing role of design. J Clean Prod 160:109–122
Donoso J (2015) Self-consumption regulation in Europe. Energetica Int 149(7):37
Drewer D, Miladinova V (2017) The BIG DATA challenge: impact and opportunity of large quantities of information under the Europol Regulation. Comput Law Secur Rev 33(3):298–308
Duso T, Szücs F (2017) Market power and heterogeneous pass-through in German electricity retail. Eur Econ Rev 98:354–372
Edens M, Lavrijssen S (2019) Balancing public values during the energy transition – how can German and Dutch DSOs safeguard sustainability? Energy Policy 128:57–65
Eid C, Hakvoort R, de Jong M (2016) Global trends in the political economy of smart grids: a tailored perspective on 'smart' for grids in transition. UNU-WIDER
Eisen JB (2013) Smart regulation and federalism for the smart grid. Harv Environ Law Rev 37:101–156
Elia Group (2016) Users group. Sophie De Baets: Project Infrastructure Communication. Elia Group, Brussels
Elia Group (2017) Press release. Avelgem-Avelin high-voltage connection upgrade project. Elia Group, Brussels
Erkman S, Ramaswamy R (2003) Applied industrial ecology: a new platform for planning sustainable societies. Aicra Publishers, Bangalore
European Commission (2017) Good practice in energy efficiency for a sustainable, safer and more competitive Europe. European Commission
European Environment Agency (2014) Country profile-Belgium. European Environment Agency, Copenhagen
European Environment Agency (EEA) (2016) Electric vehicles in Europe. EEA, Copenhagen
European Environmental Bureau (2015) Delivering resource-Efficient products: how ecodesign can drive a circular economy in Europe. European Environmental Bureau, Brussels
European Regulators' Group for Electricity and Gas (ERGEG) (2007) Smart metering with a focus on electricity regulation. ERGEG, Brussels
European Smart Grid Task Force (2016) My Energy Data. Smart Grids Task Force Ad hoc group of the Expert Group 1—Standards and Interoperability
European Technology Platform SmartGrids (2016) National and regional smart grids initiatives in Europe: cooperation opportunities among Europe's active platforms, 2nd edn
European Union Agency for Fundamental Rights (2014) Handbook on European Data Protection Law
Fernandez JMR et al (2016) Renewable generation versus demand-side management: a comparison for the Spanish Market. Energy Policy 96:458
Frosch RA, Gallopoulos N (1989) Strategies for manufacturing. Sci Am 260(3):144
Garanis-Papadatos T, Boukis D (2016) Research ethics committees in Greece. In: Beyleveld D, Townend D, Wright J (eds) Research ethics committees, data protection and medical research in European countries. Routledge, New York
Georgescu-Roegen N (1971) The entropy law and the economic process. Harvard University Press

Gissey GC, Dodds PE, Radcliffe J (2016) Regulatory barriers to energy storage deployment: the UK perspective. RESTLESS Project, London, p 2

Goldthau A (2016) Assessing Nord Stream 2: regulation, geopolitics & energy security in the EU, Central Eastern Europe & the UK. European Centre for Energy and Resource Security, London

Goldthau A, Sitter N (2015) Soft power with a hard edge: EU policy tools and energy security. Rev Int Polit Econ 22(5):941–965

Gorz A (1985) Paths to paradise: on the liberation from work. Pluto Press

Gorz A, Turner C (1994) Capitalism, socialism, ecology. Verso Books

Haas W et al (2015) How circular is the global economy? An assessment of material flows, waste production, and recycling in the European Union and the World in 2005. J Ind Ecol

Harremoes P et al (eds) (2001) Late lessons from early warnings: the precautionary principle 1896–2000. European Environment Agency

Hawken P (ed) (2017) Coming attractions: smart grids. In: Drawdown: the most comprehensive plan ever proposed to reverse global warming. Penguin, p 209

Hogan Lovells (2015) Data protection compliance in Spain. Mission impossible? Hogan Lovells

Hopkins R (2013) The power of just doing stuff: how local action can change the world

Hörter CM, Feyel N, Awad A (2015) The smart grid: energy network of tomorrow – legal barriers and solutions to implementing the smart grid in the EU and the US. Int Energy Law Rev 8:291, 297

IEA (2011) World Energy Outlook 2010. IEA, Paris

IndustRE (2016) Innovative business models for market uptake of renewable electricity unlocking the potential for flexibility in the industrial electricity use. Business models and market barriers. IndustRE

International Energy Agency, "Energy policies of IEA countries. Spain. 2015 review," International Energy Agency, Paris, 2015

International Energy Agency, "Energy policies of IEA countries. Belgium. 2016 review," International Energy Agency, Paris, 2016.

International Energy Agency (IEA) (2016) Energy Policies of IEA Countries; Italy 2016. OECD/IEA

Izyumenko E (2011) Think before you share: personal data on the social networking sites in Europe; Article 8 ECHR as a tool of Privacy Protection

Jackson T (2016) Prosperity without growth: foundations for the economy of tomorrow, 2nd edn. Routledge

Joachain H, Klopfert F (2014) Coupling smart meters and complementary currencies to reinforce the motivation of households for energy savings. Ecol Econ 105:89–96

Joskow PL et al (2008) Lessons learned from electricity market liberalization. Energy J 2(29):9–42

Kalimo H et al (2014) What roles for which stakeholders under extended producer responsibility? Rev Eur Community Int Environ Law 24(1):40–57

Kallis G (2018) Degrowth. Agenda Publishing

Katina PF et al (2016) A criticality-based approach for the analysis of smart grids. Technol Econ Smart Grids Sustain Energy 1(14):1–20

Kennedy C et al (2007) The changing metabolism of cities. J Ind Ecol 11:43–59

Kosta E (2013) Consent in European Data Protection Law

Langlet D, Mahmoudi S (2016) EU environmental law and policy. Oxford University Press

Lazar J et al (2015) Ensuring digital accessibility through process and policy

Leal-Arcas R (2016) Energy transit in the Caucasus: a legal analysis. Caucasus Int 6(2):53–74

Leal-Arcas R et al (2015a) The European Union and its energy security challenges. J World Energy Law Bus 8(4):291–336

Leal-Arcas R, Peykova M, Choudhury T, Makhoul M (2015b) Energy transit: intergovernmental agreements on oil and gas transit pipelines. Renew Energy Law Policy Rev 6(2):122–162

Leal-Arcas R et al (2015c) Multilateral, regional and bilateral energy trade governance. Renew Energy Law Policy Rev 6(1):38–87

Leal-Arcas R, Lasniewska F, Proedrou F (2017) Prosumers: new actors in EU energy security. Netherlands Yearb Int Law 48:139–172

Leal-Arcas R et al (2018) Smart grids in the European Union: assessing energy security, regulation & social and ethical considerations. Columbia J Eur Law 24(2):311–410

Leal-Arcas R et al (2019) Energy decentralization in the European Union. Queen Mary University of London, School of Law Legal Studies Research Paper No. 307/2019, pp. 1–55

Leiva J (2016) Smart metering trends, implications and necessities: a policy review. Renew Sustain Energy Rev 55

Leonard P (2014) Customer data analytics: privacy settings for "big data" business. Int Data Privacy Law 4(1):53–68

Liu J et al (2012) Cyber security and privacy issues in smart grids. IEEE Commun Surveys Tutorials 14(4):981–987

London B (1932) Ending the depression through planned obsolescence. University of Wisconsin

López Prol J, Steininger K (2017) Photovoltaic self-consumption regulation in Spain: profitability analysis and alternative regulation schemes. Energy Policy 108

Luo X, Wang J, Dooner M, Clarke J (2015) Overview of current development in electrical energy storage technologies and the application potential in power system operation. Appl Energy 137:511

Lynskey O (2014) Deconstructing data protection: the "Added-value" of a right to data protection in the EU legal order. Int Comp Law Q 63:569

Lynskey O (2015) The foundations of EU Data Protection Law

Lynskey O (2016) The Europeanisation of Data Protection Law. Camb Yearb Eur Leg Stud 18:1

Maitre-Ekern E, Dalhammar C (2016) Regulating planned obsolescence: a review of legal approaches to increase product durability and reparability in Europe. Rev Eur Community Int Environ Law 25(3):378–394

Martínez Lao J (2017) Electric vehicles in Spain: an overview of charging systems. Renew Sustain Energy Rev:77

Masera M et al (2016) The Security of Information and Communication Systems and the E+I Paradigm. In: Gheorghe AV et al (eds) Critical infrastructures at risk: securing the European Electric Power System, p 85

Masson G, Briano JI, Baez MJ (2016) International Energy Agency, Review and Analysis of PV Self-Consumption Policies 13

Mayer-Schönberger V, Padova Y (2016) Regime change? Enabling big data through Europe's New Data Protection Regulation. Colum Sci Tech Law Rev 17:315

McAfee A, Brynjolfsson E (2017) Machine, platform, crowd: harnessing our digital future. W.W. Norton

McKenna E et al (2012) Smart meter data: balancing consumer privacy concerns with legitimate applications. Energy Policy 41:807

Mihaylov M et al (2014) NRGcoin: virtual currency for trading of renewable energy in smart grids. Int Conf Eur Energy Mkt 11:58

Morgera E, Savaresi A (2013) A conceptual and legal perspective on the green economy. Rev Eur Comp Int Environ Law 22(1):14–28

Murthy Balijepalli VSK et al (2011) Review of demand response under smart grid paradigm. IEEE Innovative Smart Grid Technologies—India

Nachmany M (2015) Climate Change Legislation in Spain. An excerpt from the 2015 Global Climate Legislation Study. The London School of Economics and Political Science – Grantham Research Institute on Climate Change and Environment

Naess A (2010) The ecology of wisdom

Navarro Rodríguez P (2012) Distribución de competencias en materia de Energía en España; Pluralidad de Administraciones competentes. Actualidad Administrativa, no. 19–20

Nolan S, O'Malley M (2015) Challenges and barriers to demand response deployment and evaluation. Appl Energy 152:1

OECD (2011) Towards green growth

Ortegon K et al (2013) Preparing for end of service life of wind turbine. J Clean Prod 39:191–199

Papakonstantinou V, Kloza D (2015) Legal protection of personal data in smart grid and smart metering systems from the European Perspective. In: Goel S et al (eds) Smart grid security. SpringerBriefs in cybersecurity. Springer, London

Pearson ILG (2011) Smart grid cyber security for Europe. Energy Policy 39:5211

Pickern G (2014) Making connections between global production networks for used goods and the realm of production: a case study on e-waste governance. Global Netw 15(4):403–423

Polinpapilinho K et al (2016) A criticality-based approach for the analysis of smart grids. Technology and Economics of Smart Grids and Sustainable Energy 1(1):1–20

Proedrou F (2012) EU energy security in the gas sector: evolving dynamics, policy dilemmas and prospects. Ashgate, Farnham

Proedrou F (2016) EU energy security beyond Ukraine: towards holistic diversification. Eur Foreign Aff Rev 21(1):57–73

Purtova N (2015) Illusion of personal data as no one's property. Law Innov Techol 7:83

Roberts J, Skillings S (2015) The market design initiative: towards better governance of EU energy markets. ClientEarth and E3G – Regulatory Assistance Project publication

Rodríguez-Molina J et al (2014) Business models in the smart grid: challenges, opportunities and proposals for prosumer profitability. Energies 7:6142

Rojas A, Carreño P (2014) The reform of the Spanish electricity sector. Spanish Econ Financ Outlook 3(2)

Romain C et al (2016) Managing energy markets in future smart grids using bilateral contracts. Front Artificial Intell Appl 285:133

Roscini M (2014) Cyber operations and the use of force in international law, p 2

Rubinstein IS (2013) Big data: the end of privacy or a new beginning? Int Data Privacy Law 3(2):74–87

Rubio J (2017) Recommender system for privacy-preserving solutions in smart metering. Pervasive and Mobile Computing

Rule J, Greenleaf G (2010) Global privacy protection: the first generation

Schiavo LL, Delfanti M, Fumagalli E, Olivieri V (2011) Changing the Regulation for regulating the change; innovation-driven regulatory developments in Italy: smart grids, smart metering and emobility. The Center for Research on Energy and Environmental Economics and Policy, Bocconi University (IEFE)

Schwartz PM (2000) Beyond Lessig's Code for internet privacy: cyberspace filters, privacy control and fair information practices. Wis Law Rev 743

Shah M (2015) Big data and the Internet of Things. In: Japkowicz N, Stefanowski J (eds) Big data analysis: new algorithms for a new society, p 207 and 208

Sidi M (2017) The scramble for energy supplies to South Eastern Europe: the EU's Southern Gas Corridor, Russia's pipelines and Turkey's role. In: Turkey as an energy hub? Nomos, Baden-Baden, pp 51–66

Smart Energy Demand Coalition (2017) Explicit demand response in Europe – mapping the market 2017. SEDC, Brussels

Smedt SD (2015) Belgium – the new data protection hub? Eur Data Prot Law Rev 3:213–218

Stoppani E (2017) Smart charging and energy storage: bridging the gap between electromobility and electricity systems. Int Energy Law Rev 1

Swora M (2011) Smart grids after the third liberalization package: current developments and future challenges for regulatory policy in the electricity sector. Yearb Antitrust Regul Stud 4(4):9–22

Taefi TT et al (2016) Supporting the adoption of electric vehicles in urban road freight transport – a multi-criteria analysis of policy measures in Germany. Transport Res Part A: Policy Pract 91:61–79

The International Energy Agency (2016) World Energy Outlook. OECD/IEA, Paris. https://www.iea.org/publications/freepublications/publication/WorldEnergyOutlook2016ExecutiveSummaryEnglish.pdf

The Mediterranean Energy Regulators (2016) Study to evaluate net metering system in Mediterranean countries. MedReg, Milan

Union of the Electric Industry (Eurelectric) (2011) Regulation for smart grids. Eurelectric, Brussels, p 14

USmart Consumer Project (2016) European Smart Metering Landscape Report. Utilities and consumers. USmart Consumer Project, Madrid

Varaiya PP, Wu FF, Bialek JW (2010) Smart operation of smart grid: risk-limiting dispatch. Proc Inst Electronic Electrical Eng 99:40–57

Veldman E, Geldtmeijer M, Knigge JD, Slootweg H (2010) Smart grids put into practice: technological and regulatory aspects. Compet Regul Netw Ind 11:287–289

Verbong GP, Beemsterboer S, Sengers F (2013) Smart grids or smart users? Involving users in developing a low carbon electricity economy. Energy Policy 52:117–125

Verkade N, Hoeffken J (2018) The design and development of domestic smart grid interventions: insights from the Netherlands. J Clean Prod 202:799–805

Wang W, Lu Z (2013) Cyber security in the smart grid: survey and challenges. Comput Netw 57:1344

White R (1994) Preface. In: Allenby B, Richards D (eds) The greening of industrial ecosystems. National Academy Press, Washington, D.C.

Yergin D (2011) The prize: the epic quest for oil, money & power. Simon & Schuster, New York

Yudina A (2017) Garden City: supergreen buildings, urban skyscapes and the new plated space. Thames & Hudson

Zhou S, Brown MA (2017) Smart meter deployment in Europe: a comparative case study on the impacts of national policy schemes. J Clean Prod 144:22–32

Zulkafli Z et al (2017) User-driven design of decision support systems for polycentric environmental resources management. Environ Model Softw 88:58

Chapter 9
Innovation, Research, and Technology

9.1 Introduction[1]

Technological innovation can meaningfully help with the reduction of GHG emissions. Businesses have taken on a leadership role in climate change mitigation,[2] and cities all over the world produce innovative strategies for advancing solutions to climate change.[3] Technology appears in several of the SDGs: goal 2 on food, goal 3 on health, goal 7 on energy, goal 9 on innovation, and goal 14 on oceans.[4]

This chapter explores the various challenges and opportunities in sustainability, the options for a cleaner future, the remarkable potential contribution of sustainable companies, and the links between sustainability and spirituality. The chapter then concludes that the solution to sustainability is to reduce CO_2 emissions by decarbonizing, electrifying, making use of the circular economy (i.e., recycling and reusing products), transferring funds and technology from the West to the rest of the world, shifting the economy to services that do not use products, and sharing best practices.

---

[1] Some of the ideas developed in this chapter draw on Leal-Arcas (2017), pp. 801–877.

[2] *See Sub-Saharan Africa's First Light-Rail Train*, Sustainia, http://solutions.sustainia.me/solutions/sub-saharan-africas-first-light-rail-train/ [https://perma.cc/H56K-EEAU].

[3] *Explore 100 City Solutions for a Greener and Fairer Future*, Sustainia, https://thesustainian.com/.

[4] https://sustainabledevelopment.un.org/sdgs.

9.2 Challenges Ahead, but the Future Is Bright

The quest for sustainable technology is one of the world's and Europe's current challenges.[5] Technologies, research, and innovation are anticipated to have growing importance in Europe's pursuit of energy security.[6] The European Union remains a global leader in terms of innovation and renewable energy,[7] but this status will be at risk unless the role of technologies, research, and innovation is increased.[8] The promotion of new technologies should underlie the European Energy Union's governance.

The main challenge in the field of innovation lies in the necessity to fuse the European Union's and its member states' research programs. An integrated approach is required to complement efforts and reinforce ties between research and industry, thereby easing the emergence of new technologies in the European internal market.[9] Shared networks are expected to hasten and intensify the interplay of information between individuals and companies across the globe.[10]

Similarly, the increasing importance of digital energy[11] will require an equivalent innovation impulse in the field of cyber-security to protect the system from cyber-attacks. Indeed, an unbounded revolution in the digital exchange of information would make cyber-systems worldwide prone to new threats as digital instruments and shared networks ease intrusions into private life.[12] Therefore, the pace of innovation should be rationalized to ensure the effective safeguard of private life.

Horizon 2020 will be the European Union's principal financial means of promoting energy research and innovation in the coming years.[13] Measures in this particular

---

[5]Indeed. The EU seems to be falling behind vis-à-vis the US and China when it comes to technology innovation. The Americans have Google; the Chinese have Baidu; the EU has none of that. The largest digital firms are either American or Chinese. All of this has implications for who will be writing the rules of the new economy. The big technology decisions are made in Silicon Valley and China, not in the EU. The fact that the EU's single market still has some obstacles and the various languages in the EU do not help to achieve the goal of making the EU a competitive technology actor. Radical ideas would be most welcome for the creation of a single digital market in the EU.

[6]Leal-Arcas et al. (2016), p. 436.

[7]*See* Leal-Arcas and Minas (2016), pp. 621–666 at 650–665.

[8]*Communication from the Commission to the European Parliament, the Council, the European Economic and Social Committee, the Committee of the Regions and the European Investment Bank, A Framework Strategy for a Resilient Energy Union with a Forward-Looking Climate Change Policy,* COM (2015) 80 final (Feb. 25, 2015), http://eur-lex.europa.eu/resource.html?uri=cellar:1bd46c90-bdd4-11e4-bbe1-01aa75ed71a1.0001.03/DOC_1&format=PDF, at 3. It should also be stressed that innovation in technology (e.g., fracking) has allowed the extraction of resources that before would have been very expensive.

[9]*Id.* at 16.

[10]Leal-Arcas (2016), p. 436.

[11]Alex Molinaroli, *What Does Digital Mean for the Future of Energy?* World Econ. F. (Mar. 2, 2016), https://www.weforum.org/agenda/2016/03/perspective-distributed-digital-and-demand-side-energy-technology-implications-for-energy-security/ [https://perma.cc/95EZ-PZBJ].

[12]Sami Andoura & Jean-Arnold Vinois, "From the European Energy Community to the Energy Union: a policy proposal for the short and the long term," 125, 136 (2015).

[13]Horizon 2020 is the largest E.U. research and innovation program to date. The scheme holds €80 billion to deploy over the period 2014–2020, of which €6.6 billion will be specifically devoted to

9.2 Challenges Ahead, but the Future Is Bright

aspect of E.U. energy policy[14] will revolve around the Strategic Energy Technology (SET) Plan.[15] The SET Plan aims to foster research and development in both existing and new generations of low-carbon technologies.[16]

To start with, the Energy Union promises an updated SET Plan and a strategic transport research and innovation agenda, thereby expediting energy system transformation.[17] The Energy Union's proposal charts four goals in the area of innovation: making the European Union the world leader in developing the next generation of renewable energy technologies[18]; easing the participation of consumers in the energy transition[19]; ensuring effective energy systems[20]; and developing more sustainable transport systems that employ large-scale innovative technologies.[21]

Relatedly, another promising advancement is the Energy Union's commitment to phase out environmentally harmful subsidies altogether.[22] Continuing to fund fossil fuels within Europe would be counterproductive, not only in light of the European Union's ambitious energy and climate goals,[23] but also in that it would delay the arrival of new technologies. Therefore, redirecting these subsidies to support low-carbon technologies and digital energy innovation represents a sensible change of course.[24]

---

energy. *See What is Horizon 2020?*, European Comm'n, https://ec.europa.eu/programmes/horizon2020/en/what-horizon-2020 [https://perma.cc/MRE8-MJNC].

[14]For greater details on EU energy law and policy, see Leal-Arcas and Wouters (2017).

[15]The European Strategic Energy Technology (SET) Plan will sustain the Energy Union's pillar on technologies, research, and innovation. It outlines the long-term energy research and innovation agenda for Europe by setting strategic objectives for the future. *See The European Strategic Energy Technology Plan (SET-Plan)*, European Comm'n, http://ec.europa.eu/research/energy/eu/index_en.cfm?pg=policy-set-plan [https://perma.cc/D3NK-WLND].

[16]*Communication from the Commission to the European Parliament, the Council, the European Economic and Social Committee and the Committee of the Regions, Investing in the Development of Low Carbon Technologies (SET-Plan)*, at 2–3, COM (2009) 519 final (Oct. 7, 2009).

[17]*Communication from the Commission to the European Parliament, the Council, the European Economic and Social Committee, the Committee of the Regions and the European Investment Bank, A Framework Strategy for a Resilient Energy Union with a Forward-Looking Climate Change Policy*, COM (2015) 80 final (Feb. 25, 2015), http://eur-lex.europa.eu/resource.html?uri=cellar:1bd46c90-bdd4-11e4-bbe1-01aa75ed71a1.0001.03/DOC_1&format=PDF, at 16 n. 28.

[18]*Id.* at 16.

[19]*Id.*

[20]*Id.*

[21]*Id.*

[22]*Id.* at 10.

[23]The leaders of the G7 (Canada, France, Germany, Italy, Japan, the United Kingdom, and the United States) have the ambition to phase out fossil fuel emissions in the twenty-first century. *See* Pilita Clark & Stefan Wagstyl, *G7 Leaders Agree to Phase Out Fossil Fuels*, Fin. Times (June 8, 2015), https://www.ft.com/content/ec2c365a-0ddf-11e5-aa7b-00144feabdc0 [https://perma.cc/6LVZ-7MG8].

[24]Smart grids and microgrids may be the winners of energy innovation. For an analysis of smart grids in the EU, see Leal-Arcas et al. (2018), pp. 311–410.

Beyond the E.U. context, there is a new initiative of visionary billionaires determined to provide energy that is reliable, affordable, and carbonless. The initiative is called Breakthrough Energy Coalition.[25] The Energy Union could and should join forces with this coalition—which is currently working with a growing group of visionary countries—towards joint research to ensure that the group's vision becomes a reality.[26] Yet another initiative called Mission Innovation[27] brings together a group of 23 countries and the E.U.[28] that aim to reinvigorate and accelerate clean energy innovation throughout the world to make clean energy affordable for all by doubling investment on clean energy research and development over 5 years.[29] The Energy Union should also join forces with this initiative.[30] The benefits and possible outcomes for the European Union of joining such coalitions will only accelerate Europe's decarbonization ambitions.

Already, clean energy innovation has enjoyed success, thus there are reasons to be optimistic about further climate change mitigation because the options for the future are abundant.[31] Wind turbines and solar panels are proliferating across China,

---

[25] To access their principles, see *Reliable, affordable Energy for the World*, Breakthrough Energy, http://www.breakthroughenergycoalition.com/en/index.html [https://perma.cc/2J7T-LS2J].

[26] The Breakthrough Energy Coalition defines itself as follows: "We are a partnership committed to broad investment in new energy technologies from public and private sources. We invest our own capital as well as working with over [twenty] countries around the world who have committed to significantly increase their investments in the basic research that leads to breakthrough innovations." *See Who We Are*, Breakthrough Energy, http://www.b-t.energy/coalition/who-we-are/ [https://perma.cc/DMT9-7K6K].

[27] *Accelerating the Clean Energy Revolution*, Mission Innovation, http://mission-innovation.net/ [https://perma.cc/2BWV-4Y6Y].

[28] *Member Participation*, Mission Innovation, http://mission-innovation.net/countries/ [https://perma.cc/39BM-GH74].

[29] *The Goal*, Mission Innovation, http://mission-innovation.net/the-goal/ [https://perma.cc/DX4Y-6UNC].

[30] Efforts to do so on the part of the European Union and other Organization for Economic Co-operation and Development (OECD) partners are already visible. *See The EU and Other OECD Partners Agree on Trade Measures Supporting Cleaner Energy*, European Comm'n (Nov. 18, 2015), http://trade.ec.europa.eu/doclib/press/index.cfm?id=1401 [https://perma.cc/G2UN-YGL5].

[31] Indeed, at the international level, a relatively new initiative called the International Solar Alliance, launched by India's Prime Minister Modi and France's President Francoise Hollande, is very promising as a mechanism to mitigate climate change. It is expected to channel $300 billion in 10 years for the promotion of renewable energy projects. See T. Mishra, "Sun shines on $300-billion global fund for clean energy," The Hindu Business Line, 1 May 2017, available at http://www.thehindubusinessline.com/economy/sun-shines-on-300billion-global-fund-for-clean-energy/article9675599.ece.

9.2 Challenges Ahead, but the Future Is Bright   447

Europe, India,[32] and the United States.[33] Wind power projects[34] on a massive scale are underway in the North Sea,[35] as are projects to bring solar energy produced in the Sahara Desert to southern Europe.[36] An increasing number of developing countries are investing in renewable energy out of their own initiative, not because they are legally bound to do so.[37] There are positive examples of countries that have climate change laws in place (on GHG emissions reduction) even if they are not a party to the

---

[32] *See, e.g.*, Varun Sivaram, Gireesh Shrimali, & Dan Reicher, *Research for the Sun: How India's Audacious Solar Ambitions Could Make or Break its Climate Commitments*, Stan. Steyer-Taylor Ctr. for Energy Pol'y & Fin. (Dec. 8, 2015), https://www-cdn.law.stanford.edu/wp-content/uploads/2015/12/Reach-for-the-Sun-High-Resolution-Version.pdf [https://perma.cc/6R7N-5MR2]. *See also A New Dawn in Renewable Energy*, Gov't of India Press Info. Bureau (Dec. 18, 2016), http://pib.nic.in/newsite/PrintRelease.aspx?relid=155612 [https://perma.cc/2QQB-4C74]; Pilita Clark, *The Big Green Bang: How Renewable Energy Became Unstoppable*, Fin. Times (May 18, 2017), https://www.ft.com/content/44ed7e90-3960-11e7-ac89-b01cc67cfeec?mhq5j=e2 [https://perma.cc/FZ4R-FT29].

[33] Karl Mathiesen, *What is Holding Back the Growth of Solar Power?*, The Guardian (Jan. 31, 2016), https://www.theguardian.com/sustainable-business/2016/jan/31/solar-power-what-is-holding-back-growth-clean-energy [https://perma.cc/82VV-HF4Q].

[34] The then UK energy minister Charles Hendry gave a speech at the annual All Energy conference in Aberdeen in 2012, where he stated: 'It is shameful that with some of the strongest winds and highest tidal reaches in Europe, the UK is currently third from bottom in the whole of the EU in its use of renewables.' He was replaced soon after. Speech available at https://www.gov.uk/government/speeches/charles-hendry-speech-at-all-energy-aberdeen.

[35] *See* Chris Lo, *Offshore Wind Turbines: In Search of the Next Generation*, Power Tech. (June 19, 2012), http://www.power-technology.com/features/featureoffshore-wind-turbines-search-next-generation-renewable/ [https://perma.cc/H4VZ-RYFA]. Offshore wind energy might be a good option for the UK, given its strong maritime tradition.

[36] Morocco is currently building one of the largest solar plants in the world and hopes to export power to Europe in the future. *See Nuclear Power in the Middle East: Wasting Energy*, The Economist (Nov. 28, 2015), http://www.economist.com/news/middle-east-and-africa/21679090-egypt-and-others-alternatives-nuclear-power-hold-more-promise-why-more [https://perma.cc/U57G-YWE8]. Using the desert is an optimal option for solar energy because there is plenty of sun and space available.

[37] *See* Ian Johnston, *Developing World Invests More in Renewable Energy Than Rich Countries for First Time, New Study Says*, The Independent (May 31, 2016), http://www.independent.co.uk/environment/climate-change/renewable-energy-investment-developed-world-developing-world-ren21-report-a7058436.html [https://perma.cc/KFX2-G3R6].

U.N. Framework Convention on Climate Change (e.g., Taiwan[38]) or an Annex I country[39] (e.g., Mexico[40]).

However, challenges remain for states with fewer resources. It is now evident that technology and wealth are not just compatible with a green future[41]; they are a precondition to environmental sustainability. As Johan Norberg notes, we also know that the alarmist rhetoric of the 1960s and 1970s that envisaged a catastrophic future was not scientifically sound and turned out to be factually wrong when predicting a world without forests, with acid rain, and requiring people to use surgical masks to protect themselves from pollution.[42] In the words of the Pilot 2006 Environmental Performance Index, "Wealth emerges as a major determinant of environmental performance."[43] In other words, according to the study, the richer the country, the more it had done to clean the environment, largely due to its economic might and technological progress, which, in turn, allowed fewer workers to produce more stuff. One extraordinary example of how technological progress in wealthier nations has contributed to environmental protection is the modern car, which in motion emits less CO_2 than a 1970s car did parked (as a result of gasoline vapor leakage).[44] Therefore, it is the lack of new technology and affluence in poor countries that creates those states' worst environmental problems.[45]

But the outlook for these less affluent countries—indeed, for all countries—continues to improve. Life expectancy at birth between 1770 and the end of the

---

[38] For a list of countries that are a party to the UNFCCC, see *List of Non-Annex I Parties to the Convention*, UNFCCC, http://unfccc.int/parties_and_observers/parties/non_annex_i/items/2833.php [https://perma.cc/JL3U-Q3T3]; *List of Annex I Parties to the Convention*, UNFCCC, http://unfccc.int/parties_and_observers/items/2704.php (noting that Taiwan is absent from both lists) [https://perma.cc/G9ZE-5654]; *see also* Greenhouse Gas Reduction and Management Act, Gov't of Rep. of China (July 1, 2015), https://www.epa.gov.tw/public/Data/511181640271.pdf [https://perma.cc/Z6AQ-6ZM5].

[39] Annex I countries are developed countries and those countries in transition to a market economy. According to the Kyoto Protocol, they are legally bound to reduce their emissions of GHGs. *See Parties & Observers*, UNFCCC, http://unfccc.int/parties_and_observers/items/2704.php [https://perma.cc/4G5W-G8TJ]; Kyoto Protocol, art. 2–3.

[40] For a list of Annex I countries, see *List of Annex I Parties to the Convention* (noting that Mexico is absent from the list).

[41] See Victor and Yanosek (2017), pp. 124–131; D. Victor et al., "Transformation of the Global Energy System," World Economic Forum, January 2018.

[42] Norberg (2016), p. 109.

[43] Pilot 2006 Environmental Performance Index, Yale Ctr. for Envtl. Law and Policy, at 4, http://archive.epi.yale.edu/files/2006_pilot_epi_summary_for_policymakers.pdf [https://perma.cc/F932-79AK].

[44] Lance Ealey & Glenn Mercer, *Tomorrow's Cars, Today's Engines*, Euractiv (Oct. 8, 2002), https://www.euractiv.com/section/transport/opinion/tomorrow-s-cars-today-s-engines/ [https://perma.cc/R9ZV-6LUZ].

[45] For an analysis of how technology can help countries develop economically, see Thiel and Masters (2014).

9.2 Challenges Ahead, but the Future Is Bright

nineteenth century was only around 30 years, whereas today is over 70[46]; world GDP per capita until 1900 was just around US$1000, based on the value of the U.S. dollar in 1990[47]; and the level of illiteracy in the early nineteenth century was around 85% of the world population.[48] As a result, the international community did not have the incentive or ability to deal with environmental protection effectively. Today, however, life expectancy at birth is 70 years and, in some developed countries, it is rising to 85 years[49]; world GDP per capita has risen to about US $8000, based on 1990 US dollars[50]; and illiteracy has gone down to just 10% of the world population.[51] Additionally, levels of undernourishment between 1945 and 2015 have drastically decreased from 50% of the world population to just above 10% in 2015,[52] and poor countries today have lower poverty rates than the richest countries did in the early nineteenth century.[53] Moreover, as depicted in Fig. 9.1, the number of people living in extreme poverty has been cut in half since 1990. In fact, the percentage of people living below the threshold of extreme poverty has gone down from 80% to 8% in the past 175 years and civil rights as well as the rule of law[54] are more robust than ever before.[55] These are all excellent trends and great achievements.

The prognosis is equally excellent: in 2107, to live past 100 years of age will no longer be rare; rather, it will be the norm in at least 39 countries.[56] Figure 9.2 forecasts that worldwide extreme poverty will be cut roughly in half again by 2030.

Taking all of this into account, the way to achieve sustainability is to keep investing in a cleaner future, aimed at developing alternative and cheaper energy.

---

[46] Max Roser, *Life Expectancy*, Our World in Data, https://ourworldindata.org/life-expectancy/ [https://perma.cc/QP38-3Z8T].

[47] *See* Angus Maddison, The World Economy: Historical Statistics 262 (2003).

[48] *See* van Leeuwen and van Leeuwen-Li (2014), pp. 87 and 94.

[49] *World Health Statistics 2016: Monitoring Health for the SDGs*, WHO, http://www.who.int/gho/publications/world_health_statistics/2016/en/ [https://perma.cc/CSE3-QZUV].

[50] Madison (2003), p. 262.

[51] *See* van Leeuwen and van Leeuwen-Li (2014), pp. 87 and 94.

[52] *The State of Food and Agriculture: 1947*, Food and Agric. Org. of the United Nations (Aug. 25, 1947), http://www.fao.org/docrep/016/ap635e/ap635e.pdf [https://perma.cc/2DDS-PU8F]; *The State of Food and Agriculture in Brief: 2015*, Food and Agric. Org. of the United Nations (2015), http://www.fao.org/3/a-i4671e.pdf [https://perma.cc/VE9G-JGDQ].

[53] Martin Ravallion, *Poverty in the Rich World When it was Not Nearly so Rich*, Ctr. for Global Dev.: Blogs (May 28, 2014), http://www.cgdev.org/blog/poverty-rich-world-when-it-was-not-nearly-so-rich [https://perma.cc/R4MF-N7DB].

[54] For an analysis of the rule of law, see R. Leal-Arcas, "Essential Elements of the Rule of Law Concept in the EU," *Queen Mary School of Law Legal Studies Research Paper No. 180/2014*, pp. 1–6. The rule of law is crucial in the context of technology: non-democratic states such as China and Russia seem to limit freedom of expression via the internet or slow down the internet to the detriment of their citizens. In an ideal world, everyone should have access to the internet, in the same way that the sun shines for everyone for free.

[55] The Economist, "A Manifesto," 15 September 2018, pp. 11–12 at 11.

[56] Human Mortality Database, http://www.mortality.org/ [https://perma.cc/QA9V-K3L8].

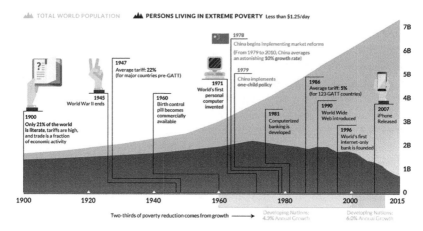

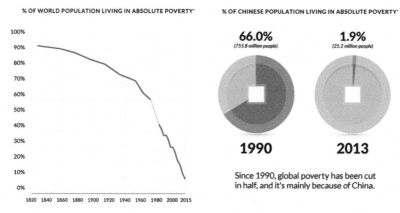

Fig. 9.1 Extreme poverty in the world. * Using a $1.90 day/PPP poverty line. Dashed line denotes trend line (Sources: ourworldindata.org; World Bank)

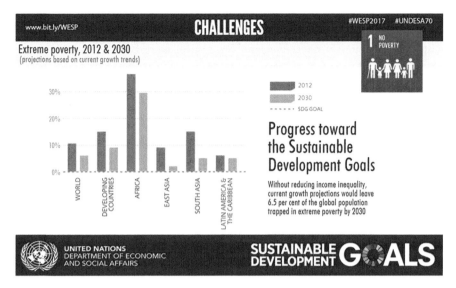

Fig. 9.2 Prognosis on world extreme poverty by 2030 (Source: World Economic Situation and Prospects 2017)

9.3 Betting on a Cleaner Future[57]

Moving forward requires creating a new energy future, accepting that we may never run out of fossil fuels[58] (otherwise, there would be no life on earth).[59] To get there, societies need to change their energy supplies and control CO_2 emissions.[60] It is a well-known fact that the use of energy impacts the environment. Both energy and the environment are global issues. As stated by energy researcher Tom Meyer, we may

---

[57] *See* Gary Brudvig, Lecture at Yale University titled Spurring Innovation (Sept. 22, 2016).

[58] But we may run out of a place where to place CO_2. The fundamental problem here is the absence of property rights for CO_2. Society needs property rights for CO_2.

[59] On that note, former U.S. Secretary of Energy Steven Chu once famously wrote that "the Stone Age did not end because we ran out of stones; we transitioned to better solutions." *Letter from Secretary Steven Chu to Energy Department Employees*, Energy.gov (Feb. 1, 2013, 11:00 AM), http://energy.gov/articles/letter-secretary-steven-chu-energy-department-employees [https://perma.cc/7399-QPP3]. The same quote is also credited to Ahmed Zaki Yamani, former Saudi minister of oil. By the same token, I would suggest that we should not need to wait until we run out of fossil fuels to make the transition to sustainable energy.

[60] See, for instance, the views of Ron Oxburgh, chairman of Shell, in 2004: "Sequestration is difficult, but if we don't have sequestration then I see very little hope for the world." He then added, "No one can be comfortable at the prospect of continuing to pump out the amounts of carbon dioxide that we are at present . . . with consequences that we really can't predict but [that] are probably not good." *Shell Boss 'Fears for the Planet,'* BBC News (June 17, 2004), http://news.bbc.co.uk/1/hi/uk/3814607.stm [https://perma.cc/3U2E-XW3V].

not speak the same language or share the same culture, but we breathe the same air.[61]

It will be necessary to utilize all energy options: clean coal, oil shale combined with CO_2 sequestration, nuclear energy,[62] hydrogen and fuel cells, renewable energy (whether wind, solar, geothermal, or biomass), inter alia. In this respect, as President Barack Obama said in 2010, "An America run solely on fossil fuels should not be the vision we have for our children and grandchildren."[63] He then added:

> [T]he only way the transition to clean energy will ultimately succeed is if the private sector is fully invested in this future, if capital comes off the sidelines and the ingenuity of our entrepreneurs is unleashed. And the only way to do that is by finally putting a price on carbon pollution.[64]

However, putting a price on carbon, so that people pay for their CO_2 emissions, will affect the poor the most negatively, and not the rich who, incidentally, are the ones to blame the most for the causes of climate change.[65] From an economic point of view, one could argue that the price of almost everything is low, which explains why there is so much of everything. Here are three examples:

– In the 1990s, very few people had a cell phone; today even students have cell phones;
– In the 1990s, very few people had laptops; today even students have them;
– In the 1920s, very few people had cars; today many people have cars.

In other words, if you increase the price of a given good, there will be less of it in the market. If the price goes up, there will be less demand. Equally, if you increase the price of carbon, you can mitigate climate change rapidly. All of this raises the question whether governments should intervene in energy markets (1) to promote competition via legal frameworks and the attraction of investments, (2) to provide energy access to citizens, and (3) as a result of environmental externalities (such as water or air pollution).

---

[61]Thomas Meyer, Lecture at the University of North Carolina titled Our Energy Future: What are the Technology Challenges of the 21st century? (Mar. 2, 2006).

[62]Although the UK government is keen to invest more on nuclear energy, most other European countries are phasing it out: for instance, Germany, Belgium, Switzerland and even France. Russia, China, and India are keen on nuclear energy. They are also accelerating renewable energy. One wonders whether the UK should learn from the experience of the 1960s, when a nuclear-powered cargo-passenger ship sailed from the US to Europe to persuade the world to embrace the 'atomic age.' See T. Thueringer and J. Parkinson, "The ship that totally failed to change the world," BBC News, 25 July 2014, available at https://www.bbc.co.uk/news/magazine-28439159.

[63]See President Barack Obama, *Remarks on the Economy at Carnegie Mellon University*, The White House (June 2, 2010), https://www.whitehouse.gov/the-press-office/remarks-president-economy-carnegie-mellon-university [https://perma.cc/25LF-59EF].

[64]*Id.*

[65]Sophie Yeo, *Who's to Blame for Climate Change?*, Climate Home (Nov. 17, 2014, 10:20 AM), available at http://www.climatechangenews.com/2014/11/17/whos-to-blame-for-climate-change/ [https://perma.cc/5UCR-EJP5].

9.3 Betting on a Cleaner Future

Investment in renewables for power generation was twice as much as it was for coal, gas, oil, and nuclear energy combined in 2017.[66] As David MacKay put it, big changes are required if we are serious about decarbonization.[67] Since it is very likely that oil demand will continue to grow, there is a need to limit footprint from fossil fuels. Renewable energy can contribute meaningfully to the reduction in fossil fuel use. For instance, solar energy could become cheaper thanks to new materials and assembly technologies.[68] At present, solar power remains expensive compared to fossil fuels as a source of energy.[69] For example, if one were to invest US$10 billion in burning gas to power a region, one could help lift 90 million people out of poverty and darkness. If, however, one were to spend the same amount on renewable energy, one could only help 20–30 million people. This means that, in this hypothetical, 60 million people would remain in poverty and darkness.[70]

New developments in solar energy show a promising future. Graphene, a new material created in 2004 at the University of Manchester, is very thin and flexible, only one carbon atom thick.[71] It is also very strong and conducts electricity and heat very efficiently.[72] Graphene could radically change the economics of solar power because most solar cells today utilize expensive indium,[73] whereas carbon atoms are abundant. If graphene is found to be a successful replacement for indium, potentially anything could be turned into a solar power station in the future.

---

[66] The Economist, "The hydrogen bombshell," *Technology Quarter: Towards zero carbon*, 1 December 2018, pp. 1–12 at 3.

[67] MacKay (2009).

[68] *Climate Change: Clear Thinking Needed*, The Economist (Nov. 28, 2015), http://www.economist.com/news/leaders/21679193-global-warming-cannot-be-dealt-using-todays-tools-and-mindsets-so-create-some-new [https://perma.cc/ZW6L-K9TA].

[69] Barbara Hollingsworth, *Study: Despite Subsidies, Solar Power More Expensive*, CNS News (Oct. 19, 2016, 10:40 AM), http://www.cnsnews.com/news/article/barbara-hollingsworth/study-solar-power-remains-considerably-more-expensive-electricity [https://perma.cc/8U8U-XSRW].

[70] Todd Moss & Benjamin Leo, Ctr. for Glob. Dev., Maximizing Access to Energy: Estimates of Access and Generation for the Overseas Private Investment Corporation's Portfolio 2 (2014), http://www.cgdev.org/sites/default/files/maximizing-access-energy-opic_1.pdf [https://perma.cc/G24X-63BC].

[71] "In simple terms, graphene[] is a thin layer of pure carbon; it is a single, tightly packed layer of carbon atoms that are bonded together in a hexagonal honeycomb lattice." Jesus de La Fuente, *Graphene – What Is It?*, Graphenea, http://www.graphenea.com/pages/graphene#.WFxPSlOLQ2w [https://perma.cc/YA3C-KYZX].

[72] *Id.*

[73] Indium is "a silvery malleable fusible chiefly trivalent metallic element that occurs especially in sphalerite ores and is used especially as a plating material, in alloys, and in electronics." *Definition of indium*, Merriam-Webster, https://www.merriam-webster.com/dictionary/indium [https://perma.cc/D6E9-GWSY].

Other interesting developments abound.[74] This raises the question of what causes transformation.[75] Others are contemplating solar power in space,[76] where nothing blocks the sun: a microwave transmitter would send energy to areas on earth that need it.[77] The energy internet, where producers and consumers can place information and power into the network, could help solve the renewable energy storage[78] issue for wind and solar power because unused energy could be stored in cars and homes.[79]

There are also scientists trying to remove surplus CO_2 from the air.[80] This process has the potential to reverse global warming and is part of the broader notion of geoengineering.[81] Other technologies would be blocking incoming solar energy in the form of planetary sunshades. Doing do will reduce temperatures. As is the case of any issue in the context of global governance, reaching consensus by all countries is a requirement that may prove very difficult to obtain.

The advantages of solar energy are that it is limitless,[82] essentially free (albeit not its technology),[83] widely dispersed,[84] and has a low environmental impact.[85] One

---

[74] For example, the WiseGRID project, funded by the EU's Horizon 2020 program. About WiseGRID, WiseGRID, http://www.wisegrid.eu/.

[75] One possible theory is innovation as a necessity. The incumbents will continue to innovate, more than the newcomers. Some of the policy instruments to get there are labor mobility (via mode 4 of the General Agreement on Trade in Services), economic incentives, investment in R&D, and intellectual property. Another possible theory is anthropological constraints. A policy instrument to get there would be social R&D (smart cities).

[76] *Space Solar Power*, Nat'l Space Soc'y, http://www.nss.org/settlement/ssp/ [https://perma.cc/KT8A-MVZW].

[77] *See id.*

[78] Large energy storage systems are usually more cost-effective than small units, but both can help balance grids. In the case of electricity, it is hard and expensive to store it, although batteries are getting better. It is far easier, and more efficient, to store energy in bulk at larger scale.

[79] An Internet for energy interconnects the energy network with the Internet, allowing units of energy (locally generated, stored, and forwarded) to be dispatched when and where it is needed. *Internet of Energy for Electric Mobility*, Internet of Energy, http://www.artemis-ioe.eu/ [https://perma.cc/E8PU-6Q4T].

[80] *See Carbon & Tree Facts*, Arbor Environmental Alliance, http://www.arborenvironmentalalliance.com/carbon-tree-facts.asp [https://perma.cc/2VG8-DTFU].

[81] Leal-Arcas and Filis-Yeloghotis (2012), pp. 128 at 128–130.

[82] For an overview of the world's total primary solar energy supply, see Matthias Loster, *Total Primary Energy Supply – From Sunlight*, EZ2C, http://www.ez2c.de/ml/solar_land_area/ [https://perma.cc/GP3E-PYBN].

[83] Yogi Goswami et al. (1999), p. 11.

[84] *Solar Panels*, Alternative Energy Primer, http://www.alternativeenergyprimer.com/Solar-Panels.html [https://perma.cc/W4TF-VY8S].

[85] *See* Vikram Aggarwal, *What is the Environmental Impact of Solar Energy?*, Mother Earth News (Dec. 2, 2015, 8:55 AM), http://www.motherearthnews.com/renewable-energy/what-is-the-environmental-impact-of-solar-energy-zbcz1512 [https://perma.cc/Z8YC-RVCU].

9.3 Betting on a Cleaner Future

great disadvantage, however, is that it is not a constant supply of energy, since the sun does not always shine.[86] So, countries will need better ways to store and trade renewable energy via large megagrids,[87] which is the key issue in solar energy implementation[88] and implicates the importance of international cooperation. Equally, carbon sequestration is important because countries are not going to stop burning carbon in the near future. Geoengineering could be further developed to mitigate climate change.[89] The greatest result of investing in low-carbon technologies will be becoming increasingly energy independent. Therefore, more research and development spending on energy technologies would be necessary to decarbonize the economy.

Shifting subsidies from fossil fuels to renewable energy is a promising policy towards clean energy support. States could provide incentives to move to renewable energy, especially in warm countries. However, there needs to be public support in the transition to a clean economy[90]: if cleaning the environment comes at the cost of higher unemployment, no democracy will accept that. Moreover, most people are interested in short-term local issues and solutions, not long-term global problems. However, it is in everyone's interest to provide the developing world with the best technology, to minimize dirty carbon technology.

The bottom line is that future climate policies must not obstruct the path to wealth creation and technology innovation, which will benefit both developing and developed countries by improving quality of life and the environment.

---

[86] A distinction needs to be made between concentrating solar power plants (which use mirrors to concentrate the energy from the sun and work on heat) and photovoltaics (which covers the conversion of light into electricity and works on light, meaning that it does not work at night).

[87] There has been discussion of an EU supergrid (connecting wind, geothermal, hydropower, biomass, and solar at a pan-European level), such as the Desertec or Medgrid projects, and an Asian supergrid, such as the Gobitec project (https://energycharter.org/what-we-do/investment/investment-thematic-reports/gobitec-and-the-asian-supergrid-for-renewable-energy-sources-in-northeast-asia-2014/). Supergrids can share power over very long distances with low losses of power.

[88] *See* David Pickup et al., Solar Trade Association, Solar + Storage = Opportunities 3–6 (2016), http://www.solar-trade.org.uk/wp-content/uploads/2016/05/Solar-storage-Opportunities-The-STAs-Position-Paper-on-Energy-Storage.pdf [https://perma.cc/Z9K4-TNDE].

[89] Leal-Arcas and Filis-Yeloghotis (2012), pp. 128–148.

[90] By this transition, we mean transport systems in transition (e.g., technologies of electrification in trucks) and industries in transition (e.g., fossil-free steel).

9.4 Sustainable Companies[91]

Three factors are necessary to reach sustainability: investors, technology, and policy.[92] Moreover, energy,[93] transportation, and agriculture are crucial industries to build a sustainable future[94] since they represent around 60% of GHG emissions.[95] Yet, the leaders in these industries[96] have been around for over 100 years and still their innovation record is irregular. In that same time period, the computer and IT industries have been remarkably innovative: computers have been transformed from mainframes to tablets; in telecommunications, phones have been transformed from landlines to smartphones. Soon there will be over six billion smartphones in circulation.[97] The owners of these smartphones will each have more computer power in their hands than the supercomputers of the 1960s, enabling them to have access to all the world's knowledge.[98]

There are positive future commitments by technology companies such as Google, which plans to buy only renewable energy in 2017 "to match the entire needs of all its data centres and offices around the world."[99] In this innovation cycle, the energy, transportation, and agriculture sectors are starting to catch up. In the energy sector, there are projects for fully integrated microgrids to replace centralized fossil fuel

---

[91] *See* Nancy Pfund, Lecture at Yale University titled Creating the Sustainable Companies of the 21st Century, Sept. 22, 2016. See also Sheffi (2018).

[92] For further details on sustainable companies, see Esty and Winston (2009).

[93] The Secretary-General's Advisory Group on energy and climate change, "Energy for a sustainable future," Summary report and recommendations, 28 April 2010, available at http://www.un.org/chinese/millenniumgoals/pdf/AGECCsummaryreport%5B1%5D.pdf.

[94] For an account of how we can reach a sustainable future, see M. Jacobson and M. Delucchi, "A Path to Sustainable Energy by 2030," *Scientific American*, November 2009, pp. 58–65, available at https://web.stanford.edu/group/efmh/jacobson/Articles/I/sad1109Jaco5p.indd.pdf.

[95] *See Sources of Greenhouse Gas Emissions*, EPA, https://www.epa.gov/ghgemissions/sources-greenhouse-gas-emissions [https://perma.cc/8U54-CZLZ]; *see also* IPCC, 5th Assessment Report, Working Group III (2014), http://www.ipcc.ch/pdf/assessment-report/ar5/wg3/ipcc_wg3_ar5_full.pdf [https://perma.cc/8K8L-FJCR].

[96] The leaders of these industries include, for instance, John Deere since 1837, General Electric since 1892, Ford since 1903, and General Motors since 1908. *See* Nancy Pfund, Lecture at Yale University titled Creating the Sustainable Companies of the 21st Century, Sept. 22, 2016.

[97] Ingrid Lunden, *6.1B Smartphone Users Globally by 2020, Overtaking Basic Fixed Phone Subscriptions*, Tech Crunch (June 2, 2015), https://techcrunch.com/2015/06/02/6-1b-smartphone-users-globally-by-2020-overtaking-basic-fixed-phone-subscriptions/ [https://perma.cc/B9UN-BVDN].

[98] Norberg (2016), p. 200.

[99] Richard Waters, *Google to Buy Only Renewable Energy for Operations in 2017*, Fin. Times (Dec. 6, 2016), https://www.ft.com/content/6794d2f0-bb6a-11e6-8b45-b8b81dd5d080?emailid=55ccb875090bff0300e78b63&segmentId=3d08be62-315f-7330-5bbd-af33dc531acb [https://perma.cc/4TUM-28LP].

9.4 Sustainable Companies

plants (such as coal plants).[100] In transportation, autonomous and connected cars are replacing gas-powered cars.[101] In agriculture, data-driven technology is replacing mechanical tools.[102] Equally importantly, renewable energy is employing many people in the energy sector.[103]

Interestingly, these three sectors are adopting many of the same themes of innovation as the computer and IT sectors. The principles on which twenty-first century companies will be built are transparency, decentralization,[104] cost reduction, personalization, and convenience, among others. Regarding energy, the demand is for energy that is cheaper, greener, more reliable, and more functional.[105] In the case of cars, thanks to technology, these principles manifest in vehicles that are electrified, autonomous, and minimized through the social tendency toward ride-sharing. Customers are asking for cars that are cheaper, greener, and safer.[106] As for agriculture, there is a demand for it to be cheaper, greener, with more choice and more farmer independence.[107] Thus, there are clear market opportunities for these three industries. The question is how to help build these companies of the twenty-first century in the energy, transportation, and agricultural sectors.[108]

---

[100] Justin Guay, *World Bank Abandons Coal, Green Light for Clean Micro-Grids*, REneweconomy (July 14, 2013), http://reneweconomy.com.au/world-bank-abandons-coal-green-light-for-clean-micro-grids-59247/ [https://perma.cc/T9CG-2YRX].

[101] Richard Viereckl et al., *Connected Care Report 2016: Opportunities, Risk, and Turmoil on the Road to Autonomous Vehicles*, Strategy& (Sept. 28, 2016), http://www.strategyand.pwc.com/reports/connected-car-2016-study [https://perma.cc/R9X3-54RG].

[102] *See* Katherine Noyes, *Cropping Up on Every Farm: Big Data Technology*, Fortune (May 30, 2014), http://fortune.com/2014/05/30/cropping-up-on-every-farm-big-data-technology/ [https://perma.cc/2SKY-T2H2].

[103] *See, e.g.*, Bill Spindle & Rebecca Smith, *Which State is a Big Renewable Energy Pioneer? Texas*, Wall St. J. (Aug. 29, 2016), http://www.wsj.com/articles/which-state-is-a-big-renewable-energy-pioneer-texas-1472414098 (describing the case of Texas, where more than 100,000 people are employed in that sector) [https://perma.cc/C9K6-LV9E].

[104] Competition creates efficiency that centralized systems do not. Hence decentralization.

[105] *See, e.g.*, Pilita Clark, Aggreko buys energy storage technology specialist, Financial Times, July 3, 2017, https://www.ft.com/content/086fe32e-6014-11e7-8814-0ac7eb84e5f1?mhq5j=e2 [https://perma.cc/C7MS-2QJS].

[106] *See* Jack Stewart, *Tesla's Cars Have Driven 140M Miles on Autopilot. Here's How*, Wired (Aug. 17, 2016, 8:00 AM), https://www.wired.com/2016/08/how-tesla-autopilot-works/ [https://perma.cc/32TD-PPNZ]; *Google Self-Driving Car Project*, WAYMO, https://www.google.com/selfdrivingcar/ [https://perma.cc/7L7H-MJ7E]; Andrew J. Hawkins, *Uber Just Completed Its Two-Billionth Trip*, The Verge (Jul. 18, 2016, 11:00 AM), http://www.theverge.com/2016/7/18/12211710/uber-two-billion-trip-announced-kalanick-china-didi [https://perma.cc/586A-6K7P].

[107] Examples of twenty-first century agriculture companies are Blue River Technology, Apeel Sciences, Farmers Business Network, and Planet. See, e.g., Blue River Technology, http://www.bluerivert.com/ [https://perma.cc/F3WP-FYSS].

[108] For an overview of how to revitalize business models to win the clean energy race, see Amory B. Lovins, Reinventing Fire: Bold Business Solutions for the New Energy Era (2013).

In 2017, a taskforce on carbon disclosure delivered recommendations[109] to the G20[110] and the Financial Stability Board,[111] a forum of global regulators. The purpose is to create a voluntary framework to allow companies to report their exposure to climate risks.[112] By creating such a framework, the market will be able to "go green" more efficiently due to smarter and more efficient allocation of capital.[113] Because business is the engine of growth and job creation, it is important that they decarbonize to achieve a sustainable future.

9.5 Sustainability and Spirituality[114]

Contemporary society is the first human community to face a comprehensive crisis, threatening ecosystems and species on a global scale.[115] To paraphrase cultural historian and scholar Thomas Berry, we have ethics for homicide, suicide, and genocide, but not for biocide or geocide.[116] While this is largely true,[117] since

---

[109]*Task Force Publishes Recommendations on Climate-Related Financial Disclosures*, Fin. Stability Board (June 29, 2017), http://www.fsb.org/2017/06/task-force-publishes-recommendations-on-climate-related-financial-disclosures/ [https://perma.cc/CA4G-C35Z].

[110]*G20 Germany 2017*, Fed. Gov't of Germany, https://www.g20.org/Webs/G20/EN/Home/home_node.html.

[111]Fin. Stability Board, http://www.fsb.org/.

[112]*See* Task Force on Climate-Related Fin. Disclosures, Recommendations of the Task Force on Climate-Related Financial Disclosures ii–v (2016), https://www.fsb-tcfd.org/wp-content/uploads/2016/12/TCFD-Recommendations-Report-A4-14-Dec-2016.pdf. See also Risky Business, "The economic risks of climate change in the United States," June 2014.

[113]For more details on how sustainable development is changing business leadership, see Bus. & Sustainable Dev. Comm'n, Better Business Better World (2017), http://report.businesscommission.org/uploads/BetterBiz-BetterWorld.pdf.

[114]See Tucker, M. "Sustainability and Spirituality," lecture given at the Yale sustainability leadership forum, September 2016, Yale University, New Haven, USA.

[115]*See* Our Common Future, From One Earth to One World, Rep. of the World Comm. on Env't & Dev. on Its Forty-Second Session, U.N. Doc. A/42/427, annex, ¶ 11 (1987).

[116]David Schenck, *The Great Teaching Work of Thomas Berry*, Ctr. for Humans & Nature, http://www.humansandnature.org/the-great-teaching-work-of-thomas-berry. One's behavior and attitudes are shaped by world views, values, and spirituality derived from world religions, environmental ethics, biophilia, humanitarian and secular values, and the arts. *See, e.g.*, Grim and Tucker (2014), pp. 1–12; Swimme and Tucker (2011), pp. 103–119.

[117]One exception is that of Tajikistan, whose criminal code stipulates in Article 399 (on biocide) the following: "Using of nuclear, neutron, chemical, biological (bacteriological), climatic or other kind of mass destruction weapons with the intent of destruction of people and environment is punishable by imprisonment for a period of [fifteen] to [twenty] years, or death penalty." Criminal Code of the Republic of Tajikistan, art. 399 (unofficial).

2015 the international community officially committed to the SDGs as a roadmap for a sustainable future.[118]

Well-known obstacles make sustainability a challenge: population growth,[119] increased consumption, urbanization, alienation from land, climate change, biodiversity loss,[120] to name but a few. In addition to the means discussed throughout this chapter, additional ways to reach sustainability include social and cultural changes such as greater reverence for the earth community, respect for species, restraint in the use of natural resources, redistribution of technology and aid, responsibility for the future of the planet, and restoration of ecosystems and the human spirit.

Two foundational socio-cultural *principles* may serve as roadmaps to achieve sustainability. First, acknowledgement of the intrinsic value of nature, the idea that nature is a source, not a resource, and acceptance of environmental degradation as an ethical issue. Second, honoring humans through awareness of environmental rights for present and future generations and the notion of distributive justice as part of our environmental responsibilities.

In the context of sustainability and spirituality, then, two socio-cultural *strategies* emerge to tackle the challenge of a sustainable future. First, society must think consequentially, both short- and long-term; and second, society must integrate solutions in the context of energy and technology via renewable energy and technology transfer.[121]

Finally, two interrelated *tactics* may achieve a way forward in the context of sustainability and spirituality: ensuring restraint on global consumption and population growth, and creating law in the context of global governance and global ethics.

9.6 Conclusion and a Future Research Agenda

What we had yesterday is not what we need tomorrow. Moreover, "the future ain't what it used to be."[122] Society has seen a tremendous transformation from mechanization to electrification to computerization to artificial intelligence (AI). The last transformation (namely AI) will need to be carefully regulated to avoid abusive situations towards humans, such as, in the words of Yuval Harari, humans becoming "useless" and irrelevant in the twenty-first century as a result of the rise of AI,

---

[118]*Sustainable Development Goals*, UN, http://www.un.org/sustainabledevelopment/sustainable-development-goals/.

[119]There is scholarly work that states that demographic change is an unavoidable force that can change the course of history. See Morland (2019) and Bricker and Ibbitson (2019).

[120]Think for instance of the gradual destruction of biodiversity in Colombia as a result of high levels of consumption by many importers of Colombian coffee throughout the world. The higher the demand of coffee, the greater the destruction of biodiversity by the coffee supplier. For further analysis on this issue, see Wiedmann and Lenzen (2018), pp. 314–321.

[121]*See* TRIPS Agreement, art. 66(2).

[122]This quote is often attributed to Yogi Berra.

whereas other humans are digitally enhanced "gods."[123] According to David Victor, history has taught us that: (1) new technological systems interact with old systems, and the old systems fight back; (2) successful new technologies tend to radically raise performance; and (3) profound effects of technological transformation (e.g., email or the internet) take around one generation to affect the rest of society.[124] Technologies may provide opportunities that we cannot imagine today. So we need to be prepared to take aspiration into action.

Yet, there is little clarity of where the world is going. There are still many pending questions: How rapidly can new technologies and infrastructure unfold (think, for instance, of the substitution of horses for cars in the transportation sector)? What are the sources of technological transformation?[125] How can we have access to energy without climate change? Will energy be more or less traded with the digital economy? How can technology help for economic and social benefits? How can it contribute to the achievement of the SDGs, which are fantastic tools for sustainability? What does technology mean for trade? Why not have new rules on technology transfer in the WTO? All these questions need an answer for a sustainable and integrated global economy. The international community needs to think long-term and embrace renewable energy as well as the impact of trade on (child) labor. Globally, renewable scales are still small and the zero-carbon technological development remains slow.

Climate change is a technological, political, and economic challenge. In the eighteenth century, biomass and waste were the main natural resources used for energy generation; in the nineteenth century, coal was the main natural resource used for energy generation[126]; in the twentieth century, it was oil.[127] In the twenty-first century, the expectation is that it will be renewable energy, although industries still use large amounts of coal for energy production and coal may remain the most used fossil fuel for years to come, with a great possibility that gas may supplant coal because it is cleaner.[128] So energy seems to move from black (oil and coal) to gray (gas) to gradually green (renewable energy) and renewable energy should be used without damaging nature. That said, businesses are increasingly interested in becoming more environmentally aware, but action is still largely missing. Ideally, one should make use of fossil fuels only to create the next technology, not to generate energy. Renewable energy is currently not cost-competitive compared to fossil fuels, so much so that, economically, it might make little sense to move to, say, solar

---

[123]Harari (2018).

[124]David Victor, Lecture given at Harvard University, 2 October 2018.

[125]An explanation is innovation as a necessity and as a public good.

[126]*History of Fossil Fuels Usage Since the Industrial Revolution*, Mitsubishi Heavy Industries, https://www.mhi-global.com/discover/earth/issue/history/history.html.

[127]Dr. Jean-Paul Rodrigue, *Evolution of Energy Sources*, The Geography of Transport Systems, https://people.hofstra.edu/geotrans/eng/ch8en/conc8en/evolenergy.html.

[128]*Are Fossil Fuel Companies Using IEA Report To Talk Up Demand?*, The Guardian (Oct. 25, 2015), https://www.theguardian.com/environment/2015/oct/23/are-fossil-fuel-companies-using-iea-reports-to-talk-up-demand.

9.6 Conclusion and a Future Research Agenda

energy.[129] There are options to reduce net emissions of GHGs: switching fuels, changing industrial production processes, carbon capture and storage, reducing energy demand, increasing the efficiency of energy generation, and reducing deforestation.

Following the title of the famous novel *What Is to Be Done?* by Nikolai Chernyshevsky, a balance must be struck between fossil fuels and renewables, and carbon needs to be challenged. A credible solution for the energy mix is combining renewable energy with natural gas because generating energy based solely on renewables is not yet feasible—for many of the reasons explored herein—and because natural gas is the least destructive of all the fossil fuels.[130] Moreover, decarbonization is possible not only via renewable energy—for which investing in innovation will be necessary—but also via clean electricity and by decarbonizing fossil fuels, namely through carbon capture and storage, which will be necessary in the future.[131] In the future, the goal is that renewable energy will shift from being a complement to a substitute for fossil fuels.

Two mega-trends are happening in the world economy that will impact climate change mitigation: (1) artificial intelligence and (2) the circular economy (which is a game changer). History has taught us that reshaping the economy provides more job creation than destruction. So the solution would be to reduce CO_2 emissions by decarbonizing, electrifying, making use of the circular economy (i.e., recycling and reusing products),[132] transferring funds and technology from the West to the rest of the world, shifting the economy to services that do not use products, and sharing best practices. The concept of a circular economy is also an opportunity for innovation in how we produce and consume. For instance, the concept of eco-design will help us with the design of products to minimize waste and maximize resources.

We are the first generation who is aware of the dangers of climate change and probably the last generation able to do something about it, which is why international cooperation/agreements are crucial to provide long-term certainty and clarity. Since business-as-usual is no longer an option, here are some of the ingredients to reach sustainability: (1) regulation, namely targets for reducing GHG emissions; (2) innovation; (3) mobilizing finance; and (4) education, starting with children, so that they are aware of the current situation and making sure that that they incrementally change the way they consume and behave.

Through effective regional and global collaboration on the decarbonization of the economy, the European Union (and the rest of the world) can pave the way for a sustainable and secure future for generations to come. Cooperation on renewable

---

[129] *But see* Zachary Shahan, *Advantages & Disadvantages of Solar Power*, Clean Technica (Oct. 8, 2013), https://cleantechnica.com/2013/10/08/advantages-disadvantages-solar-power/.

[130] *Natural Gas, the Cleanest or Less Dirty Fossil Fuel*, Energy News (Sept. 19, 2016), http://www.energynews.es/english/natural-gas-the-cleanest-or-less-dirty-fossil-fuel/.

[131] Stuart Haszeldine (2009), p. 1647 at 1647–1652 (2009).

[132] According to The Economist, "in 2009–15 the number of biogas plants in the EU grew from 6,000 to 17,700—heating houses with old banana skins and uneaten porridge." See The Economist, "A load of rubbish," Special report: Waste, pp. 1–12 at 9, 29 September 2018.

energy will enable E.U. member states to reduce their GHG emissions, in line with their commitments under the Paris Agreement[133] and obligations under the European Union's Sustainable Development Strategy.[134] By enhancing sustainable energy, the European Union and the international community are mitigating climate change. In addition, effective cooperation will culminate in the spread of global renewable energy security, a global public good that can only be supplied through collective efforts. Among other issues, regional and global cooperation on decarbonization will enable the European Union to tackle some of the most pressing human rights issues in the region, boost the economy by encouraging investment, and generate employment.

It is possible to achieve global renewable energy security. In 2011, the Intergovernmental Panel on Climate Change (IPCC) argued that "as infrastructure and energy systems develop, in spite of the complexities, there are few, if any, fundamental technological limits to integrating a portfolio of renewable energy technologies to meet a majority share of total energy demand in locations where suitable renewable resources exist or can be supplied."[135] The IPCC has further said that if governments are supportive, and the full complement of renewable energy technologies are deployed, renewable energy supply could account for almost 80% of the world's energy use within 40 years, namely by 2050.[136]

This is an era of changes and challenges. Historian Yuval Harari notes that "for the first time in history, more people die today from eating too much than from eating too little; more people die from old age than from infectious diseases; and more people commit suicide than are killed by soldiers, terrorists and criminals combined."[137] The challenge of the third millennium will be a sustainable future, where common people understand common concerns and public goods are taken seriously. Conservation is the biggest source of GHG emissions reduction.[138] The challenge is not technological (with the exception of carbon capture and storage), nor is it financial; it is political, namely lack of political will to cooperate internationally to solve such issues. The challenge is therefore that we are asking countries to do something internationally that they do not agree to do domestically.

There is a knowledge gap on the links between four major global concerns: trade, energy, climate change, and sustainability. Each one of them has its own culture; for

---

[133] Paris Agreement, art. 4(4).

[134] *Commission Proposal to the Gothenburg European Council, A Sustainable Europe For a Better World: A European Union Strategy for Sustainable Development*, COM (2001) 264 final.

[135] Intergovernmental Panel on Climate Change, *Special Report on Renewable Energy Sources and Climate Change Mitigation*, at 17–18 (2012), https://www.ipcc.ch/pdf/special-reports/srren/SRREN_FD_SPM_final.pdf.

[136] *See* Renewable Energy Can Power the World, Says Landmark IPCC Study, The Guardian (May 9, 2011), https://www.theguardian.com/environment/2011/may/09/ipcc-renewable-energy-power-world (citing the Special Report on Renewable Energy Sources and Climate Change Mitigation, of the Intergovernmental Panel on Climate Change, finalized in 2011).

[137] Harari (2016).

[138] *See* Gary Brudvig, Lecture at Yale University titled Spurring Innovation (Sept. 22, 2016).

9.6 Conclusion and a Future Research Agenda

instance, both trade and climate are similar in that they are global in scope, but they differ in institutional structure and governance in that trade is more developed due to its dispute settlement system, which is absent in climate change, whereas the climate regime operates more with persuasion than punishment. From this point of view, the trade regime is exclusive because punishment will take place if one is not in compliance with regulations. With the threat of climate change looming, and energy increasingly important to all aspects of human and economic development, learning more about these links is extremely timely. Specifically, it is necessary to do more research into the use of trade as a tool to achieve sustainable energy and therefore reduce poverty, while also addressing climate change.

An open trading system in all its three aspects (political, legal, and economic) is crucial for sustainable development to take shape. Pending questions remain, such as: What can citizens do to be more empowered in inter-state trade agreements? How can they be better informed? How should future trade and environmental agreements be designed to be socially acceptable and more inclusive of civil society? How can trade agreements be modernized to help climate change? How can we reach social sustainability?

Politically, taking the Paris Agreement forward with its implementation is imperative to make sure no one is left behind, given that the Paris Agreement is as much about economic and social transformation as it is about climate change. The Agreement may not be perfect, but it is better than previous legal instruments. The concept of *in dubio pro natura*, advocated by Brazil's National High Court Justice Antonio Benjamin, is the strongest legal form of environmental protection.[139] And providing concessional financing for CO_2 to incentivize countries to decarbonize their economies would assist in the transition to clean energy. While the transition to clean energy has discouraged oil and gas imports throughout the world, it has brought about an increase in coal use.[140]

Regarding clean energy, the potential of solar energy is phenomenal: solar energy today represents only around 0.3% of global energy[141]; one hour of sun can generate energy for the whole Earth for an entire year[142]; in 14 and a half seconds, the sun

---

[139] Superior Court of Justice (Brazil), Recurso Especial No. REsp 883.656/RS, Rel. Herman Benjamin, available at http://www.planetaverde.org/arquivos/biblioteca/arquivo_20131123 195922_9398.pdf [https://perma.cc/CL9D-5GF2].

[140] However, see the views by Tom Randall on coal's prognosis, "The latest sign that coal is getting killed," Bloomberg, 13 July 2015, available at https://www.bloomberg.com/news/articles/2015-07-13/the-latest-sign-that-coal-is-getting-killed.

[141] Information gathered from the roundtable on "Science, Law and Climate Change – Innovating sustainable solutions" of the London Energy Forum. The event was held in The Law Society, London, UK, on 4 February 2016.

[142] *Solar Frontiers*, The Economist (Dec. 1, 2015), https://www.youtube.com/watch?v=4-m9OR9vcaM.

emits enough energy to power the Earth for an entire day[143]; and "we could power the entire world if we covered less than 3% of the Sahara Desert with solar panels."[144] So there is hope[145] and great research and business opportunities.[146] Moreover, predictions are that the fastest-growing occupation until 2028 will be that of solar installer.[147]

Equally important is to study the pivotal role that cities will play in becoming new platforms to help mitigate climate change and use sustainable energy more effectively. Energy consumption has increased exponentially since the 1960s due to world population growth and prosperity. With world population growth, with the rise of the middle class mainly in developing countries, and with people moving to cities looking for better job opportunities, cities are becoming crucial platforms for climate change mitigation. In terms of new approaches to governance resulting from the Paris Agreement, there is renewed focus on mayors and citizens acting at the center of analysis for climate change mitigation and sustainable energy. To that, one should add that young people tend to believe in the importance of sustainability, which means that, moving forward, transformation of our way of life to a more sustainable one is likely to prevail. Making use of such innovative methodologies will bridge an important knowledge gap and, in doing so, open the door to an entirely new research agenda.

No solution to the above big challenges is possible without cooperation among governments, companies, researchers (whose role is to provide good information to create good policy), and mobilization of individuals. Business may have a role to play when politicians fall short and help decarbonize the economy at large. While elected politicians may be too shy to risk failure and seem to suffer from short-termism, entrepreneurs seem to be riskophiles and persistent, with a long-term commitment, especially multibillionaire entrepreneurs—for instance, Elon Musk's companies Space X and Tesla. Change may come sooner than later thanks to them. Technology seems to be the resource for success. To that, one should add the optimism of Steven Pinker that things will only get better in the future because people generally think reasonably and logically[148] and that the geopolitics of clean energy may make the world more peaceful and stable. Peace is certainly a key condition for sustainable development.

A number of policy changes may make a difference moving forward:

---

[143]Z. Shahan, "In 14 and a half seconds, the sun provides as much energy to Earth as humanity uses in a day," *CleanTechnica*, 18 April 2012, available at https://cleantechnica.com/2012/04/18/in-14-and-a-half-seconds/.

[144]*See Global Lessons: The GPS Road Map for Powering America*, CNN Transcripts (Oct. 21, 2012), http://transcripts.cnn.com/TRANSCRIPTS/1210/21/cp.01.html.

[145]See Hawken (2017).

[146]For more details, see Sivaram (2018).

[147]The Economist, "Trade tariffs: Duties call," 27 January 2018, p. 12.

[148]Pinker (2018).

9.6 Conclusion and a Future Research Agenda

- Decreasing/cutting fossil-fuel subsidies and promoting energy efficiency to reduce fossil-fuels emissions. The international community keeps subsidizing unsustainability, which needs to stop. Let's focus on what is unsustainable and leave it behind, if we want to get to sustainability;
- Educating consumers and corporations on sustainable energy/climate change to reduce the demand and supply, respectively, of non-green goods and services and to embrace a minimalist approach to life.[149] For change to happen, we need to understand people's values;
- Implementing new regulation that favors green trade as a policy objective. In the technology field, it is well known that technology evolves quicker than regulation can keep up. Yet, regulation seems necessary to provide guidance, but that can gradually change to de-regulation and let the market dictate;
- Boosting renewable-energies development by increasing renewable-energy subsidies, promoting investment in climate-friendly technologies, and by gradually prohibiting the use of fossil fuels to generate energy; and
- Providing financial mechanisms (such as emissions trading schemes and carbon taxes/border carbon measures) and eco-labelling schemes, leading consumers to buy green goods/services such as electric cars. For electric-vehicle projections to be fulfilled, better batteries are necessary as well as economies of scale. Equally, making a policy shift from taxing labor to taxing natural resources.

Regarding the energy transition, as the world reduces its oil dependence, the winners in this race will be those that will be able to produce and export green technology and rely on clean energy, whereas the losers will be those that will continue to depend mainly on fossil fuels. Two ingredients may help move forward the energy transition: international collaboration and energy decentralization. Potential international collaboration can be achieved in the field of technology, for which international trade will certainly play a major role. Initiatives such as "Breakthrough Energy Coalition"[150] of visionary billionaires determined to provide energy that is reliable, affordable and carbon-less are an excellent way forward. Even more ambitious would be to replicate the Breakthrough Energy Coalition from the bottom up so that every consumer feels they are part of this energy transformation.

Another initiative called Mission Innovation[151] brings together a group of 23 countries and the EU[152] that aim to reinvigorate and accelerate clean energy innovation throughout the world to make clean energy affordable for all. As for energy decentralization, the emergence of micro-/minigrids dealing with locally produced wind and solar energy, as well as electric-vehicle batteries is the way forward for sustainability. All of this will not only help to have better access to energy, but it will also decentralize economies.

---

[149]H. Yusof, "The Pursuit of Less," The Business Times, 19 January 2018, available at https://www.businesstimes.com.sg/lifestyle/feature/the-pursuit-of-less.
[150]To access their principles, see http://www.breakthroughenergycoalition.com/en/index.html.
[151]http://mission-innovation.net/.
[152]http://mission-innovation.net/countries/.

When it comes to the fight against climate change, winning slowly is the same as losing the fight. However, when opportunity meets willingness, action takes place. Change in behavior by citizens (and businesses) is key to make the economy more sustainable because policy targets come from governments, but policy implementation will be done by citizens. Change in behavior implies enabling people the choices of change. There is evidence that young people want to consume in a sustainable manner, which is a positive change and will make the future brighter.[153] Equally, as pointed out by David Korten, changing the story will change the future.[154] So citizens need to have a voice to change the story of their future. Being pragmatic and practical at city/company level will help.

Two counter-intuitive trade-related points deserve to be mentioned. First, that trade agreements may be more effective legal instruments than environmental agreements for environmental-protection purposes is both counter-intuitive and surprising. Just as the huge improvement in quality of life after World War II was largely due to the expansion of world trade by lowering technical barriers, one can use the international trading system (whether regionally, bilaterally, plurilaterally, multilaterally or in any other form) to help mitigate climate change and enhance sustainable energy. If multilateralism is currently in crisis, plurilateralism might be an effective platform to work on the links between trade and climate action. How? By making sure that major GHG emitters conclude mega-FTAs with each other to liberalize green goods and services.

Second, on the trade-climate change nexus, whether clean-energy technology eventually triggers a healthy competition or geopolitical friction will depend on international trade. If the Trump administration ends up creating a trade war, there will be less trade and, therefore, less international shipping for the transnational movement of goods. Thus, fewer emissions of GHGs will result, which is good for climate change. In conclusion, a trade war would be beneficial for climate change from the point of view of GHG emission reduction, but it will make the world poorer. So if climate change mitigation is about money, how can a trade war help fix the climate change problem? Moreover, a trade war may help with the reduction of GHG emissions, but would prevent global access to clean goods.

All of this would need to be implemented in terms of bottom-up governance. Recent examples of citizens' discontent in EU governance show the apathy among voters for supranational parliamentary elections, whose participation has decreased in each and every election since the first in 1979. Instead, there is an increasing interest in national/sub-national parliamentary politics, as exemplified by Brexit and the Catalonian independent movement, which are closer to the citizens than metanational/supranational/international entities. Greater use of social media (Twitter, Facebook, videos on YouTube) could be a very effective platform to educate

---

[153]"Green generation: Millennials say sustainability is a shopping priority," available at https://www.nielsen.com/uk/en/insights/news/2015/green-generation-millennials-say-sustainability-is-a-shopping-priority.html.
[154]Korten (2015).

9.6 Conclusion and a Future Research Agenda

youth—which is the segment of society that makes most use of it—on the links between trade and climate change, raise awareness at local level, and involve them in parliamentary elections. These ideas will make it possible to have sustainable and economically advantageous trade.

Many of today's big changes are demographics, a shift in power from the West to the East, rapid urbanization, technology, health and well-being, and climate change and natural resources. The last two points are crucial to the arguments made in this chapter in the broader context of inclusive prosperity. Access to affordable and clean energy as well as climate action are two of the seventeen UN Sustainable Development Goals, which the international community is committed to meeting by 2030. The Earth is our home and common inheritance. We need to make sure it is sustainably managed. We now have enough scientific knowledge to know that climate change is a problem. But the policies in place are wrong and good leadership is essential to meet the agreed targets. More specifically, collective action by all leaders would make a difference: we see that leaders are good at individual goals for their own company/country; what is required is collective vision of a dream to share among leaders.

In the past, electricity has been generated centrally, running across vast physical grids. One can think of five 'D's when analyzing what is shaping the economy and the energy transition:

- Decentralization;
- Democratization;
- Decarbonization;
- Digitalization; and
- De-regulation.

Moving forward, several key challenges to the energy transition are likely to emerge, including:

- A modern and clean energy economy: The energy transition trend has been from a centralized system in the past to a currently decentralized system and smart technologies, to a future smart, data-centric system and electrification of transportation. Mass electrification from zero-carbon sources could stimulate new industries and decentralize the global economy even further. This transition will happen with innovation (in energy, architecture, agriculture, transportation, government) and if the appropriate investment takes place to empower prosumers and renewable energy cooperatives and to manage data;
- A fair energy system with access to energy for all: No one and no country should be left behind. In other words, the energy transition must be designed in a fair manner;
- The enhancement of existing regional cooperation at all levels of governance: The current normative complexity would need further cooperation between various parties involved in the energy-transition process and at all levels, whether

it is the EU, national level, regulators, distributors, stakeholders, or transmission system operators (TSOs)[155];
- Digitalization: Cyber security in energy will inevitably have cascading effects in other sectors such as finance and transport.[156] Smart grids may be linked to artificial intelligence (AI) via digitalization. Digitalization may help achieve the 17 Sustainable Development Goals and is a major potential force to bring about technology to climate change mitigation and the energy sector. Equally, it is presumed that digitalization will bring down the cost of energy for consumers; and
- A global level playing field: The Paris Climate Agreement is a case in point. The objectives of the Paris Agreement would need to be in alignment with the objectives of future legislation on clean energy. But what about international trade and investment?[157] How can the objectives of the Paris Agreement be aligned with those of future trade agreements?

In a nutshell, technology (which is absolutely crucial for full renewable energy) and leadership from the bottom up, not just top-down, are the essential ingredients for sustainable development. The solutions to climate change are all technological. This technology will come from the market.[158] Technology and politics may not yet allow full potential of renewable energy (mainly solar). Most likely, we will continue with fossil fuels for years to come despite the fact that there are very good reasons to make a solid transition to renewable energy: it is good for climate action because it reduces CO_2; it is good for energy security because there is less dependence on fossil fuels; it is good for consumers thanks to the creation of renewable-energy cooperatives; it is good for job creation thanks to green jobs; and it is good for trade creation thanks to renewable-energy exports.

So how can we use technology (cybersecurity, cloud computing, the internet-of-things, artificial intelligence, blockchain), evidence-based policymaking, and big data to enable cities, companies, and citizens mitigate climate change? Will policy intervention be necessary to shape the benefits of AI and blockchain? Should robots be taxed because they have replace people who were taxed, and that would be revenue for the state? Can AI and blockchain transform the processes around trade flows, facilitating interactions between distant actors and reducing logistical friction? How do we measure sustainability? The cluster of different technologies is what makes artificial intelligence. Artificial intelligence is based on data. The amount of data we have today has never existed before. All of this will change climate change, energy security, and trade. These are the big challenges of tomorrow.

---

[155] A lot of these TSOs are naturally regional, not national.

[156] An example is the potential risks of cyber-attacks associated with autonomous vehicles. See The Economist, "Reinventing wheels," Special Report, p. 6, 3rd March 2018.

[157] For the specific case of energy trade, see Leal-Arcas et al. (2017), pp. 520–549; Leal-Arcas (2016), pp. 53–74; Leal-Arcas (2015), pp. 202–219; Leal-Arcas et al. (2015), pp. 38–87.

[158] On the role of the market, see Posner and Weyl (2018) (who argue that the way out of the current impasse is to expand the role of markets).

One final personal point of view: one simple solution to sustainability is to become a minimalist by reducing our unsustainable production and consumption patterns. We, as consumers, have a lot of power. Before you buy anything, ask yourself the question: do I really need this? How was it made? Was there an abuse of human rights when producing this good? Another simple solution is to reduce population growth, whose implication is the reduction of energy consumption and less construction.[159] The reduction of population growth can be done by educating girls on family planning. The impacts of educating girls will have a multiplying effect across the board. Finally, localism in conjunction with globalism can be a very powerful for rapid change. In other words, social innovation, changing the mindset of people, and thinking out of the box as possible solutions to sustainability.

References

Bricker D, Ibbitson J (2019) Empty planet: the shock of global population decline. Signal
Esty D, Winston A (2009) Green to Gold: how smart companies use environmental strategy to innovate, create value, and build competitive advantage. Wiley
Grim J, Tucker ME (2014) Ecology and religion, pp 1–12
Harari YN (2016) Homo Deus: a brief history of tomorrow
Harari Y (2018) 21 lessons for the 21st century. Jonathan Cape, London
Hawken P (ed) (2017) Drawdown: the most comprehensive plan ever proposed to reverse global warming. Penguin Books
Korten D (2015) Change the story, change the future: a living economy for a living earth. Berrett-Koehler Publishers
Leal-Arcas R (2015) How governing international trade in energy can enhance EU energy security. Renew Energy Law Policy Rev 6(3):202–219
Leal-Arcas R (2016) Energy transit in the Caucasus: a legal analysis. Caucasus Int 6(2):53–74
Leal-Arcas R (2017) Sustainability, common concern and public goods. George Wash Int Law Rev 49(4):801–877
Leal-Arcas R, Filis-Yeloghotis A (2012) Geoengineering a future for humankind: some technical and ethical considerations. Carbon Climate Law Rev 6:128
Leal-Arcas R, Minas S (2016) Mapping the international and European governance of renewable energy. Oxf Yearb Eur Law 35(1):621–666
Leal-Arcas R, Wouters J (eds) (2017) Research handbook on EU energy law and policy. Edward Elgar
Leal-Arcas R et al (2015) Multilateral, regional and bilateral energy trade governance. Renew Energy Law Policy Rev 6(1):38–87
Leal-Arcas R et al (2016) Energy security, trade and the EU: regional and international perspectives. Edward Elgar, Cheltenham, p 436
Leal-Arcas R et al (2017) Energy trade in the MENA Region: looking beyond the Pan-Arab electricity market. J World Energy Law Bus 10(6):520–549
Leal-Arcas R et al (2018) Smart grids in the European Union: assessing energy security, regulation & social and ethical considerations. Columbia J Eur Law 24(2):311–410
MacKay D (2009) Sustainable energy – without the hot air. UIT Cambridge Ltd.
Madison A (2003) The world economy: historical statistics. OECD, Paris, p 262

---

[159]The construction sector is responsible for large GHG emissions from cement and steel.

Morland P (2019) The human tide: how population shaped the modern world. John Murray
Norberg J (2016) Progress: ten reasons to look forward to the future. Oneworld Publications
Pinker S (2018) Enlightenment now: the case for reason, science, humanism, and progress. Viking
Posner E, Weyl EG (2018) Radical markets: uprooting capitalism and democracy for a just society. Princeton University Press
Sheffi Y (2018) Balancing green: when to embrace sustainability in a business (and when not to). MIT Press
Sivaram V (2018) Taming the Sun: innovations to harness solar energy and power the planet. MIT Press
Stuart Haszeldine R (2009) Carbon capture and storage: how green can black be? Science 325:1647
Swimme BT, Tucker ME (2011) Journey of the Universe, pp 103–119
Thiel P, Masters B (2014) Zero to one: notes on startups, or how to build the future. Virgin Books
van Leeuwen B, van Leeuwen-Li J (2014) Education since 1820. In: How was life? Global well-being since 1820, pp 87 and 94
Victor D, Yanosek K (2017) The next energy revolution: the promise and peril of high-tech innovation. Foreign Aff 96(4):124–131
Wiedmann T, Lenzen M (2018) Environmental and social footprints of international trade. Nat Geosci 11:314–321
Yogi Goswami D, Kreith F, Kreider JF (1999) Principles of solar engineering, 2nd edn, p 11